Organic and Inorganic Nanostructures

Organic and Inorganic Nanostructures

Edited by **Lindy Bowman**

Willford Press
New York

Published by Willford Press,
118-35 Queens Blvd., Suite 400,
Forest Hills, NY 11375, USA
www.willfordpress.com

Organic and Inorganic Nanostructures
Edited by Lindy Bowman

International Standard Book Number: 978-1-68285-238-5 (Hardback)

Printed in China.

Contents

Preface

The world is advancing at a fast pace like never before. Therefore, the need is to keep up with the latest developments. This book was an idea that came to fruition when the specialists in the area realized the need to coordinate together and document essential themes in the subject. That's when I was requested to be the editor. Editing this book has been an honour as it brings together diverse authors researching on different streams of the field. The book collates essential materials contributed by veterans in the area which can be utilized by students and researchers alike.

This book unfolds the innovative aspects of organic as well as inorganic nanostructures which will be crucial for the progress of their applications in the future. It studies, analyses and upholds the pillars of this discipline and its utmost significance in modern times. Synthesis, processing, modelling and simulation of organic, inorganic and hybrid nanosized materials, etc. are some of the key concepts that have been discussed herein. It will serve as a valuable source of reference for graduate and post graduate students pursuing nanosciences and allied fields of study.

Each chapter is a sole-standing publication that reflects each author´s interpretation. Thus, the book displays a multi-facetted picture of our current understanding of applications and diverse aspects of the field. I would like to thank the contributors of this book and my family for their endless support.

Editor

Sulfonic acid functionalized ordered nanoporous Na^+ montmorillonite as an efficient and recyclable catalyst for the chemoselective methoxymethylation of alcohols

Farhad Shirini*, Manouchehr Mamaghani and Seyyed Vahid Atghia

Abstract

Sulfonic acid functionalized ordered nanoporous sodium montmorillonite was easily prepared by the reaction of sodium montmorillonite with chlorosulfonic acid. The new catalyst demonstrated efficient and chemoselective methoxymethylation reaction of alcohols with formaldehyde dimethyl acetal in chloroform under reflux conditions. This reaction affords corresponding ethers in good to excellent yields. The present method offers several advantages such as short reaction times, high yields, simple procedure, mild conditions, heterogeneous nature, and reusability of the catalyst.

Keywords: Methoxymethylation, Alcohols, Formaldehyde dimethyl acetal, Na^+ montmorillonite, Solid acid catalyst, Heterogeneous catalysis

Background

Selective protection and deprotection of hydroxyl groups have occupied a unique position in organic synthesis because of the fundamental importance of hydroxyl groups in multistep synthesis of complex natural products [1]. Between several methods available for the protection of the alcoholic hydroxyl groups, methoxymethylation has attracted the attention of many organic chemists. This considerable attention can be attributed to the stability of the produced methoxymethyl (MOM) ethers against reagents such as strong bases, butyl lithium, lithium aluminum hydride, and Grignard reagents as well as easy removal of the MOM moiety by acid treatment. Generally, the formation of MOM ethers is carried out by the alkylation of alcohols with excess amounts of chloromethyl methyl ether (CME) in alkaline solution [2]. However, the potent carcinogenic property of CME limits the use of this method, and MOM ethers are prepared by the reaction of alcohols or phenols with formaldehyde dimethyl acetal (FDMA), as a cheap and commercially available compound. Even though the handling of this reagent does not need special precautions and workup of the reaction mixture is not time-consuming, the low methoxymethylating power of FDMA is the main drawback for its application. To overcome this limitation, a number of catalysts such as *p*-toluenesulfonic acid [3], Nafion-H [4], TMSI [5], Envirocat [6], expansive graphite [2], sulfated metal oxides [7], silica sulfuric acid [8], $Sc(OTf)_3$ [9], $Bi(OTf)_3$ [10], $Al(HSO_4)_3$ [11], $MoO_2(acac)_2$ [12], $H_3PMo_{12}O_{40}$ [13], anhydrous $FeCl_3$ dispersed on 3A molecular sieve [14], TiO_2/SO_4^{2-} [15], $H_3PW_{12}O_{40}$ [16,17], melamine trisulfonic acid [18], benzyltriphenylphosphonium tribromide [19], and high-valent $[Sn^{IV}(Br_8TPP)(OTf)_2)]$ [20] have been reported. Although these methods are improvements, most of them suffer from disadvantages such as harsh reaction conditions, long reaction times, poor yields, poor selectivity, use of toxic or expensive reagents, and use of large amounts of FDMA. Thus, the searches for new reagents and methods that utilize eco-friendly protocols are still in demand.

In recent years, clays as nanostructured materials have been widely used in organic transformations as solid acid catalysts [21-23]. The main reasons for use of clays are accessibility, easy modification, cheapness, non-

* Correspondence: shirini@guilan.ac.ir
Department of Chemistry, College of Science, the University of Guilan, PO Box 1914, Rasht 41335, Iran

corrosiveness, and recyclability. Montmorillonite (MMT) is one of the most widely used clays. Montmorillonite minerals have very small micron-sized particles, and they are extremely fine-grained and thin-layered [24]. Layers of MMT have a thickness of about 1 nm and a length of 100 nm or a little more. Broken bonds on the edge of MMT layers are common phenomena for layered silicates and lead to the free formation of hydroxyl groups [25,26], which can be utilized for chemical modification. The first attempt may be traced back in 1941 when Berger found that hydroxyl groups of montmorillonite could be methylated with diazomethane [27]. Many scientists then used this property for different modifications on the surface of montmorillonite [28,29].

Results and discussion

Recently, preparation of new catalysts for organic reactions, by modification of hydroxyl groups of various hetero- and homogenous compounds by sulfonic moiety, became an important part of our research program [30-33]. In continuation of these studies, we have found that chlorosulfonic acid can be used for the modification of Na^+-MMT until -OH groups on the surface of this clay be converted to $-SO_3H$ (Scheme 1). The reaction is easy and clean and needs no special precautions because HCl gas evolved from the reaction vessel immediately.

On the basis of the structure of sulfonic acid-functionalized ordered nanoporous Na^+-MMT (SANM), we anticipated that this reagent would act as an efficient catalyst in the reactions that need the use of acidic reagents to speed up. In our first communication, we used this reagent as an efficient catalyst for *N-tert*-butoxycarbonylation of amines with di-*tert*-butyl dicarbonate [34], our procedure provided better activity with high yields compared to the heterogeneous ones in the *N-tert*-butoxycarbonylation of amines. Next, we have shown the catalytic application of our synthesized catalyst in the trimethylsilylation of alcohols and phenols with high chemoselectivity and yields [35]. Herein, we are reporting the promoting effect of SANM in the conversion of alcohols to their corresponding MOM ethers. All reactions were performed in $CHCl_3$ at reflux temperature under completely heterogeneous reaction

$$\mathrm{ROH} \xrightarrow[\mathrm{CHCl_3,\ reflux}]{\mathrm{CH_2(OMe)_2\ /\ SANM}} \mathrm{ROCH_2OMe}$$

Scheme 2 Methoxymethylation of alcohols catalyzed by SANM.

conditions in good to high yields (Scheme 2, Table 1). Phenols remain intact under the same reaction conditions (entry 18 of Table 1). Therefore, the method can be useful for the chemoselective methoxymethylation of alcohols in the presence of phenols and primary alcohols in the presence of tertiary alcohols (entry 19 of Table 1).

Investigation in the reusability of the catalyst showed that SANM is reusable four times (entries 1 and 2 of Table 1).

The possible mechanism for the methoxymethylation of various alcohols in the presence of SANM as a promoter is shown in Scheme 3. On the basis of this mechanism, SANM catalyzes the reaction by electrophilic activation of FDMA and by making the central carbon of FDMA susceptible to nucleophilic attack by the alcohol. Successive elimination of MeOH results in the formation of MOM-ether derivatives and regenerates SANM in the reaction mixture [7].

To illustrate the efficiency of the present method, Table 2 compares our results with some of those reported in the literature [8-12].

Conclusions

In conclusion, we have developed a simple and efficient protocol for the methoxymethylation of various alcohols using SANM as a novel heterogeneous catalyst. Good yields of the products, short reaction times, heterogeneous nature of reaction conditions, use of relatively small amounts of FDMA, ease of preparation, stability of the reagent, recyclability, and easy workup procedure are important features of the reported method. We are exploring further applications of SANM for the other types of functional group transformations in our laboratory.

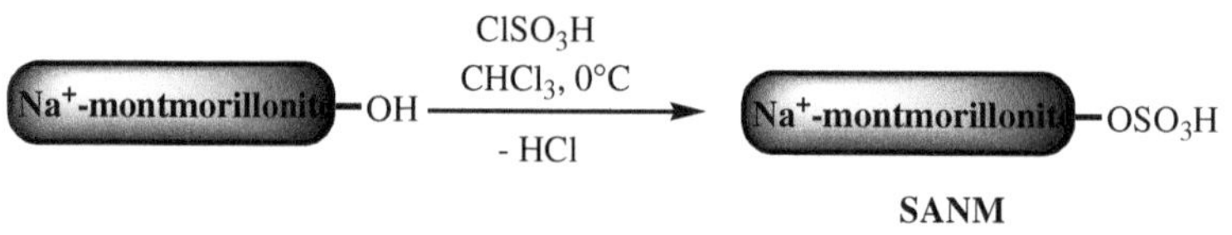

Scheme 1 Preparation of SANM.

Table 1 Methoxymethylation of alcohols

Entry	Alcohol	MOM ether	Time(h)	Yield (%)
1	$PhCH_2OH$	$PhCH_2OMOM$	2	92 (90, 85, 80[a])
2	$4\text{-}ClC_6H_4CH_2OH$	$4\text{-}ClC_6H_4CH_2OMOM$	1.5	95 (92, 88, 86[a])
3	$2\text{-}ClC_6H_4CH_2OH$	$2\text{-}ClC_6H_4CH_2OMOM$	3	96
4	$4\text{-}BrC_6H_4CH_2OH$	$4\text{-}BrC_6H_4CH_2OMOM$	2	94
5	$2\text{-}BrC_6H_4CH_2OH$	$2\text{-}BrC_6H_4CH_2OMOM$	4.5	92
6	$4\text{-}MeOC_6H_4CH_2OH$	$4\text{-}MeOC_6H_4CH_2OMOM$	0.5	95
7	4-*tert*-butyl$C_6H_4CH_2OH$	4-*tert*-butyl$C_6H_4CH_2OMOM$	3	90
8	$4\text{-}NO_2C_6H_4CH_2OH$	$4\text{-}NO_2C_6H_4CH_2OMOM$	7.5	93
9	$PhCH_2CH_2OH$	$PhCH_2CH_2OMOM$	4	95
10	PhCH(OH)Me	PhCh(OMOM)Me	5	91
11	$PhCH(OH)CH_2CH_3$	$PhCH(OMOM)CH_2CH_3$	2	96
12	$CH_3(CH_2)_5CH_2OH$	$CH_3(CH_2)_5CH_2OMOM$	3.5	91
13	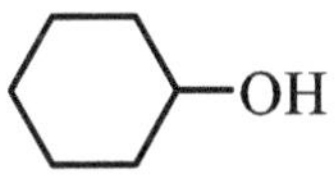	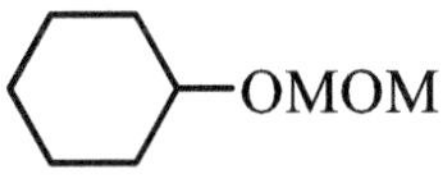	8	95
14	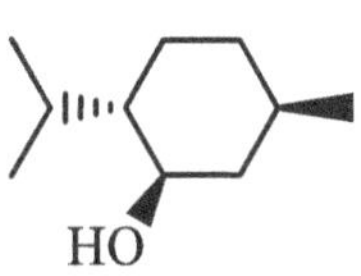	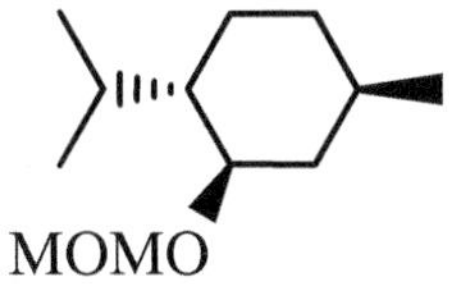	7.5	97
15	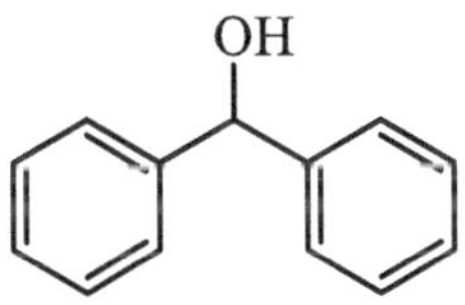	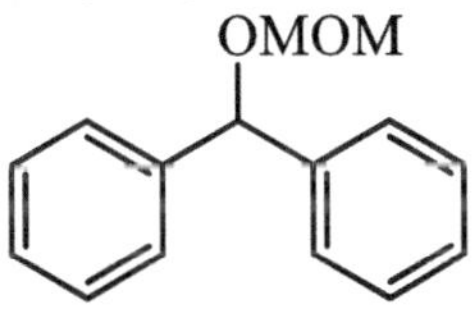	0.75	95
16	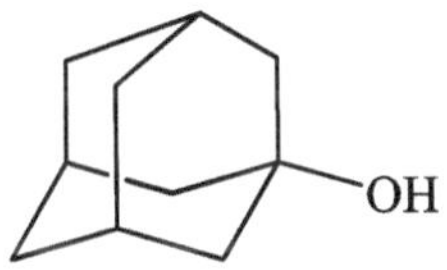	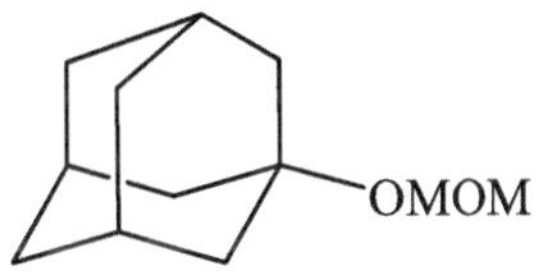	8	90
17	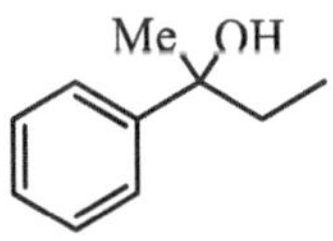	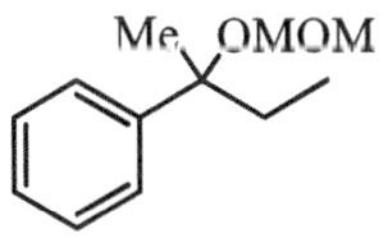	4	85
18	PhOH	PhOMOM	11	0
19	$PhCH_2OH$+PhOH	$PhCH_2OMOM$+PhOMOM	2.5	100[b]+0[b]

Products were characterized by comparison of their physical and spectral data with authentic samples [9,10,18,20] and isolated yields. [a]Results obtained using recycled catalyst; [b]Conversion.

Methods

General

Chemicals were purchased from Southern Clay Products (Gonzales, TX, USA), Fluka (Buchs, Switzerland), Merck (Darmstadt, Germany), and Sigma-Aldrich (St. Louis, MO, USA) chemical companies. All of the products are known compounds and were characterized by spectral analyses, comparisons with authentic samples (IR and NMR), and regeneration of the corresponding alcohols. All yields refer to the isolated products. The purity determination of the substrate and reaction monitoring was accompanied by gas chromatography or thin-layer chromatography (TLC) on silica-gel polygram SILG-UV 254 plates.

Catalyst preparation

A 500-mL suction flask charged with 2.5 g Na^+-montmorillonite (Southern Clay Products) and 10 mL $CHCl_3$ was equipped with a constant pressure dropping funnel containing chlorosulfonic acid (0.50 g, 9 mmol) and a gas inlet tube for conducting HCl gas into water as adsorbing solution. Chlorosulfonic acid was added dropwise over a period of 30 min while the reaction mixture

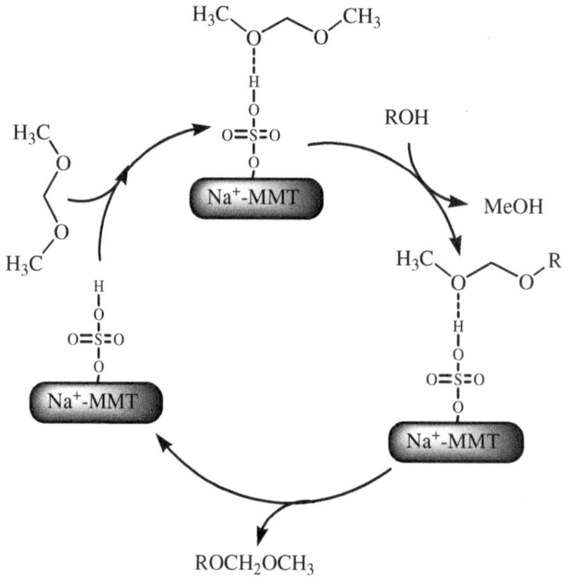

Scheme 3 Proposed mechanism.

was stirred slowly in an ice bath (0°C). After addition was completed, the mixture was stirred for additional 30 min to remove all HCl. The mixture was then filtered, and the solid residue was washed with methanol (20 mL) and dried at room temperature to obtain SANM as white powder (2.58 g) [34,35].

Catalyst characterization

The synthesized catalyst was characterized by Fourier transform infrared spectroscopy, transmission electron microscopy, nitrogen sorption technique, X-ray diffraction, thermogravimetric analysis, and elemental analysis [34,35].

General procedure

A mixture of the substrate (1 mmol), FDMA (6 mmol, 0.456 g), and SANM (3 mg) in $CHCl_3$ (3 mL) was stirred at reflux temperature. The progress of the reaction was monitored by TLC. After completion of the reaction, the solvent was evaporated, and Et_2O (5 mL) was added. The mixture was filtered, and the solid residue was washed with Et_2O (5 mL). The filtrate was washed with a saturated solution of $NaHCO_3$ and H_2O and dried over

Table 2 Comparison of some results obtained by our method (I) with those reported using various catalysts

Entry	Substrate	Time (h)/yield (%)/mmol of FDMA					
		I	II	III	IV	V	VI
1	$PhCH_2OH$	2/92/6	1.5/85/10	-	1/95/8	1.67/83/30	3/90/10
2	$4\text{-}NO_2C_6H_4CH_2OH$	7.5/93/6	12.5/80/10	5/94/50	-	12/75/30	4.5/83/10
3	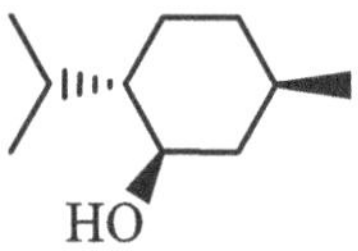	7.5/97/6	16.5/78/10	7/80/50	4/85/8	17.5/85/30	-
4	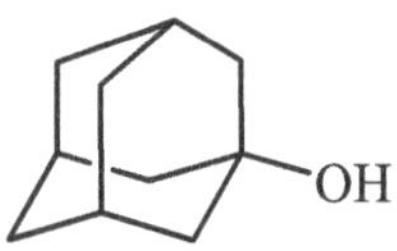	8/90/6	8/70/10	7/77/50	-	8.5/75/30	5/75/30

Various catalysts used silica sulfuric acid (II) [8], $Sc(OTf)_3$(III) [9], $Bi(OTf)_3$ (IV) [10], $Al(HSO_4)_3$ (V) [11], and $H_3Mo_{12}O_4$ (VI) [12].

$MgSO_4$. Evaporation of the solvent afforded the requested MOM ether in high purity.

Competing interests

The authors declare that they have no competing interests.

Acknowledgments

We are thankful to the University of Guilan Research Council for the partial support of this work.

References

1. Schelhaas, M, Waldmann, H: Protecting group strategies in organic synthesis. Angew Chem Int Ed Engl **35**, 2056–2083 (1996)
2. Jin, TS, Li, TS, Gao, YT: A facile preparation of methoxymethyl ethers of primary and secondary alcohols with dimethoxymethane catalyzed by expansive graphite. Synth Commun **28**, 837–841 (1998)
3. Yardley, JP, Fletcher, H: Introduction of the methoxymethyl ether protecting group. Synthesis **1976**, 244–245 (1976)
4. Olah, GA, Husain, A, Gupta, BGB, Narang, SC: Catalysis by solid superacids 1; 15. Facile preparation of methoxymethyl ethers. Synthesis (6), 471–472 (1981)
5. Olah, GA, Husain, A, Narang, SC: Iodotrimethylsilane-catalyzed preparation of methoxymethyl ethers of primary and secondary alcohols with dimethoxymethane. Synthesis **1983**(11), 896–897 (1983)
6. Bandgar, BP, Hajare, CT, Wadgaonkar, PP: Envirocat EPZG catalyzed methoxymethylation of alcohols. J Chem Res, 90–91 (1996)
7. Lin, CH, Wan, MY, Huang, YM: Methoxymethylation of alcohols catalyzed by sulfated metal oxides. Catal Let. **87**, 253–256 (2003)
8. Niknam, K, Zolfigol, MA, Khorramabadi-Zad, A, Zare, R, Shayegh, M: Silica sulfuric acid as an efficient and recyclable catalyst for the methoxymethylation of alcohols under solvent-free conditions. Catal Commun **7**, 494–498 (2006)
9. Karimi, B, Ma'mani, L: Scandium trifluoromethanesulfonate as a recyclable catalyst for efficient methoxymethylation of alcohols. Tetrahedron Lett **44**, 6051–6053 (2003)
10. Sreedhar, B, Swapna, V, Sridhar, C: Bismuth triflate and microencapsulated bismuth triflate: Efficient catalysts for methoxymethylation of alcohols and carboxylic acids. Catal Commun **6**, 293–296 (2005)
11. Niknam, K, Zolfigol, MA, Shayegh, M, Zare, R: Metal hydrogen sulfates catalyzed methoxymethylation of alcohols under solvent free conditions. J Chin Chem Soc **54**, 1067–1073 (2007)
12. Kantam, ML, Santhi, PL: Molybdenum catalyzed methoxymethylation of alcohols. Synlett, 429–430 (1993)
13. Zolfigol, MA, Shiri, M: Molybdatophosphoric acid as a catalyst for the methoxymethylation of alcohols under solvent-free conditions. Mendeleev Commun **15**, 165–166 (2005)
14. Patney, HK: Anhydrous iron(III) chloride dispersed on 3A molecular sieves, a mild and efficient catalyst for methoxymethylation of alcohols. Synlett, 567–568 (1992)
15. Jin, TS, Guo, JJ, Yin, YH, Zhang, SL, Li, TS: TiO_2/SO_4^{2-}, an efficient catalyst for the methoxymethylation of alcohols. J Chem Res, 188–189 (2002)
16. Mohammadpoor-Baltork, I, Moghadam, M, Tangestaninejad, S, Mirkhani, V, Mirjafari, A: $H_3PW_{12}O_{40}$ - a selective, environmentally benign, and reusable catalyst for the preparation of methoxymethyl and ethoxymethyl ethers and their deprotections under mild conditions. Can J Chem **86**, 831–840 (2008)
17. Mohammadpoor-Baltork, I, Moghadam, M, Tangestaninejad, S, Mirkhani, V, Mirjafari, A: 12-Tungstophosphoric acid supported on inorganic oxides as heterogeneous and reusable catalysts for the selective preparation of alkoxymethyl ethers and their deprotections under different reaction conditions. Polyhedron **27**, 2612–2624 (2008)
18. Shirini, F, Zolfigol, MA, Albadi, J: Melamine trisulfonic acid (MTSA): a new efficient catalyst for the chemoselective methoxymethylation of alcohols. Synth Commun **40**, 910–914 (2010)
19. Shirini, F, Imanzadeh, GH, Mousazadeh, SAR, Mohammadpoor-Baltork, I, Abedini, M: A mild and efficient method for the methoxymethylation and acetylation of alcohols promoted by benzyltriphenylphosphonium tribromide. Chin Chem Lett **21**, 1187–1190 (2010)
20. Moghadam, M, Tangestaninejad, S, Mirkhani, V, Mohammadpoor-Baltork, I, Khajehzadeh, M, Kosari, F, Araghi, M: High-valent $[Sn^{IV}(Br_8TPP)(OTf)_2]$ as a highly efficient and reusable catalyst for selective methoxymethylation of alcohols and phenols: the effect of substituted bromines on the catalytic activity. Polyhedron **29**, 238–243 (2010)
21. Bergaya, F, Lagaly, G: Surface modification of clay minerals. Appl Clay Sci **19**, 1–3 (2001)
22. Vaccari, A: Preparation and catalytic properties of cationic and anionic clays. Catal Today **41**, 53–71 (1998)
23. Laszlo, P: Chemical reactions on clays. Science **235**, 1473–1477 (1987)
24. Giannelis, EP, Krishnamoorti, R, Manias, E: Polymer-silicate nanocomposites: model systems for confined polymers and polymer brushes. Adv Polym Sci **138**, 107–147 (1999)
25. Shi, H, Lan, T, Pinnavaia, TJ: Interfacial effects on the reinforcement properties of polymer-organoclay nanocomposites. Chem Mater **8**, 1584–1587 (1996)
26. Yariv, S, Cross, H: Organo-clay complexes and interactions, 1st edn. CRC, New York (2002)
27. Berger, G: The structure of montmorillonite. Preliminary communication on the ability of clays and clay minerals to be methylated. Chem Weekblad **38**, 42–43 (1941)
28. Deuel, H, Huber, G, Iberg, R: Organische derivate von tonmineralien. Helv Chim Acta **33**, 1229–1232 (1950)
29. Gieseking, JE: The clay minerals in soils. Adv Agrononmy **1**, 159–204 (1949)
30. Shirini, F, Zolfigol, MA, Mohammadi, K: Silica sulfuric acid as a mild and efficient reagent for the acetylation of alcohols in solution and under solvent free conditions. Bull Korean Chem Soc **25**, 325–327 (2004)
31. Shirini, F, Zolfigol, MA, Mohammadi, K: Silica sulfuric acid as an efficient reagent for the synthesis of symmetrical ethers under mild and heterogeneous conditions. Phosphorous Sulfur Silicon **178**, 2357–2361 (2003)
32. Shirini, F, Sadeghzadeh, P, Abedini, M: Silica sulfuric acid: a versatile reagent for oxathioacetalyzation of carbonyl compounds and deprotection of 1,3-oxathiolanes. Chin Chem Lett **20**, 1457–1460 (2009)
33. Salehi, P, Zolfigol, MA, Shirini, F, Baghbanzadeh, M: Silica sulfuric acid and silica chloride as efficient reagent for organic reactions. Curr Org Chem **10**, 2171–2189 (2006)
34. Shirini, F, Mamaghani, M, Atghia, SV: Sulfonic acid-functionalized ordered nanoporous Na^+-montmorillonite (SANM): A novel, efficient and recyclable catalyst for the chemoselective *N*-Boc protection of amines in solventless media. Catal Commun **12**, 1088–1094 (2011)
35. Shirini, F, Mamaghani, M, Atghia, SV: A mild and efficient method for the chemoselective trimethylsilylation of alcohols and phenols and deprotection of silyl ethers using sulfonic acid-functionalized ordered nanoporous Na^+-montmorillonite. Appl Clay Sci **58**, 67–72 (2012)

2

Green synthesis of densely dispersed and stable silver nanoparticles using myrrh extract and evaluation of their antibacterial activity

Ibrahim M El-Sherbiny[1,2*], Ehab Salih[3] and Fikry M Reicha[3]

Abstract

In the present study, green synthesis of densely dispersed and stable silver nanoparticles (Ag NPs) using myrrh extract as green-reducing and stabilizing agent was investigated. The Ag NPs were synthesized by exposing a mixture of $AgNO_3$ and myrrh extract to UV irradiation for different time intervals. The effects of parameters such as concentration of $AgNO_3$ and reaction time onto the physicochemical characteristics of the developed Ag NPs were studied. Formation of the Ag NPs was noted upon changing the solution color from pale yellow into brown and was confirmed by the appearance of surface plasmon resonance peaks at about 445 nm. The resulting Ag NPs were characterized by UV-vis spectrophotometry, transmission electron microscopy, Fourier transform infrared spectroscopy, and X-ray diffraction. The developed NPs demonstrated a relatively high antibacterial activity against *Bacillus thuringiensis* and *Pseudomonas aeruginosa* bacteria as compared to that of myrrh extract. The antibacterial activity increased with increasing the Ag NP concentration.

Keywords: Myrrh, Green synthesis, Silver, Nanoparticles, Antibacterial activity

Background

Nanotechnology comprises the science and technology at dimensions below 100 of nanometers. At this scale, the physicochemical characteristics of materials are significantly different from those found in larger scales and bulk materials due to the quantum effects [1].

Synthesis of metal nanoparticles (NPs) is an expanding research area due to the wide range of their applications in various fields such as medicine, electronics, and energy [2]. One of the most important classes of metal NPs is that made of noble metals such as gold (Au), silver (Ag), and platinum (Pt) [3]. Among the noble metal NPs, Ag NPs are perhaps the most widely recognized for their applications in photonics [4-6], micro-electronics [7,8], photocatalysis [9,10], and lithography [11,12]. The widespread use of Ag NPs in medicine, for instance, can be attributed to their potent antimicrobial activity against a wide range of pathogenic microorganisms [13].

* Correspondence: imelsherbiny@gmail.com
[1]Zewail University, Zewail City of Science and Technology, 6th October City, Giza 12588, Egypt
[2]Chemistry Department, Faculty of Science, Mansoura University, Mansoura ET-35516, Egypt
Full list of author information is available at the end of the article

The production of significant amounts of Ag NPs is achievable using various physical and chemical techniques including laser ablation [14], lithography [15], and the photochemical reduction [16]. However, synthesis techniques remain relatively expensive and occasionally involve the use of some hazardous moieties [17].

Biosynthesis of Ag NPs using microorganisms like bacteria [18], fungi [19], and yeast [20] has been investigated. However, exploration of the plant extracts as the potential nanofactories for the green synthesis of Ag NPs has a growing interest due to the major benefit of this synthesis technique in protecting the environment [21].

In this study, new series of stable densely dispersed Ag NPs were obtained using an entirely green synthesis approach. In this approach, the natural non-toxic ingredients extracted from the myrrh plant were used, instead of common hazardous chemicals, in combination with UV light irradiation to create the Ag NPs. Myrrh extract has been used as reducing and capping agent for the synthesis of NPs which could be advantageous over microbial synthesis as there is no need of the elaborated process of culturing and maintaining the cells.

Myrrh has been used for long time as a medicine and wound dressing, and has also been closely associated with the health and purification rituals of women. The medical potential of myrrh was first described in China in 600 AD during the Tang Dynasty, and it is still used to date in Chinese medicine to treat wounds, relieve painful swelling, and menstrual pain associated with blood stagnation. Myrrh consists of water-soluble gum, alcohol-soluble resins, and volatile oil [22]. The gum constituent contains polysaccharides and proteins, whereas the volatile oils are composed of steroids, sterols, and terpenes; the myrrh's characteristic odor is attributed to the presence of furanosesquiterpenes [22].

Results and discussion

Formation of the Ag NPs

The myrrh can be considered a potential source of hydrocarbons, and it consists of water-soluble gum, alcohol-soluble resins, and volatile oil [22]. The gum constituent contains various polysaccharides and proteins, whereas the volatile oil is composed of steroids, sterols, and terpenes [22]. Many of these compounds in the myrrh extract, especially the polysaccharides and saccharide moieties illustrated in Figure 1, are expected to be responsible for the partial reduction process of $AgNO_3$ into metallic Ag. During the synthesis process, the aqueous Ag^+ ions were exposed to the myrrh extract which initiated the reduction process of the Ag^+ ions in solution, leading to the formation of Ag hydrosol. Then, a complete reduction of the Ag^+ ions into atomic Ag then to Ag NPs was achieved through UV irradiation reduction of the reaction mixture for different time intervals from 10 to 120 min.

The formation of Ag NPs was observed upon the color change of the myrrh extract from transparent yellow into brown, as shown in Figure 2, due to the coherent oscillation of electrons at the surface of NPs, resulting in surface plasmon resonance (SPR) [23]. The color change into brown was noted within 10 min, and the color intensity increased significantly with increasing the $AgNO_3$ concentration at a fixed volume of myrrh extract. The UV-vis spectrophotometry was also used to confirm the formation of the Ag NPs as shown in Figure 3. From Figure 3a, the SPR band steadily increased in intensity with a prominent peak at about 480 nm at 1 mM concentration and then made a blue shift up to about 445 nm with increasing UV irradiation reduction time. The change of color and intensity of the SPR band with increasing irradiation time might be due to the variation in concentration, size, and shape of the resulting Ag NPs [24]. The peak that appeared at around 390 nm is likely to arise from the myrrh extract. Increasing the concentration of the Ag NP precursor (Ag NO_3) from 1 to 15 mM also led to the increasing intensity of the SPR band with a blue shift from 480 to about 450 nm (Figure 3b).

FTIR spectroscopy

FTIR analysis was carried out in order to identify the possible reducing and stabilizing biomolecules in the myrrh extract. The FTIR spectra of myrrh extract before and after the development of Ag NPs are shown in Figure 4. Both spectra are very similar and showed bands around 1045, 1445, 1438, 1635, 2855, 2925, and 3449 cm^{-1}. The band around 1,025 to 1,200 cm^{-1} corresponds to the C-O stretching, while the weak bands in the 1,340 to 1,450 cm^{-1} range can be attributed to aliphatic CH_2 and CH_3 groups, CH_2 groups of aldehydes and ketones, and the bending modes of O-H bonds in alcohols, phenols, and carboxylic acids [25]. The band around 1,620 to 1,650 cm^{-1} corresponds to aromatic rings, while the band that appeared around 2,920 to 2,930 cm^{-1} corresponds to the asymmetric stretching of the C-H bonds. The strong broad band appearing at about 3,440 cm^{-1} in both FTIR spectra can be assigned to the stretching vibrations of O-H groups in alcohols and phenols [25]. This band in particular confirms the presence of various O-H groups in the myrrh extract (Figure 1) which is expected to play a good role in the reduction process of Ag^+ ions into Ag then to Ag NPs. This band at 3,340 cm^{-1} in the Ag NP spectrum has also showed a slight shift when compared to that of the myrrh extract, indicating the formation of Ag NPs.

X-ray diffraction analysis

The X-ray diffraction pattern of the myrrh-Ag NPs developed after 60 min of UV irradiation of 6 mM $AgNO_3$ is illustrated in Figure 5. As apparent from the figure, there is a broad peak that appeared at $2\theta = 20°$ which

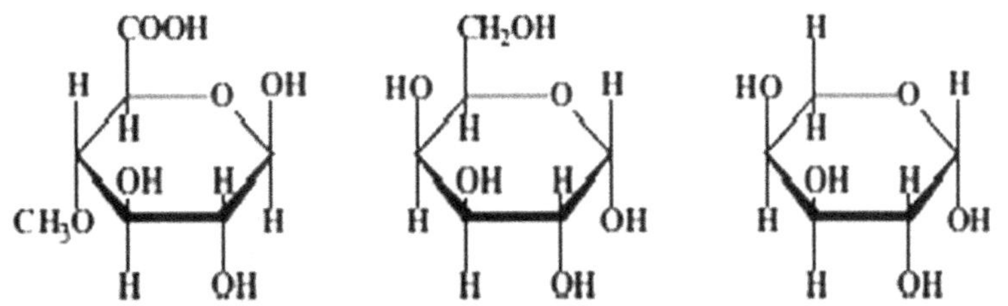

Figure 1 Some saccharides in the myrrh extract (4-methyl D-glucuronic acid, D-galactose, and L-arabinose) [22].

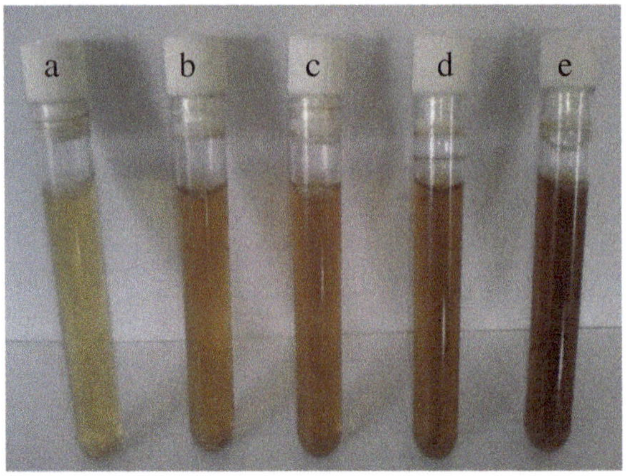

Figure 2 The color change of the myrrh extract. (**a**) Upon the formation of Ag NPs from different concentrations of $AgNO_3$: (**b**) 1 mM, (**c**) 4 mM, (**d**) 10 mM, and (**e**) 15 mM.

can be attributed to the amorphous phase of myrrh in addition to a number of Bragg reflections with 2θ values of 38°, 44°, 64.6°, and 77°. These bands correspond to the (111), (200), (220), and (311) sets of lattice planes, which may be indexed as the bands for face centered cubic structures of Ag [26]. The XRD pattern, thus, clearly demonstrates that the Ag NPs synthesized by the present green method are crystalline in nature.

Transmission electron microscopy

TEM has been employed to characterize the size, shape, and morphologies of the formed Ag NPs. The typical TEM micrographs and the corresponding particle size distribution of the myrrh-Ag NPs developed after 60 min of UV irradiation reduction of different concentrations of $AgNO_3$ are shown in Figure 6. As apparent from the figure, the obtained myrrh-Ag particles are in the nano range with uniform and spherical shapes. The myrrh-Ag NPs obtained using 1 mM of $AgNO_3$ followed by 60 min of UV irradiation (Figure 6a,c) showed average particle diameters from 40 to 100 nm. This particle size range was decreased to 10 to 50 nm by increasing the $AgNO_3$ concentration to 6 mM followed by 60 min of UV irradiation (Figure 6b,d). This result is in agreement with the blue shift observed in the UV-vis spectra of these samples (Figure 3). The TEM micrographs demonstrated also that the myrrh-Ag NPs are well and densely dispersed which indicates a good stabilization effect of the myrrh extract. The stability of the developed myrrh-Ag NPs was also confirmed through the absence of any visible changes for more than 6 months. Figure 6e shows the diffraction pattern of the developed NPs, which demonstrates the characteristic crystal planes of the elemental Ag.

Antibacterial studies

The inhibition zone values were determined for the Ag NPs synthesized using myrrh extract tested against two types of bacteria, Gram positive (*B. thuringiensis*) and Gram negative (*P. aeruginosa*). The results of the inhibition zones were presented as average values (cm) as shown in Figure 7. The obtained data (Figure 7) demonstrate that all the investigated Ag NPs had high antibacterial activity against the Gram-positive (*B. thuringiensis*) bacteria and even much better antibacterial activity against the Gram-negative (*P. aeruginosa*) bacteria as compared to that of the control (myrrh extract). It is also apparent from Figure 7 that increasing the concentration of the Ag NPs has led to a significant increase in the antibacterial activity. This promising antibacterial activity may be related to the small size and the high surface area of the developed Ag NPs, which allow them

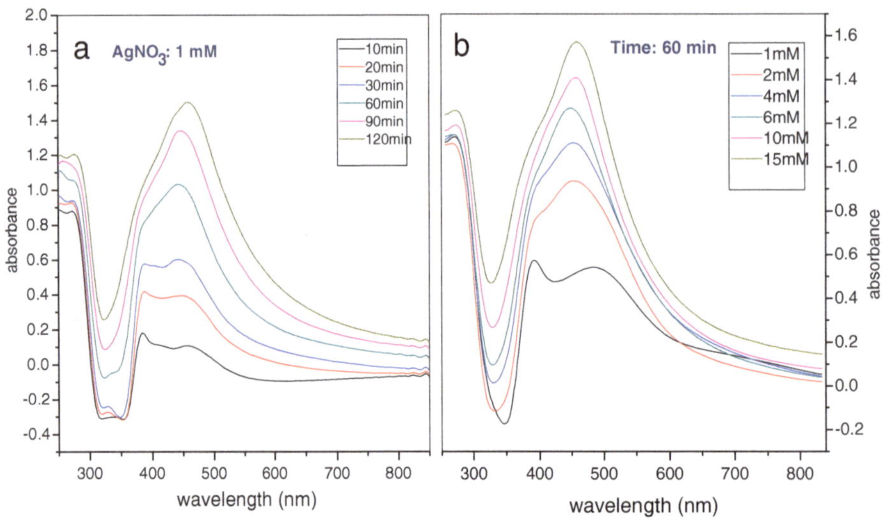

Figure 3 UV-vis spectra. UV-vis spectra of the formed Ag NPs recorded as a function of (**a**) time of UV irradiation and (**b**) $AgNO_3$ concentration.

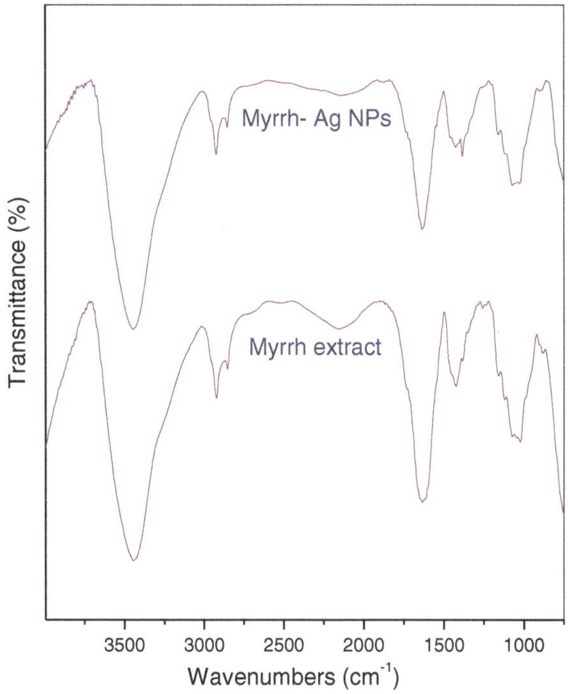

Figure 4 FTIR spectra of the myrrh extract before and after the synthesis of Ag NPs.

to reach easily the nuclear content of bacteria [27,28]. Also, it has been reported that the pathogenic effect of the NPs can be attributed to their stability in the medium as colloids, which modulates the phosphotyrosine pattern of the pathogen proteins and consequently arrests their growth. Moreover, it was proposed that the variation in the growth inhibition of bacteria by NPs may be attributed to the presence of peptidoglycan, which has a strong negative complex structure [29]. This negative structure may participate in the sequestration of free Ag^+ ions. Therefore, Gram-positive bacteria may allow less Ag NPs to reach the cytoplasmic membrane than the Gram-negative bacteria, which tends to explain the founded higher antibacterial activity of the synthesized Ag NPs towards the Gram-negative (*P. aeruginosa*) bacteria than towards the Gram-positive one. These obtained preliminary results indicate a potential of the prepared myrrh-Ag NPs as antibacterial agents; however, further investigations are required to explore their bactericidal effects on other types of bacteria.

Conclusion

The current study demonstrated that the Myrrh extract can be used as a green reducing and capping agent for the eco-friendly synthesis of Ag NPs. The developed Ag NPs showed good stability, and no visible changes were observed even after 6 months. The NPs also demonstrated high bactericidal activity against both *B. thuringiensis* and *P. aeruginosa* bacteria as compared to that of myrrh extract, and the antibacterial activity increased with increasing Ag NP concentration. In conclusion, this preliminary study revealed that the developed Ag NPs can be tailored and used, after more investigations, as potential antibacterial agents for various biological and biomedical applications.

Methods

Materials

Myrrh was purchased from a local market located at Mansoura, Egypt. Silver nitrate, ($AgNO_3$, AR grade, 99.5% purity) was obtained from Carl Roth GmbH & Co. KG (Schoemperlenstrasse, Karlsruhe, Germany). All of the other chemicals and solvents were of analytical grade and used as provided.

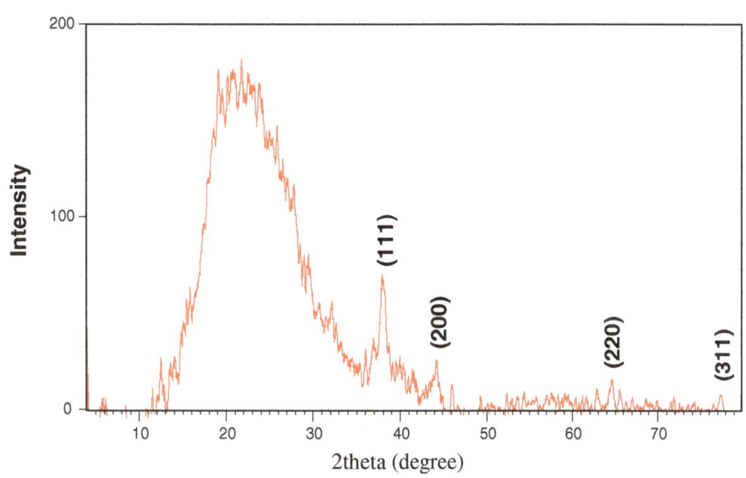

Figure 5 XRD patterns. XRD patterns of the Ag NPs developed after 60 min of UV irradiation of 6 mM $AgNO_3$.

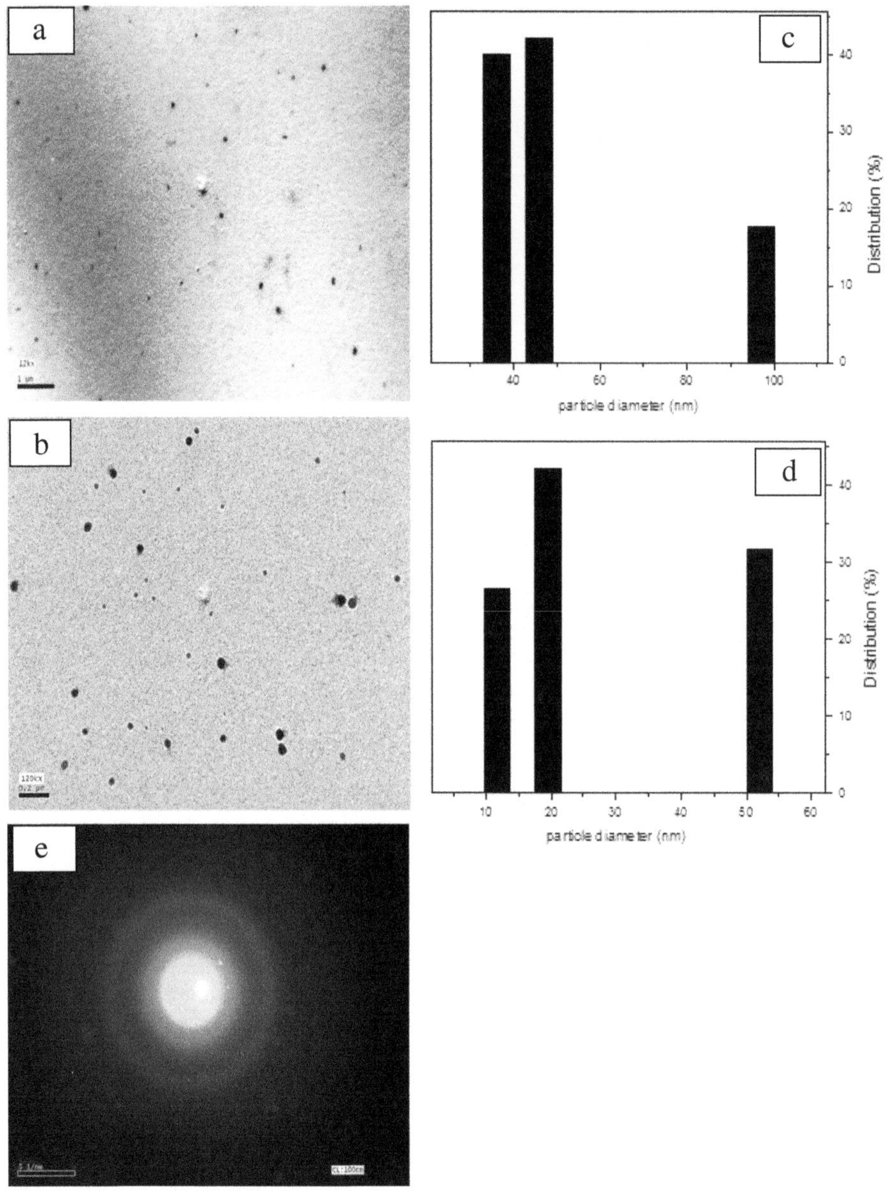

Figure 6 TEM micrographs. TEM micrographs of the Ag NPs developed in presence of the myrrh extract and 60 min of UV irradiation using different $AgNO_3$ concentrations. (**a**) 1 mM and (**b**) 6 mM. (**c**, **d**). The histograms of size distribution of the corresponding synthesized Ag NPs and (**e**) selected area of electron diffraction. The scale bars are (**a**) 1 μm, (**b**) 0.2 μm, and (**e**) μm.

Preparation of myrrh extract

Myrrh parts were collected and washed thoroughly with distilled water and grinded into fine pieces. About 10 g of thoroughly washed finally grinded myrrh was weighted and transferred into a 300-ml Erlenmeyer flask containing 200-ml deionized water, mixed well and heated for 7 h in water bath at 80°C. The resulting solution was centrifuged at 4,000 rpm for 15 min and then filtered using Whatman number 1 filter paper, and the filtrate was stored at 4°C for further use.

Synthesis of Ag NPs

Aqueous $AgNO_3$ solution (6 mM) was prepared and used for the synthesis of Ag NPs. Typically, 15 ml of myrrh aqueous extract was mixed with 60 ml of the $AgNO_3$ solution, and the resulting mixture was then irradiated by UV light (Philips UV lamp, 36 W; Koninklijke Philips Electronics N.V., Amsterdam, Netherlands) for different time intervals from 10 to 120 min. For studying the effect of $AgNO_3$ concentration, 5 ml of myrrh extract was added to 20 ml of different concentrations of the aqueous $AgNO_3$ solution

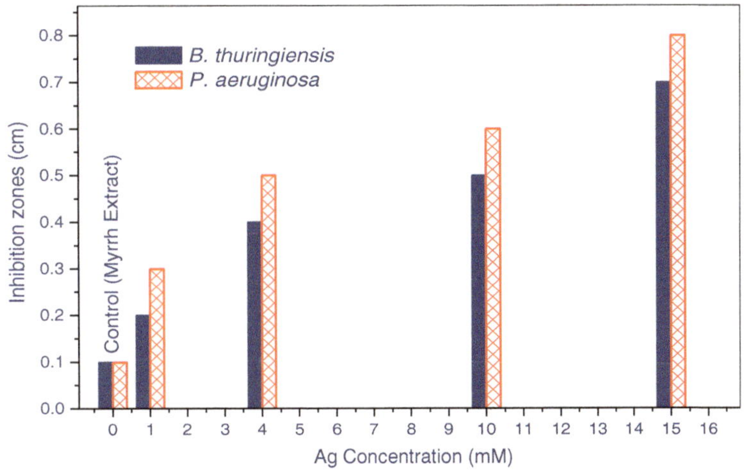

Figure 7 The bacterial inhibition zones of the developed myrrh-Ag NPs of different $AgNO_3$ concentrations.

(1 to 15 mM) followed by UV irradiation reduction of the mixture for 60 min.

Analysis of Ag NPs

The reduction of pure Ag^+ ions was monitored by measuring the UV-vis spectrum of the reaction medium at different time intervals. The UV-visible spectra of the formed Ag NPs were recorded in the range of 200 to 800 nm using ATI Unicom UV-vis spectrophotometer (UNICOM Systems, Inc., England) with the aid of ATI Unicom UV-vis. vision software V 3.20. The analysis was performed at room temperature using quartz cuvettes (1-cm optical path), and the blank was the corresponding myrrh aqueous solution. Fourier transform infrared spectroscopy (FTIR) measurements were carried out to identify the possible biomolecules in the aqueous extract of myrrh, which are responsible for the reduction of the Ag^+ ions and capping of the resulting Ag NPs. The sample was dried and grinded with KBr pellets and analyzed on Mattson 5000 FTIR spectrometer (Mattson Instruments, Inc., WI, USA) in the range of 400 to 4,000 cm^{-1} at a resolution of 8 cm^{-1} at 25°C. The size and morphology of the synthesized Ag NPs were investigated by transmission electron microscopy (TEM; JEOL TEM-1230, JEOL Ltd., Tokyo, Japan) attached to a CCD camera at an accelerating voltage of 120 kV. The samples were prepared by placing few drops of the Ag NP suspension on carbon-coated copper grids, followed by allowing the solvent to slowly evaporate before recording the TEM images. The X-ray diffraction patterns of the Ag NP samples were recorded using Philips PW 1390 X-ray diffractometer (Netherland). The X-ray diffraction was provided with a beam monochromator and Cu $K\alpha$ radiation at λ = 1.5406 Å. The applied voltage was 40 kV, and the current intensity was 40 mA sec. The 2θ angles were recorded in the range of 40° to 60°.

Assessment of the antibacterial activity

The antibacterial activity of myrrh aqueous extract and the Ag NPs developed at different $AgNO_3$ concentrations (1, 4, 10, and 15 mM) was evaluated against two types of bacteria, *Bacillus thuringiensis* and *Pseudomonas aeruginosa*, isolated from patients in Mansoura University Hospital, Mansoura, Egypt. The antibacterial assessment was performed using the nutrient agar diffusion method and measuring the inhibition zones (cm). In brief, sterile paper discs (6 mm) were impregnated overnight in the sample solutions and then left to dry for 24 h at 37°C in sterile conditions. The bacterial suspensions were obtained by making a saline suspension of isolated colonies selected from 18 to 24 h of nutrient agar plating. The suspensions were then adjusted to match the tube of 0.5 McFarland turbidity standards using spectrophotometry at λ = 600 nm, which equals 1.5.108 colony-forming units per milliliter. The surface of the nutrient agar was completely inoculated using a sterile swab, which was steeped in the prepared bacterial suspension. Then, the impregnated discs were placed on the inoculated agar and incubated at 37°C for 24 h. After incubation, the diameters of the growth inhibition zones were determined.

Competing interests

The authors declare that they have no competing interests.

Authors' contributions

IME participated in the idea of the study, the design of the study, interpretation of the results, and writing the manuscript for publication. ES performed the green synthesis and the physicochemical characterization of the myrrh-Ag nanoparticles and early drafted the manuscript. FMR suggested the myrrh for the green synthesis of the Ag nanoparticles, conceived of the study, and participated in its coordination. All authors read and approved the final manuscript.

Authors' information

IME achieved the following in his academic years: B.Sc. and M.Sc. in Polymer Chemistry, Mansoura University, Egypt; Ph.D. in Drug Delivery from IFS,

Massey University, New Zealand; Postdoctoral Fellow from the College of Pharmacy, University of New Mexico, USA; and Postdoctoral Associate from the College of Pharmacy, University of Texas, Texas, USA. He is an Associate Professor from the Zewail University of Science and Technology, Egypt and an Associate Professor from Mansoura University, Egypt. He is a Research Assistant Professor in Pharmaceutics, Texas University at Austin, USA, and a Fellow of BMES, AAPS and NZ Institute of Chemistry. His research area interests cover smart nano-biomaterials, controlled drug delivery, wastewater treatment, tissue engineering, and gene therapy. ES obtained his BS in Physics from the Faculty of Science, Mansoura University, Egypt, with a graduation project on Fractional calculus. He earned his MSc in Physics with the thesis entitled 'Green synthesis of metal nanoparticles in plant extracts and evaluation of their physicochemical and biological characteristics', from the Faculty of Science, Mansoura University, Egypt. His research area interests include nanomaterials and biomaterials. FMR earned his BSc in Physics & Chemistry from the Faculty of Science, Alexandria University, Egypt. He is an MSc degree holder of Chemistry and Physics. He obtained his PhD in Thin Films from Magaryans. He was a Professor of Experimental Solid State Physics. He is the Head of Advanced Biomaterials Laboratory, Physics Department, Mansoura University. His research area covers nano-biotechnology, electrochemical polymerization and complexation, biomaterials, tissue engineering, AC-dielectric studies, and crystal science.

Author details
[1]Zewail University, Zewail City of Science and Technology, 6th October City, Giza 12588, Egypt. [2]Chemistry Department, Faculty of Science, Mansoura University, Mansoura ET-35516, Egypt. [3]Biological Advanced Materials, Physics Department, Faculty of Science, Mansoura University, Mansoura ET-35516, Egypt.

References

1. Matos, RA, Cordeiro, TS, Samad, RE, Vieira Jr, ND, Courrol, LC: Green synthesis of stable silver nanoparticles using Euphorbia milii latex. Colloid Surface A **389**, 134–137 (2011)
2. Saxena, A, Tripathi, RM, Zafar, F, Singh, P: Green synthesis of silver nanoparticles using aqueous solution of Ficus benghalensis leaf extract and characterization of their antibacterial activity. Mater. Lett. **67**, 91–94 (2012)
3. Kaviya, S, Santhanalakshmi, J, Viswanathan, B, Muthumary, J, Srinivasan, K: Biosynthesis of silver nanoparticles using *Citrus sinensis* peel extract and its antibacterial activity. Spectrochim Acta A **79**, 594–598 (2011)
4. Inacio, PL, Barreto, BJ, Horowitz, F, Correia, RRB, Pereira, MB: Silver migration at the surface of ion-exchange waveguides: a plasmonic template. Opt Mater Express **3**, 390–399 (2013)
5. Hosel, M, Krebs, FC: Large-scale roll-to-roll photonic sintering of flexo printed silver nanoparticle electrodes. J. Mater. Chem. **22**, 15683–15688 (2012)
6. Gould, IR, Lenhard, JR, Muenter, AA, Godleski, SA, Farid, S: Two-electron sensitization: a new concept for silver halide photography. J. Am. Chem. Soc. **122**, 11934–11943 (2000)
7. Reddy, VR, Currao, A, Calzaferri, G: Gold and silver metal nanoparticle-modified AgCl photocatalyst for water oxidation to O_2. J. Phys. Conf. Ser. **61**, 960–965 (2007)
8. Deheer, WA: The physics of simple metal clusters: experimental aspects and simple models. Rev. Mod. Phys. **65**, 611–676 (1993)
9. Bawendi, MG, Steigerwald, ML, Brus, LE: The quantum mechanics of larger semiconductor clusters ("Quantum Dots"). Annu. Rev. Phys. Chem. **41**, 477–496 (1990)
10. Bar, H, Bhui, DK, Sahoo, GP, Sarkar, P, Pyne, S, Misra, A: Green synthesis of silver nanoparticles using seed extract of Jatropha curcas. Colloid Surface A **348**, 212–216 (2009)
11. Xia, Y, Rogers, JA, Paul, KE, Whitesides, GM: Unconventional methods for fabricating and patterning nanostructures. Chem. Rev. **99**, 1823–1848 (1999)
12. Shipway, AN, Lahav, M, Willner, I: Nanostructured gold colloid electrodes. Adv. Mater. **12**, 993–998 (2000)
13. Klaine, SJ, Alvarez, PJ, Batley, GE, Fernandes, TF, Handy, RD, Lyon, DY: Nanomaterials in the environment: behavior, fate, bioavailability and effects. Environ. Toxicol. Chem. **27**, 1825–1851 (2008)
14. Tsuji, T, Kakita, T, Tsuji, M: Preparation of nano-size particles of silver with femtosecond laser ablation in water. Appl. Surf. Sci. **206**, 314–320 (2003)
15. Jensen, TR, Malinsky, MD, Haynes, CL, Duyne, RPV: Nanosphere lithography: tunable localized surface plasmon resonance spectra of silver nanoparticles. J. Phys. Chem. B **104**, 10549–10556 (2000)
16. Lu, HW, Liu, SH, Wang, XL, Qian, XF, Yin, J, Zhu, ZK: Silver nanocrystals by hyperbranched polyurethane-assisted photochemical reduction of Ag^+. Mat Chem Phys **81**, 104–107 (2003)
17. Lukman, AI, Gong, B, Marjo, CE, Roessner, U, Harris, AT: Facile synthesis, stabilization, and anti-bacterial performance of discrete Ag nanoparticles using Medicago sativa seed exudates. J Colloid Interf Sci **353**, 433–444 (2011)
18. Saifuddin, N, Wong, CW, Nur Yasumira, AA: Rapid biosynthesis of silver nanoparticles using culture supernatant of bacteria with microwave irradiation. Eur. J. Chem. **6**, 61–70 (2009)
19. Duran, N, Marcato, PD, De Souza, GIH, Alves, OL, Esposito, E: Mechanistic aspects of biosynthesis of silver nanoparticles by several Fusarium oxysporum strains. J. Biomed. Nanotechnol. **3**, 1–8 (2009)
20. Kowshik, M, Ashtaputre, S, Kharrazi, S, Vogel, W, Urban, J, Kulkarni, SK, Paknikar, KM: Extracellular synthesis of silver nanoparticles by a silver-tolerant yeast strain MKY3. Nanotechnology **14**, 95–100 (2003)
21. Jacob, SJP, Finub, JS, Narayanan, A: Synthesis of silver nanoparticles using Piper longum leaf extracts and its cytotoxic activity against Hep-2 cell line. Colloid Surfaces B **91**, 212–214 (2012)
22. Hanusa, LO, Rezankab, T, Dembitskya, VM, Moussaieff, A: Myrrh—commiphora chemistry. Biomed Pap **149**, 3–28 (2005)
23. Narayanan, KB, Sakthivel, N: Extracellular synthesis of silver nanoparticles using the leaf extract of Coleus amboinicus Lour. Mater. Res. Bull **46**, 1708–1713 (2011)
24. Rani, PU, Rajasekharreddy, P: Green synthesis of silver-protein (core–shell) nanoparticles using Piper betle L. leaf extract and its ecotoxicological studies on Daphnia magna. Colloid Surface A **389**, 188–194 (2011)
25. Yilmaz, M, Turkdemir, H, Kilic, MA, Bayram, E, Cicek, A, Mete, A, Ulug, B: Biosynthesis of silver nanoparticles using leaves of Stevia rebaudiana. Mater. Chem. Phys. **130**, 1195–1202 (2011)
26. Sathishkumar, M, Sneha, K, Won, SW, Cho, CW, Kim, S, Yun, YS: Cinnamon zeylanicum bark extract and powder mediated green synthesis of nano-crystalline silver particles and its bactericidal activity. Colloid Surfaces B **73**, 332–338 (2009)
27. Reicha, FM, Sarhan, A, Abdel-Hamid, MI, El-Sherbiny, IM: Preparation of silver nanoparticles in the presence of chitosan by electrochemical method. Carbohydr. Polym. **89**, 236–244 (2012)
28. Lok, CN, Ho, CM, Chen, R, He, QY, Yu, WY, Sun, H, Tam, PK, Chiu, JF, Che, CM: Proteomic analysis of the mode of antibacterial action of silver nanoparticles. J. Proteome Res. **5**, 916–924 (2006)
29. Ahmad, N, Sharma, S, Singh, VN, Shamsi, SF, Fatma, A, Mehta, BR: Biosynthesis of silver nanoparticles from Desmodium trifolium: a novel approach towards weed utilization. Biotech Res Int **8**, 1–8 (2011)

3

Synthesis and identification of carvedilol nanoparticles by ultrasound method

Moayad Hossaini Sadr and Hafezeh Nabipour*

Abstract

Nanoparticles of carvedilol are prepared by ultrasonic method. Particle sizes and morphology of nanoparticles are dependent on the power of ultrasound irradiation and sonication time. Results show an increase in the particle size with increasing the power of ultrasound irradiation. The particle size increased with increasing sonication time. The produced carvedilol nanoparticles were characterized by X-ray diffraction, infrared spectroscopy, and scanning electron microscopy.

Keywords: Preparation, Carvedilol, Nanoparticles, Sonochemical method

Background

One of the major causes of new drug molecules failing to reach the market is poor bioavailability. Indeed, many of the newly discovered drug candidates suffer from low to extremely low water solubility and hence are prone to limited bioavailability [1-3]. To cope with these issues, drug formulation scientists are developing formulation strategies that enhance the dissolution and thus the bioavailability of these poorly water-soluble drugs [4]. The most important technologies are applied in the formulation development to enhance solubility and dissolution which include particle size reduction, salt formation, and amorphous systems, as well as co-solvents, complexants, surfactants, and lipid formulations. Among these techniques, particle size reduction is highly promising. Micronization of drug crystals by various mechanisms to the lower micron range improves drug dissolution [5]. According to the Noyes-Whitney equation, the dissolution rate increases by increasing the specific surface area. The saturation solubility increases with decreasing particle size according to the Ostwald-Freundlich equation [6,7]. Therefore, reducing particle size to the nanorange has been investigated through several approaches. Current methods to prepare drug nanocrystals can be divided into top-down and bottom-up methods [8,9]. At present, most industrially feasible methods are the top-down technologies; all the products on the market are made by size reduction. The major two processes are ball milling and high-pressure homogenization, either in water or in non-aqueous water-reduced media. Disadvantages of the ball milling process are the relatively long milling times in the case of hard drug material, the limited scaling up ability (the weight of the ball mill and the milling material limit maximum size), and potential contamination of the milling material by erosion from the milling pearls [10]. Nanocrystal production by high-pressure homogenization requires typically 10 to 20 homogenization cycles at pressures of, e. g., 1,000 to 1,500 bar. Metal contamination from the homogenizer is obviously not a problem. However, despite when using a high-capacity homogenizer with 1 ton product per hour, the acceleration of production by reduction of the homogenization cycle number would be desirable. Currently, the utilization of ultrasound for material synthesis has been extensively examined over many years, and it is now positioned as one of the most powerful tools in nanostructured material synthesis. Acoustic cavitation provides a unique interaction of energy and matter, and ultrasonic irradiation of liquids causes high-energy chemical reactions to occur, often accompanied by the emission of light [11]. Bubbles generated during sonication can effectively accumulate the diffuse energy of ultrasound, and upon collapse, an enormous concentration of energy is released to heat the contents of the bubble. These transient, localized hot spots with extremely high temperatures and pressures are primarily responsible for the chemical effects of ultrasound [12].

* Correspondence: ha.nabipour@gmail.com
Chemistry Department, Faculty of Science, Azarbaijan Shahid Madani University, Tabriz 53714-161, Iran

The usefulness of sonochemical synthesis as a synthetic tool resides in its versatility. With a simple modification in the reaction conditions, various forms of nanostructured materials can be synthesized, including metals [13,14], carbides [15] and sulfides [16] which are nanostructured supported catalysts.

Carvedilol, (+)–1-(Carbazol-4-yloxy)-3-f[2-(*o*-methoxyphe noxy)ethyl]aminog-2-propanol, is a β-receptor blocker and antihypertensive drug with a multiple-action spectrum [17], and it also has vasodilating properties that are attributed mainly to its blocking activity at α1 receptors. The use of carvedilol has been shown to provide additional morbidity and mortality benefits in severe congestive heart failure [18]. Previously, Varshosaz et al. reported the preparation of carvedilol nanoparticles by emulsification method [19]. Kalimuthu et al. reported the formulation of carvedilol-loaded Eudragit e 100 nanoparticles [20]. In this paper, we have developed a simple ultrasonic method to prepare nanocarvedilol and studied some parameters, such as sonication time and ultrasonic device power in the size and morphology of the carvedilol nanoparticles. The result shows that different particle diameters of carvedilol were prepared, and infrared studies show that the bonds do not change.

Figure 1 SEM images of carvedilol nanoparticles in (A) sample 1 and (B) sample 2.

Results and discussion

Various conditions for the preparation of carvedilol nanoparticles were summarized in Table 1. In series 1, the reactions were studied in 20 and 40 min. Results show that the particle size in sample 1 (Figure 1B) is smaller than that of 2 (Figure 1A). This point shows that an increase in the sonication time led to an increase in the particle size. In the same way, an increase in the sonication time led to the increase of the particle sizes in sample 4 (Figure 2B). Various ultrasound powers led to different morphologies. Samples 1 and 2 have amorphous structures, but the morphology of samples 3 and 4 changed to spherical structures, and their particle distribution is uniform. Particle size in sample 3 is smaller than that of 4. Therefore, the best morphology with smaller particles and good distribution was obtained from sample 4 with the sonication time of 40 min and the ultrasound device power of 80 W (Figure 2B).

Table 1 Experimental conditions for the preparation of carvedilol nanoparticles

	Sample	Sonication time (min)	Ultrasound power (W)	Particles size (nm)
Series 1	1	20	60	21
	2	40	60	48
Series 2	3	20	80	64
	4	40	80	96

Figure 2 SEM images of carvedilol nanoparticles in (A) sample 3 and (B) sample 4.

Figure 3 shows the typical Fourier transform-infrared (FTIR) spectrum pattern of the nanocarvedilol in the range of 400 to 4,000 cm^{-1}. The IR spectrum of pure carvedilol showed the peak at 3,344.88 cm^{-1} (N-H, str), which corresponds to N-H stretching. The hetero-aromatic structure shows the presence of the C-H stretching vibrations in the region 2995.87 cm^{-1} (C-H, Sp^2), also in 2923.56 cm^{-1}, which is the characteristic region for the ready identification of the C-H stretching vibrations. The bands corresponding to the in-plane C-H deformations are observed in the regions 1,000 to 1,300 cm^{-1}. The bands are sharp but of weak to minimum intensity. The medium- and strong-intensity bands are in the regions 1,099.72, 1,119.95, 1,156.24, and 1,177.15 cm^{-1} in the FT-IR spectrum. A medium experimental peak around 1252 to 1,402 cm^{-1} in the FT-IR was assigned to the C-C stretching vibrations. The bands in the regions 1,502 to 1,607 cm^{-1} observed in both FT-IR were assigned to the C = C stretching vibrations. In the present case, the bands observed at 849.50 and 783.90 cm^{-1} in the FT-IR are assigned to the ring breathing mode. The identification of C-N vibrations is a very difficult task since mixing of several bands is possible in the region assigned for the C-N stretching vibrations which is in the regions 1,347.95 to 1,285.41 cm^{-1}. In the present work, a weak band at 1,402.71 cm^{-1} and a strong band at 1,453.97 cm^{-1} were seen in the FT-IR. It is observed that the ultrasonic wave has no effect on nanoscale materials and the structure does not change.

X-ray diffraction (XRD) technique was used to determine the ingredients of the sample. Figure 4 shows the XRD patterns of typical sample 4 prepared by the sonochemical process. The mean size of carvedilol nanoparticles was calculated to be about 96 nm using Scherrer's equation: $D = K\lambda/(\beta \cos \theta)$, where D is the mean crystalline size (nm), λ is the wavelength of Cu Kα

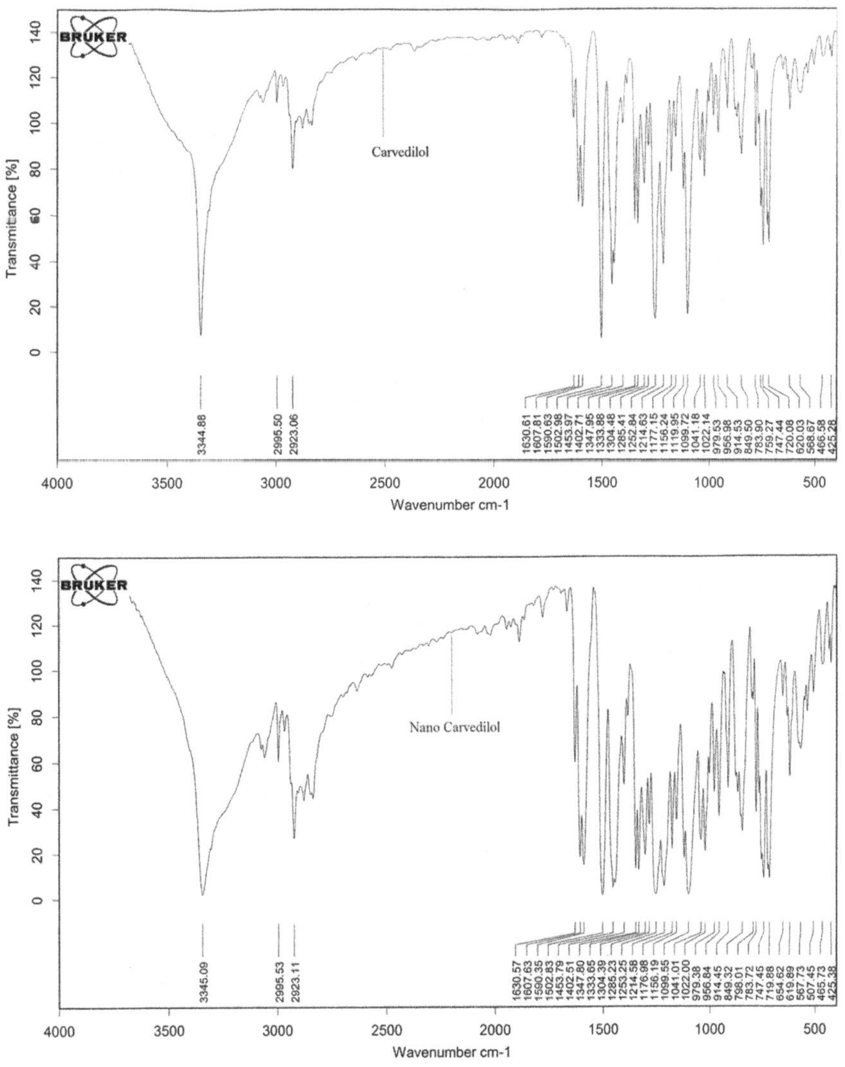

Figure 3 Room-temperature FT-IR spectra of carvedilol and nanocarvedilol (sample 4).

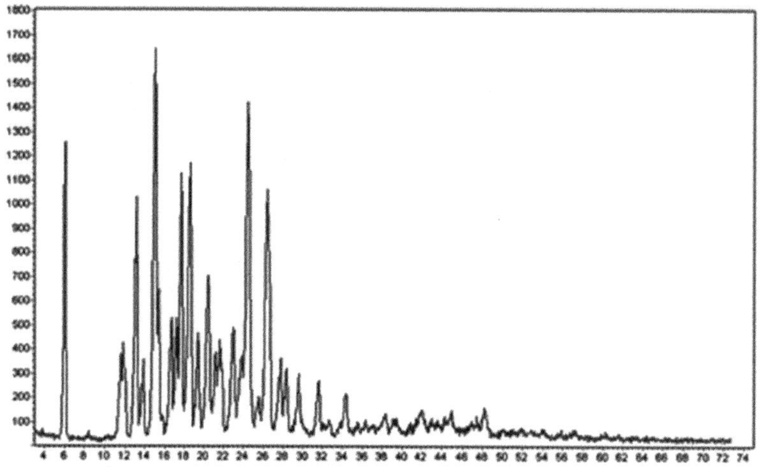

Figure 4 X-ray diffraction spectrum of nanocarvedilol in (A) sample 4.

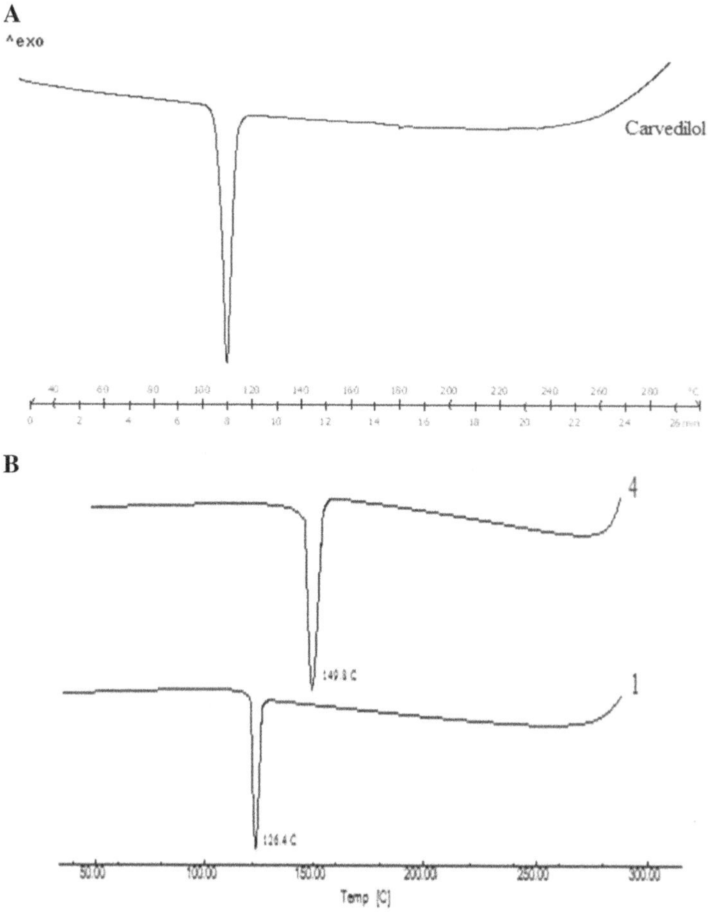

Figure 5 DSC spectra of (A) carvedilol and (B) carvedilol nanoparticles.

(0.154 nm), β is the full width at half maximum intensity in radian, and θ is the Bragg angle (°) [21].

The differential scanning calorimetry (DSC) curve of carvedilol showed a sharp endothermic peak (T_{peak} = 115°C) corresponding to its melting point, indicating its crystalline nature (Figure 5A). In DSC studies, the characteristic endothermic peaks, corresponding to drugs melting point, were shifted toward higher temperature. This could be attributed to the bigger particle sizes and uniform distribution of drugs (Figure 5B). Now, it is found that the melting temperature of carvedilol nanoparticles decreases with decreasing their particle size. The melting temperature increases as its size increases. In other words, their melting temperature is lower than the corresponding bulk materials.

It is also noteworthy that one of the problems of carvedilol is its very low aqueous solubility. The importance of solubility enhancement is for the absorption of drug from the site of absorption. The size of the solid particle influences the solubility because as a particle becomes smaller, the surface area-to-volume ratio of the particle increases. The larger surface area allows a greater interaction with the solvent. Increase aqueous solubility with the nanoparticle size increases the efficiency and/or reduces the side effects for certain drugs.

Conclusions

The described method for preparing carvedilol nanoparticles can potentially be used at industrial scale because it does not need special conditions, such as high temperature, long times, or high-temperature and high-pressure controlling. The advantage of using ultrasound radiation is that it yields smaller particles. The same results have been shown for recent reports too. This approach can be used to improve the therapeutic efficacy of poorly soluble drugs.

Methods

Material and methods

Carvedilol was purchased from Alborz Darou Pharmaceutical Company in Tehran, Iran. A multiwave ultrasonic generator (Sonicator_3000, Misonix, Inc., Farmingdale, NY, USA), equipped with a converter/transducer and titanium oscillator (horn), 12.5 mm in diameter, operating at 20 kHz with a maximum power output of 600 W, was employed. The morphology of synthesized samples was studied using scanning electron microscopes (KYKY-EM3200, KYKY Technology Development Ltd., Beijing, China; Philips XL 30 and S-4160, Philips, Amsterdam, The Netherlands) and by a sputtering technique with gold as covering contrast material. The thermal analysis was carried out on DSC Q10 V9.4 Build 287 (TA Instruments, New Castle, DE, USA). The XRD measurements of synthesized samples were carried out using a Philips X-pert PRO powder diffractometer (Philips, Amsterdam, The Netherlands) with Cu Kα radiation (λ = 1.54 Å) in the scan range of 0° to 100°. The IR spectra were recorded using a Bruker spectrometer (Bruker AXS, Inc., Madison, WI, USA) with KBr pellets in the range of 400 to 4,000 cm^{-1}.

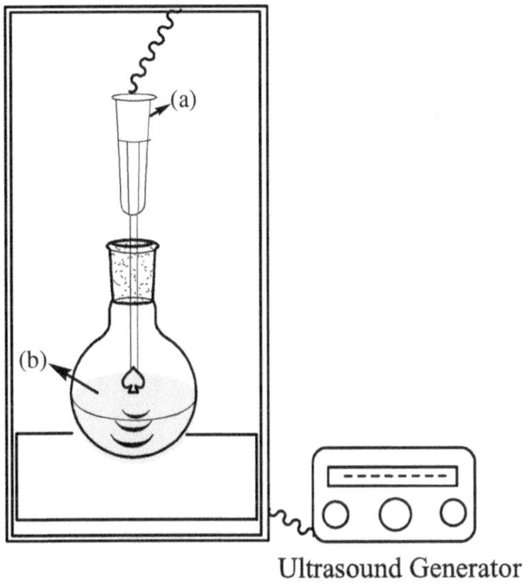

Figure 6 Schematic of the experimental setup used for the sonochemical reactions. (**a**) Carvedilol and carbon tetrachloride, (**b**) sonochemical probe.

Synthesis of nanoparticles

Carvedilol and carbon tetrachloride were mixed to prepare nanoparticles by ultrasonic method. The sonochemical reactions were carried in the sonochemical probe, as shown in Figure 6. The ultrasonic device power and sonication time were the parameters which were changed to reach the optimum condition.

Then, the suspension was ultrasonically irradiated with a high-density ultrasonic probe immersed directly into the solution under various conditions (Table 1). The precipitate was separated from the mother liquor by a centrifuge at 4,000 rpm for 3 min. The final product was dried at 50°C in a vacuum system.

Competing interests

The authors declare that they have no competing interests.

Authors' contributions

Both authors participated in writing the manuscript. Both authors read and approved the final manuscript.

Acknowledgements

We acknowledge Dr. Ali Afshar from Alborz Darou Pharmaceutical Company, who supported this study through a grant.

References

1. Alsenz, J, Kansy, M: High throughput solubility measurement in drug discovery and development. Adv. Drug Deliv. Rev. **59**, 546–567 (2007)
2. Li, P, Zhao, LW: Developing early formulations: practice and perspective. Int. J. Pharmaceut. **341**, 1–19 (2007)
3. Gardner, CR, Walsh, CT, Almarsson, O: Drugs as materials: valuing physical form in drug discovery. Nat. Rev. Drug Discov. **3**, 926–934 (2004)
4. Horn, D, Rieger, J: Organic nanoparticles in the aqueous phase-theory, experiment, and use. Angew. Chem. Int. Ed Engl. **40**, 4331–4361 (2001)
5. Patravale, VB, Date, AA, Kulkarni, RM: Nanosuspensions: a promising drug delivery strategy. J. Pharm. Pharmacol. **56**, 827–840 (2004)
6. Pillay, V, Fassihi, R: A new method for dissolution studies of lipid-filled capsules employing nifedipine as model drug. Pharm. Res. **16**, 333–337 (1999)
7. Müller, RH, Peters, K: Nanosuspensions for the formulation of poorly soluble drugs: I. Preparation by a size-reduction technique. Int. J. Pharm. **160**, 229–237 (1998)
8. Merisko-Liversidge, E, Liversidge, GG, Cooper, ER: Nanosizing: a formulation approach for poorly-water-soluble compounds. Eur. J. Pharm. Sci. **18**, 113–120 (2003)
9. Chen, X, Young, TJ, Sarkari, M, Williams 3rd, RO, Johnston, KP: Enhanced drug dissolution using evaporative precipitation into aqueous solution. Int. J. Pharm. **242**, 3–14 (2002)
10. Bang, JH, Suslick, KS: Applications of ultrasound to the synthesis of nanostructured. Materials Adv. Mater. **22**, 1039–1059 (2010)
11. Suslick, KS: in Kirk-Othmer Encyclopedia of Chemical Technology, 26, 4th edn, pp. 517–541. Wiley, New York (1998)
12. Koltypin, Y, Katabi, G, Prozorov, R, Gedanken, A: Sonochemical preparation of amorphous nickel. J. Non-Cryst. Solids **201**, 159 (1996)
13. Okitsu, K, Mizukoshi, Y, Bandow, H, Maeda, Y, Yamamoto, T, Nagata, Y: Formation of noble metal particles by ultrasonic irradiation. Sonochem **3**, 249 (1996)
14. Hyeon, T, Fang, M, Suslick, KS: Nanostructured molybdenum carbide: sonochemical synthesis and catalytic properties. J. Am. Chem. Soc. **118**, 5492 (1996)
15. Wu, GS, Yuan, XY: A simple synthesis route to CdS nanomaterials with different morphologies by sonochemical reduction. J. Mat. Lett. **58**, 794–797 (2004)
16. Wu, GS, Yuan, XY, Xie, T, Xu, GC, Zhang, LD, Zhuang, YL: A simple synthesis route to CdS nanomaterials with different morphologies by sonochemical reduction. Mater. Lett. **58**, 794 (2004)
17. Ruffolo Jr, RR, Feuestein, GZ: Pharmacology of carvedilol: rationale for use in hypertension, coronary artery disease, and congestive heart failure. Cardiovasc. Drugs Ther. **11**, 247–256 (1997)
18. Packer, M, Fowler, MB, Roecker, EB, Coats, AJ, Katus, HA, Krum, H, Mohacsi, P, Rouleau, JL, Tendera, M, Staiger, C, Holcslaw, TL, Amann-Zalan, I, DeMets, DL: Carvedilol Prospective Randomized Cumulative Survival (COPERNICUS) Study Group. Effect of carvedilol on the morbidity of patients with severe chronic heart failure: results of the carvedilol prospective randomized cumulative survival (COPERNICUS) study. Circulation **106**(17), 2194–2199 (2002)
19. Varshosaz, J, Moazen, E, Fathi, M: Preparation of carvedilol nanoparticles by emulsification method and optimization of drug release: surface response design versus genetic algorithm. J. Dispersion Sci. Tech. **33**, 1480–1491 (2012)
20. Kalimuthu, S, Yadav, AV: Formulation and evaluation of carvedilol loaded udragit e 100 nanoparticles. Int. J. Pharm. Res. **1**(2), 179–183 (2009)
21. Darezereshki, E: Synthesis of maghemite (α-Fe_2O_3) nanoparticles by wet chemical method at room temperature. Mater. Lett. **64**, 1471–1472 (2010)

Synthesis of uniform MgO/CNT nanorods by precipitation method

Farshad Taleshi[1*] and Ali A Hosseini[2]

Abstract

In this work, we used carbon nanotubes (CNTs) as a support for synthesis, size control, and morphology of magnesium oxide using the precipitation method. The morphology of MgO nanoparticles varies by changing the weight percent of carbon nanotubes in the solution. Experimental results indicate that, at optimum condition in a mixture of materials, the surface of purified CNTs is covered by MgO nanoparticles completely. The purified CNTs and MgO/CNTs nanorods had average diameters of about 35 and 65 nm, respectively, and their length was in the order of a few micrometers. The structure of CNTs and MgO/CNTs nanopowders has been characterized by analyzing the X-ray diffraction pattern, Fourier transform infrared spectra, and scanning electron microscopy images.

Keywords: Carbon nanotubes, Magnesium oxide, Nanoparticles, MgO/CNT nanorods, Precipitation method

Background

Carbon nanotubes (CNTs) have been extensively studied and known as materials with outstanding properties. Several applications were proposed for CNTs, many of which are concerned with conductive or high-strength composites, in which the inclusion of CNTs in a ceramic matrix is expected to produce composites possessing high stiffness and improved mechanical properties [1-6]. They have also been widely used as support for the controlled synthesis of nanostructures [7].

However, there are major obstacles to overcome in using CNTs as support additives, mainly the difficulty of the uniform dispersion of CNTs, wettability of the CNTs surface, and the load transfer from the matrix to the CNTs [8]. Due to the chemically 'inert' characteristic and the highly hydrophobic nature of the CNT surface, it is difficult to make an attachment with metal precursors [7]. Then, surface modification is required so that efficient tube-matrix interactions are established. The extensive dispersion of various functional groups such as hydroxyl, carboxyl, and carbonyl on the CNT surface could provide active sites for metal ions to attach [9]. Unfortunately, some of the previous studies indicated that these chemical treatments resulted in an extreme degradation of the tubes and the decrease in Young's modulus [10]. The creation of stable nanotube coatings, which do not significantly alter the tube surface, is expected to exhibit improving mechanical properties when compared with heavily degraded nanotubes [11]. Hernadi et al. [7] have obtained homogeneous coverage of multi-walled carbon nanotubes (MWCNTs) using inorganic materials implementing organometallic compounds. SiO_x-coated CNTs [12] have also been fabricated through a sol–gel technique at room temperature. Chen et al. [13] obtained SnO-CNT composites, using a sol–gel method, as anode-active material for lithium-ion batteries. Sun et al. [14] have coated CNTs with a thin SnO_2 layer using a chemical solution route.

Nanoscale magnesium oxide exhibits unique physical and chemical properties due to its structural characteristic [15]. Several novel methods have been reported in literature for the preparation of nanosized MgO particles [15,16]. The particle size and surface morphology of synthesized MgO by wet chemical procedure depend upon several factors such as the rate of hydrolysis of magnesium salts, temperature, type of base, concentration of the salt, and drying and calcination steps. Proper choice of these parameters can lead to particles of uniform morphology and size [16,17].

The coated nanotubes may improve the adhesion between the nanotubes and the matrix when CNTs are used as fillers in the composite materials. However, the

* Correspondence: far.taleshi@gmail.com
[1]Department of Applied Science, Qaemshahr Branch, Islamic Azad University, Qaemshahr 163, Iran
Full list of author information is available at the end of the article

mechanical behavior of CNTs is more complicated. Whether or not the strength and stiffness of CNTs can be transferred to the MgO depends on the amount of interfacial bonding between the two phases, which is affected by the CNTs' wettability and interfacial area [18]. Thus, it is important to synthesize CNT-MgO composites and to characterize the interfacial bonding action between the CNTs and the MgO without disruption of the local order of CNTs [9].

In this work, we describe our successful attempt with a simple precipitation method to achieve homogeneous coverage of CNTs with MgO nanoparticles, during which the CNTs were well dispersed in an aqueous solution. The results show that carbon nanotubes can play an important role in controlling the morphology and size of MgO nanoparticles.

Results and discussion

Morphology

The morphology and microstructure of CNTs and MgO/CNT nanopowders were carried out by analyzing scanning electron microscopy (SEM) images. The SEM images of the CNT sample before and after purification are shown in Figure 1. The images show the large amount of impurities attached to the CNT surface (Figure 1a). After the purification steps and the acid treatment, the impurity materials (metal catalyst particles and amorphous carbon) are efficiently removed (Figure 1b). The purification disclosed intertwined CNTs which form a dense 3D-network, extending for several micrometers.

We synthesized nanocomposite powders with different concentrations of CNTs to determine the optimized condition. For now, we discuss the obtained results for the following four concentrations of CNTs in the solution: 0, 0.02, 0.04, and 0.06 g. Figure 2a shows the SEM images of the as-synthesized MgO nanoparticles without any use of CNTs. These nanoparticles are like tiny aggregated nanoparticles which make them bulky. For the 0.02- and 0.04-g concentrations of CNTs, because of the high concentration of $Mg(NO_3)_2{\cdot}6H_2O$, the CNTs are invisible (Figure 2b,c). By comparing Figure 2a with Figure 2b,c, we can see that nanoparticles have different morphologies due to the presence of CNTs. Though larger MgO aggregates still exist, many individual nanoparticles are attached onto the CNT surfaces. Thus, the presence of CNTs in the solution can cause the change of MgO nanoparticle morphology.

The SEM image of MgO/CNT nanorods with the 0.06-g concentration of CNTs in the solution is shown in Figure 2d. The images show the development of a large amount of MgO nanoparticles attached to the surfaces of CNTs, making dense and entangled MgO/CNT hybrid nanostructures forming a three-dimensional network structure. For the prepared composites, all of the MgO nanoparticles were firmly deposited on the surface of CNTs, and all of the CNTs are covered by well-dispersed MgO nanoparticles.

In order to obtain the average diameters of CNTs and MgO/CNT nanorods and their range of distribution, we have selected more than 30 per samples and measured their outer diameters from their respective SEM images. It can be seen that the purified CNTs have diameters in the range of 20 to 50 nm (Figure 1b), and MgO/CNT nanorods have diameters within 45 to 150 nm (Figure 2d); their average diameters are about 35 and 65 nm, respectively. Then, the thickness of the MgO layer is about 30 nm.

FT-IR spectroscopy

The nanoparticle loading process was dependent on the functionalization of CNTs. The positive Mg ions may be adsorbed on some sites due to imperfections of the CNT surface and act as preferential nucleation sites for the formation of larger crystals. The functionalization of CNTs using oxidizing acid is one of the most crucial factors, as prerequisite, which would introduce a large number of carboxyl groups on the surface of CNTs, and the high density of functional groups in these restrictive positions would provide great opportunity for further position-selective modification of CNTs with MgO.

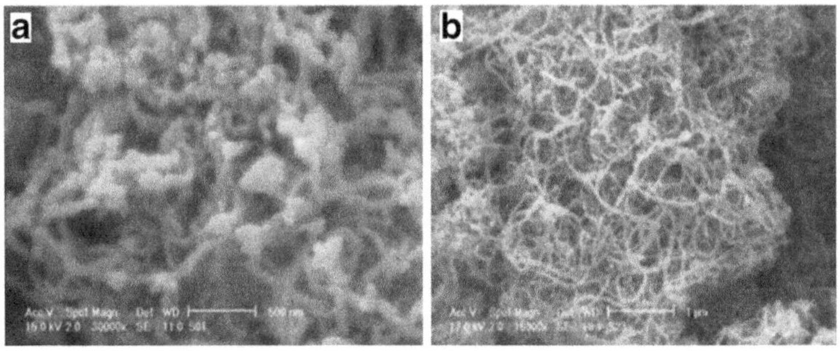

Figure 1 SEM images of CNTs. (**a**) Before and (**b**) after purification, showing the long intertwined CNTs.

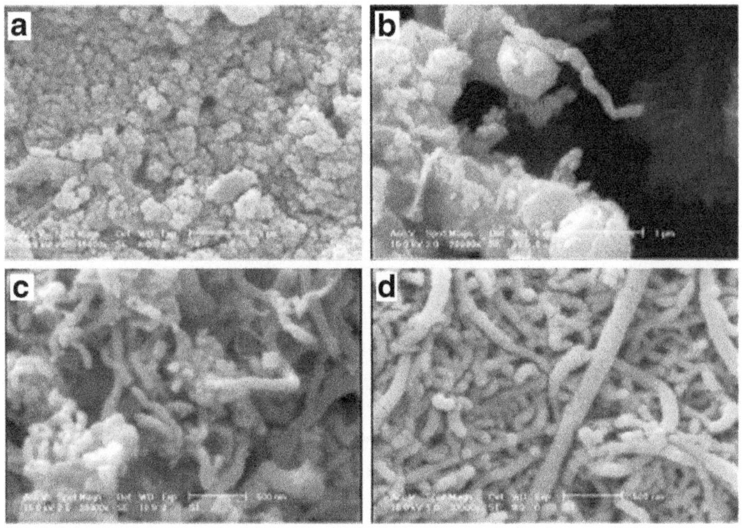

Figure 2 SEM images of MgO nanoparticles and MgO/CNT nanopowders. (**a**) MgO nanoparticles without using CNTs. MgO/CNT nanopowders synthesized by precipitation method with different concentrations of CNTs in the solution: (**b**) 0.0 2, (**c**) 0.04, and (**d**) 0.06 g.

A Fourier transform infrared (FT-IR) spectrum was mainly used to identify the presence or absence of functional groups. Figure 3 shows the FT-IR spectra of initial and functionalized CNTs using HNO_3. As shown on the FT-IR spectra, acid treatment can result in the appearance of some important absorption peaks approximately at 1,210.9, 1,461.8, and 1,514.4 cm^{-1} which correspond to O-C=O, C-O, and C=O, respectively. These peaks are, however, barely noticeable in the functionalization of the CNTs using HNO_3 [6,19-21]. The FT-IR spectra indicate that oxygen with high density is generated from functionalized groups on the surface of CNTs during the acid treatment.

As a result, we herein conjecture that the MgO/CNT nanorods form under precipitation conditions; the redox reaction between Mg ions will take place on the surface of CNTs. Consequently, MgO nucleated heterogeneously via precipitation deposition because CNTs have prevented MgO from growing into aggregated nanoparticles [22,23].

XRD analysis

The multi-walled carbon nanotubes, MgO nanoparticles, and MgO/CNT nanopowders were analyzed using X-ray diffraction (XRD) analyses. The XRD spectra of CNTs before and after purification with the 2θ range of 20° to 80° are shown in Figure 4. The purified CNTs present a characteristic diffraction peak at about 26.5°, corresponding to the interlayer spacing of multi-walled CNTs,

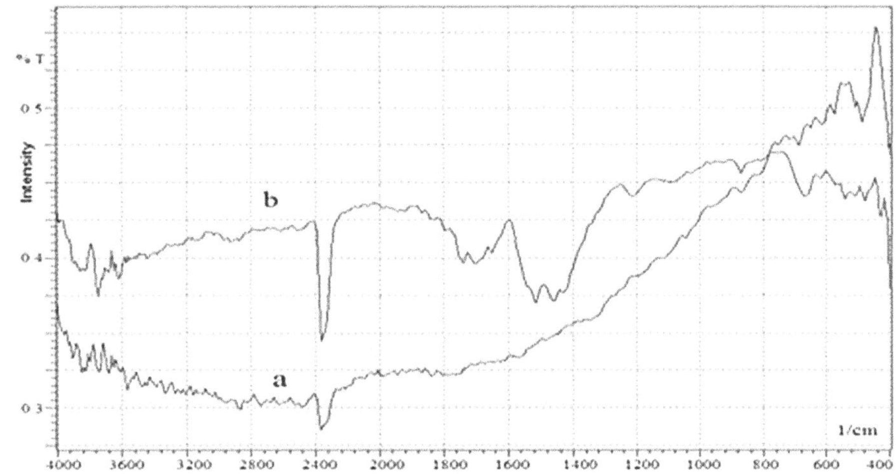

Figure 3 FT-IR spectra of CNTs before (a) and after (b) functionalization.

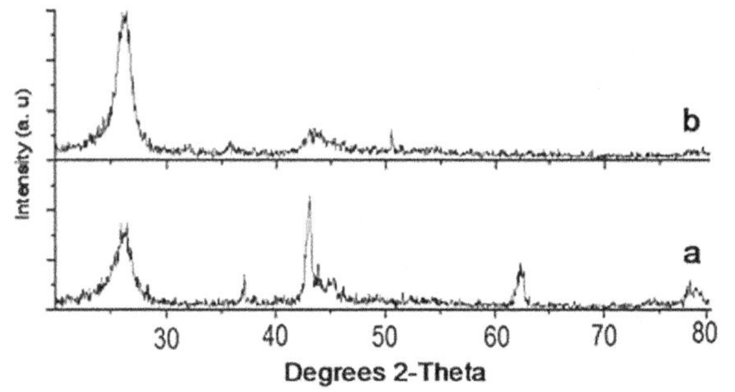

Figure 4 XRD pattern of MWCNT. Before (**a**) and after (**b**) purification in the 2θ range of 20° to 80°.

which indicates that the structure of CNTs was not destroyed during the purification process.

Also, the XRD pattern of MgO nanoparticles and MgO/CNT nanopowders (for 0.06 g of CNTs in the solution), in the 2θ range of 20° to 100°, is shown in Figure 5. For both XRD patterns (Figure 5a,b), the intense peaks were observed at 36.8°, 42.7°, 62.1°, and 75.5°, indicating the crystalline nature of the synthesized MgO nanoparticles with fcc structure [16,24]. MgO nanoparticles are composed of pure crystalline MgO since no other impurity peaks are observed in the XRD patterns. In addition, we can see that the characteristic peak of CNTs still exists at 2θ = 26.5° (Figure 5b), but the peak intensity decreases after decorating. The average MgO nanoparticle size for both syntheses (with and without the use of CNTs) was determined from the X-ray diffraction line broadening using the Scherrer formula [25], and their average values are about 11 and 7.1 nm, respectively. The results indicated that the presence of carbon nanotubes in the solution can not only change the morphology of MgO nanopowders, but also decreased the average size of the MgO nanoparticles. We believe that the nucleation and growth processes of MgO were significantly influenced by the introduction of CNTs, which have altered the size and shape of MgO.

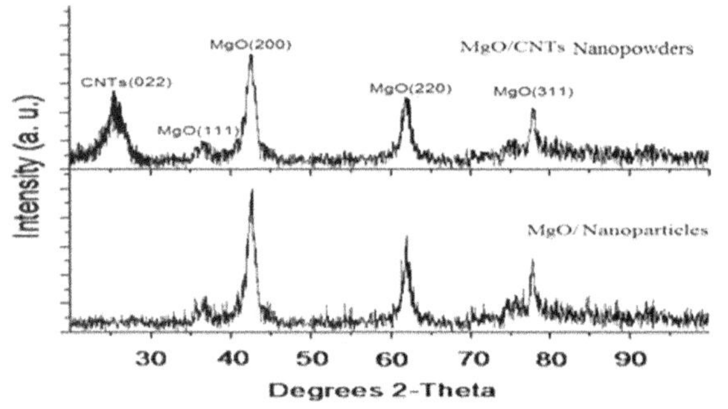

Figure 5 XRD pattern of MgO nanoparticles and MgO/CNT nanopowders in the 2θ range of 20° to 100°.

Conclusions

The results of the synthesis of carbon nanotubes decorated with MgO nanoparticles using the precipitation method were demonstrated in this work. High efficiency of the proposed method was confirmed as well as the possibility of the coating of MgO nanoparticles onto CNTs, without aggregation of these particles. Magnesium ions were attached on CNTs due to the coordination reaction between magnesium ions and polar oxygenated functional groups in the precipitation process. Calculations show that MgO particles appear to have a smaller average size due to the presence of CNTs, and they are durably attached to the nanotubes.

Methods

Synthesis, purification, and functionalization of CNTs

CNTs were synthesized by the catalytic decomposition of ethylene at 930°C over commercial Fe nanoparticles (Sigma-Aldrich, St. Louis, MO, USA; purity was 99% and the average size of nanoparticles was about 25 nm), using catalytic chemical vapor deposition method. It is well known that most produced CNTs contain some impurities such as amorphous carbon, fullerenes, and catalyst particles that could significantly affect the CNT properties, which are a serious impediment for CNTs to be directly used as functional filler in composites. Therefore, such impurities of carbon and metal catalyst particles are encapsulated in the CNT closed tips and walls and need to be removed, trying to avoid a concurrent damage of the tubular structures. In order to remove the impurity, at first, the as-produced CNTs were heated in a horizontal furnace under air flow condition at 500°C for 2 h, sonicated in a mixed solution of HNO_3 (5 M) and HCl (5 M) for 30 min, and stirred for 4 h. This suspension was filtered using a membrane having a 0.2-μm pore size, then washed several times with distilled water until acid-free, and finally dried at 120°C for 2 h. Under these treatments, the product was purified from the impurities, and the surfaces of CNTs were modified, which could alter the surface interactions significantly during the coating procedure.

Synthesis of MgO/CNT nanopowders

The MgO/CNT nanopowders were synthesized by precipitation of the magnesium hydroxide gels in aqueous solution containing CNTs using $Mg(NO_3)_2{\cdot}6H_2O$ as a salt and liquid ammonia as the precipitating agent. At first, certain amounts of purified CNTs (0, 0.02, 0.04, and 0.06 g) were dispersed into the distilled water solution of ammonia (50 cm^3; pH = 10.5) by ultrasonication for 10 min. Then 1.8 g of $Mg(NO_3)_2{\cdot}6H_2O$ was dissolved in 100 cm^3 of distilled water. With constant magnetic stirring, the solution of $Mg(NO_3)_2{\cdot}6H_2O$ was added dropwise to the above solution at 50°C through a dropping funnel. The rate of addition of the salt solution was kept approximately at 20 ml/h. During the addition, the pH of the mixture decreased due to hydrolysis of the salt. The pH was maintained at 10.5 by controlling the addition of the liquid ammonia solution. After completion of the precipitation procedure, the mixture was stirred at room temperature for 12 h, washed and filtered continually in distilled water, and dried at 120°C followed by calcination at 450°C for 2 h. During the calcination procedure, the temperature of the furnace was steadily raised from room temperature to 450°C with an increment of 10°C/min. At 450°C, the temperature was maintained for 2 h to achieve the final material.

Characterization of CNTs and MgO/CNT nanopowders

The resulting CNTs and MgO/CNT nanopowders were characterized by SEM images (Philips MAG 15 kV, ×30,000; FEI Co., Hillsboro, OR, USA), FT-IR spectra (8400 S, Shimadzu Corporation, Nakagyo-ku, Kyoto, Japan), and XRD pattern (CuKα spectra, λ = 1.54 Å; GBC Scientific Equipment LLC, Hampshire, IL, USA). Also, the MgO nanoparticle size was determined using the Debye-Scherrer equation.

Competing interests

The authors declare that they have no competing interest.

Authors' contributions

All authors provided the same contributions in this article. All authors read and approved the final manuscript.

Acknowledgment

The authors would like to acknowledge the University of Mazandaran, Azad University of Qaemshahr and the Iranian National Nanotechnology Initiation Council (INNIC) for their financial support of this project.

Author details

[1]Department of Applied Science, Qaemshahr Branch, Islamic Azad University, Qaemshahr 163, Iran. [2]Department of Physics, University of Mazandaran, Babolsar 47416-95447, Iran.

References

1. Thostenson, ET, Ren, Z, Chou, TW: Advances in the science and technology of carbon nanotubes and their composites: a review. Composites Science and Technology **61**, 1899–1912 (2001)
2. Paradise, M, Gogswami, T: Carbon nanotubes – production and industrial applications. Materials and Design **28**, 1477–1489 (2007)
3. Raffaelle, RP, Landi, BJ, Harris, JD, Bailey, SG, Hepp, AF: Carbon nanotubes for power applications. Materials Science and Engineering B **116**, 233–243 (2005)
4. Wei, T, Fan, Z, Luo, G, Wei, F: A new structure for multi-walled carbon nanotubes reinforced alumina nanocomposite with high strength and toughness. Mater Lett **62**, 641–644 (2008)
5. Xia, ZH, Lou, J, Curtin, WA: A multiscale experiment on the tribological behavior of aligned carbon nanotube/ceramic composites. Scr Mater **58**, 223–236 (2008)
6. Chen, L, Zhang, BL, Qu, MZ, Yu, ZL: Preparation and characterization of CNTs–TiO_2 composites. Powder Technology **154**, 70–72 (2005)
7. Hernadi, K, Ljubovic, E, Seo, JW, Forro, L: Synthesis of MWNT-based composite materials with inorganic coating. Acta Mater **51**, 1447–1452 (2003)
8. Esawi, AMK, Borady, MAE: Carbon nanotube-reinforced aluminum strips. Composites Science and Technology **68**, 486–492 (2008)
9. Zhao, L, Gao, L: Novel in situ synthesis of MWNTs–hydroxyapatite composites. Carbon **42**, 423–426 (2004)
10. Andrews, R, Jacques, D, Minot, M, Rantell, T: Fabrication of carbon multi-wall nanotube/polymer composites by shear mixing. Macromolecular Materials and Engineering **287**, 395–403 (2002)
11. Jia, ZJ, Wang, ZY, Xu, CL, Liang, J, Wei, BQ, Wu, DH, Zhu, SW: Study on poly (methyl methacrylate)/carbon nanotube composite. Mater Sci Eng A **271**, 395–400 (1999)
12. Seeger, T, Redlich, P, Grobert, N, Terrones, M, Walton, DRM, Kroto, HW, Rühle, M: SiOx-coating of carbon nanotubes at room temperature. Chemical Physics Letters **339**, 41–46 (2001)
13. Chen, MH, Huang, ZC, Wu, GT, Zhu, GM, You, JK, Lin, ZG: Synthesis and characterization of SnO–carbon nanotube composite as anode material for lithium-ion batteries. Mater Res Bull **38**, 831–836 (2003)
14. Sun, Z, Liu, Z, Han, B, An, G: Supercritical carbon dioxide-assisted deposition of tin oxide on carbon nanotubes. Mater Lett **61**, 4565–4568 (2007)
15. Wang, W, Qiao, X, Chen, J, Li, H: Facile synthesis of magnesium oxide nanoplates via chemical precipitation. Mater Lett **61**, 3218–3220 (2007)
16. Kumar, D, Reddy, VB, Mishra, BG, Rana, RK, Nadagouda, MN, Varma, RS: Nanosized magnesium oxide as catalyst for the rapid and green synthesis of substituted 2-amino-2-chromenes. Tetrahedron **63**, 3093–3097 (2007)
17. Wang, JA, Novaro, O, Bokhimi, X, Lopez, T, Gomez, R, Navarrete, J, Llanos, ME, Lopez-Salinas, E: Characterizations of the thermal decomposition of brucite prepared by sol–gel technique for synthesis of nanocrystalline MgO. Mater Lett **35**, 317–323 (1998)
18. White, AA, Best, SM: Hydroxyapatite-carbon nanotube composites for biomedical applications: a review. International Journal of Applied Ceramic Technology **4**, 1–13 (2007)
19. Osorio, AG, Silveria, ICL, Bueno, VL, Bregmann, PC: $H_2SO_4/HNO_3/HCl$—functionalization and its effect on dispersion of carbon nanotubes in aqueous media. Appl Surf Sci **255**, 2485–2489 (2008)
20. He, B, Wang, M, Sun, W, Shen, Z: Preparation and magnetic property of the MWNT-Fe^{2+} composite. Mater Chem Phys **95**, 289–293 (2006)
21. Sun, W, Huang, Z, Zhang, L, Zhu, J: Luminescence from multi-walled carbon nanotubes and the Eu(III)/multi-walled carbon nanotube composite. Carbon **41**, 1685–1687 (2003)
22. Ma, SB, Ahn, KY, Lee, ES, Oh, KH, Kim, KB: Synthesis and characterization of manganese dioxide spontaneously coated on carbon nanotubes. Carbon **45**, 375–382 (2007)
23. Yue, H, Huang, X, Yang, Y: Preparation and electrochemical performance of manganese oxide/carbon nanotubes composite as a cathode for rechargeable lithium battery with high power density. Mater Lett **62**, 3388–3390 (2008)
24. Soumitra, K, Subhadra, C: Synthesis and characterization of one-dimensional MgO nanostructures. J Nanosci Nanotechnol **6**, 1447–1452 (2006)
25. Bergeret, G, Gallezot, P, Ertl, G, Knözinger, H, Weitkamp, J: Handbook of Heterogeneous Catalysis. Wiley, Weinheim (1997)

5

Fixed-bed adsorption dynamics of Pb (II) adsorption from aqueous solution using nanostructured γ-alumina

Zahra Saadi, Reyhane Saadi and Reza Fazaeli*

Abstract

In the present study, removal of metal ions Pb (II) using nanostructured γ-alumina was investigated in a fixed-bed column. Comparison of Thomas, Yoon-Nelson, and Adams-Bohart models with experimental kinetic results was done, and model parameters were evaluated by linear regression analysis for Pb^{2+} adsorption in different bed heights, initial concentrations, and flow rates. The obtained experimental data are in good agreement with Thomas and Yoon-Nelson models, but in the case of Adams-Bohart model, low correlation coefficient is observed.

Keywords: Lead; Nanostructured γ-alumina; Adsorption kinetic models; Fixed-bed column

Background

Industrial wastewater polluted with heavy metals has endangered the human health and environment. Lead is one of the most extensive toxic metals in the environment [1], so this study has discussed on the continuous flow removal of lead from aqueous solution.

A number of conventional methods for the removal of heavy metals from aqueous solutions have been reported such as precipitation as hydroxides, carbonates, or sulfides; subsequent liquid–solid separation by gravity settling; flotation; ultra-filtration; coagulation; reverse osmosis; ion exchange; electrodialysis; liquid-liquid extraction; membrane processes; evaporation; and biological treatment.

Most of these methods have disadvantages such as high-cost devices, toxic waste production, low efficiency, and require high energy and space. Adsorption process has been shown as an alternative economic method for the removal of heavy metals from aqueous solutions [2-9].

A number of materials have also been used to remove heavy metals from wastewater, such as activated carbon [10], silica [11], titanium dioxide [11], calcium carbonate [11], alumina [11,12], and recently, various nanomaterials such as nanometal oxides [13], carbon nanotubes [14], and nanozeolite composites [15].

Nanoalumina has several crystalline form and mainly used as γ-alumina form.

By considering the γ-alumina, surface area is in the range of 150 to 500 m^2/g and pore radius is in the range of 1.5 to 6 nm, so γ -alumina is anticipated to be more adsorptive active than α-alumina. γ-Al_2O_3 nanoparticles is a promising material as an adsorbent because of its large specific surface area, high adsorption capacity, mechanical strength, and low-temperature modification. Nanoalumina is used widely as electrical insulator, catalyst, and base catalyst in many chemical reactions, membrane processes, microelectronics, and water and wastewater purification [16-21].

Unuabonah et al. studied in polymer-clay-based composite adsorbent to remove lead (II) ions from aqueous solution in a fixed-bed mode. The increase in bed height and initial metal ion concentration increased the adsorption capacity of lead (II). The Yoon-Nelson, Thomas, and Clark models were found to give good fit to adsorption data. On the other hand, Adams-Bohart was found to be a poor predictor for the column operation [22].

Kundu and Gupta studied in continuous fixed bed to evaluate the efficiency of iron oxide-coated cement (IOCC) as an adsorbent for the removal of As(III) from

* Correspondence: r_fazaeli@azad.ac.ir
Department of Chemical engineering, Faculty of engineering, South Tehran Branch, Islamic Azad University, P.O. Box 11365–4435, Tehran, Iran

aqueous solution under the effect of various process parameters like bed depth, flow rate, and initial As(III) concentrations. The dynamics of the adsorption process was modeled by bed depth service time (BDST), mass transfer, Thomas, and Yoon-Nelson models. The Thomas and Yoon-Nelson model predictions were in very good agreement with the experimental results at all the process parameters studied, indicating that they were very suitable for IOCC column design [23].

Rao et al. studied in *Syzygium cumini* L leaf powder as a biosorbent for generating adsorption data in a fixed-bed minicolumn. Effect of flow rate, initial Cd (II) concentration, and bed height was the experimental parameters chosen to obtain breakthrough curves. Adams-Bohart, BDST, Thomas, and Yoon-Nelson models were applied to the data for predicting breakthrough curves and to determine the characteristic parameters. Prominent and unique characteristic features of the respective models like service time (Hutchins BDST model), adsorption capacity (Thomas model), and time required for 50% breakthrough (Yoon-Nelson model) were determined [24].

Han et al. studied in continuous adsorption in a fixed-bed column using phoenix tree leaf powder as an adsorbent for the removal of methylene blue (MB) from aqueous solution. The effect of flow rate, influent MB concentration, and bed depth on the adsorption characteristics of adsorbent was investigated at pH 7.4. Four kinetic models, Thomas, Adams-Bohart, Yoon-Nelson, and Clark, were applied to experimental data to predict the breakthrough curves by nonlinear regression and to determine the characteristic parameters of the column that are useful for process design [25].

Purpose of this study is removal of Pb^{2+} by nanostructured γ-alumina using fixed bed with upward flow. Effects of parameters such as flow rate, initial concentration of lead, and fixed bed height on lead adsorption using Thomas, Yoon-Nelson, and Adams-Bohart kinetic models were considered.

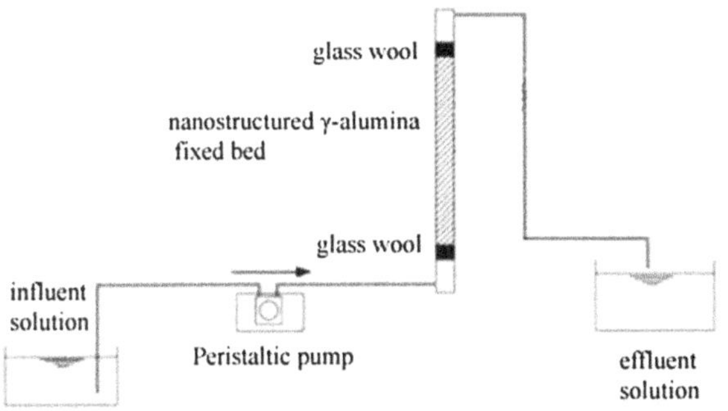

Figure 1 Schematic diagram of the experimental column.

Table 1 Characterization of nanostructured γ-alumina

Step number	Characteristics	Value
1	Particle form	Powder
2	Particle size, nm	20
3	Surface area(minimum), m^2/g	138
4	Bulk density, g/cm^3	0.9
5	Porosity	0.5246
6	**Chemical analysis**	
	Al_2O_3(minimum) %	99
	Ca(maximum), ppm	25
	V(maximum), ppm	7
	CL(maximum), ppm	315
	Na(maximum), ppm	70
	Mn(maximum), ppm	3
	Co(maximum), ppm	2

Materials

Nanostructured γ-alumina prepared from nano Pars Lima Co. (Tehran, Iran) as adsorbent and lead nitrate as adsorbate were obtained from Merck C. (Germany). All materials were prepared with high purity. Double distilled water was used. Dilute HNO_3 and NaOH were used to adjust pH.

Atomic absorption spectrophotometer with acetylene flame (PG 990) was used for the measurement of lead ions; digital pH meter (Sartorius (PB-11)) was used to measure pH. Different flow rates were performed using a peristaltic pump ETATRON DS (Rome, Italy) that is capable to adjust flow rates in milliliter-per-minute range.

Continuous flow studies

Thomas model

The Thomas model is widely used in column performance modeling. Its derivation assumes Langmuir kinetics of adsorption-desorption and no axial dispersion. The

Table 2 Thomas model parameters using linear regression analysis for Pb2+ adsorption under various operating conditions

Q (mL/min)	L (cm)	C_0(mg/L)	$q_{e/max}$ (mg/g)	k_{Th} (mL/min.mg)	R^2	$q_{e,max(exp)}$ (mg/g)
3	10	100	49.635	0.031	0.9109	47.79
6	10	100	46.607	0.052	0.9112	42.6
9	10	100	44.457	0.094	0.9308	38.15
6	5	100	42.962	0.118	0.9182	38.62
6	15	100	48.546	0.038	0.9147	44.99
6	10	65	43.288	0.064	0.9401	39.89
6	10	150	49.181	0.049	0.9319	43.59

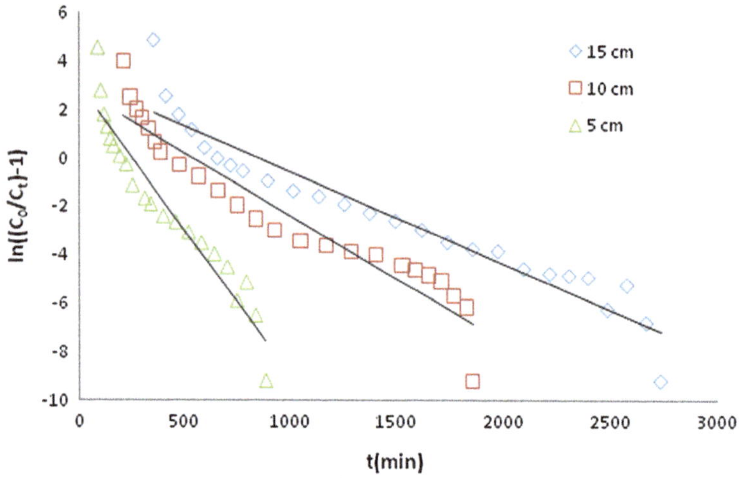

Figure 2 Linear plot of Thomas model with experimental data at different heights. C_0 = 100 mg/L, Q_0 = 6 mL/min.

expression for the Thomas model for adsorption column is given as Equation 1 [26]:

$$\frac{C_t}{C_0} = \frac{1}{1+\exp\left[\left(\frac{k_{\mathrm{Th}}q_e x}{Q}\right) - k_{\mathrm{Th}}C_0 t\right]}, \quad (1)$$

where k_{Th} (mL/min.mg) is the Thomas model constant, q_e (mg/g) is the predicted adsorption capacity, x is mass of adsorbent (g), Q is influent flow rate (mL/min), C_0 is initial solution concentration (mg/L), and C_t is effluent solution concentration (mg/L). The linear form of Thomas model is expressed as Equation 2:

$$\ln\left(\frac{C_0}{C_t} - 1\right) = \frac{k_{Th}q_e x}{Q} - k_{Th}C_0 t \quad (2)$$

Yoon-Nelson model

This model is simpler than other models and also requires no data about the characteristics of the system such as the type of adsorbent and the physical properties of the adsorption bed.

The Yoon-Nelson equation is expressed as Equation 3 [27]:

$$\frac{C_t}{C_0 - C_t} = \exp(k_{\mathrm{YN}}t - \tau k_{\mathrm{YN}}), \quad (3)$$

where k_{YN} (1/min) is the rate constant and τ (min) is the time required for 50% adsorbate breakthrough. The linear form of Yoon-Nelson model is expressed as Equation 4:

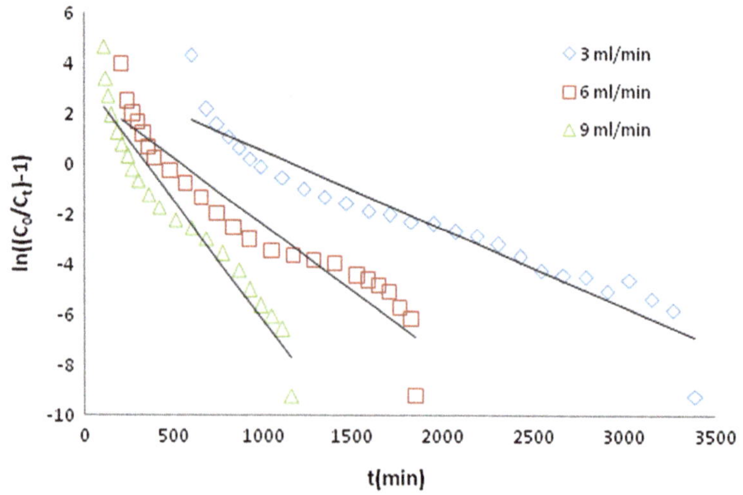

Figure 3 Linear plot of Thomas model with experimental data at different flow rates. C_0 = 100 mg/L, L_0 = 10 cm.

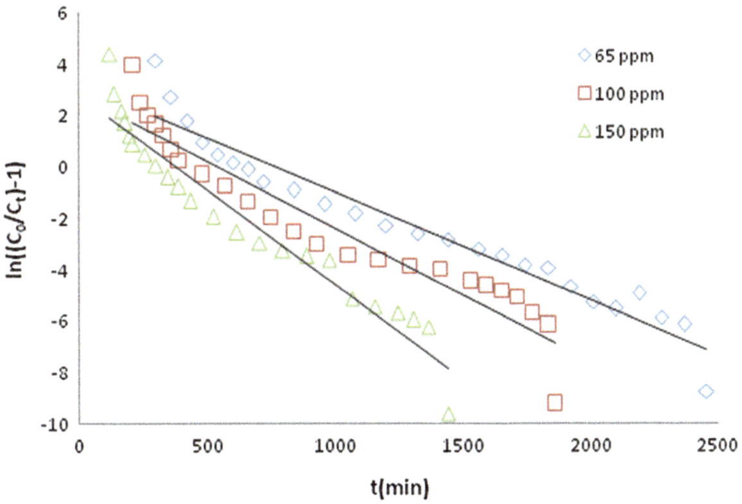

Figure 4 Linear plot of Thomas model with experimental data at different concentrations. Q_0 = 6 mL/min, L_0 = 10 cm.

$$\ln\frac{C_t}{C_0-C_t} = k_{YN}t-\tau k_{YN} \quad . \tag{4}$$

Adams-Bohart model

The Adams-Bohart model is based on the assumption that the rate of adsorption is proportional to both the concentration of the adsorbing species and the residual capacity of the adsorbent. The Adams-Bohart model is only used for the description of the initial part of the breakthrough curve and is expressed as Equation 5 [28]:

$$\frac{C_t}{C_0} = \exp\left(k_{AB}C_0t-k_{AB}N_0\frac{z}{U_0}\right), \tag{5}$$

where k_{AB} (l/min.mg) is rate constant of Adams-Bohart model, z (cm) is the bed depth, N_0 (mg/L) is maximum ion adsorption capacity per unit volume of adsorbent column, and U_0 (cm/min) is the linear velocity of influent solution. The linear form of Adams-Bohart model is expressed as Equation 6:

$$\ln\left(\frac{C_t}{C_0}\right) = k_{AB}C_0t-k_{AB}N_0\frac{z}{U_0}. \tag{6}$$

Methods

A Pyrex column (Sigma-Aldrich, St. Louis, MO, USA) 30 cm in height and 1 cm in diameter was used with upward flow. In order to avoid the exiting adsorbent from column, glass wool at the top and bottom of the column was applied. Schematic of experimental set-up is shown in Figure 1.

In order to investigate the effect of flow rate on the behavior of the adsorbent column, flow rates 3, 6, and 9 mL/min with constant concentrations 100 ppm, constant bed height 10 cm were used. In order to investigate the effect of height on the behavior of the adsorbent column, heights 5, 10, and 15 cm (equivalent with 3.5342, 7.0685, and 10.6028 gr) with constant concentrations 100 ppm and constant flow rate 6 mL/min were used, and in order to investigate the effect of initial concentration of lead ion on behavior of the adsorbent column, concentrations 65, 100, and 150 ppm with constant bed height 10 cm and constant flow rate 6 mL/min were used. Effluent samples from the column were collected at specified time intervals, and remaining lead concentrations in the solution were measured by atomic absorption technique.

In all experiments, the temperature and pH values were adjusted to 25°C and 4.5, respectively.

Maximum adsorption capacity of the adsorbent in a fixed-bed column is calculated by Equations 7 and 8 [22]:

$$q_e = \frac{Q}{1000W}\int_0^{t_{total}}(C_0-C_t)\,dt \tag{7}$$

Table 3 Yoon-Nelson model parameters using linear regression analysis for Pb2+ adsorption under various operating conditions

Q(mL/min)	L(cm)	C_0(mg/L)	τ (min)	k_{YN}(1/min)	R^2
3	10	100	1,196.516	0.003	0.9109
6	10	100	549.076	0.005	0.9112
9	10	100	349.17	0.009	0.9308
6	5	100	253.067	0.011	0.9182
6	15	100	857.894	0.003	0.9147
6	10	65	777.119	0.004	0.9401
6	10	150	383.662	0.007	0.9319

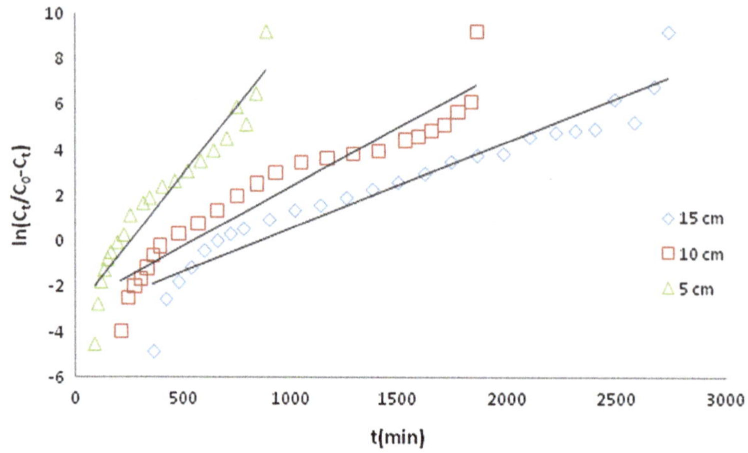

Figure 5 Linear plot of Yoon-Nelson model with experimental data at different heights. C_0 = 100 mg/L, Q_0 = 6 mL/min.

$$q_{e,\max} = \frac{q_{\text{total}}}{w}, \quad (8)$$

where $q_{e,\max}$ (mg/g) is maximum adsorption capacity of the adsorbent, Q (mL/min) is flow rate, C_0 (mg/L) is initial concentration, C_t (mg/L) is concentration in specified time, and w (g) is mass of the adsorbent.

Results and discussions

Characterization of adsorbent

Characterization of nanostructured γ-alumina is presented in Table 1.

Evaluation of Thomas model parameters

To determine maximum adsorption capacity of the adsorbent ($q_{e,\max}$) and kinetic coefficient (k_{Th}) in Thomas model, experimental data were fit into the Equation 2. Linear regression results and values of R^2 are listed in Table 2, where values of R^2 ranged from 0.9109 to 0.9401. According to Table 2, the calculated $q_{e,\max}$ from Thomas model is greater than experimental $q_{e,\max}$. It is also seen in Table 2 that as C_0 and adsorbent weight increased, the value of $q_{e,\max}$ and k_{Th} increased and decreased, respectively. This is because the driving force for adsorption is the difference in concentration between the lead on the adsorbent and in the solution [29]. As flow rate increased, the value of $q_{e,\max}$ decreased. This is because of unavailability of reaction sites, but the value of k_{Th} increased. Linear form of Thomas model in different heights, flow rates, and concentrations is shown in Figures 2, 3, and 4, respectively. As can be observed, the experimental data are in good agreement with the theoretical results.

Evaluation of Yoon-Nelson model parameters

Linear regression results and values of R^2 are presented in Table 3. It was observed that rate constant (k_{YN}) and

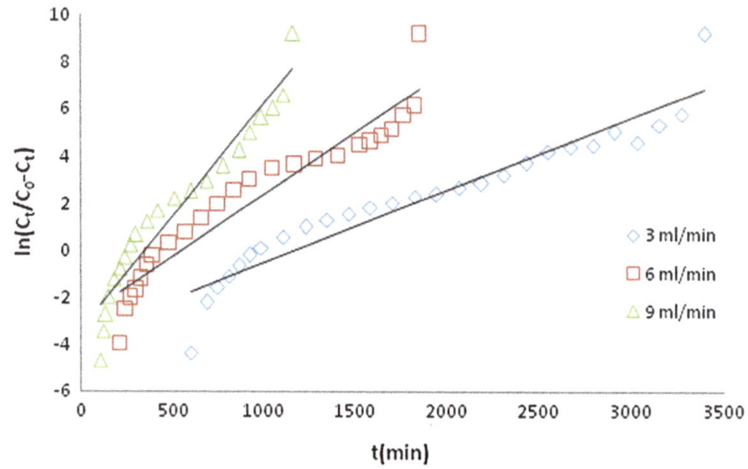

Figure 6 Linear plot of Yoon-Nelson model with experimental data at different flow rates. C_0 = 50 mg/L, L_0 = 15 cm.

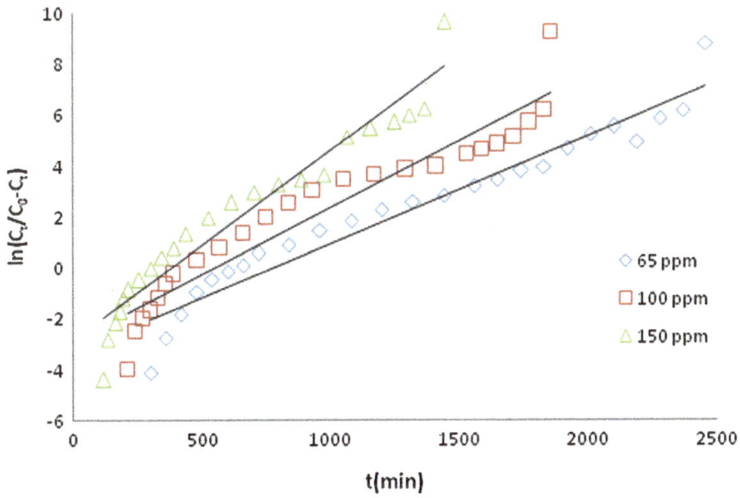

Figure 7 Linear plot of Yoon-Nelson model with experimental data at different concentrations. Q_0 = 6 mL/min, L_0 = 10 cm.

time required for 50% adsorbate breakthrough (τ) decreased and increased, respectively, with increasing adsorbent dose. Also, it was found that the rate constant (k_{YN}) increases as much as flow rate increases. Because of less residence time of metal ion in adsorbent bed, time required for 50% adsorbate breakthrough (τ) decreased with increasing flow rate (Table 3). Linear plot of Yoon-Nelson model in different heights, flow rates, and concentrations is shown in Figures 5, 6, and 7, respectively. As can be observed, the experimental data are fit well with the model. It can be concluded that both Thomas and Yoon-Nelson models are appropriate models to describe fixed-bed operations.

Evaluation of Adams-Bohart model parameters

Linear regression results (Table 4) are shown that adsorption capacity of the adsorbent (N_0) and kinetic constant of the model (k_{AB}) increased and decreased, respectively, with increasing adsorbent dose and initial lead concentration. Because of more saturation of adsorbent sites, adsorption capacity of the adsorbent (N_0) decreased with increasing flow rate. Also, kinetic constant of the model (k_{AB}) increased with increasing flow rate. Linear plot of Adams-Bohart model in different heights, flow rates, and concentrations is shown in Figures 8, 9, and 10, respectively.

The lower R^2 values relative to the other models can be interpreted that the Adams-Bohart model is not as appropriate a predictor for the breakthrough curve.

Conclusions

Studies on continuous adsorption using nanostructured γ-alumina to remove Pb (II) indicate that this adsorbent has high ability to remove heavy metal ions from aqueous solutions.

Adsorption capacity strongly depends on flow rate, influent concentration, and bed height.

Comparison of Thomas, Yoon-Nelson, and Adams-Bohart kinetic models with experimental data was performed, and model parameters were determined by linear regression analysis for Pb^{2+} adsorption under various operating conditions. The experimental data fit well with Thomas and Yoon-Nelson models, but the Adams-Bohart model predicted poor performance of fixed-bed column.

Results showed that with increasing initial concentration of lead, adsorption capacity of the adsorbent and kinetic constant of model increased and decreased, respectively, in the Thomas model. Fifty percent adsorbate breakthrough and rate constant decreased and increased, respectively, in the Yoon-Nelson model. Capacity of the adsorbent and kinetic constant increased and decreased, respectively, in the Adams-Bohart model.

Table 4 Adams-Bohart model parameters using linear regression analysis for Pb2+ adsorption under various operating conditions

Q(mL/min)	L(cm)	C_0(mg/L)	N_0(mg/L)	k_{AB}(mL/min.mg)	R^2
3	10	100	37,182.271	0.099	0.8699
6	10	100	30,835.733	0.206	0.8669
9	10	100	27,371.123	0.334	0.8813
6	5	100	29,346.952	0.367	0.8488
6	15	100	32,464.054	0.15	0.8733
6	10	65	30,136.07	0.207	0.9096
6	10	150	35,097.074	0.118	0.7767

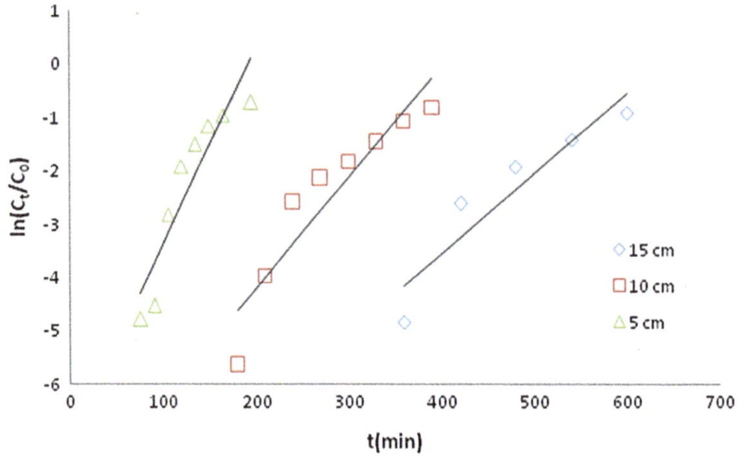

Figure 8 Linear plot of Adams-Bohart model with experimental data at different height. C_0 = 100 mg/L, Q_0 = 6 mL/min.

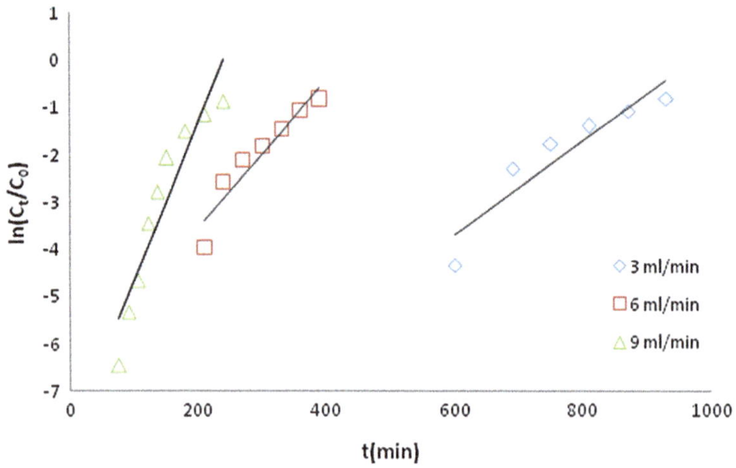

Figure 9 Linear plot of Adams-Bohart model with experimental data at different flow rates. C_0 = 100 mg/L, L_0 = 10 cm.

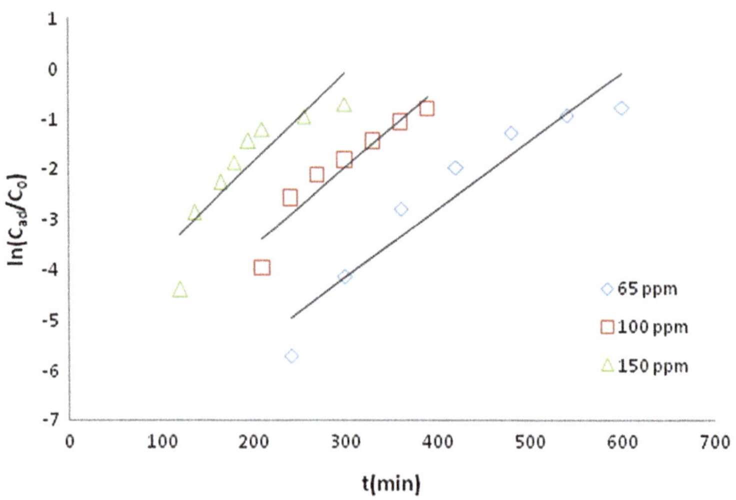

Figure 10 Linear plot of Adams-Bohart model with experimental data at different concentration. Q_0 = 6 mL/min, L_0 = 10 cm.

Competing interests
The authors declare that they have no competing interest.

Authors' contributions
ZS carried out the preparation of experimental elements. ZS and RS participated in doing the experiments and analyzed, interpreted, and organized the data. RF designed and coordinated the overall project. ZS, RS and RF carried out the manuscript preparation. All authors read and approved the final manuscript.

Acknowledgments
The authors wish to thank Mr. Bagher Biralvand for his great help in the atomic absorption analysis.

References
1. Paulino, AT, Santos, LB, Nozaki, J: Removal of Pb^{2+}, Cu^{2+}, and Fe^{3+} from battery manufacture wastewater by chitosan produced from silkworm chrysalides as a low-cost adsorbent. React. Funct. Polym. **68**, 634–642 (2008)
2. Marder, L, Sulzbach, GO, Bernardes, AM, Ferreira, JZ: Removal of cadmium and cyanide from aqueous solutions through electrodialysis. J. Electrochem. Soc. **14**, 610–615 (2003)
3. Bradl, HB: Adsorption of heavy metal ions on soils and soils constituents. J Colloid Interf Sci **277**, 1–18 (2004)
4. Zouboulis, AI, Matis, KA: Removal of metal ions from dilute solutions by sorptive flotation. Crit. Rev. Sci. Tech. **27**, 195–235 (1997)
5. Srivastava, SK, Gupta, VK, Mohan, D: Removal of lead and chromium by activated slag-A blast-Furnace waste. J. Envir. Eng. **123**, 461–468 (1997)
6. Wang, J, Chen, C: Biosorbents for heavy metals removal and their future. Biotech. Adv. **27**, 195–226 (2009)
7. Hota, G, Kumar, BR, Ng, WJ, Ramakrishna, S: Fabrication and characterization of a boehmite nanoparticles impregnated electrospun fiber membrane for removal of metal ions. J. Mater. Sci. **43**, 212–217 (2008)
8. Afkhami, A, Conway, BE: Investigation of removal of Cr(VI), Mo(VI), W(VI), V (IV), and V(V) oxy-ions from industrial waste-waters by adsorption and electrosorption at high-area carbon cloth. J. Colloid Interface Sci. **251**, 248–255 (2002)
9. Afkhami, A, Madrakian, T, Amini, A, Karimi, Z: Effect of the impregnation of carbon cloth with ethylenediaminetetraacetic acid on its adsorption capacity for the adsorption of several metal ions. J. Hazard. Mater. **150**, 408–412 (2008)
10. Mohan, D, Singh, KP: Single- and multi-component adsorption of cadmium and zinc using activated carbon derived from bagasse–an agricultural waste. J. Water Res. **36**, 2304–2318 (2002)
11. Kim, MS, Chung, JG: Removal of copper(II) ion by kaolin in aqueous solutions. Environ. Eng. Res. **7**, 49–57 (2002)
12. Genc-Fuhrman, H, Mikkelsen, PS, Ledin, A: Simultaneous removal of As, Cd, Cr, Cu, Ni and Zn from stormwater: experimental comparison of 11 different sorbents. Water Res. **41**, 591–602 (2007)
13. Zhang, L, Huang, T, Zhang, M, Guo, X, Yuan, Z: Studies on the capability and behavior of adsorption of thallium on nano-Al_2O_3. J. Hazard. Mater. **157**, 352–357 (2008)
14. Ahmedna, M, Marshall, WE, Husseiny, AA, Rao, RM, Goktepe, I: The use of nutshell carbons in drinking water filters for removal of trace metals. Water Res. **38**, 1062–1068 (2004)
15. Ngomsik, AF, Bee, A, Draye, M, Cote, G, Cabuil, V: Magnetic nano- and microparticles for metal removal and environmental applications: a review. CR Chimie. **8**, 963–970 (2005)
16. Turker, AR: New sorbents for solid-phase extraction for metal enrichment. Clean **35**, 548–557 (2007)
17. Li, J, Shi, Y, Cai, Y, Moua, S, Jiang, G: Adsorption of di-ethyl-phthalate from aqueous solutions with surfactant-coated nano/microsized alumina. Chem. Eng. J. **140**, 214–220 (2008)
18. Hiraide, M, Iwasawa, J, Hiramatsu, S, Kawaguchi, H: Use of surfactant aggregates form on alumina for the preparation of chelating sorbents. Anal. Sci. **11**, 611–615 (1995)
19. Srivastava V, Weng CH, Singh VK, Sharma YC: Adsorption of Nickel Ions from Aqueous Solutions by Nano Alumina: Kinetic, Mass Transfer, and Equilibrium Studies. J. Chem. Eng. Data. **56**, 1414–1422 (2011)
20. Sharma, YC, Srivastava, V, Upadhyay, SN, Weng, CH: Alumina nanoparticles for the removal of Ni(II) from aqueous solutions. Ind. Eng. Chem. Res. **47**, 8095–8100 (2008)
21. Treybal, RE: Mass Transfer Operations, 3rd edn. Mc Graw Hill Book Company, Japan (1990)
22. Unuabonah, EI, El-Khaiary, MI, Olu-Owolabi, BI: Predicting the dynamics and performance of a polymer-clay based composite in a fixed bed system for the removal of lead(II) ion. Chem. Eng. Res. Design. **90**, 1105–1115 (2012)
23. Kundu, S, Gupta, AK: As(III) removal from aqueous medium in fixed bed using iron oxide-coated cement (IOCC): experimental and modeling studies. Chem. Eng. J. **129**, 123–131 (2007)
24. Rao, KS, Anand, S, Venkateswarlu, P: Modeling the kinetics of Cd(II) adsorption on *Syzygium cumini* L leaf powder in a fixed bed mini column. J. Ind. Eng. Chem. **17**, 174–181 (2011)
25. Han, R, Wang, Y, Zhao, X, Wang, Y, Xie, F, Cheng, J, Tang, M: Adsorption of methylene blue by phoenix tree leaf powder in a fixed-bed column: experiments and prediction of breakthrough curves. Desalination **245**, 284–297 (2009)
26. Thomas, HC: Heterogeneous ion exchange in a flowing system. J. Am. Chem. Soc. **66**, 1466–1664 (1944)
27. Yoon, YH, Nelson, JH: Application of gas adsorption kinetics. I. A theoretical model for respirator cartridge service time. Am. Ind. Hyg. Assoc. J. **45**, 509–516 (1984)
28. Loderio, P, Herreo, R, Sastre de Vicentes, ME: The use of protonated *Sargassum muticum* as biosorbent for cadmium removal in a fixed-bed column. J. Hazard. Mater. **137**, 244–253 (2006)
29. Aksu, Z, Gonen, F: Biosorption of phenol by immobilized activated sludge in a continuous packed bed: prediction of breakthrough curves. Process. Biochem. **39**, 599–613 (2004)

An improved method for the purification of fullerene from fullerene soot with activated carbon, celite, and silica gel stationary phases

Hassan Keypour[1*], Mohammad Noroozi[1,2†] and Alimorad Rashidi[3†]

Abstract

Extracting pure fullerene (C_{60}) from fullerene soot was accomplished in two steps. In the first step, extraction was carried out using a Soxhlet extractor. Different solvents, such as toluene, were used in this method. In this process, light and heavy fullerenes were separated from impurities such as soot and other compounds. The applied fullerene soot was produced using an electric arc method, and the obtained purity was close to 7%. In the next step, fullerene C_{60} can be separated from other fullerenes (particularly C_{70}) by column chromatography. The stationary phase in the chromatography column contained a mixture of activated carbon, silica gel, and celite, and in order to achieve a high efficiency for separation, the column was filled with a combination of these compounds. Organic solvents such as chlorobenzene and 1,2-dichlorobenzene were used as the mobile phase. Several experiments were designed and conducted to obtain optimal conditions, and consequently, the fullerene C_{60} was purified with proper efficiency. In this method, parameters including efficiency, low cost, and simplicity are of great importance in the aim of obtaining an optimized method. As a result, a suitable method with low cost and high efficiency was used in this research.

Keywords: Fullerene, Purification, Chromatography, Celite

Correspondence/Findings

Background

Fullerenes can be produced through several developing methods which have their own advantages and disadvantages. Fullerenes can be synthesized by ablating graphite with a laser [1], burning graphite in a furnace [2], and by producing an arc across two graphite electrodes in an inert atmosphere [3]. Other methods include using negative ion/desorption chemical ionization techniques [4-13], benzene flame [14,15], combustion method [16], and soot-free synthesis of C_{60} [17]. At the present time, the most common method for synthesizing fullerenes is the Huffman-Kratschmer carbon arc technique [18]. In this method, pure carbon in the form of graphite was heated under plasma temperature by means of graphite electrodes in an inert atmosphere (usually helium). This method leads to the production of soot, in which the fullerenes have to be separated from. Until recently, only a few methods have been developed to separate fullerenes successfully. Most known separation methods employ a column chromatography technique [19]. Column chromatography uses a tube that is open at the top with a valve at the bottom to control the flow of liquid through the tube. The column is filled with a substance that has some affinity with the material that one is trying to separate (stationary phase). After the material to be separated is placed inside the top of the column, a solvent is passed through the column. The material to be separated has different interactions with the solvent and stationary phase; consequently, it moves through the tube slower than the solvent. The components will be carried along with the mobile phase to varying degrees, and a separation will be achieved. Most attempts to separate fullerenes involved using large amounts of stationary phase and solvent and an extended period of time for separation. The most common stationary phase for column chromatography is silica gel which is not able to separate the compounds effectively, and consequently, silica gel is used with alumina. Fullerenes have low

* Correspondence: haskey1@yahoo.com
†Equal contributors
[1]Faculty of Chemistry, Bu-Ali Sina University, Hamedan 65174, Iran
Full list of author information is available at the end of the article

solubility in common solvents, such as hexane. Their low solubility in inexpensive common solvents leads to ineffective separations with common stationary phases such as silica gel. An area of potential interest lies in the development of effective materials to separate the compounds properly for isolating gram quantities of 25 fullerenes. Recently, it was discovered that the use of an activated carbon (AC) and silica gel mixture can provide adequate and cost-effective separation of gram quantities of C_{60} and C_{70} [20,21].

One of the current interests in fullerenes is directed toward large-scale production and commercial applications. Although more than a decade has passed since fullerenes have been separated, an increasing number of researchers in both academic and industrial environment are still working on this subject [2]. Quite recently, industrial-scale fullerene extraction (a mixture of C_{60}, C_{70}, and C >70) began and lowered the price to less than one-fourth compared to the former prices [22]. With respect to the purification of fullerenes [19,23], chromatographic methods have been commonly employed [24-28]. Although these methods are quite useful for laboratory-scale purification, a few drawbacks were observed in the case of larger-scale production such as limited column loadings, duration of the process with constant attention, and the amounts of both required stationary and mobile phases. Non-chromatographic methods such as selective complexation [29-33] and fractional crystallization [34-37] are not always efficient especially for large-scale purification since they require a long stirring time, sometimes at elevated temperatures, and repeating the precipitation-filtration process a few times. Therefore, there is an increasing demand for a more efficient purification method that is applicable for the large-scale production of pure fullerenes. There are also several methods for the purification of fullerene derivatives [38]. In the present paper, we report a novel method for purifying C_{60} with a Soxhlet extractor, column choromatography, and the characterization of purified fullerene by Fourier transmittance infrared (FT-IR), ultraviolet–visible (UV–vis) spectroscopy, nuclear magnetic resonance (NMR) spectroscopy, X-ray diffraction (XRD), and high-performance liquid chromatography (HPLC). As a result, these methods lead us to methods with low cost and high-efficiency purification.

Results and discussion

A process has been developed to separate the compounds effectively and recover pure amounts of C_{60} from the fullerene soot containing higher and lower molecular weight fullerenes. This process first extracts the fullerenes from carbon soot and then separates the C_{60} molecules sequentially with high-performance or low-adsorption chromatography packed column containing a silica gel, AC, or celite.

In this process, the first step of fullerene extraction is preferably accomplished by the Soxhlet method. Several solvents were examined on small scales using the Soxhlet method. The choice of solvent is important and is based on the solubility of C_{60} fullerene. Experimental solubilization of C_{60} has been carried out via different solvents. The used solvents are classified in the following groups: polars (*N*-methyl-2-pyrrolidone (NMP)), benzenes (toluene, chlorobenzene (CB), orthodichlorobenzene (ODCB), 1,2,4-trichlorobenzene (TCB)), naphthalenes (MN, dimethylnaphthalenes (DMN), 1-phenylnaphthalene (PN), 1-chloronaphthalene (CN)), and miscellaneous (carbon disulfide (CDS)). The measured solubilities confirm that C_{60} is poorly soluble in polar and H-bonding solvents and has low solubility in alkanes (the higher the number of carbons, the better the solubility), even though declining that solubilized C_{60} is more soluble in halogenated derivatives than in analog alkanes. In general, aromatics are proper solvents; increasing the number of rings increases solubility. As shown in Table 1, the most applicable solvent is CN (51 g/L, at room temperature) [39]. In addition to the high solubility, the solvent cost is also of great importance. As shown in Table 2, in the constant conditions of extraction, the highest efficiency is observed in run numbers of 4 to 9. The best solvent for extracting fullerenes from soot at a reasonable price that also showed high performance is ODCB. As illustrated in Figure 1, some of the solvents yield more than 100%, which means that the actual amount of fullerenes is 7% higher than the theoretical value. In the second step, fullerenes were extracted using ODCB solvent, and about 20 g of fullerenes were obtained by the Soxhlet extraction process. The solvent was then recovered using a rotary evaporator. The fullerenes were used for the purification in the second stage. Qualitative and quantitative tests were performed on fullerenes, such as IR, HPLC, and UV–vis spectroscopy. Analysis of the results showed that

Table 1 Solubility of C_{60} in various solvents

Solvents	Solubility of C_{60} (mg/mL)
Toluene	2.8
N-methyl-2-pyrrolidone	0.89
Chlorobenzene	7.0
1,2-Dichlorobenzene	27
1,2,4-Trichlorobenzene	8.5
1-Methylnaphthalene	33
Dimethylnaphthalenes	36
1-Phenylnaphthalene	50
1-Chloronaphthalene	51
Carbon disulfide	7.9

Table 2 Extraction of fullerenes from soot with the Soxhlet method

Number of run	Solvent	Temperature (°C)	Fullerenes extracted (g)	Fullerenes extracted (%)	Efficiency (%)
1	Carbon disulfide	57	0.34	6.8	97.14
2	NMP	215	0.08	1.6	22.86
3	Chlorobenzene	142	0.352	7.04	100.0
4	1,2-Dichlorobenzene	190	0.357	7.14	100.0
5	1,2,4-Trichlorobenzene	225	0.357	7.14	102.0
6	1-Methylnaphthalene	255	0.357	7.14	100.0
7	Dimethylnaphthalenes	280	0.357	7.14	100.0
8	1-Phenylnaphthalene	335	0.357	7.14	100.0
9	1-Chloronaphthalene	275	0.357	7.14	100.0
10	Toluene	120	0.16	3.2	45.71

a mixture of fullerenes was extracted by the Soxhlet method.

In the second step, C_{60} was purified from the mixture of fullerenes by using column chromatography. As shown in Table 3, the purification efficiency can be optimized by changing parameters such as the combination of stationary phase and mobile phase. The various chromatographic methods used for fullerene separation have been investigated in several investigations, including stationary and mobile phases, conditions of detection, separation yields in some case, and a general discussion for each method. Stationary phases that have been reviewed include alumina [40], silica [41], molecular sieve, and AC [42-44]. In comparison with single materials, combinations of materials are more applicable and have been investigated more. In order to examine changes in the composition of the stationary phase material, 33 tests were designed. In addition to changing the mobile phase, four solvents were considered separately as a mobile phase. Large numbers of experiments have to be performed in order to evaluate all factors. The effects of all the parameters can be incorporated into designing the experiments. In runs 1 to 11, the effect of increasing AC and celite in the silica gel are shown. By adding these compounds to the silica gel, the

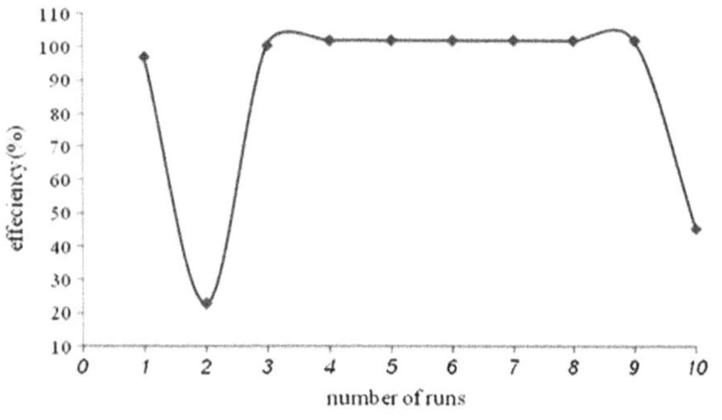

Figure 1 Efficiency of fullerene extraction from soot by the Soxhlet method.

purification efficiency increases significantly. Similarly, runs 12 to 22 show the effect of increasing the silica gel and celite on the AC. In runs 23 to 33, adding silica gel and AC to celite at first shows an increase in efficiency and purification, and then a reduction was observed. This trend of increase and decrease is shown in Figure 2. The increased efficiency in the first and second series of runs indicates that the exchange molecules of silica gel with celite and AC, and AC with celite and silica gel lead to increases in the separation process in columns. The reduced efficiency of separation in the third series of runs indicates that unlike previous series, the exchange of celite with silica gel and AC reduces the performance of the separation. The results in Table 3 and the slope of the curve in Figure 2 show that the purification efficiency of C_{60} is increased by the addition of celite to the stationary phase. A comparison of efficiency in runs 1, 12, and 23 indicates that when celite is used individually, high separation performance will be observed.

Comparing the results shown in Table 3, it can clearly be seen that celite is essential for the separation of fullerene C_{60}. For the stationary phase containing 100% of silica gel, AC, and celite, the efficiency of C_{60} purification is 33.21, 82.36, and 90.14, respectively.

The maximum peak of four curves is related to run number 29 in which the mobile phase consisted of 30%, 30%, and 40% of silica gel, AC and celite, respectively. Four solvents, CB, ODCB, CDS, and toluene, were used as the mobile phase in this method. As shown in Figure 2, the purification efficiency is as follows: ODCB > CB > toluene > CDS. The main impurity in fullerene C_{60} is C_{70}. The highest obtained purity in these experiments was 99.34%. High purity can be confirmed by FT-IR, NMR, UV–vis spectroscopy, and XRD patterns.

XRD is another technique for the determination of fullerene purity. The XRD pattern of pure fullerene C_{60} (shown in Figure 3) is clear and appears at $2\theta = 10.8$, 17.6, and 20.7. Figure 3 shows the XRD patterns of a mixture of

Table 3 Separation of C_{60} (%) with different mobile phases and stationary phases

Number of run	Combination of stationary phases (%)			Efficient separation of C_{60} (%) with different mobile phases			
	Silica gel	AC	Celite	ODCB	CB	CDS	Toluene
1	100	0	0	33.21	20.46	18.12	25.25
2	90	5	5	46.15	43.58	37.33	40.25
3	80	10	10	52.31	50.05	42.54	55.24
4	70	15	15	65.47	62.79	58.21	60.05
5	60	20	20	76.32	73.86	68.95	70.33
6	50	25	25	87.58	84.37	79.01	82.10
7	40	30	30	89.12	87.44	80.44	83.72
8	30	35	35	90.32	87.92	79.43	83.90
9	20	40	40	92.43	88.67	80.98	85.42
10	10	45	45	93.05	90.04	82.65	86.32
11	0	50	50	94.06	92.83	86.11	90.11
12	0	100	0	82.36	80.33	72.37	75.44
13	5	90	5	84.32	82.68	72.49	75.10
14	10	80	10	85.07	83.25	75.30	81.86
15	15	70	15	86.12	84.76	77.61	82.06
16	20	60	20	86.84	83.53	79.05	82.76
17	25	50	25	87.32	84.26	79.74	83.43
18	30	40	30	87.99	85.03	80.43	83.87
19	35	30	35	88.09	85.36	81.36	84.02
20	40	20	40	93.12	87.27	83.02	85.71
21	45	10	45	94.32	90.62	83.95	87.54
22	50	0	50	97.12	92.65	84.52	88.72
23	0	0	100	90.14	88.51	79.04	81.04
24	5	5	90	94.34	90.39	79.47	82.43
25	10	10	80	96.02	94.72	83.64	86.54
26	15	15	70	97.32	95.21	86.75	90.06
27	20	20	60	98.06	95.41	89.05	92.83
28	25	25	50	99.32	96.33	90.27	93.74
29	30	30	40	99.34	96.73	90.82	94.83
30	35	35	30	98.94	96.04	90.54	93.75
31	40	40	20	94.15	93.22	83.06	88.93
32	45	45	10	87.26	90.42	79.43	82.54
33	50	50	0	85.38	85.73	77.34	81.04

fullerenes extracted by the Soxhlet method. Solid C_{60} forms a face-centered-cubic structure at room temperature [45,46]. The presence of impurities in the extracted product by the Soxhlet system is observable by broadening obtained peaks and observing other peaks that are related to C_{60}.

The ^{13}C NMR spectrum of purified C_{60} shows one signal at 143.2. The amount of fullerene in the NMR samples is low, and the spin–lattice relaxation times are quite long (≥20 s for C_{60}) [47]. The FT-IR spectra of fullerene soot and purified C_{60} are shown in Figure 4. Four absorption areas at 527, 576.5, 1,182.5, 1,429 cm^{-1} are related to C-C bond stretching and bending frequencies [48].

The electronic absorption spectra (UV–vis) of C_{60} and fullerene soot are characterized by stronger absorptions between 190 and 410 nm and the forbidden transitions in the visible region of the spectrum. For C_{60}, the number and the type of transitions were confirmed by means of theoretical calculation results. The assignment of the transitions has been carried out using the results of theoretical calculations [49,50]. The absorptions between 190 and 410 nm are related to the allowed $^{1}T_{1u}$ - $^{1}A_g$

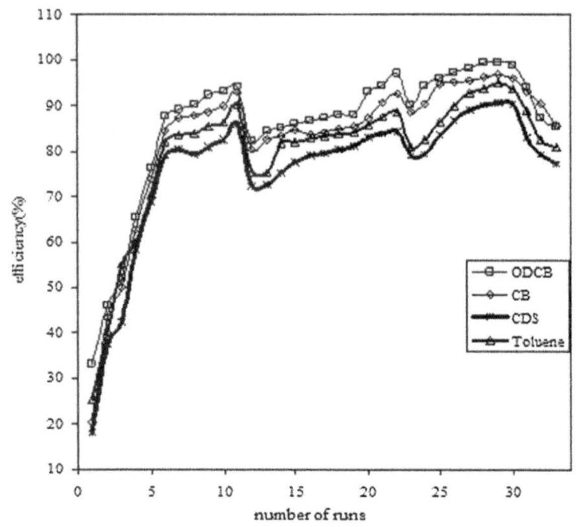

Figure 2 Efficiency of C_{60} purification by column chromatography.

transitions, while absorptions between 410 and 620 nm are related to orbitally forbidden singlet-singlet transitions. These absorptions that appeared in the visible part are responsible for the purple color of C_{60} and the red color of C_{70} [51].

Raw and pure samples of fullerene were also analyzed by HPLC (Figure 5). Chromatography was performed using Agilent 1200 HP HPLC (Agilent Technologies, Santa Clara, CA, USA) with Develosil ODS-5 column (Nomura Chemical Co., Ltd., Seto, Japan) at room temperature (*ca.* 15°C to 20°C) with mobile phase toluene/acetonitrile = 50/50, flow rate of 2 mL/min, and UV detector at 325 nm. Clearly, the peaks corresponding to fullerenes separated by means of an appropriate method of analysis.

Methods

Materials and equipment

Fullerene soot was purchased from Sigma-Aldrich Chemise GmbH Company (purity 7%; Steinheim, Germany); celite R-630, silica gel 40, and powdered AC Norite A (particle size 1 to 150 μm) were purchased from the Merck company (Dramstadt, Germany). Toluene, ODCB, other chemical materials, and solvents were purchased from both Merck and Aldrich. These materials were used without further purification. IR spectra were recorded using a Shimadzu FT-IR 8400 (Kyoto, Japan); UV–vis spectrophotometers Scinco S-2100 (SCINCO, Twin Lakes, WI, USA), NMR Jeol 90 MHz (JEOL Ltd., Tokyo, Japan), HPLC Agilent 1200 HP, and rotary evaporator Heidolph Laborota 4000 (Schwabach, Germany) were used.

Soxhlet extraction procedure

Fullerene-containing carbon soot (5 g) was directly added to a cellulose extraction thimble. The extractions were performed using 50 mL of toluene for more than 1 h. Then, solids were separated from the solution by a rotary evaporator. The obtained solid was dried in an oven at 110°C (0.16 g). The extraction of fullerenes from soot was carried out by different solvents such as NMP, CB, ODCB, TCB, 1-ethylnaphthalene, DMN, PN, CN, or CDS. The fullerenes obtained were analyzed with FT-IR, NMR, UV–vis, XRD, and HPLC.

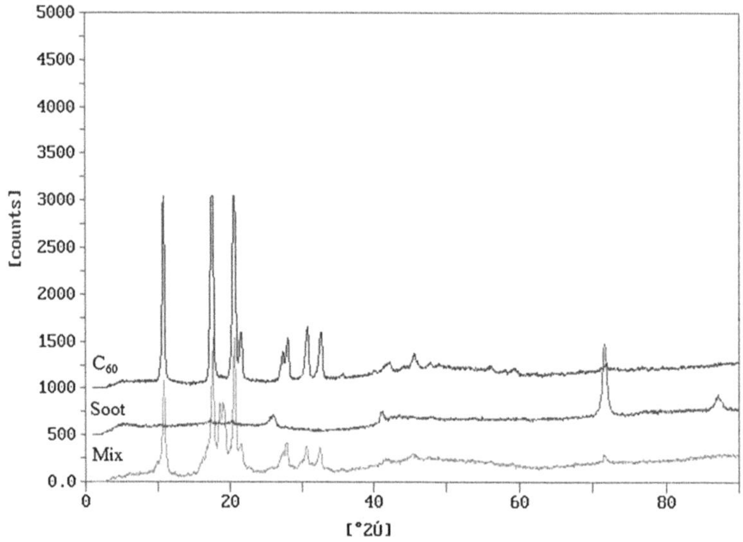

Figure 3 Comparison between XRD patterns of soot, mixture of fullerenes, and extracted fullerene (C_{60}).

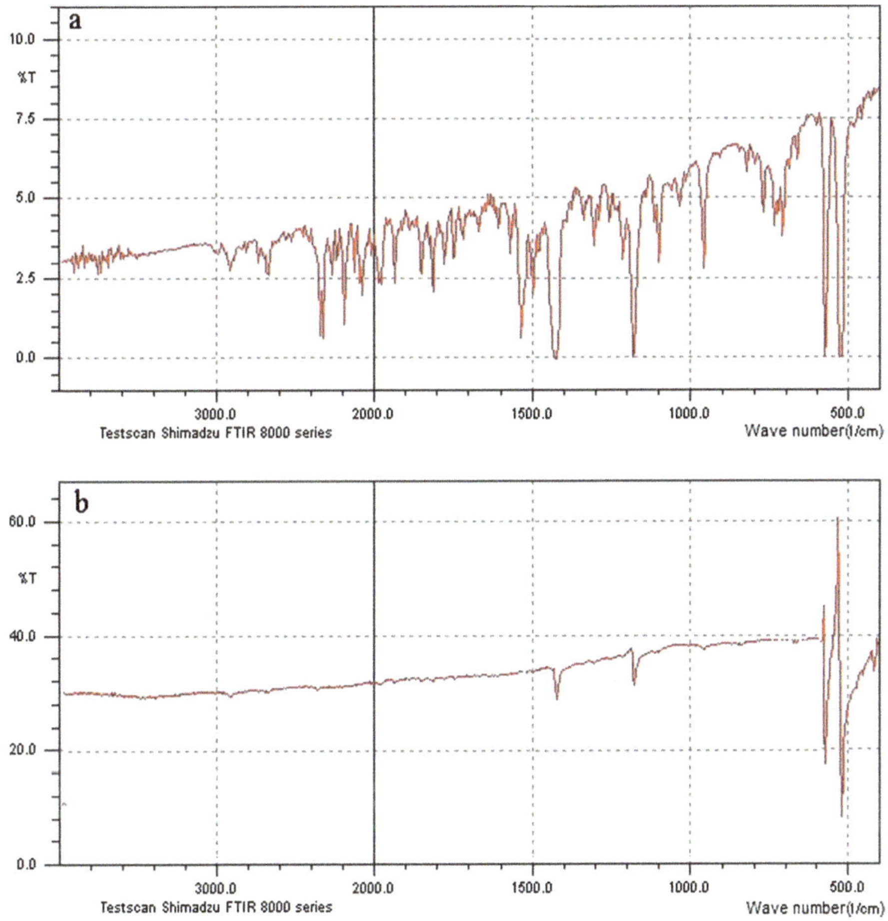

Figure 4 FT-IR spectra of fullerene soot and purified C_{60}. (**a**) Fullerenes extracted by Soxhlet (toluene) and (**b**) fullerenes purified by column chromatography are related to samples no. 29.

Preparing the column

The following steps were performed to prepare the column. In this step, a clean column (1.5 × 50 cm) with a glass wool at the bottom was used. It is notable that the applied glass container and column have to be washed with soapy water, rinsed with distilled water and a small amount of acetone, and ultimately dried in an oven. A small amount (0.5 cm) of dry sand was added to the column on top of the glass wool. The column was filled with the stationary phase including different percentages of silica gel, AC, and celite. Hexan was added to the beaker containing stationary phase as a solvent. The contents of the beaker were stirred to remove air bubbles, and solid particles were allowed to get wet. A funnel was placed on top of the column, and the stationary phase was poured quickly into the column. If necessary, the columns' stopcock will be opened, and some of the hexane was drained away to the stationary phase. This process was repeated until most of the stationary phase transferred to the column. The column was then gently tapped in order to remove the bubbles and compress the stationary phase; 1 cm of sand was smoothly added to the top of the stationary phase. The stopcock was opened, and the excess hexane was drained until its level reaches above the sand level.

Running the column

The extracted fullerene (50 mg) was poured into an Erlenmeyer flask, and then 20 mL of the solvent was added to the flask for purification purpose. The mixture was stirred properly and transferred to the top of the sand layer in the column. The opened stopcock in the column allows the fullerene to pass through the sand and the stationary phase. Subsequent to this step, 5 mL of the solvent was added to the solution. The components were collected as the solvent drips from the bottom of the column, and a rotary was used to evaporate the solvent and isolate the fullerenes. The obtained solid fullerenes were analyzed. This process was repeated for different stationary phases, and the efficiency of each test was also evaluated.

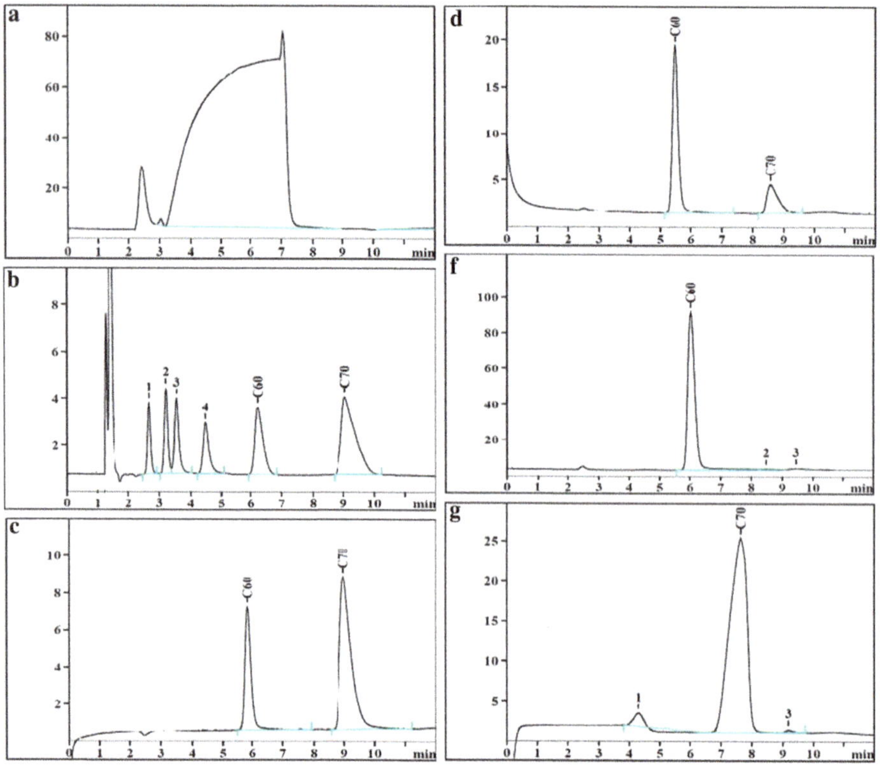

Figure 5 HPLC chromatograms. (**a**) Fullerene soot, (**b**) fullerenes extracted by Soxhlet (toluene), and (**c**) fullerene purified by column chromatography are related to samples no. 2 (toluene); those in (**d**, **f**, **g**) are related to samples no. 29.

Conclusions

Celite is a naturally occurring, soft, chalk-like sedimentary rock that is easily crumbled into a fine white to off-white powder. This powder has an abrasive characteristic, similar to pumice powder, and is very light due to its high porosity. The typical chemical composition of diatomaceous earth is 86% silica, 5% sodium, 3% magnesium, and 2% iron. It is used as a filtration aid, mild abrasive, mechanical insecticide, absorbent for liquids, cat litter, activator in blood clotting studies, and heterogeneous catalyst [52-55].

Celite is a cheap material, which together with activated carbon and silica gel increases the efficiency of purification. This material with good pores traps the heavier fullerenes of C_{60} and allows fullerene C_{60} to pass through the column. The use of this material with silica gel and activated carbon can increase the efficiency significantly. Here are a series of chemical and physical interactions of fullerene with celite and also celite with AC and silica gel. This method is a simple and yet efficient method to use. In addition to the laboratory base, this method can be used in the bench base. The applied solvents are easily recyclable by rotary evaporators for use in the next stages. The mobile phase can be recovered using a suitable solvent. In this method, parameters including efficiency, low cost, and simplicity are of great importance in the aim of obtaining an optimized method. As a result, a suitable method with low cost and high efficiency was used in this research.

Competing interests

The authors declare that they have no competing interests.

Authors' contributions

MN participated in the experimental section. MN and AR participated in the design of the study and performed the statistical analysis. HK, as the supervisor, participated in this work. HK, MN, and AR carried out the manuscript preparation. All authors read and approved the final manuscript.

Acknowledgements

We are grateful to the Department of Chemistry of Bu-Ali Sina University and the Ministry of Science, Research Institute of Petroleum Industry (RIPI) for their financial support.

Author details

[1]Faculty of Chemistry, Bu-Ali Sina University, Hamedan 65174, Iran. [2]Center for Research and Development of Petroleum Technologies at Kermanshah, Research Institute of Petroleum Industry (RIPI), Kermanshah, Iran. [3]Nanotechnology Research Center, Research Institute of Petroleum Industry (RIPI), Tehran, Iran.

References

1. Kroto, HW, Heath, JR, O'Brien, SC, Curl, RF, Smalley, RE: C_{60}: buckminsterfullerene. Nature **318**(6042), 162–163 (1985)
2. Kratschmer, W, Lamb, LD, Fostiropoulos, K, Huffman, DR: Solid C_{60}: a new form of carbon. Nature **347**(6291), 354–358 (1990)

3. Haufler, RE, Conceicao, J, Chibante, LPF, Chai, Y, Byrne, NE, Flanagan, S, Haley, MM, O'Brien, SC, Pan, C, Xiao, Z, Billups, WE, Ciufolini, MA, Hauge, RH, Margrave, JL, Wilson, LJ, Curl, RF, Smalley, RE: Efficient production of C_{60} (buckminsterfullerene), $C_{60}H_{36}$, and the solvated buckide ion. J. Phys. Chem **94**(24), 8634–8636 (1990)
4. Gerhardt, PH, Loffler, S, Homann, KH: Polyhedral carbon ions in hydrocarbon flames. Chem. Phys. Lett **137**(4), 306–310 (1987)
5. Gerhardt, PH, Loffler, S, Homann, KH: The formation of polyhedral carbon ions in fuel-rich acetylene and benzene flames. Proc. Combust. Inst. **22**, 395–401 (1988)
6. Gerhardt, PH, Homann, KH: Ions and charged soot particles in hydrocarbon flames. J. Phys. Chem. **94**, 5381–5391 (1990)
7. Baum, T, Loffler, S, Weilmunster, P, Homann, KH: Polycyclic aromatics, fullerenes and soot particles as charged species in flames. Prepr. Pap. Am. Chem. Soc. Div. Fuel. Chem. **36**, 1533–1540 (1991)
8. Baum, T, Loffler, S, Loffler, P, Weilmunster, P, Homann, KH: Fullerene ions and their relation to PAH and soot in low-pressure hydrocarbon flames. Ber. Bunsen. Ges. Phys. Chem. **96**, 841–857 (1992)
9. Ahrens, J, Kovacs, R, Shafranovskii, EA, Homann, KH: On-line multi-photon ionization mass spectrometry applied to PAH and fullerenes in flames. Ber. Bunsen. Ges. Phys. Chem. **98**, 265–268 (1994)
10. Ahrens, J, Bachmann, M, Baum, T, Griesheimer, J, Kovacs, R, Weilmunster, P, Homann, KH: Fullerenes and their ions in hydrocarbon flames. Int. J. Mass. Spectrom. Ion. Proc. **138**, 133–148 (1994)
11. Bachmann, M, Wiese, W, Homann, KH: Fullerenes versus soot in benzene flames. Combust. Flame. **101**, 548–550 (1995)
12. Bachmann, M, Wiese, W, Homann, KH: PAH and aromers: precursors of fullerenes and soot. Proc. Combust. Inst. **26**, 2259–2267 (1996)
13. Ahrens, J, Keller, A, Kovacs, R, Homann, KH: Large molecules, radicals, ions, and small soot. Particles in fuel-rich hydrocarbon flames. Ber. Bunsenges. Phys. Chem **102**, 1823–1839 (1998)
14. Howard, JB, McKinnon, JT, Makarovsky, Y, Lafleur, AL, Johonson, ME: Production and characterization of fullerenes in flames. Prepr. Pap.-Am. Chem. SOC., Div. Fuel Chem **36**(3), 1022–1025 (1991)
15. McKinnon, JT, Bell, WL, Barkley, RM: Combustion synthesis of fullerenes. Combust. Flame. **88**, 102–112 (1992)
16. Hiroaki, T, Masashi, F, Mineyuki, A, Michael, DDJ, Michael, A: Experimental study of industrial scale fullerene production by combustion synthesis. Carbon **43**, 311–319 (2005)
17. Ryan, PR, Peter, TAR, William, B, Whitten, J, Michael, R: Soot-free synthesis of C_{60}. Carbon **41**, 1469–1475 (2003)
18. Howard, JB, Lafleur, AL, Makarovsky, Y, Mitra, S, Pope, CJ, Yadav, TK: Fullerenes synthesis in combustion. Carbon **30**(8), 1183–1201 (1992)
19. Fetzer, JC: The preparative separation of fullerenes. In: Jinno, K (ed.) Separation of fullerenes by liquid chromatography, pp. 25–48. The Royal Society of Chemistry, Cambridge (1999)
20. Govindraj, A, Rao, CNR: Convenient procedures for obtaining pure C_{60} and C_{70} in relatively large quantities. Fullerene. Sci. Tec. **1**(4), 557–562 (1993)
21. James, M: Purification of fullerenes. (1997). US 55662876
22. Howard, JB, McKinnon, JT, Johnson, ME, Makarovsky, Y, Lafleur, AL: Production of C_{60} and C_{70} fullerenes in benzene-oxygen flames. J. Phys. Chem **96**(16), 6657–6662 (1992)
23. Theobald, J: Extraction and purification of fullerenes: a comprehensive review. Sep. Sci. Technol. **30**(14), 2783–2819 (1995)
24. Scrivens, WA, Cassell, AM, North, BL, Tour, JM: Single column purification of gram quantities of C_{70}. J. Am. Chem. Soc **116**(15), 6939–6940 (1994)
25. Darwish, AD, Kroto, HW, Taylor, R, Walton, DRM: Improved chromatographic separation of C_{60} and C_{70}. J. Chem. Soc. Chem. Commun. **1**, 15–16 (1994)
26. Isaacs, L, Wehrsig, A, Diederich, F: Improved purification of C_{60} and formation of sigma- and pai-homoaromatic methano-bridged fullerenes by reaction with alkyl diazoacetates. Helv. Chim. Acta. **76**(3), 1231 (1993)
27. Scrivens, WA, Bedworth, PV, Tour, JM: Purification of gram quantities of C_{60}. A new inexpensive and facile method. J. Am. Chem. Soc **114**(20), 7917–7919 (1992)
28. Wilson, MA, Pang, LSK, Vassallo, AM: C_{60} separation on coal. Nature **355**, 117–118 (1992)
29. Komatsu, N: Preferential precipitation of C_{70} over C_{60} with phalohomooxacalix [3] arenes. Org. Biomol. Chem. **1**(1), 204–209 (2003)
30. Komatsu, N: New synthetic route to homooxacalix[n]arenes via reductive coupling of diformylphenols. Tetrahedron. Lett. **42**(9), 1733–1736 (2001)
31. Atwood, JL, Koutsantonis, GA, Raston, CL: Purification of C_{60} and C_{70} by selective complexation with calixarenes. Nature **368**, 229–231 (1994)
32. Suzuki, T, Nakashima, K, Shinkai, S: Very convenient and efficient method for fullerrene (C_{60}) with calix [8]arene. Chem. Lett. **4**, 699–702 (1994)
33. Bucsi, I, Aniszfeld, R, Shamma, T, Prakash, GKS, Olah, GA: Convenient separation of high-purity C_{60} from crude fullerene extract by selective complexation with $AlCl_3$. Proc. Natl. Acad. Sci. USA **91**(19), 9019–9021 (1994)
34. Doome, RJ, Fonseca, A, Richter, H, Nagy, JB, Thiry, PA, Lucas, AA: Purification of C_{60} by fractional crystallization. J. Phys. Chem. Solids **58**(11), 1839–1843 (1997)
35. Zhou, X, Gu, Z, Wu, Y, Sun, Y, Jin, Z, Xiong, Y: Separation of C_{60} and C_{70} fullerenes in gram quantities by fractional crystallization. Carbon **32**(5), 935–937 (1994)
36. Wu, Y, Sun, Y, Gu, Z, Zhou, X, Xiong, Y, Sun, B: A combined recrystallization and preparative liquid chromatographic method for the isolation of pure C_{70} fullerenes. Carbon **32**(6), 1180–1182 (1994)
37. Coustel, N, Bernier, P, Aznar, R, Zahab, A, Lambert, J-M: Purification of C_{60} by a simple crystallization procedure. J. Chem. Soc. Chem. Commun. **19**, 1402–1403 (1992)
38. Juan, L, Mingyi, Z, Baoyun, S, Gengmei, X, Yan, S, HaiLi, G, Chang, Y, Ge, Y, Zhao, Y: Separation and purification of fullerenols for improved biocompatibility. Carbon **50**, 460–469 (2012)
39. Ruoff, RS, Tse, DS, Malhotra, R, Lorents, DC: Solubility of C_{60} in a variety of solvents. J. Phys. Chem. **97**, 3379–3383 (1993)
40. Sivaraman, N, Dhamodaran, R, Kaliappan, I, Srinivasan, TG, Vasuveva, Rao, PR, Mathews, CK: Solubility of C60 in organic solvents. J. Org. Chem **57**, 6077–6079 (1992)
41. Taylor, R, Hare, JP, Abdul-Sada, A, Kroto, H: Isolation, separation and characterisation of the fullerenes C_{60} and C_{70}. J. Chem. Soc. Chem. Commun. **20**, 1423–1425 (1990)
42. Diack, M, Compton, RN, Guiochon, G: Evaluation of stationary phases for the separation of buckminsterfullerenes by high-performance liquid chromatography. J. Chrom. **639**, 129–140 (1993)
43. Steven, PW, Thomas, P, Jamey, L, Michele, AW, Christopher, SF: Extraction, isolation and characterization of fullerene C_{60}. J. Chem. Educ **74** (30), 311–312 (1997)
44. Manolova, N, Rashkov, I, Legras, D, Delpeux, S, Beguin, F: Separation of C_{60}/C_{70} mixture on activated carbon and activated carbon fibers. Carbon **33**, 209–213 (1995)
45. Naoki, K, Toshiyuki, O, Kazumi, M: A highly improved method for purification of fullerenes applicable to large-scale production. Carbon **42**, 163–167 (2004)
46. Kortan, AR, Kopylov, N, Glarum, S, Gyorgy, EM, Ramirez, AP, Fleming, RM: Superconductivity at 8.4 K in calcium-doped C_{60}. Nature **355**, 529 (1992)
47. Gasyna, Z, Schatz, PN, Hare, JP, Dennis, TJ, Kroto, HW, Taylor, R: The magnetic circular dichroism and absorption spectra of C_{60} isolated in argon matrices. Chem. Phys. Lett. **183**, 283–291 (1991)
48. Taylor, R, Langley, GJ, Dennis, TJS, Kroto, HW, Walton, DRM: A mass spectrometric/NMR study of fullerene-78 isomers. J. Chem. Soc. Chem. Commun. **15**, 1043–1046 (1992)
49. Hare, JP, Dennis, TJ, Kroto, HW, Taylor, R, Allaf, AW, Balm, S, Walton, DRM: The IR spectra of fullerene-60 and −70. J. Chem. Soc. Chem. Commun. **6**, 412–413 (1991)
50. Hare, JP, Kroto, HW, Taylor, R: Preparation and UV/VIS spectra of fullerenes C_{60} and C_{70}. Chem. Phys. Lett. **177**, 394–398 (1991)
51. Leach, S, Vervloet, M, Desprès, A, Brèheret, E, Hare, JP, Dennis, T, Kroto, HW, Taylor, R, Walton, DRM: Electronic spectra and transitions of the fullerene C_{60}. Chem. Phys. **160**, 451–466 (1992)
52. Krätschmer, W, Fostiropoulos, K, Huffman, DR: The infrared and ultraviolet absorption spectra of laboratory-produced carbon dust evidence for the presence of the C_{60} molecule. Chem. Phys. Lett. **170**, 167 (1990)
53. Arthur, B: Diatomite in industrial minerals and rocks, 3rd edn, pp. 303–319. American Institute of Mining, Metallurgical, and Petroleum Engineers, Dove Valley (1960)

54. Fatemeh, T, Alireza, N, Somayeh, F: CsF–celite as an efficient heterogeneous catalyst for sulfonylation and desulfonylation of heteroatoms. Catal. Commun. **12**, 1477–1482 (2011)
55. Shakeel, AA, Qayyum, H: Lactose hydrolysis from milk/whey in batch and continuous processes by concanavalin A-celite 545 immobilized *Aspergillus oryzae* β galactosidase. Food. Bioprod. Process. **90**, 351–359 (2012)

7

Eco-friendly synthesis and characterization of gold nanoparticles using *Klebsiella pneumoniae*

Chelladurai Malarkodi, Shanmugam Rajeshkumar, Mahendran Vanaja, Kanniah Paulkumar, Gnanadhas Gnanajobitha and Gurusamy Annadurai*

Abstract

The thrust to develop eco-friendly procedures for the production of nanoparticles arises from the extremely recent nanotechnology research. Extracellular biosynthesis of gold nanoparticles was achieved by an easy biological procedure using *Klebsiella pneumoniae* as the reducing agent. After exposing the gold ions to *K. pneumoniae*, rapid reduction of gold ions is observed foremost to the formation of gold nanoparticles in colloidal solution. UV-vis spectrum of the aqueous medium containing gold nanoparticles showed a peak around 560 nm. The crystalline nature of the particles was confirmed from an X-ray diffractometer. Transmission electron microscopy (TEM) micrograph analysis of the gold nanoparticles indicated that they were well dispersed and ranged in sizes 35 to 65 nm. The high crystalline in the FCC phase is evidenced by bright circular spots in a selected-area electron diffraction pattern and clear lattice fringes in the high-resolution TEM image. Fourier transform infrared spectroscopy revealed possible involvement of reductive groups on the surface of nanoparticles. The method exploits a cheap and easily available biomaterial not explored so far for the synthesis of metallic nanoparticles.

Keywords: Gold nanoparticles, Extracellular synthesis, Characterization, *K. pneumoniae*, TEM

Background

Currently, nanoparticles (NPs) have drawn marvelous consideration because of their valuable properties on various fields such as medical, sensor, optical, electronic, and catalytic application [1]. Green-mediated synthesis and characterization of nanoparticles have emerged as a significant division of nanotechnology in the last decade, particularly for noble metals such as gold, silver, platinum, and palladium. Moreover, the chemical or physical approaches are usually employed to synthesize metal nanoparticles because of their intrinsic advantage in producing well-defined NPs with quite controllable shapes and sizes [2]. However, the former methods involve tedious treatments, such as inert gas condensation, pyrolysis [3], laser ablation, and hydrothermal and solvothermal synthesis [4]. Sol-gel method is based on inorganic polymerization reactions. The sol-gel process includes four steps: hydrolysis, polycondensation, drying, and thermal decomposition [5]. Ambitious by the growing momentum of green nanotechnology, the synthesis of NPs utilizing biological materials could be an improved alternative to toxic chemicals and the expensive physical methods. In general, biological organisms, such as bacteria and fungi majorly referred to as microorganisms [6], plants [7], and algae [8] represent a development of an eco-friendly and cost-effective approach [9]. Using of bacteria for the biosynthesis of metal nanoparticle process is the easiest method.

Following the revolutionary study of [10] on bacteria-mediated biosynthesis of silver NPs, many efforts have been directed toward this area with the use of bacteria, such as *Thermomonospora* sp. [11], *Lactobacillus strains* [12], *Shewanella algae* [13], *Actinobacter* sp. [14], *Plectonema boryanum* [15], *Rhodopseudomonas capsulate* [16], *Shewanella algae* [17], *Geobacter sulfurreducens* [18], *Morganella* sp. [19], *Bacillus subtilus* [20], *Staphylococcus aureus* [21], *Escherichia coli* [22], and *Serratia nematodiphila* [6]. These bacteria were demonstrated to be responsible for the bacterial synthesis and stabilization of metal nanoparticles. One of the chief applications of nanotechnology is the use of gold nanoparticles in biomedical research like X-ray computed tomography

* Correspondence: annananoteam@gmail.com
Environmental Nanotechnology Division, Sri Paramakalyani Centre for Environmental Sciences, Manonmaniam Sundaranar University, Alwarkurichi, Tamil Nadu 627 412, India

and magnetic resonance imaging [23], cancer nanotechnology [24], drug delivery applications [25], and its optical properties for cancer diagnosis and photo thermal therapy.

Herein, an eco-friendly method for the bacteria-mediated synthesis of gold nanoparticles by the reduction of $HAuCl_4$ ions using the broth of *Klebsiella pneumoniae* is reported. The bioreduction process was monitored by the UV-visible spectroscopy, and the crystalline structure was investigated by the X-ray diffraction (XRD) technique. The nanostructure and size of the synthesized gold nanoparticles were characterized by transmission electron microscopy (TEM) and Fourier transformation infrared spectroscopy (FTIR) was used to understand the biomolecules responsible for the biosynthesis.

Results and discussion

Isolation and identification

The bacterial strain is used for the synthesis of gold nanoparticles which is isolated from the saltpan soil from Tuticorin. The isolate strain MAA was morphologically and biochemically identified as *K. pneumoniae* that produces a red pigment. *K. pneumoniae* MAA was a gram-positive rod-shaped non-motile bacterium that has to be identified and maintained at the Microbial Type Culture Collection and Gene Bank (MTCC), Chandigarh.

Synthesis of gold nanoparticles

Visual identification of gold nanoparticles

The detailed study on extracellular biosynthesis of gold nanoparticle by *K. pneumoniae* was carried out in this work. Figure 1a shows the biomass of *K. pneumoniae* without the addition of $HAuCl_4$, and Figure 1b,c exhibits the $HAuCl_4$-treated biomass incubated at 6 to 24 h. In the biomass of *K. pneumoniae*, the gold nanoparticle synthesis process was completed at 24 h. The color change from yellow to dark purple indicates the formation of gold nanoparticles at 24 h incubation time. The appearance of a dark-purple color at 24 h of incubation time confirms the reduction of gold chloride into gold NPs using the *K. pneumoniae* culture supernatant. The pink color in the solution at 6 h indicates that the gold nanoparticle synthesis process has started. After 24 h, there was no absorbance. The color of the colloidal solution reaction turned from yellow to purple which indicated the formation of gold nanoparticles upon using *K. pneumoniae* observed in the UV-vis spectra and also confirmed the completion of gold nanoparticle synthesis reaction in both bacteria. The color change is dependent upon the incubation time (6 to 24 h) and the shape of the nanoparticles. Similarly, the gold nanoparticles were synthesized by *P. aeruginosa* [26] and *R. capsulate* [27]. The colloidal solutions of gold nanoparticles show a very intense color, which is absent in the bulk material as well as for individual atoms. The origin is attributed to the collective oscillation of free conduction electrons by an interacting electromagnetic field. These resonances are called surface plasmon resonances [28].

UV-vis spectra of gold nanoparticles

The production and stabilization of the reduced gold nanoparticles in the colloidal solution was monitored by

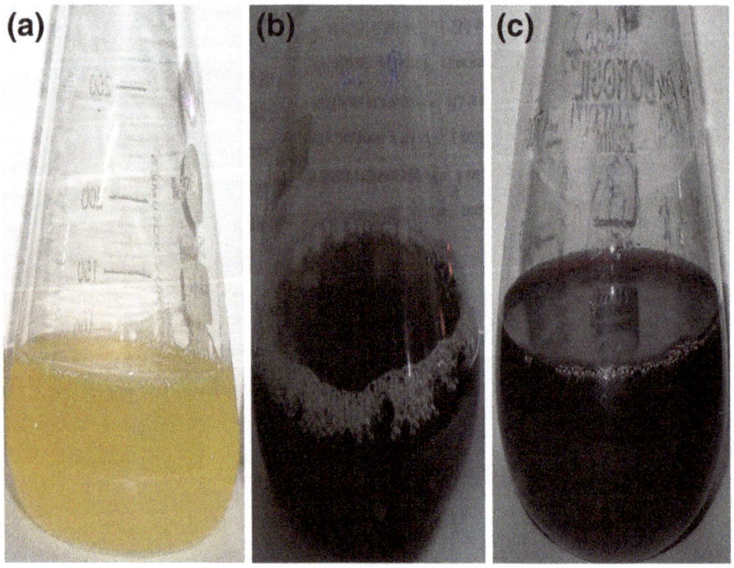

Figure 1 Extracellular synthesis of *K. pneumoniae* biomass and gold chloride mixed with biomass. *K. pneumoniae* biomass (**a**), 1 mM gold chloride mixed with biomass at the beginning of reaction showing a greenish-brown color change (**b**), and after 1 day of reaction showing a dark-purple in color change (**c**).

UV-vis spectrophotometer analysis. UV-vis spectroscopy is one of the most important techniques to identify the formation and stability of the gold nanoparticles in aqueous solution. Gold nanoparticles are known to exhibit at maximum in the range of 400 to 700 nm. The synthesis of gold nanoparticles was monitored at different time intervals such as 6, 12, 18, 24, 36, and 48 h. The gold nanoparticles synthesized by *K. pneumoniae* are positioned at 560 nm (Figure 2). Initially, at 6 h, the gold nanoparticle was absorbed slowly, and the absorbance was gradually increased up to 48 h. After 48 h, there was no absorbance which indicated that the gold NP synthesis process was completed. Similarly, Skirtach et al. (2005) reported that the gold NPs were absorbed at 560 nm, and it was synthesized using *P. aeruginosa.* The reduction of gold ions occurs comparatively slowly, but the gold nanoparticles are found to be very stable in the colloidal suspension. The spectrum shows three bands: one in the visible region and the other two in the NIR region, which is characteristic for the formation of anisotropic nanoparticles [29,30]. It is well known that spherical nanoparticles of Au should exhibit single-surface plasmon bands whereas anisotropic particles should exhibit two or three bands, corresponding to the quadrupole and higher multipole plasmon excitations [31,32].

XRD spectrum of gold nanoparticle

The crystalline nature of the gold nanoparticles was confirmed from X-ray diffraction analysis. The XRD pattern clearly shows that the extracellular synthesis of gold nanoparticles formed by the reduction of gold chloride ions using *K. pneumoniae.* Gold nanoparticles exhibited four prominent Bragg reflections at around 38.2°, 44.7°, 64.5°, and 78.3° (Figure 3) for *K. pneumoniae.* The fraction between the intensity of the (200), (220), and (311) diffraction peaks is much lower, suggesting that the (111) plane is the predominant orientation. The XRD facets of the gold nanoparticle match with standard gold which was published by JCPDS (file no. 04-0784). The mean size of gold nanoparticles was calculated using the Debye-Scherer equation by determining the width of the (111) and the similar Bragg reflection [33,34]. *R. capsulata* was found to have an average size of 38 nm which matched with the particle sizes obtained from TEM results discussed later.

FTIR

The FTIR measurements were carried out to identify the possible biomolecules responsible for the reduction and capping of the reduced gold nanoparticles synthesized by *S. nematodiphila* and *K. pneumoniae.* The FTIR spectra of gold nanoparticles with absorption peaks 1,716, 1,670, 1,630, 1,526, 1,389, 1,239, and 1,037 cm^{-1} for *K. pneumoniae* are seen in Figure 4. The strong peak at 1,630 cm^{-1} are identified as C=O stretching vibrations due to the carbonyl stretch in proteins of the respective amide I group of proteins [35,36]. The peak at 1,526 cm^{-1} shows the characteristic of N-O asymmetric stretching vibrations of nitro compounds. The small band at 1,670 cm^{-1} arises from the -C=C stretching vibrations corresponding to the -C=O stretching vibrations due to carboxylic acids and carbonyl groups [37,38]. The small peak at 1,239 cm^{-1} can be assigned to the C-N stretching vibrations of aliphatic amine groups. Gold nanoparticles can bind to proteins through free amine groups or carboxylate groups in the protein. The presence of the intense

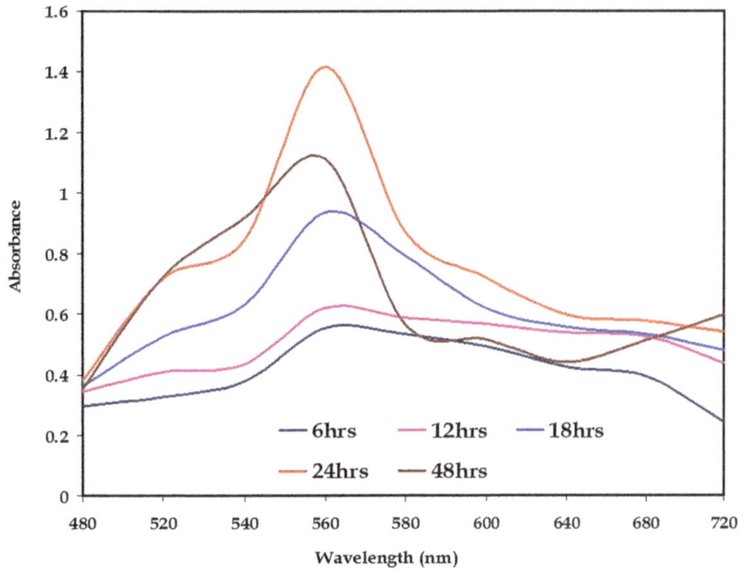

Figure 2 UV-visible absorption spectra of reduction of gold ions to gold nanoparticles using *K. pneumoniae.*

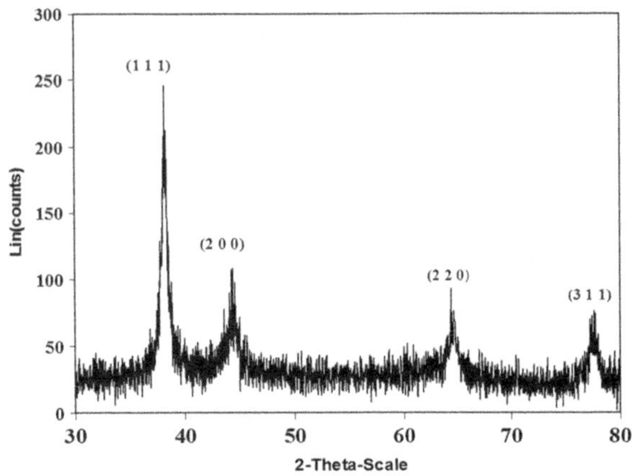

Figure 3 X-ray diffraction pattern of gold nanoparticles using *K. pneumoniae*.

peak at C=O stretching mode indicates the presence of carboxylic groups in the material bound to gold nanoparticles [39].

TEM and SAED pattern of gold nanoparticle

TEM analysis is performed to examine the size and shape of the biosynthesized gold nanoparticles using *K. pneumoniae*. Interestingly, the *K. pneumoniae*-derived gold nanoparticles exhibit a predominantly spherical shape (Figure 5). Such different sizes of gold nanoparticles are extracellularly synthesized by the bacteria *R. capsulate*, [34], fungus *Sclerotium rolfsii*, and plant *Coriandrum sativum*. The spherical-shaped nanoparticles are formed at the beginning of the reaction, and after that, the spherical-shaped nanoparticles are aggregated with each other. Due to the presence of enough amount of reducing agent in the biomass, the initially synthesized gold nanoparticles are stablilized in spherical shape, similar to the shape of nanoparticles synthesized by the bacteria *K. planticola* [40].

The single-crystalline nature of *K. pneumoniae* of these nanoparticles was further confirmed by their

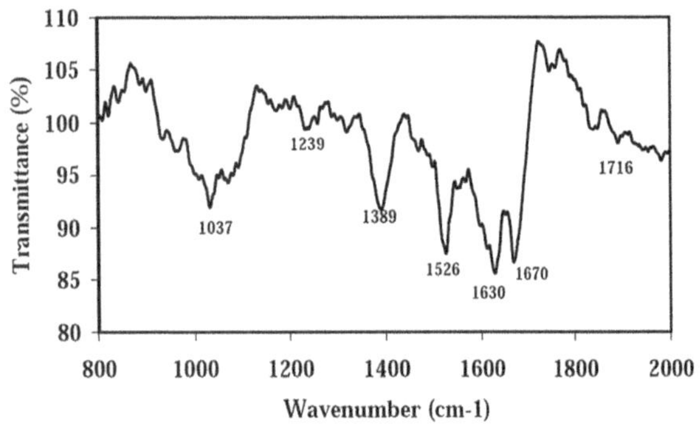

Figure 4 Typical FTIR absorption spectra of the bio-moieties of the macerated extracellular solution.

corresponding selected-area electron diffraction (SAED) analysis for *K. pneumoniae*. The SAED pattern of the spherical nanoparticle showed clear lattice fringes with bright circular rings corresponding to (111), (200), (220), and (311) planes. There are three sets of spots that could be identified from this diffraction pattern as shown in Figure 6. *K. pneumoniae*-derived gold nanoparticles also showed strong intensity at (111) plane of gold nanoparticles. Similarly, four concentric rings are observed in the gold nanoparticles indicating their crystalline nature, as reported by [39]. TEM analysis revealed that the synthesized gold nanoparticles are stable in solution.

Conclusion

The microbe-mediated biosynthesis of gold nanoparticles is non-toxic compared to all other reported methods. Also, the method has advantages: the process is easy since it can be scaled, and it is also economically possible. Reduction and surface accretion of metals may be processed, by which bacteria keep themselves from the toxic effects of metallic ions. In the present investigation, the gold nanoparticles were synthesized using *K. pneumoniae* which was isolated from saltpan soil and identified by MTCC. The synthesis of gold nanoparticles was confirmed by color change of the liquid medium from yellow to intense dark purple, and it exhibited its maximum absorbance at 560 nm which played a prominent role in the reduction of gold chloride to gold nanoparticles. The X-ray diffractometer showed the crystalline nature of nanoparticles. The Fourier transform infrared spectroscopy showed that the functional groups of gold nanoparticles can bind to proteins through free amine groups or carboxylate groups in the protein. The morphology of the gold nanoparticles are found to be spherical in shape and stable in water for 3 months that can be attributed to the surface binding of stabilization materials secreted by the bacteria. Gold colloidal solution is biologically well suited and has the potential to be used in medical and pharmaceutical applications due to their homologous size distribution.

Methods

Isolation and identification of microorganism

The organism was isolated from saltpan soil, and the sample was collected from Tuticorin. The collection of samples was serially diluted, and the isolates were morphologically and microbiologically characterized as *Klebsiella pneumoniae*. The isolated culture was identified as *Klebsiella pneumoniae* (MAA) from MTCC and maintained it by a subculture process for the synthesis of gold nanoparticles.

Extracellular synthesis of gold nanoparticle

K. pneumoniae (MAA) was grown in 100 ml nutrient broth and incubated at 35°C for 24 h. The overnight

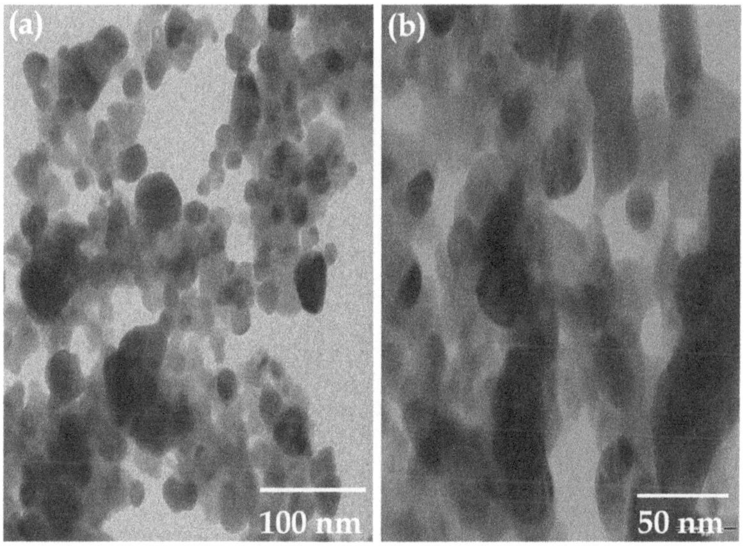

Figure 5 The Transmission electron microscopy and photo of gold nanoparticles prepared by using K. pneumoniae (a) 100 nm and (b) 50 nm. TEM images of gold nanoparticles using *K. pneumoniae* (**a**) and (**b**).

culture broth was centrifuged at 6,000 rpm for 10 min. The cell-free supernatants were collected, 1 mM of gold chloride was added and mixed, and the solution was incubated at 35°C for 24 h. The spectrum of the sample was measured using a UV-visible spectrophotometer.

Characterization of gold nanoparticle

UV-vis spectrophotometer

The bioreduction of pure $HAuCl_4$ are monitored using UV-vis spectroscopy at regular intervals. During the reduction, 0.1 ml of samples was taken and diluted several times with millipore water. After dilution, it was centrifuged at 800 rpm for 5 min. The supernatant was scanned using a UV-300 spectrophotometer (UNICAM, York Street, Cambridge, Cambridgeshire) for a UV-vis 1601 Schimodzu spectrophotometer (Kyoto, Japan), operated at a resolution of 420 nm.

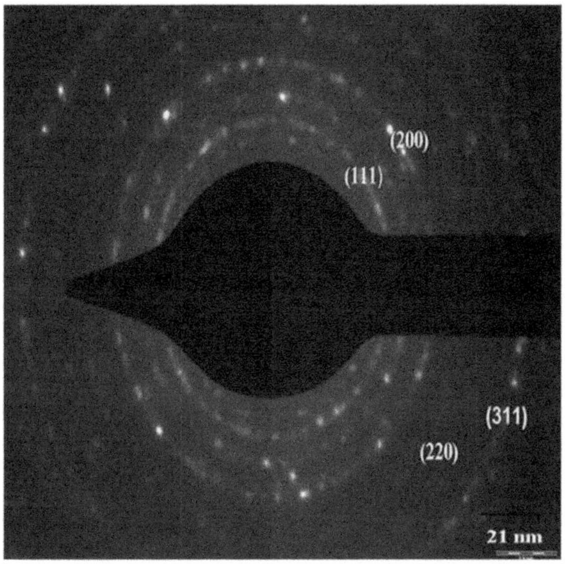

Figure 6 SAED pattern of gold nanoparticles synthesized using *K. pneumoniae.*

X-ray diffractometer

X-ray diffraction (Bruker, Karlsruhe, Germany) is one of the important techniques for the structure characterization of the crystalline material. Prepared nanoparticles can be analyzed by this instrument using the lynx eye detector (silicon strip detector technology).

Transmission electron microscopy

Transmission electron microscopy (TEM) analysis of the sample was done using a Philips CM 200 instrument (Philips, Amsterdam, The Netherlands) operated at an accelerating voltage of 200 kV with a resolution of 0.23 nm. A drop of the solution was placed on carbon-coated copper grid and later exposed to infrared light (45 min) for solvent evaporation.

FTIR

A known weight of sample (1 mg) was taken in a mortar and pestle and ground with 2.5 mg of dry potassium bromide (KBr). The powder so obtained was filled in a 2-mm internal diameter microcup and loaded onto the FTIR set at 26°C ± 1°C. The samples were scanned in the range of 4,000 to 400 cm^{-1} using a Fourier transform infrared spectrometer (Thermo Nicolet Model-6700, Waltham, MA, USA). The spectral data obtained were compared with the reference chart to identify the functional groups present in the sample.

Competing interests

The authors declare that they have no competing interests.

Authors' contributions

CM and SR carried out the nanoparticle synthesis. MV and KP carried out the manuscript preparation. GG carried out the antimicrobial activity. All authors read and approved the final manuscript.

Authors' information

CM completed her B.Sc. in Zoology and M.Sc. Environmental Biotechnology in Sri Paramakalyani Centre for Environmental Sciences, Manonmaniam Sundaranar University, Alwarkurichi, Tirunelveli. She achieved her Ph.D. degree under the guidance of GA in the field of nanotechnology. Her research interests include synthesis of semiconductor nanomaterials and metal nanoparticles and its biomedical applications. SR completed his M.Sc. Biotechnology in Periyar University and obtained his Ph.D. degree in Nanotechnology under the guidance of GA. He published seven research articles in nanoparticle synthesis using algae. He is interested in the metallic nanoparticle synthesis using algae and algae-derived compounds and their potential applications in the biomedical field. VM obtained her M.Sc. in Environmental Biotechnology in Sri Paramakalyani Centre for Environmental Sciences, Manonmaniam Sundaranar University, Alwarkurichi, Tirunelveli. She obtained her Ph.D. degree under the guidance of GA in the field of nanotechnology. Her research interests include eco-friendly synthesis of nanoparticles and its application in the environment. KP received his B.Sc. (2005) and M.Sc. in Biotechnology (2007) from Madurai Kamaraj University and Periyar University, India, respectively. He is currently working to obtain his Ph.D. degree at Manonmaniam Sundaranar University in India. He is interested in the green and immobilized microbe-mediated synthesis process of nanoparticles and nanocomposites and their biomedical and textile industry applications. GG completed her M.Sc. in Biotechnology in Periyar University and obtained her Ph.D. degree in Nanotechnology under the guidance of GA. Her research interests include green-meditated synthesis of nanoparticles and its agricultural applications for controlling plant diseases. GA received his M.Sc. in Applied Chemistry (1992) and Ph.D. in Environmental Biotechnology (1997) from Anna University, India. He had ten years (1999 to 2008) post-doctoral experiences from the National Taiwan University in Taiwan, National Institute of Advanced Industrial Science and Technology in Japan, and National Central University in Taiwan. He has received many research awards from Indian and other country governments. He is an associate editor in five international journals. At present, he is an associate professor of the Environmental Biotechnology and the leader of Environmental Nanobiotechnology Division at Sri Paramakalyani Centre for Environmental Sciences, Manonmaniam Sundaranar University, India. His main research interest is on the biosynthesis of nanoparticles and nanomaterials, nanobiocatalyst, and environmental chemistry.

Acknowledgements

CM thanks UGC, India for the financial support under the Rajiv Gandhi National Fellowship (Sanction Letter no. F. 14-2(SC)/2009 (SA-III) -09/2010) and IIT Bombay University for the TEM analyses.

References

1. Mohanpuria, P, Ran, KN, Yadav, SK: Biosynthesis of nanoparticles: technological concepts and future applications. J. Nanopart. Res. **10**, 507–517 (2008)
2. Yuqing, M, Sun, K, Qiu, J, Liu, J: Preparation and characterization of gold nanoparticles using ascorbic acid as reducing agent in reverse micelles. J. Mater. Sci. **44**, 754–758 (2009)
3. Rao, CNR, Muller, A, Cheetham, AK: The Chemistry of Nanomaterials: Synthesis. Properties and Applications, Wiley VCH, Weinheim (2004)
4. Moreno-Álvarez, SA, Martínez-Castañón, GA, Niño-Martínez, N, Reyes-Macías, JF, Patiño-Marín, N, Loyola-Rodríguez, JP, Ruiz, F: Preparation and bactericide activity of gallic acid stabilized gold nanoparticles. J. Nanopart. Res. **12**, 2741–2746 (2010)
5. Oshavsky, MA, Allcock, HR: Small scale system for in-vivo drug delivery. Chem. Mater. **9**, 1367 (1997)
6. Malarkodi, C, Annadurai, G: A novel biological approach on extracellular synthesis and characterization of semiconductor zinc sulfide nanoparticles. Appl. Nanosci. doi:10.1007/s13204-012-0138-0 (2012).
7. Gnanajobitha, G, Annadurai, G, Kannan, C: Green synthesis of silver nanoparticle using *Elettaria cardamomom* and assesment of its antimicrobial activity. Int. J. Pharma Sci. Res. (IJPSR) **3**, 323–330 (2012)
8. Rajeshkumar, S, Kannan, C, Annadurai, G: Synthesis and characterization of antimicrobial silver nanoparticles using marine brown seaweed *Padina tetrastromatica*. Drug Invention Today **ISSN**, 0975–7619 (2012)
9. Karthiga, P, Soranam, R, Annadurai, G: Alpha-mangostin, the major compound from *Garcinia mangostana*linn. responsible for synthesis of Ag nanoparticles: its characterization and evaluation studies. Research Journal of Nanoscience and Nanotechnology **2**, 46–57 (2012)
10. Klaus, T, Joerger, R, Olsson, E, Granqvist, CG: Silver-based crystalline nanoparticles, microbially fabricated. Proc. Natl. Acad. Sci. U. S. A. **96**, 13611–13614 (1999)
11. Ahmad, A, Senapati, S, Khan, MI, Kumar, R, Sastry, M: Extracellular biosynthesis of monodisperse gold nanoparticles by a novel extremophilic actinomycete. Thermomonospora sp. Langmuir **19**, 3550–3553 (2003)
12. Nair, B, Pradeep, T: Coalescence of nanoclusters and formation of submicron crystallites assisted by *Lactobacillus* strains. Crystal Growth and Design **2**(4), 293–298 (2002)
13. Konishi, Y, Ohno, K, Saitoh, N, Nomura, T, Nagamine, S: Microbial synthesis of gold nanoparticles by metal reducing bacterium. Transactions of the Materials Research Society of Japan **29**, 2341–2343 (2004)
14. Bharde, A, Wani, A, Shouche, Y, Pattayil, A, Bhagavatula, L, Sastry, M: Bacterial aerobic synthesis of nanocrystalline magnetite. J. Am. Chem. Soc. **127**(26), 9326–9327 (2005)
15. Lengke, MF: Ravel, B, Fleet, ME, Wanger, G, Gordon, RA, Southam, G: Mechanisms of gold bioaccumulation by filamentous *cyanobacteria* from gold (III)-chloride complex. Environ. Sci. Technol. **40**(20), 6304–6309 (2006)
16. Shiying, H, Zhirui, G, Zhanga, Y, Zhanga, S, Wanga, J, Ning, G: Biosynthesis of gold nanoparticles using the bacteria *Rhodopseudomonas capsulate*. Material Letters **61**(18), 3984–3987 (2007)
17. Yasuhiro, K, Kaori, O, Norizoh, S, Toshiyuki, N, Shinsuke, N, Hajime, H, Yoshio, T, Tomoya, U: Bioreductive deposition of platinum nanoparticles on the bacterium *Shewanella* algae. J. Biotechnol. **128**, 648–653 (2007)
18. Law, N, Ansari, S, Livens, FR, Renshaw, JC, Lloyd, JR: The formation of nano-scale elemental silver particles via enzymatic reduction by *Geobacter sulfurreducens*. Appl. Environ. Microbiol. **74**, 7090–7093 (2008)
19. Parikh, RP: Singh, S, Prasad, BLV, Patole, MS, Sastry, M, Shouche. YS: Extracellular synthesis of crystalline silver nanoparticles and molecular evidence of silver resistance from Morganella sp: towards understanding biochemical synthesis mechanism. Chembiochem **9**(9), 1415–1422 (2008)
20. Saifuddin, N, Wong, CW, Nuryasumira, AA: Rapid biosynthesis of silver nanoparticles using culture supernatant of bacteria with microwave irradiation. E-Journal of Chemistry **6**, 61–70 (2009)
21. Nanda, A, Saravanan, M: Biosynthesis of silver nanoparticles from *Staphylococcus aureus* and its antimicrobial activity against MRSA and MRSE. Nanomedicine: Nanotechnology, Biology and Medicine **5**, 452–456 (2009)
22. Bao, H, Lu, Z, Cui, X, Guo, J, Anderson, JM, Li, CM: Extracellular microbial synthesis of biocompatible CdTe quantum dot. Acta Biomater. **6**(9), 3534–3541 (2010)
23. Alric, C, Taleb, J, Le Duc, G, Mandon, C, Billotey, C, Meur-Herland, AL, Brochard, T, Vocanson, F, Janier, M, Perriat, P, Roux, S, Tillement, O: Gadolinium chelate coated gold nanoparticles as contrast agents for both X-ray computed tomography and magnetic resonance imaging. J. Am. Che. Soc. **138**, 5908–5915 (2008)
24. Cail, W, Gao, T, Hong, H, Sun, J: Applications of gold nanoparticles in cancer nanotechnology. Nanotechnol. Sci. Appl. **1**, 17–32 (2008)
25. Ghosh, P, Han, G, De, M, Kim, CK, Rotello, VM: Gold nanoparticles in delivery applications. Adv. Drug Deliv. Rev. **60**, 1307–1315 (2008)
26. Husseiny, MI, Abd El-Aziz, M, Badr, Y, Mahmoud, MA: Biosynthesis of gold nanoparticles using *Pseudomonas aeruginosa*. Spectrochimica Acta Part A **67**, 1003–1006 (2007)
27. Wang, H, Xu, K, Liu, L, Tan, JP, Chen, Y, Li, Y, Fan, W, Wei, Z, Sheng, J, Yang, YY, Li, L: The efficacy of self-assembled cationic antimicrobial peptide nanoparticles against *Cryptococcus neoformans* for the treatment of meningitis. Biomaterials **31**(10), 2874–2881 (2010)

28. Smitha, SL, Philip, D, Gopchandrana, KG: Green synthesis of gold nanoparticles using *Cinnamomum zeylanicum* leaf broth. Spectrochimica Acta Part A **74**, 735–739 (2009)
29. Liz, LM, Tailoring, M: Surface plasmons through the morphology and assembly of Metal nanoparticles. Langmuir **22**, 32 (2006)
30. Skirtach, AG, Dejugnat, C, Braun, D, Susha, AS, Rogach, AL, Parak, WJ, Mohwald, H, Sukhorukov, GB: The role of metal nanoparticles in remote release of encapsulated. Nano Lett. Materials **5**, 1371–1377 (2005)
31. Murphy, TK, Sau, AM, Gole, CJ, Orendorff, J, Gao, L, Gou, SE, Hunyadi, T, Li: Measurement of lactose crystallinity using Raman spectroscopy. J. Pharm. Biomed. **38**, 186–190 (2005)
32. Xu, W, Wang, Y, Xu, R, Liang, S, Zhang, G, Yin, D: Synthesis and fluorescence spectrum analysis of CdS nanocrystals. J. Mat. Sci. **42**, 6942–6945 (2007)
33. Kasthuria, Veerapandian, S, Rajendiran, N: Biological synthesis of silver and gold nanoparticles using apiin as reducing agent. Colloids Surf. B Biointerfaces **68**, 55–60 (2009)
34. He, S, Guo, Z, Zhang, Y, Zhang, S, Wang, J, Gu, N: Biosynthesis of gold nanoparticles using the bacteria *Rhodopseudomonas capsulate*. Mater. Lett. **61**, 3984–3987 (2007)
35. Narayanan, KB, Sakthivel, N: Coriander leaf mediated biosynthesis of gold nanoparticles. Mater. Lett. **62**, 4588–4590 (2008)
36. Rajeshkumar, S, Kannan, C, Annadurai, G: Green synthesis of silver nanoparticles using marine brown algae *Turbinaria conoides* and its antibacterial activity. J. Pharm. Bio. Sci. **3**(4), 502–510 (2012)
37. Rajeshkumar, S, Kannan, C, Annadurai, G: Synthesis and characterization of antimicrobial silver nanoparticles using marine brown seaweed *Padina tetrastromatica*. Drug Invention Today **4**, 511–513 (2012)
38. Gnanajobitha, G, Vanaja, M, Paulkumar, K, Rajeshkumar, S, Malarkodi, C, Annadurai, G, Kannan, C: Green synthesis of silver nanoparticles using *Millingtonia hortensis* and evaluation of their antimicrobial efficacy. International Journal of Nanomaterials and Biostructures **3**(1), 21–25 (2013)
39. Philip, D: Biosynthesis of Au, Ag and Au–Ag nanoparticles using edible mushroom extract. Spectrochimica Acta Part A **73**, 374–381 (2009)
40. Paulkumar, K, Rajeshkumar, S, Gnanajobitha, G, Vanaja, M, Malarkodi, C, Annadurai, G: Eco-friendly synthesis of silver chloride nanoparticles using *Klebsiella planticola* (MTCC 2277). International Journal of Green Chemistry and Bioprocess **3**(1), 12–16 (2013)

8

The effect of silver and aluminum oxide nanoparticles on thermophysical properties of nanofluids

Heydar Maddah[1], Mahdokht Rezazadeh[2*], Mojtaba Maghsoudi[3] and Syamak NasiriKokhdan[4]

Abstract

The present paper describes experimental and theoretical aspects of the effective thermal conductivity, electrical conductivity, and viscosity of nanofluids. The thermal conductivity, electrical conductivity, and viscosity of nanofluids increase with the nanoparticle volume fraction. The nanofluid was prepared by synthesizing Al_2O_3 and Ag nanoparticles using microwave-assisted chemical precipitation method and then dispersed in distilled water using a sonicator. Water nanofluid with nominal diameters of 20 and 40 nm at various volume concentrations (0.25% to 5%) at a temperature of 15°C was used for the investigation. The thermal conductivity, electrical conductivity, and viscosity of nanofluids were measured, and it was found that the viscosity and electrical conductivity increase is substantially higher than the increase in thermal conductivity. The pure base fluid thermal conductivity displayed a Newtonian behavior at 15°C; it transformed to a non-Newtonian fluid with the addition of a small amount of nanoparticles $\varphi > 3\%$.

Keywords: Nanofluids, Aluminum oxide, Silver, Viscosity, Thermal conductivity, Electrical conductivity

Background

Nanoparticles have many remarkable properties because of their small size and very large specific surface areas [1]. Nanofluids have gained significant attention due to their enhanced thermal properties. Pak and Cho reported that convection increases transfer coefficient by 75% for an Al_2O_3 particle concentration of 2.7% at a fixed Reynolds number. Most published studies on nanofluids deal with viscose properties of nanoparticle suspensions at low temperature. Praveen and Namburu reported that with the heat transfer behavior including thermal conduction, phase change, heat transfer, and convection heat transfer from an application point of view, the ideal nanofluid should possess not only high thermal conductivity but also low viscosity. Until now, only few studies have addressed the viscosity of nanofluids which increases when the volume concentration of nanoparticles increases, for example, the viscosity of 6.12% copper oxide volume concentration is about four times the value of the base fluid at −35°C [2]. Although copper oxide nanofluids exhibit Newtonian behavior in ethylene glycol and water mixture for concentrations varying from 0% to 6.12%, they found that TiO_2 particles of 27-nm average diameter at a volumetric loading of 4.3% increased the viscosity of water by 60%. Pak found that for 10% volume concentration of nanoparticles, the viscosity of Al_2O_3 (13 nm) water- and TiO_2 (17 nm) water-based nanofluids is several times higher than that of water. The viscosity of nanofluids is an important transport property for applications of nanofluids as a new class of heating or cooling medium in thermal devices or systems such as heat exchangers or cooling systems. Determining the viscosity of nanofluids is essential to establishing adequate pumping power as well as the convective heat transfer coefficient, because the Prandtl and Reynolds numbers (functions of viscosity) will be influenced. Einstein proposed a viscosity correlation for particle suspensions in base fluid when the volume concentration is lower than 5% [3]:

$$\mu_s = \mu_f(1 + 2.5\varphi), \quad (1)$$

* Correspondence: mahdokht_r62@yahoo.com
[2]Young Research Club, Saveh Branch, Islamic Azad University, Saveh 39187/366, Iran
Full list of author information is available at the end of the article

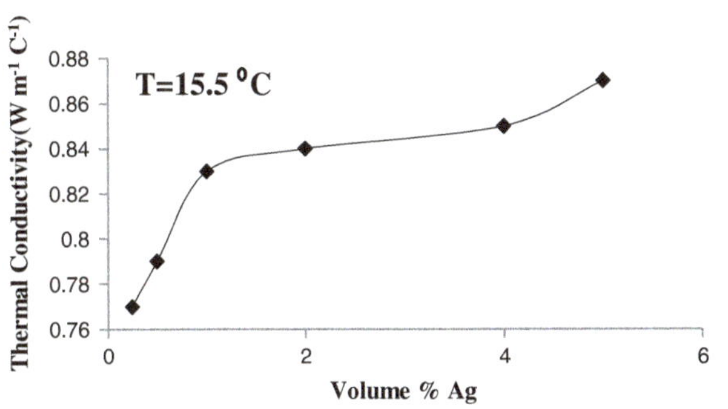

Figure 1 Thermal conductivity ratio of Ag/water nanofluid as a function of volume concentration.

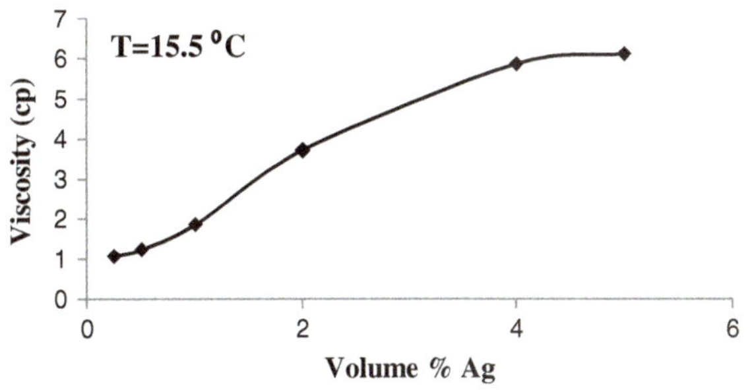

Figure 3 Viscosity of Ag/water nanofluid as a function of volume concentration.

where μ_s is the suspension's viscosity, μ_f is the viscosity of the base fluid, and φ is the volume percentage of particles in base fluid. Brinkman extended Einstein's formula for use with moderate particle concentration as follows [4]:

$$\mu_{nf}/\mu_{bf} = (1-\varphi)^{-2.5}. \tag{2}$$

Batchelor considered the effect of the Brownian motion of a particle on the bulk stress of an isotropic suspension of spherical particles and derived the viscosity expression as [5] follows:

$$\mu_{nf}/\mu_{bf} = \left(1 + 2.5\varphi + 6.5\varphi^2\right). \tag{3}$$

Various recent studies [6] suggest that the high viscosity of nanofluids is likely to be associated with the aggregation of nanoparticles for which the Krieger-Dougherty equation has been used by other authors [7]. The Krieger-Dougherty model has a form:

$$\frac{\mu_{nf}}{\mu_{bf}} = \left(1 - \frac{\varphi_n}{\varphi_m}\right)^{-(\eta)\varphi_m}, \tag{4}$$

where φ_m is the maximum concentration at the flow, φ_a the effective volume fraction of aggregates, and η is the intrinsic viscosity wherein for a monodisperse system, it has a typical value of 2.5.

Notice that all four correlations (Equations 1 to 4) were developed to relate viscosity as a function of volume percentage only. There is no consideration of temperature dependence. A correlation for pure fluids between viscosity and temperature is presented as follows [8]:

$$\mathrm{Ln}\frac{\mu_{nf}}{\mu_{bf}} \approx a + b\left(\frac{T_0}{T}\right) + C\left(\frac{T_0}{T}\right)^2, \tag{5}$$

where μ_{bf} and T_0 are reference values, and c is the dimensionless curve fitting constant. In this research, we report the rheological behavior of silver nanoparticles (40 nm) and Al_2O_3 (20 nm) suspended in water over a wide range of volumetric solid concentrations (φ = 0.25% to 5%). The nanoparticle suspensions generally exhibited a yield pseudoplastic behavior. We conducted a detailed investigation on their effects on the viscosity as a function of Al_2o_3 and Ag volume concentrations, shear rate, and temperature. Results have been discussed both in terms of several classical models and the theoretical prediction for viscosity of nanofluids proposed recently.

The research work on the measurement of thermophysical property, so far, has been done mostly on the effective thermal conductivity. Inspite of the scientific and technological importance of electrical conductivity

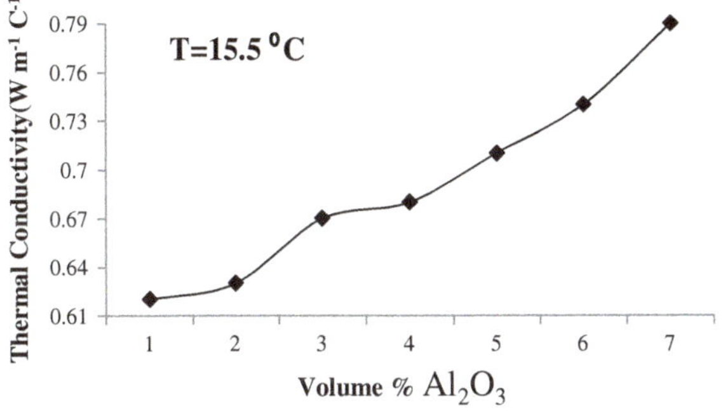

Figure 2 Thermal conductivity ratio of Al_2O_3/water nanofluid as a function of volume concentration.

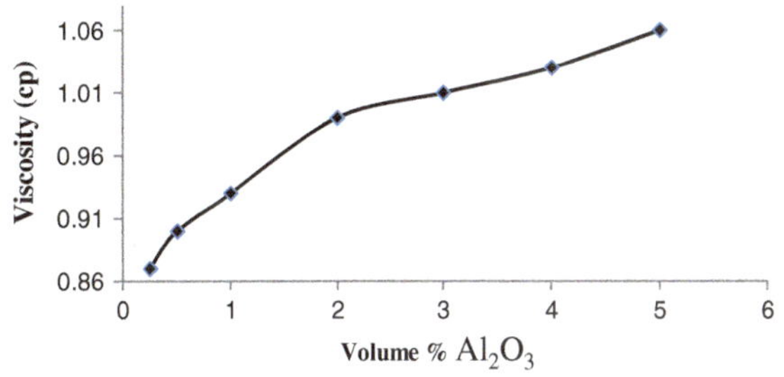

Figure 4 Viscosity of Al_2O_3/water nanofluid as a function of volume concentration.

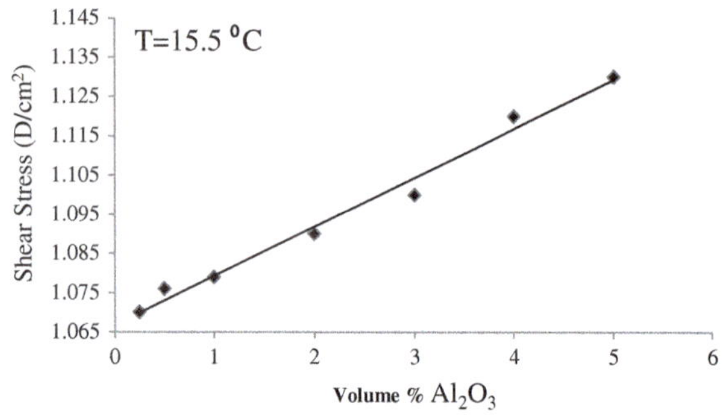

Figure 5 Shear stress as a function of volume concentration for Al_2O_3.

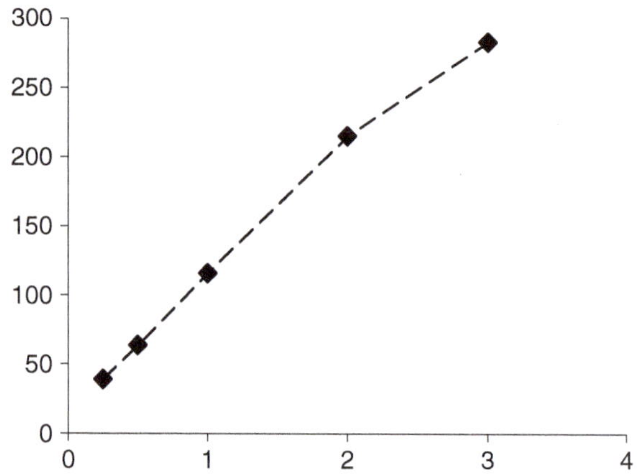

Figure 7 Measured effective electrical conductivity of alumina nanofluid.

characteristics of nanoparticle suspensions, the effective electrical conductivities of nanofluids have not been investigated thoroughly, and there are very few data which presented the electrical properties of nanofluids. In addition, among the transport properties, electrical conductivity may help to attain the understanding of dispersion state and stability of the particulate suspension [9].

Suvankar Ganguly et al. [10] investigated the effective electrical conductivity of alumina nanofluid at different volume fractions. It is seen that the electrical conductivity of alumina nanofluid increases almost linearly with increase in the volume fraction of the alumina nanoparticles. The experimental data also indicate that for a given volume fraction, the electrical conductivity of the suspension will increase with increases in the temperature. The highest value of electrical conductivity, 351 μS/cm, was recorded for a volume fraction of 0.03 at a temperature of 45°C; the corresponding value at room temperature (T = 24°C) was 258 μS/cm.

Results and discussion

In this paper, we have experimentally investigated the effective thermal conductivities, electrical conductivity, and viscosities of the water-based nanofluids containing Al_2O_3 and Ag nanoparticles (Al_2O_3 and Ag /water nanofluids). The thermal conductivity and viscosity of the nanofluid were measured using a KD2 Pro thermal property analyzer and Brookfield viscometer. Figures 1 and 2 show the thermal conductivity of Al_2O_3 and Ag/water nanofluids as a function of particle volume concentration. The viscosity of nanofluids increases when the volume concentration of nanoparticles increases.

The results show that the thermal conductivity of nanofluids significantly increases linearly with increasing particle volume concentration. In contrast, Figures 3 and 4 show that the viscosity was observed to have a nonlinear relation at a volume concentration greater than 2%.

Figures 5 and 6 show the shear stress (τ) for various nanofluids with different volume concentrations of Al_2O_3 and Ag nanoparticles. We show the figures with a correlation coefficient $R^2 > 0.98$.

Figures 7 and 8 show the effective electrical conductivity of alumina and silver nanofluids at different volume fractions and at 15°C. It can be found that the electrical conductivity of nanofluids increases linearly with an increase in the volume fraction of the alumina

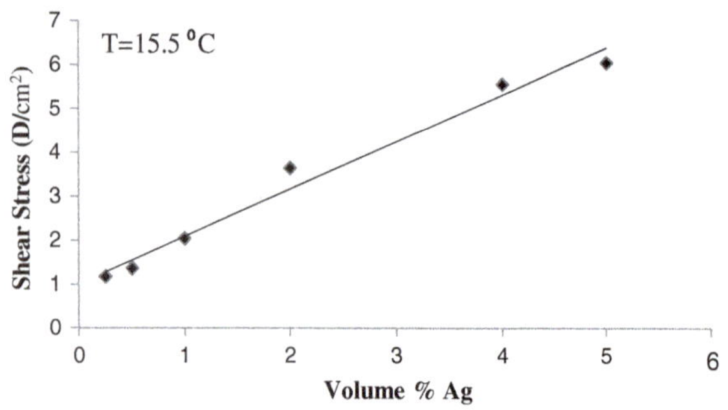

Figure 6 Shear stress as a function of volume concentration for Ag.

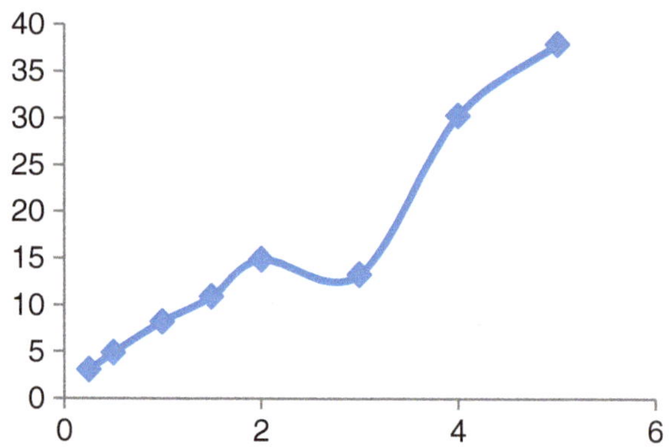

Figure 8 Measured effective electrical conductivity of silver nanofluid.

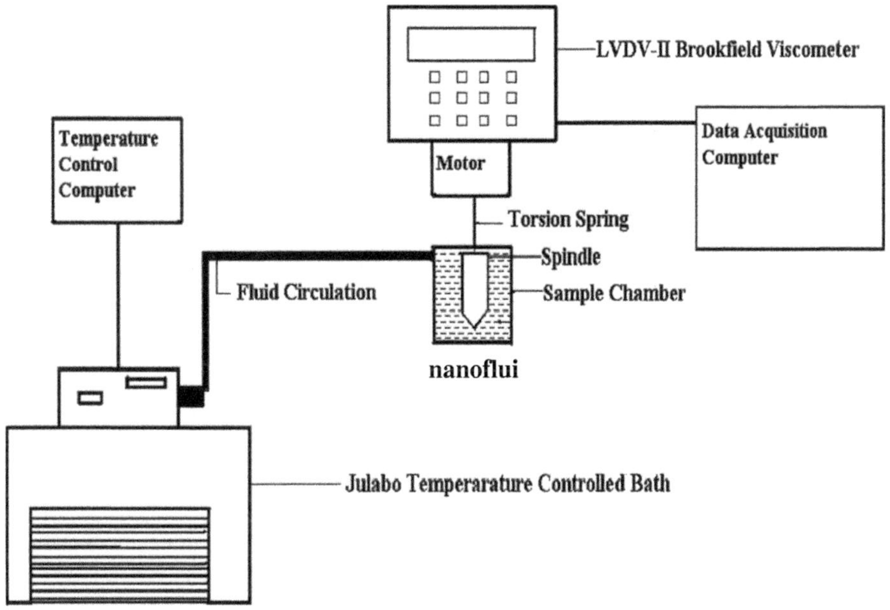

Figure 9 Experimental setup for viscosity measurement of nanofluids.

nanoparticles. The enhancement in the electrical conductivity of the alumina and silver nanofluids is compared to that of the base fluid.

Conclusions

In this paper we have experimentally investigated the effective thermal conductivities and viscosities of water-based nanofluids containing Al_2O_3 and Ag nanoparticles:

1. Ideal nanofluids should not only possess high thermal conductivity but also should have low viscosity.
2. Al_2O_3 and Ag nanofluids exhibit Newtonian behavior in water for concentrations varying from 0.25% to 2% at 15°C.
3. The viscosity and thermal conductivity of nanofluids increase when the volume concentration of nanoparticles increases.
4. The viscosity of nanofluids is dependent on volume percentage. The higher the concentration of nanofluids, the higher viscosity.
5. The pure base fluids display Newtonian behavior at 15°C.
6. The viscosity is as critical as thermal conductivity in engineering systems that employ fluid flow.
7. Nanofluids (Ag and Al_2O_3) with different volume concentrations (0.25% to 5%) are appropriate for pumping in a heat exchanger setup.
8. The results show that the thermal conductivity of Ag nanofluids significantly increases linearly with increasing particle volume concentration.
9. The results indicate considerable enhancement of electrical conductivity with increase in volume fraction.
10. The experimental results show that the electrical conductivity of alumina and silver nanofluid is significantly greater than that of the base fluid.

Methods

In our experiments, we used Al_2O_3 and Ag nanoparticles with an average diameter of 20 nm for Al_2O_3 and 40 nm for Ag. Nanofluids with different volume concentrations (0.25%, 0.5%, 1%, 2%, 3%, 4%, 5%) were dispersed in water. The suspension of nanofluids was then stirred and agitated thoroughly for 1 h with an ultrasonic agitator similar to a sonicator. This ensures uniform dispersion of nanoparticles in the base fluid. The viscosity of the nanofluids is measured using the Brookfield programmable

Table 1 Conductivity meter properties

	Description
Accuracy	5% Thermal conductivity/resistivity
Measurement speed	90 s
Range of measurement	*K*: 0.02 to 2 Wm^{1} C^{1}
	R: 0.5 to 50 mCW^{1}
Operating environment	−10°C to 60°C
KD2 sensor	Model: KD2-S
	Needle material: stainless steel
	Needle length: 60 mm
	Needle diameter: 1.27 mm

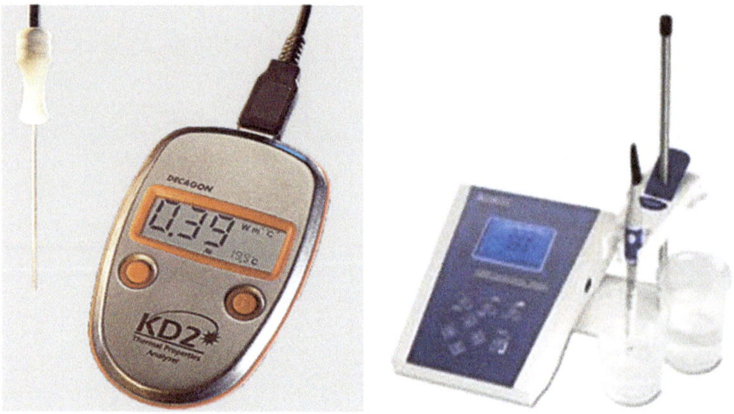

Figure 10 Thermal conductivity meter and electrical conductivity meter.

viscometer (model: LVDV-II+, Brookfield Engineering Labs., Inc, Middleboro, MA, USA) connected to a PC-controlled Julabo temperature-controlled bath (JULABO GmbH, Seelbach, Germany) which can vary the fluid temperature control of 15°C. The viscometer drives a spindle immersed in nanofluids. When the spindle is rotated, the viscous drag of the fluid against the spindle is measured by the deflection of the calibrated spring. This viscometer has a viscosity measurement range of 1.5 and 30,000 mPas and can handle the viscosity measurement results within the temperature ranges of this experiment. The total volume of nanofluid required in a UL adapter is ≈20 ml. The viscometer contains a sample chamber and is carefully monitored using a RTD temperature sensor during the viscosity measurements. The spindle type and speed combinations will produce satisfactory results when the applied torque is between 10% and 100%; therefore, spindle types and speeds are selected in such way that the torque values lie in this prescribed range. A wide range of spindle speeds are available in this viscometer (0 to 200 RPM). Viscosity measurements were started at 15°C. The temperature of the fluid is measured by a calibrated pt100 temperature sensor. The viscometer is connected to another computer as shown in Figure 9. The operation of the viscometer and data collection, viscosity, shear stress, shear rate, RPM, torque, and temperature is performed using Wingather software. All the viscosity measurements were recorded at steady state conditions (Figure 9, Table 1).

Thermal conductivity measurement

The thermal conductivity of Al_2O_3 and Ag/water nanofluid was measured by using a KD2 Pro (Figure 10) thermal property analyzer (Decagon Devices, Inc., Pullman, WA, USA). It consists of a handheld microcontroller and sensor needles. The KD2 sensor needle contains both a heating element and a thermistor. The controller module contains a battery, a 16-bit microcontroller/AD converter, and power control circuitry. The sensor needle can be used for measuring thermal conductivity of fluids in the range of 0.2 to 2 W/mk with an accuracy of ±5%. Each measurement cycle consists of 90 s. During the first 30 s, the instrument will equilibrate which is then followed by heating and cooling of the sensor needle for 30 s each. At the end of the reading, the controller computes the thermal conductivity using the change in temperature (ΔT) time data from

$$k = \frac{q(lnt_2 - lnt_1)}{4\pi(\Delta T_2 - \Delta T_1)}, \tag{6}$$

where q is the constant heat rate applied to an infinitely long and small 'line' source, and ΔT_1 and ΔT_2 are the changes in the temperature at times t_1 and t_2, respectively (Figure 10).

Measurements of the effective electrical conductivity of nanofluids

The 4510 is easy to use with the flexibility to meet the broadest range of applications, and for those where greater accuracy is required, the 4510 has automatic conductivity standard recognition and endpoint detection (Figure 10). Setup options include cell constant, temperature coefficient, and reference temperature to measure the electrical conductivity of nanofluids; a precision conductivity cell with an application range of 0.01 μS/cm to 2 S/cm has been used. Measurements of the electrical conductivity were carried out both as a function of nanoparticle volume fraction and temperature. The typical response time for measurement was about 20 s. The electrical conductivity of the Al_2O_3-water nanofluid was measured at a temperature of 15°C.

Competing interests

The authors declare that they have no competing interests.

Authors' contributions

HM carried out the effective thermal conductivity, electrical conductivity, and viscosity of nanofluid studies and participated in the sequence alignment. MR carried out the nanoparticle studies. MM participated in the design of the study and performed the statistical analysis. SN participated in its design and coordination. All authors read and approved the final manuscript.

Acknowledgments

Generally, Mahdokht Rezazadeh wishes to give her grateful acknowledgments to all the members of the Institute of Chemistry, Shahrood University of Technology, who provided a pleasant atmosphere where she spent 1 year of her happy time. Special thanks also go to R. Aghayari, A. R. Moghadassi, and A. A. Rohani for being the reviewers of this thesis.

Author details

[1]Shahrood Branch, Islamic Azad University, Shahrood 3619995161, Iran. [2]Young Research Club, Saveh Branch, Islamic Azad University, Saveh 39187/366, Iran. [3]Sama-Saveh Branch, Islamic Azad University, Saveh 39187/366, Iran. [4]Kashan University, Kashan 87314, Iran.

References

1. Gleiter, H: Nanocrystalline materials. Prog. Mater. Sci. **33**, 223–35 (1989)
2. Praveen, K, Namburu, D, Devdatta, P: Viscosity of copper oxide nanoparticles dispersed in ethylene glycol and water mixture. Exp. Therm. Fluid. Sci. **32**, 398–410 (2007)
3. Choi, US: Enhancing thermal conductivity of fluid with nanoparticles. ASME FED **66**, 99–105 (1995)
4. Brinkman, HC: The viscosity of concentrated suspensions and solution. J. Chem. Phys. **20**, 571–581 (1952)
5. Batchelor, GK: The effect of Brownian motion on the bulk stress in a suspension of spherical particles. J. Fluid Mechanics **83**, 97–117 (1977)
6. Wang, BX, Zhou, LP, Peng, XP: A fractal model for predicting the effective thermal conductivity of liquid with suspension of nanoparticles. Int. J. Appl. Phys **81**, 6692–6699 (1977)
7. Krieger, IM, Dougherty, TJ: A mechanism for non-Newtonian flow in suspension of rigid spheres. Trans. J. Soc. Rheology **3**, 055501 (1959)
8. Hosseini, MS, Mohebi, A, Ghader, S: Correlation of shear viscosity of nanofluids using the local composition theory. J. Therm. Chem. Eng. Data **18**, 102–107 (2010)
9. Glory, J, Bonetti, M, Helezen, M, Hermite, M, Reynaud, C: Thermal and electrical conductivities of water-based nanofluids prepared with long multiwalled carbon nanotubes. J. Appl. Phys. **103**(Article number 094309), 1–7 (2008)
10. Barrau, S, Demont, P, Perez, E, Peigney, A, Laurent, C, Lacabanne, C: Effect of palmitic acid on the electrical conductivity. Macromolecules **36**, 9678–9680 (2003)

Reduced graphene oxide: a promising electrode material for oxygen electrodes

Michail O Danilov*, Ivan A Slobodyanyuk, Igor A Rusetskii and Gennadii Ya Kolbasov

Abstract

A method for the chemical synthesis of reduced graphene oxide has been developed. It is based on the chemical oxidation of multiwalled carbon nanotubes by means of heptavalent manganese with subsequent reduction. Sodium hypophosphite and sulfite were used as reductants. It has been confirmed by different methods that reduced graphene oxide is obtained. Current-potential curves for electrodes based on reduced graphene oxide in an oxygen fuel cell mock-up with alkaline electrolyte have been studied. It has been found that the characteristics of the electrode of graphene oxide reduced with sodium hypophosphite in oxygen reduction reaction are three to four times higher than those of the electrode of starting multiwalled carbon nanotubes. The electrodes based on reduced graphene oxide are superior to multiwalled carbon nanotube-MnO_2 composite in characteristics. The characteristics of oxygen electrodes based on reduced graphene oxide were stable during 6-month tests. The reduced graphene oxide obtained promises much as a material for the oxygen electrodes of electrochemical power sources.

Keywords: Reduced graphene oxide; Electrocatalysis; Oxygen electrode; Electrochemical power sources

Background

The use of air or oxygen electrode in devices generating electrical energy is very promising since it does not give rise to environmental problems and makes it possible to save natural resources, such as oil and gas. Air and oxygen electrodes in current sources are a three-phase electrode-electrolyte-gas system, where the electric current generation processes are localized at the interface between these phases. The current magnitude generated at such a gas diffusion electrode depends on the size of the zone of the triple contact of these three phases. The electrode consists of a catalyst and a support; the interaction between them determines mainly the value of generated current. Platinum is at present the most efficient oxygen reduction catalyst, but it has a considerable demerit - high price. There are a large number of papers dealing with the investigation of other efficient catalysts [1]. Another important problem is catalytically active and stable support. The advantage of carbon nanotubes as a catalyst support is shown in [2-5]. At the present time, in view of the advent of a new carbon nanomaterial such as graphene, a number of papers have appeared which deal with its investigation as an electrode material for lithium-ion batteries [6] and as a support for catalysts in fuel cells [7-11]. Graphene is an atom-thick carbon layer, which consists of condensed six-membered rings.

The carbon atoms in graphene are bonded by sp^2 bonds into a hexagonal two-dimensional lattice. Ideal graphene consists solely of six-membered rings; the appearance of defects leads to the formation of a number of five- or seven-membered rings in the graphene structure and hence to the curving of flat surface. At the same time, the extended π-system of conjugated aromatic rings makes graphene fairly stable in comparison with other nanosized objects. The structure peculiarities of graphene sheet are such that the charge carriers, having unlimited freedom to move in-plane, are confined in a narrow space between 'walls' (Figure 1), which are approximately 0.3 nm (the shortest atomic distance) apart, which gives rise to unique electrophysical characteristics and other extraordinary properties of graphene.

In view of this, the study of the dependence of electrochemical properties on the method for the preparation of reduced graphene oxide (RGO), which is used as a catalyst support for fuel cell oxygen electrodes, is of great interest.

* Correspondence: danilovmickle@rambler.ru
V.I. Vernadskii Institute of General and Inorganic Chemistry of the Ukrainian NAS, prospekt Palladina 32-34, 03680, Kyiv 142, Ukraine

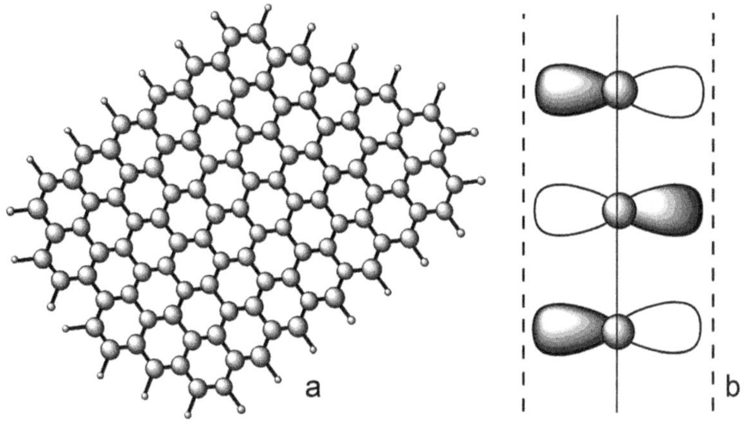

Figure 1 The schematic structure of graphene: (a) top view, (b) side view.

Results and discussion

To synthesize oxidized graphene, a modified Hammers method [12,13] using multiwalled carbon nanotubes (MCNTs) with a bulk density of 25 to 30 g/dm^3, purified of catalyst impurities by treatment with a hydrofluoric acid solution, was employed. The outside diameter of nanotubes was approximately 10 to 30 nm, and the specific surface area was 126 m^2/g (determined by the BET method).

One gram of MCNTs was dispersed for 1 h by stirring in 300 mL of concentrated sulfuric acid. Then 5 g of $KMnO_4$ was added under stirring for 1 h at a temperature of not over 17°C. After that, the mixture was heated on a water bath to 55°C for 30 min and held for 1 h. Then the temperature of the solution was brought to 65°C, and it was left to cool to room temperature. To prevent the formation of manganese dioxide, the resulting mixture was diluted in 400 mL of bidistilled water with ice, which contained 5 mL of 30% H_2O_2. Then the diluted mixture was filtered on a fine paper filter. The filtered off precipitate was dispersed in bidistilled water.

A typical reductant for graphene oxide is hydrazine [12,14-16], whose standard redox potential of electrochemical reaction in the alkaline solution (1) $E = -1.16$ V [17]. It follows that reductants with close redox potentials can be used for the reduction of graphene oxide. Such reductants can be hypophosphite (2) ($E = -1.51$ V) and sulfite (3) ($E = -0.936$ V) [17]. These reductants are nontoxic, which is their advantage.

$$N_2 + 4H_2O + 4e = N_2H_4 + 4OH^- \quad (1)$$

$$HPO_3^{2-} + 2H_2O + 2e = H_2PO_2^- + 3OH^- \quad (2)$$

$$SO_4^{2-} + H_2O + 2e = SO_3^{2-} + 2OH^- \quad (3)$$

Two samples were selected for reduction. One sample of oxidized product was reduced with 0.14 mol of an alkaline solution of sodium hypophosphite (pH = 11) and the other with 0.14 mol of an alkaline solution of sodium sulfite (pH = 11). The reduced substance was filtered off on a fine paper filter, after which it was separated from the filter and dried in a desiccator at 140°C for 3 h. The solid filtrate has been investigated by electron microscopy and an X-ray phase analysis. The diluted solution of reduced graphene oxide has been investigated by spectral photometry. Figure 2 shows X-ray diagrams of products obtained

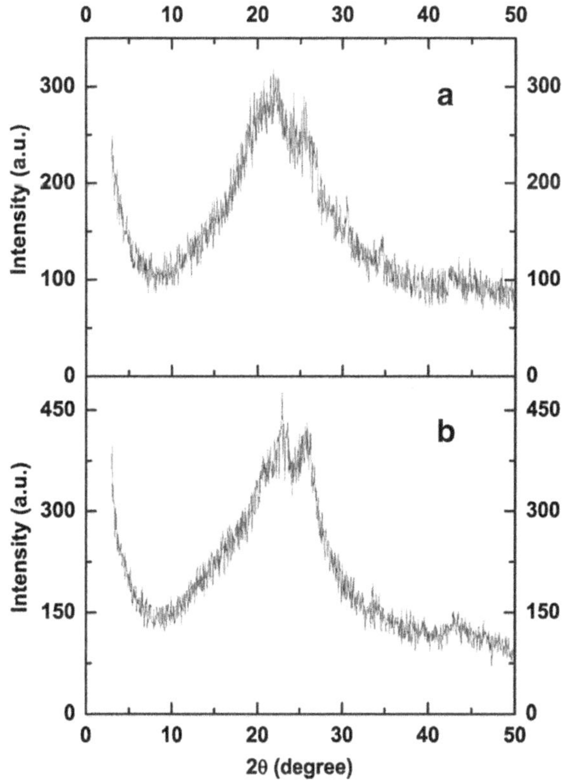

Figure 2 X-ray diagram samples of RGO reduced with (a) sodium hypophosphite and (b) sodium sulfite.

by reduction with sodium hypophosphite (Figure 2a) and sodium sulfite (Figure 2b). In the X-ray diffraction (XRD) pattern of RGO (which is exfoliated into individual platelets and then agglomerated into a powder form), the major peak is observed at about 23° to 24° [14]. This gives an interlayer spacing of approximately 3.7 to 3.8 Å [14,18,19]. Figures 3 and 4 show micrographs of RGO, which was obtained using different reductants. Figure 5 shows an absorption spectrum (1-cm path length) of an alkaline dispersion (pH = 1) of RGO in the range 200 to 800 nm. As can be seen, the absorption peak of RGO dispersion is in the range of 231 to 270 nm [20-28]. One main peak at approximately 250 nm comes from the π-π* transition of C-C and C=C bonds in sp^2 hybrid regions, and a shoulder peak at approximately 300 nm is due to the n-π* transition of the C=O bond in sp^3 hybrid regions [25,29,30]. Figure 6 shows fluorescence emission spectra (1-cm path length) of a RGO dispersion in an alkaline medium (pH = 11) with the excitation wavelength λ = 285 to 400 nm. The fluorescence peaks lie in the short-wavelength region (λ = 430 to 450 nm), which corresponds to the aqueous dispersion of RGO [23-25,31].

Thus, when studying absorption spectra and photoluminescence spectra of RGO dispersion, analyzing electron micrographs, and comparing peaks in X-ray diagrams with literature data, it may be concluded that RGO nanosheets can be fabricated by the above procedure.

Oxygen electrodes, where RGO was used as an active layer, have been made by pressing. The electrodes obtained were investigated on an oxygen fuel cell mock-up with alkaline electrolyte. Electrodes with an active layer of starting MCNTs have been made for comparison.

On the oxygen electrodes made, current-potential curves were measured. Figure 7 shows plots of potential against current density for oxygen electrodes based on graphene oxide reduced with sodium hypophosphite (RGO-SH; curve 1 in Figure 7) and graphene oxide obtained by reduction with sodium sulfite (curve 2 in Figure 7). In this figure, characteristics of electrodes with starting MCNTs are presented for comparison (curve 3). As is seen from Figure 7, RGO-SH-based electrodes give the best results.

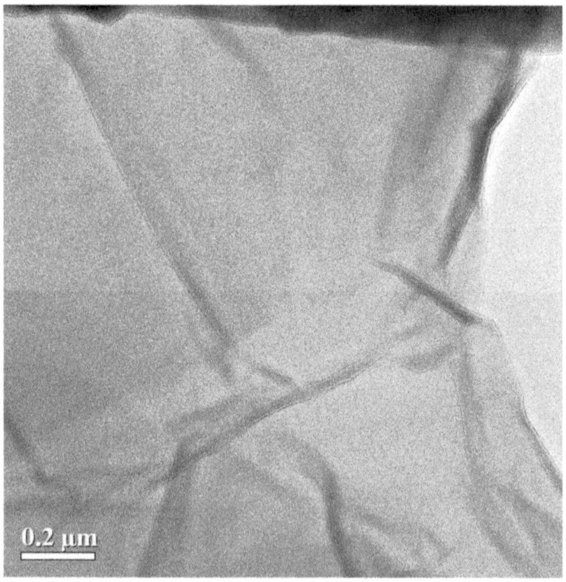

Figure 4 Micrograph of a sample of graphene oxide reduced with sodium hypophosphite.

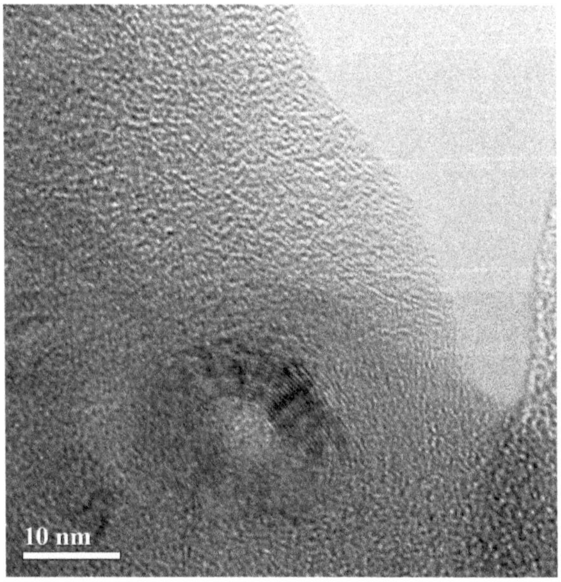

Figure 3 Micrograph of a sample of graphene oxide reduced with sodium sulfite.

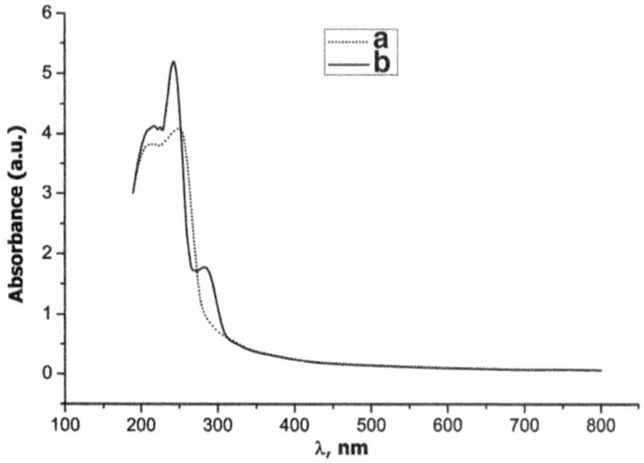

Figure 5 Absorption spectra of RGO with sodium hypophosphite (a) and sodium sulfite (b).

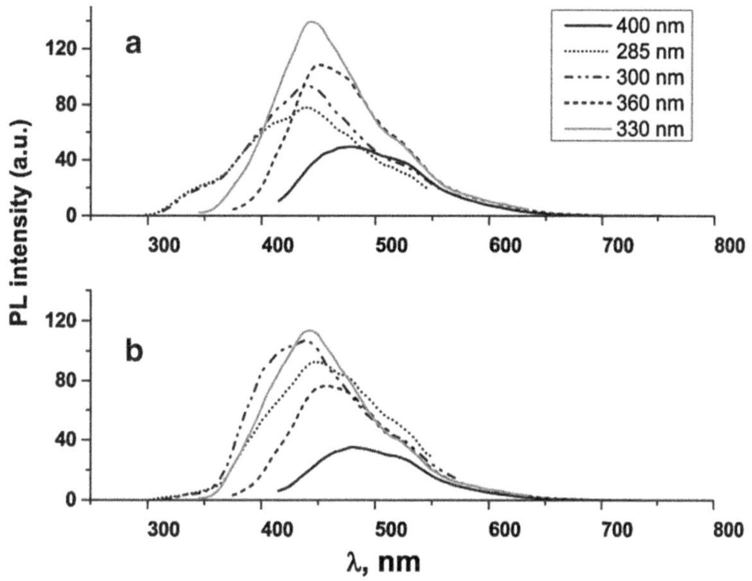

Figure 6 Photoluminescence spectra of graphene oxide reduced with sodium sulfite (a) and sodium hypophosphite (b).

It has been found that the characteristics of the RGO-SH electrode in oxygen reduction reaction are three to four times higher than those of the electrode of starting MCNTs. RGO-based electrodes are also superior to MCNT-MnO_2 nanocomposite in characteristics, which follows from comparison with the data presented in [2]. The characteristics of RGO-based oxygen electrodes were stable in the course of 6-month tests.

Conclusions

The proposed procedure for the reduction of graphene oxide allows one to obtain highly efficient materials for the oxygen electrodes of electrochemical power sources. The synthesis procedure is simple and does not require the use of highly toxic reagents.

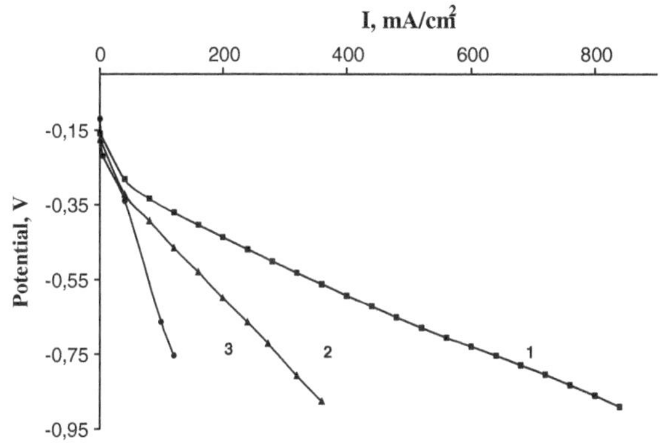

Figure 7 Curves for RGO reduced with (1) sodium hypophosphite, (2) sodium sulfite, and (3) starting nanotubes.

Methods

Preparation of reduced graphene oxide

MCNTs were chosen as a precursor, for they resemble, in structure, several graphene layers rolled into a tube; therefore, using a strong oxidant, one can 'break' nanotubes to form oxidized graphene nanostrips [12]. To this end, a procedure for the oxidation of MCNTs with heptavalent manganese [12] with subsequent reduction was chosen. We used alkaline solutions of sodium hypophosphite and sodium sulfite as oxidized-graphene reductants [32,33].

Electrochemical measurements

Two-layer oxygen electrodes were prepared by pressing. The hydrophobic layer contained 0.07 g/cm^2 acetylene black with 25% polytetrafluoroethylene, and the active layer contained 0.02 g/cm^2 RGO, with 5% polytetrafluoroethylene. The investigations were carried out on a fuel cell mock-up, a zinc electrode being used as the anode. A mock-up for the testing of gas diffusion electrodes is described in [34]. The electrolyte was a solution of 5 M KOH with 1 M LiOH. A silver chloride electrode connected through a salt bridge was used as a reference electrode. The electrochemical characteristics were recorded under galvanostatic conditions. The oxygen source was a U-shaped electrolyzer with alkaline electrolyte. Oxygen was supplied to the gas electrodes under an excess pressure of 0.01 MPa. Before measurements, the oxygen electrode was blown through with oxygen for an hour. The electron micrographs have been obtained with the aid of a JEM-100 CXII electron microscope (JEOL Ltd., Akishima-shi, Japan). The X-ray phase analysis was performed with the aid of a DRON-4 X-ray diffractometer with CuK_α radiation (Bourevestnik, Inc., Saint Petersburg, Russia). The electronic absorption spectra were recorded on a PerkinElmer UV/VIS Lambda 35 spectrophotometer (Waltham, MA, USA) and the fluorescence spectra on a PerkinElmer LS 55 spectrophotometer in a range of 300 to 700 nm.

Competing interests
The authors declare that they have no competing interests.

Authors' contributions
MOD carried out the electrochemical investigations and drafted the manuscript, IAS carried out the synthesis of reduced graphene oxide, IAR carried out the electronic absorption spectra, and GYK carried out the XRD characterization of samples and rechecked the whole manuscript. All authors read and approved the final manuscript.

Acknowledgements
The authors are most grateful to the STCU (no. 5335) for their financial support.

References
1. Bidault, F, Brett, D, Middleton, P, Brandon, N: Review of gas diffusion cathodes for alkaline fuel cells. J. Power Sources **187**, 39–48 (2009)
2. Soehn, M, Lebert, M, Wirth, T, Hofmann, S, Nicoloso, N: Design of gas diffusion electrodes using nanocarbon. J. Power Sources **176**, 494–498 (2008)
3. Danilov, M, Melezhyk, A: Carbon nanotubes modified with catalyst—promising material for fuel cells. J. Power Sources **163**, 376–381 (2006)
4. Hsieh, C, Lin, J, Wei, J: Deposition and electrochemical activity of Pt-based bimetallic nanocatalysts on carbon nanotube electrodes. Int. J. Hydrogen Energy **34**, 685–693 (2009)
5. Wang, X, Waje, M, Yan, Y: CNT-based electrodes with high efficiency for PEMFCs. Electrochem. Solid-State Lett. **8**, A42–A44 (2005)
6. Wang, G, Shen, X, Yao, J, Park, J: Graphene nanosheets for enhanced lithium storage in lithium ion batteries. Carbon **47**, 2049–2053 (2009)
7. Xin, Y, Liu, J, Jie, X, Liu, W, Liu, F, Yin, Y, Gu, J, Zou, Z: Preparation and electrochemical characterization of nitrogen doped graphene by microwave as supporting materials for fuel cell catalysts. Electrochim. Acta **60**, 354–358 (2012)
8. Lin, Z, Waller, G, Liu, Y, Liu, M, Wong, C: Facile synthesis of nitrogen-doped graphene via pyrolysis of graphene oxide and urea and its electrocatalytic activity toward oxygen reduction reaction. Adv. Energy Mater. **2**, 884–888 (2012)
9. Qu, L, Liu, Y, Baek, J, Dai, L: Nitrogen-doped graphene as efficient metal-free electrocatalyst for oxygen reduction in fuel cells. ACS Nano **4**, 1321–1326 (2010)
10. Lin, Z, Song, M, Ding, Y, Liu, Y, Liu, M, Wong, C: Facile preparation of nitrogen-doped graphene as a metal-free catalyst for oxygen reduction reaction. Phys. Chem. Chem Phys. **14**, 3381–3387 (2012)
11. Shao, Y, Zhang, S, Wang, C, Nie, Z, Liu, J, Wang, Y, Lin, Y: Highly durable graphene nanoplatelets supported Pt nanocatalysts for oxygen reduction. J. Power Sources **195**, 4600–4605 (2010)
12. Kosynkin, D, Higginbotham, A, Sinitskii, A, Lomeda, J, Dimiev, A, Price, B, Tour, J: Longitudinal unzipping of carbon nanotubes to form graphene nanoribbons. Nature **458**, 872–876 (2009)
13. Hummers, W, Offeman, R: Preparation of graphitic oxide. J. Am. Chem. Soc. **80**, 1339 (1958)
14. Park, S, An, J, Potts, J, Velamakanni, A, Murali, S, Ruoff, R: Hydrazine-reduction of graphite- and graphene oxide. Carbon **49**, 3019–3023 (2011)
15. Lee, Y, Song, M, Lee, K, Kim, I, Hwang, S: Synthesis and electrochemical characterization of reduced graphene oxide-manganese oxide nanocomposites. J. Electrochem. Sci. Technol. **2**, 1–7 (2011)
16. Pei, S, Cheng, H: The reduction of graphene oxide. Carbon **50**, 3210–3228 (2012)
17. Steven, G: Standard electrode potentials and temperature coefficients in water at 298.15 K. J. Phys. Chem. **18**, 1–21 (1989)
18. Dubin, S, Gilje, S, Wang, K, Tung, V, Cha, K, Hall, A, Farrar, J, Varshneya, R, Yang, Y, Kaner, R: A one-step, solvothermal reduction method for producing reduced graphene oxide dispersions in organic solvents. ACS Nano **4**, 3845–3852 (2010)
19. Moon, I, Lee, J, Ruoff, R, Lee, H: Reduced graphene oxide by chemical graphitization. Nat. Commun. (2010). doi:10.1038/ncomms1067
20. Krishnamoorthy, K, Veerapandian, M, Mohan, R, Kim, S: Investigation of Raman and photoluminescence studies of reduced graphene oxide sheets. Appl. Phys. A **106**, 501–506 (2012)
21. Wang, G, Wang, B, Park, J, Yang, J, Shen, X, Yao, J: Synthesis of enhanced hydrophilic and hydrophobic grapheme oxide nanosheets by a solvothermal method. Carbon **47**, 68–72 (2009)
22. Sun, Z, Popa, D, Hasan, T, Torrisi, F, Wang, F, Kelleher, E, Travers, J, Nicolosi, V, Ferrari, A: A stable, wideband tunable, near transform-limited, graphene-mode-locked, ultrafast laser. Nano. Res. **3**, 653–660 (2010)
23. Eda, B, Lin, Y, Mattevi, C, Yamaguchi, H, Chen, H, Chen, I, Chen, C, Chhowalla, M: Blue photoluminescence from chemically derived graphene oxide. Adv. Mater. **22**, 505–509 (2010)
24. Loh, K, Bao, Q, Eda, G, Chhowalla, M: Graphene oxide as a chemically tunable platform for optical applications. Nat. Chem. **2**, 1015–1024 (2010)
25. Shang, J, Ma, L, Li, J, Ai, W, Yu, T, Gurzadyan, G: The origin of fluorescence from graphene oxide. Sci. Reports (2012). doi:10.1038/srep00792
26. Hu, Y, Wang, K, Zhang, Q, Li, F, Wu, T, Niu, L: Decorated graphene sheets for label-free DNA impedance biosensing. Biomaterials **33**, 1097–1106 (2012)
27. Li, D, Muller, M, Gilje, S, Kaner, R, Wallace, G: Processable aqueous dispersions of graphene nanosheets. Nat. Nanotech. **3**, 101–105 (2008)
28. Galande, C, Mohite, A, Naumov, A, Gao, W, Ci, L, Ajayan, A, Gao, H, Srivastava, A, Weisman, R, Ajayan, P: Quasi-molecular fluorescence from graphene oxide. Sci. Reports (2011). doi:10.1038/srep00085
29. Cuong, T, Pham, V, Tran, Q, Hahn, S, Chung, J, Shin, E, Kim, E: Photoluminescence and Raman studies of graphene thin films prepared by reduction of graphene oxide. Mater. Lett. **64**, 399–401 (2010)
30. Luo, Z, Lu, Y, Somers, L, Johnson, A: High yield preparation of macroscopic graphene oxide membranes. J. Am. Chem. Soc. **131**, 898–899 (2009)
31. Chien, C, Li, S, Lai, W, Yeh, Y, Chen, H, Chen, I, Chen, L, Chen, K, Nemoto, T, Isoda, S, Chen, M, Fujita, T, Eda, G, Yamaguchi, H, Chhowalla, M, Chen, C: Tunable photoluminescence from graphene oxide. Angew. Chem. Int. Ed. **51**, 6662–6666 (2012)
32. Danilov, M, Kolbasov, G, Rusetskii, I, Slobodyanyuk, I: Electrocatalytic properties of reduced graphene oxide in oxygen electrode [abstract]. ABAF **13**, 18 (2012)
33. Danilov, M, Kolbasov, G, Rusetskii, I, Slobodyanyuk, I: Electrocatalytic properties of reduced graphene oxide in oxygen electrode. In: ABAF'2012: Advanced Batteries, Accumulators and Fuel Cells. 13th International Conference, August 2012; Brno, pp. 17–22. Brno University of Technology, Brno (2012)
34. Danilov, M, Kolbasov, G, Rusetskii, I, Slobodyanyuk, I: Electrocatalytic properties of nanocomposites for electrochemical power sources from the standpoint of localization of reactions. ECS Trans. **40**, 133–137 (2012)

Visible-light-activated nanocomposite photocatalyst of Cr_2O_3/SnO_2

Rajendra Bhosale[1], Sidram Pujari[1], Gajanan Muley[2], Bathuwell Pagare[3] and Anil Gambhire[3*]

Abstract

Visible-light-activated Cr_2O_3/SnO_2 nanocomposite photocatalyst was prepared by coprecipitation method and characterized by X-ray diffraction, transmission electron microscopy, X-ray photoelectron spectroscopy, N_2 adsorption-desorption measurement, and UV–vis diffuse reflectance spectroscopy. The results show that phase composition, crystallite size, Brunauer-Emmett-Teller surface area, and optical absorption of samples varied significantly with the heat treatment temperatures. The Cr_2O_3/SnO_2 photocatalyst (the molar ratio Cr to Sn is 1:2) calcined at 400°C for 2 h exhibited maximum photocatalytic activity because it has a smaller particle size of 10.05 nm and a higher surface area of 38.75 m^2/g. Under visible-light ($\lambda > 400$ nm) irradiation, the degradation rate of Rhodamine B reached 98.0% in 60 min, which is about 3.5 times higher than that of the standard P25 photocatalyst.

Keywords: Coprecipitation method, Coupled photocatalyst, Photocatalytic activity, Photoelectron spectroscopy

Background

In recent years, photocatalytic degradation of various kinds of organic–inorganic pollutants using semiconductor powder as a photocatalyst has been extensively studied [1,2]. Among various oxide semiconductor photocatalysts, TiO_2 was intensively investigated because of its biological and chemical inertness, strong oxidizing power, nontoxicity, and long-term stability [3-5]. However, the photocatalytic activity of TiO_2 (the bandgap is 3.2 eV, and it can be excited by photons with wavelengths below 387 nm) is limited to irradiation wavelengths in the UV region so that the effective utilization of solar energy is limited to about 3% to 5% of the total solar spectrum. Furthermore, the fast recombination of photo-generated electron–hole pairs hinders the commercialization of this technology [6]. The decomposition of adsorbed organic compound is closely correlated with the density of space charge-separated electron–hole pair on TiO_2. Therefore, it is of great interest to separate the electron–hole pairs effectively to increase the photon efficiencies and develop new visible-light photocatalyst to extend the absorption wavelength range into the visible-light region. In this sense, an interesting approach to deal with the issue is carried out by coupled semiconductor technique. Recently, there are a number of studies related to the photocatalytic activity of coupled semiconductor photocatalyst such as ZnO/SnO_2 [6], SnO_2-TiO_2 [7], Au/TiO_2-CeO_2 [8], and TiO_2-CdS [9]. The results show that nearly all the composite semiconductors have presented higher photocatalytic activity than single ones. Indeed, this effect could be due to the synergistic effect of the coupled semiconductor photocatalyst. It was also reported that the chromium-doped TiO_2 has been found to exhibit superior photocatalytic activity under visible-light irradiation because chromium atom can effectively narrow the energy bandgap of TiO_2 [10,11]. Chromium-based catalysts have been widely examined for polymerization, partial oxidation, and aromatization reaction because of the peculiar characteristics of Cr oxide species on the surface of the support, including oxidation state and coordination environment [12]. However, there have been few studies clearly elucidating the roles of the surface chromate species and supporting the oxidation reaction, particularly the decomposition of environmental pollutants.

In this study, a series of Cr_2O_3/SnO_2 photocatalysts with different calcination temperatures were prepared by coprecipitation method, and their phase compositions, crystalline structures, and particle sizes were studied. We also explored their photocatalytic degradation performance

* Correspondence: abg_chem@ymail.com
[3]Department of Chemistry, Shri Anand College of Science, Pathardi, Ahmednagar, Maharashtra 414102, India
Full list of author information is available at the end of the article

in treating Rhodamine B solution under visible-light (λ > 400 nm) irradiation.

Results and discussion

XRD analysis

Figure 1 shows X-ray diffraction (XRD) patterns of the Cr_2O_3/SnO_2 photocatalysts calcined at different temperatures for 2 h. It can be seen that the diffraction pattern of the sample prepared at 300°C is indistinct because of the amorphous configuration. While the sample calcined at 400°C, the diffraction peaks clearly show that the Cr_2O_3 and SnO_2 nanocrystals coexisted in the samples. With the increase of calcination temperature (at 500°C), the phase composition is still the mixture of Cr_2O_3 and SnO_2 phase, but the peak intensity of the sample obviously increases due to the growth of crystallites and enhancement of crystallization. Table 1 summarizes the

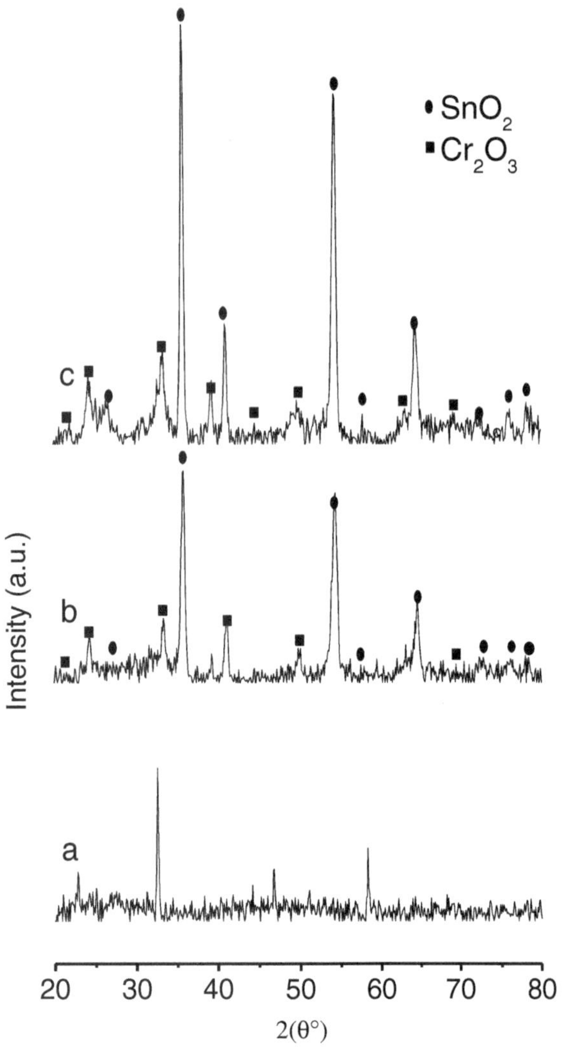

Figure 1 XRD patterns of the Cr_2O_3/SnO_2 photocatalysts calcined at (a) 300°C, (b) 400°C, and (c) 500°C.

Table 1 Effect of calcination temperature on average particle size and BET surface area of Cr_2O_3/SnO_2 samples

Calcination temperature (°C)	BET surface area (m^2/g)	Grain size (nm)
300	–	6.80
400	38.75	10.05
500	19.02	23.15
600	8.35	31.19

Brunauer-Emmett-Teller (BET) surface area and grain size of the samples. The average grain sizes estimated from the highest intensity diffraction peaks of SnO_2 (2θ = 33.37) using Scherrer's equation were 6.80, 10.05, and 23.15 nm for samples a, b, and c, respectively. It is obvious that the grain sizes of the samples improved with the increase of heat treatment temperature, while the BET surface areas decreased.

TEM analysis

The grain sizes of the Cr_2O_3/SnO_2 nanocomposite photocatalysts varied with calcination temperature, which has been confirmed in the transmission electron microscopy (TEM) images (Figure 2a,b,c,d). It can be seen that the sample calcined at 300°C appeared amorphous because of its weak crystallization, and the grain edge of the sample was a little dim. The TEM images also indicated that the grain size was homogenous and fairly small (about 10 nm) when the sample was calcined at 400°C. In contrast, the sample calcined at 500°C displayed greater particle size due to the enhancement of crystallization or sintering between smaller particles, which is estimated to be about 25 nm. The increase of heat treatment temperature caused the catalyst grains to become larger. The result was in agreement with that of XRD. Figure 2d shows a high-resolution (HR) TEM image of Cr_2O_3/SnO_2 powder calcined at 400°C and its corresponding fast Fourier transform (FFT) pattern. The space between adjacent lattice planes is 0.32 nm, corresponding to 110 planes of rutile SnO_2. Combined with the results of FFT analysis, the preferential growth direction is (112).

UV–vis DR spectral analysis

Figure 3 gives the UV–vis diffuse reflectance spectra (DRS) of the Cr_2O_3/SnO_2 photocatalysts calcined at 300°C, 400°C, and 500°C. A progressive red shift in the bandgap absorption is noticed with the increase of calcination temperature. This implies that the absorption edges of the samples shift to the longer wavelength range. When calcined at 300°C, the sample has a little absorption in the range of visible light (λ > 400 nm) because the catalyst is mainly composed of amorphous SnO_2 powder (shown in XRD patterns), which is only effective under ultraviolet irradiation. With the increase

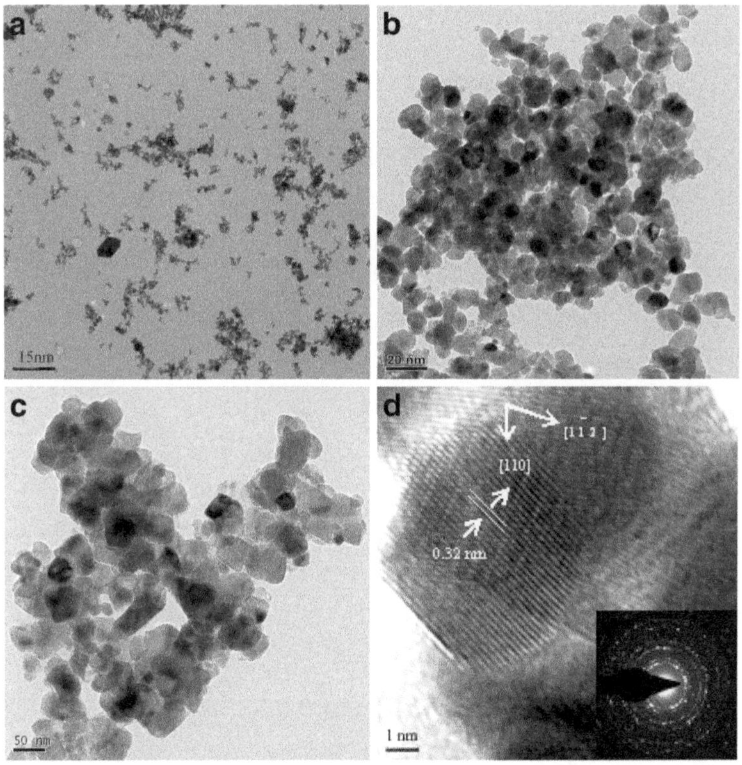

Figure 2 TEM images of Cr_2O_3/SnO_2 photocatalyst calcined at different temperatures, HRTEM image, and inset of FFT. TEM images at (**a**) 300°C, (**b**) 400°C, and (**c**) 500°C; (**d**) HRTEM image of photocatalyst (**b**); and inset of corresponding FFT of (**d**).

of calcination temperature (reaching 400°C), the absorption edge of the sample has some red shift. The red shift is presumably ascribed to the formation of the homogeneous Cr_2O_3 nanocrystals. The absorption onsets were determined by linear extrapolation from the infection point of the curve to the baseline. The edges of the absorption of the Cr_2O_3 incorporated SnO_2 samples were shifted to approximately 500 nm, corresponding to a bandgap energy of 2.25 eV.

The bandgap energy of Cr_2O_3 is 2.5 eV and can be activated by the light below 560 nm [13], when it couples with SnO_2 semiconductor, the conduction band of SnO_2 acts as a sink for photo-generated electrons. The photo-generated holes move in the opposite direction and accumulate in the valence band of the Cr_2O_3 particle, which leads to increasing the charge separation efficiency and extending the photo-responding range to visible light. When the heat treatment temperature reaches 500°C, the crystallite sizes become larger, and the absorption edge shows more red shift.

XPS studies

Figure 4a,b,c shows the high-resolution X-ray photoelectron spectroscopy (XPS) spectra of the $Cr2p_{3/2}$, O 1s, and Sn 3d regions, taken on the surface of Cr_2O_3/SnO_2 photocatalyst calcined at 400°C. The $Cr2p_{3/2}$ level (Figure 4a) is broad and showed asymmetry towards higher binding-energy side of the main peak. The deconvolution of the $Cr2p_{3/2}$ signals based upon the previous studies [14] and the comparison to the reference samples including CrO_3 and Cr_2O_3 indicate that spectra for the present catalytic systems contain two major peaks at 579.1 eV for Cr(VI) and 576.1 eV for Cr(III). The change in oxidation state of chromium from sixth to third might be due to oxide/oxide support interaction. From high-resolution XPS spectra of O 1s region (Figure 4b), it can be seen that oxygen on the sample surface exists at least in three forms: at the binding energy of 530.4 eV which corresponds to oxygen in SnO_2 lattice; 531.7 eV which corresponds to oxygen in SnO_2 surface adsorption of (–OH); and 533.5 eV which corresponds to oxygen in SnO_2 surface adsorption of H_2O. It is just because so much adsorption oxygen exists on the SnO_2 surface that they become captives of photo-generated electron–hole pairs directly or indirectly. So, the recombination of photo-generated electron–hole pairs is suppressed, and therefore, the quantum efficiency of photocatalytic reaction is improved. This may be the possible reason for the enhancement of photocatalytic activity. The binding energy of $Sn3d_{5/2}$ (Figure 4c) was 486.36 eV, slightly lower than that of SnO_2 (>486.4 eV), might be due to oxide/oxide support interaction.

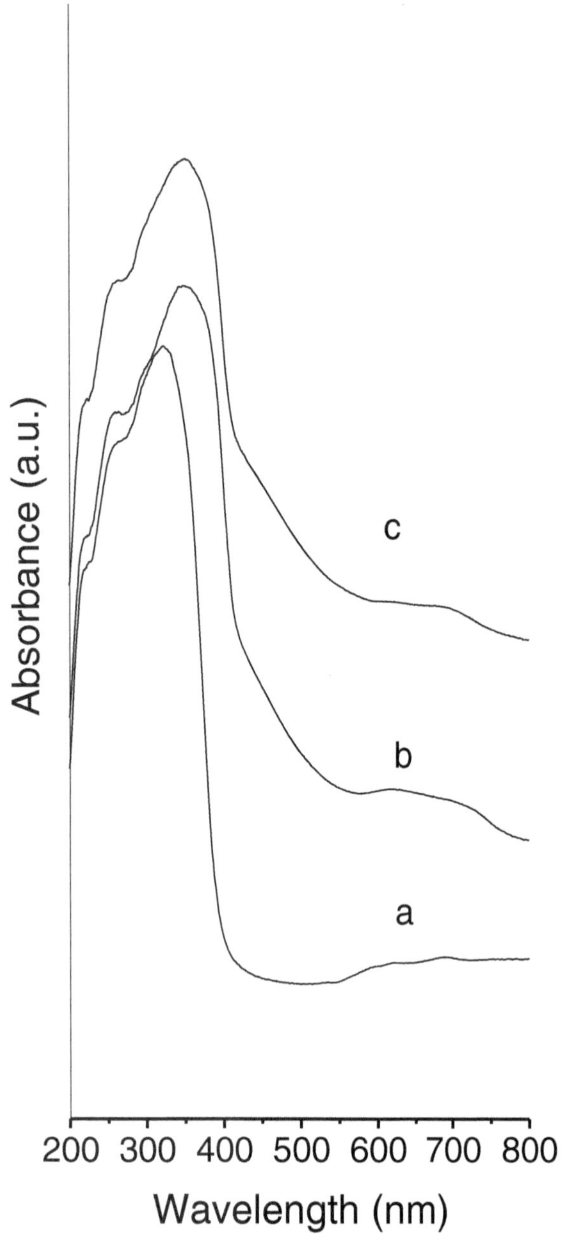

Figure 3 **UV–vis DRS of the Cr_2O_3/SnO_2 photocatalyst calcined at (a) 300°C, (b) 400°C, and (c) 500°C.**

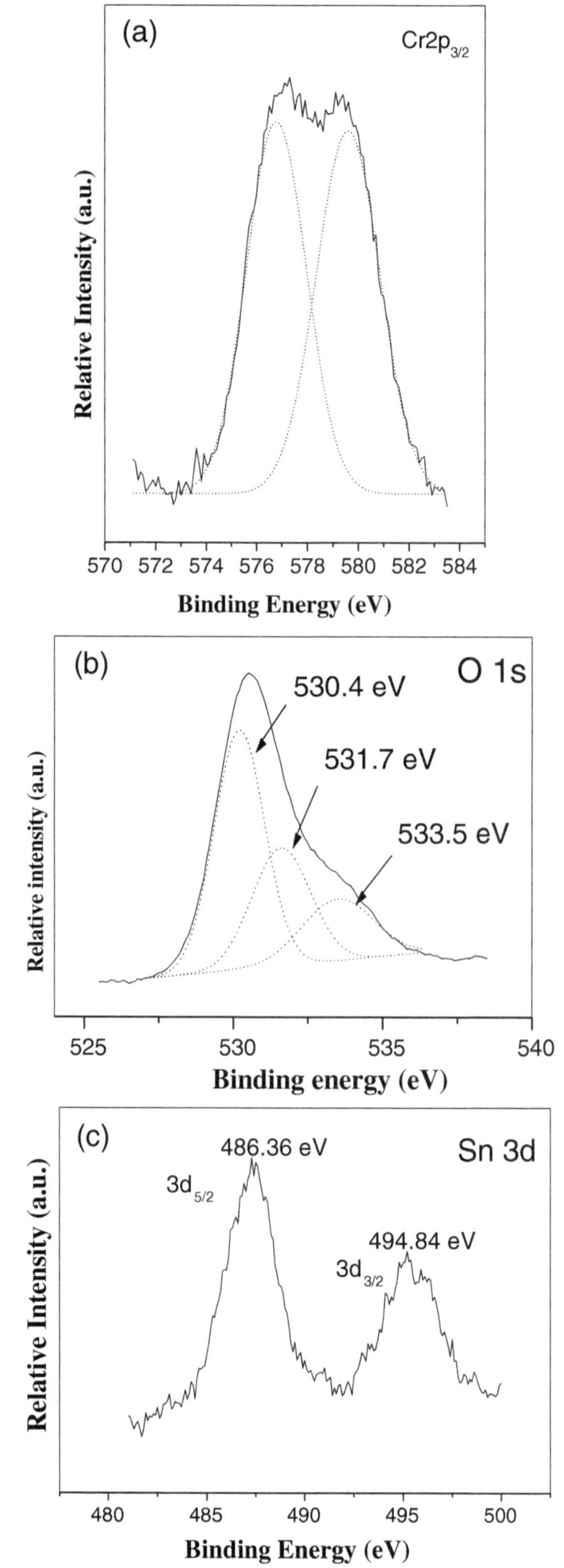

Figure 4 **High-resolution XPS spectra of Cr_2O_3/SnO_2 photocatalyst calcined at 400°C for 2 h in air.** (**a**) $Cr2p_{3/2}$ XPS spectra, (**b**) O 1s XPS spectra, and (**c**) Sn 3d XPS spectra.

Photocatalytic activity studies

The characteristic patterns of XRD, TEM, and UV–vis DRS displayed that the Cr_2O_3/SnO_2 photocatalyst calcined at 400°C for 2 h (the molar ratio of Cr to Sn is 1:2) has better crystallization, smaller crystal size, and stronger response to visible light. In order to explore the photocatalytic activity of the catalyst in the degradation of Rhodamine B solution (0.5 g/L) under visible light ($\lambda > 400$ nm), the photocatalytic behavior of the standard photocatalyst Degussa P25 was also measured as a reference. Results indicate that the Cr_2O_3/SnO_2

photocatalyst displayed significantly high degradation ability for Rhodamine B, the degradation rate reached 98.0% in 60 min, which is obviously about 3.5 times higher than the value of 27.5% over Degussa P25 (Figure 5). The enhancement may be explained in terms of the synergetic effect on the specific adsorption property and efficient electron–hole separation at the Cr_2O_3/SnO_2 nanocomposite photocatalyst interfaces and surfaces. Cr_2O_3-doped SnO_2 photocatalyst have stronger reduction/oxidation capabilities, probably due to the high polymerized chromium oxide species on the catalyst surface [15]. As shown in the XPS results, formed polychromate species can substitute Ti to form a stable solid that creates a charge imbalance and must be satiated therefore; more hydroxide ions would be adsorbed on the surface to balance the charge. These hydroxide ions on the surface can accept holes generated by visible-light illumination to form hydroxyl radicals which oxidize adsorbed organic molecule.

Factors influencing the photocatalytic activity

Effect of catalyst loading

The effect of catalyst loading on the photocatalytic degradation of Rhodamine B was studied by varying the amount of Cr_2O_3/SnO_2 nanocomposite calcined at 400°C. Figure 6 showed the degradation profile of Rhodamine B with an initial concentration of 0.5 mg/L under various catalyst loadings from 0.2 to 4 g/L. It could be seen that photocatalytic degradation efficiency has increased up to 0.5 g/L and then declined with increasing catalyst loading. This could be attributed to the shadowing effect, wherein high turbidity due to high Cr_2O_3/SnO_2 nanocomposite concentration decreased the penetration depth of solar radiation. Hence, the optimal catalyst loading of 0.5 g/L was employed throughout the present study.

Effect of pH

Figure 7 presented the effect of pH value in the suspension on photocatalytic activity. The variation of pH value showed its strong influence on the Rhodamine B photodegradation. The photodegradation efficiency as a function of pH value decreased in the order of 10 > 7 > 3. It was generally accepted that the pH-dependent photodecomposition was mainly ascribed to the variation of surface charge properties of a photocatalyst. Consequently, this changed the adsorption behavior of a dye on the catalyst surface. Since Rhodamine B had a cationic configuration, its adsorption was favored in alkaline solution as demonstrated in Figure 8. The increase of pH value resulted in a higher adsorption amount of Rhodamine B on the Cr_2O_3/SnO_2 nanoparticle surface.

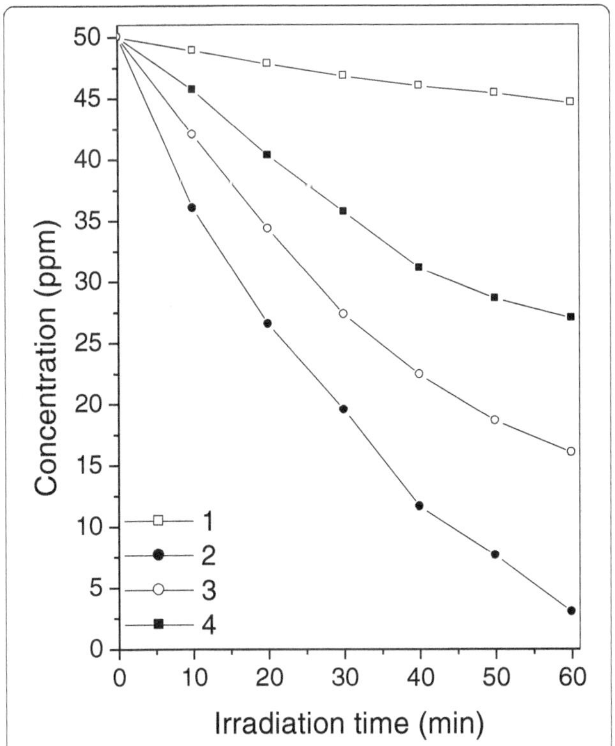

Figure 5 Photocatalytic degradations for 50 ml Rhodamine B by Cr_2O_3/SnO_2 photocatalyst calcined at different temperatures and standard P25. Calcined at (1) 300°C, (2) 400°C, (3) 500°C, and (4) standard P25.

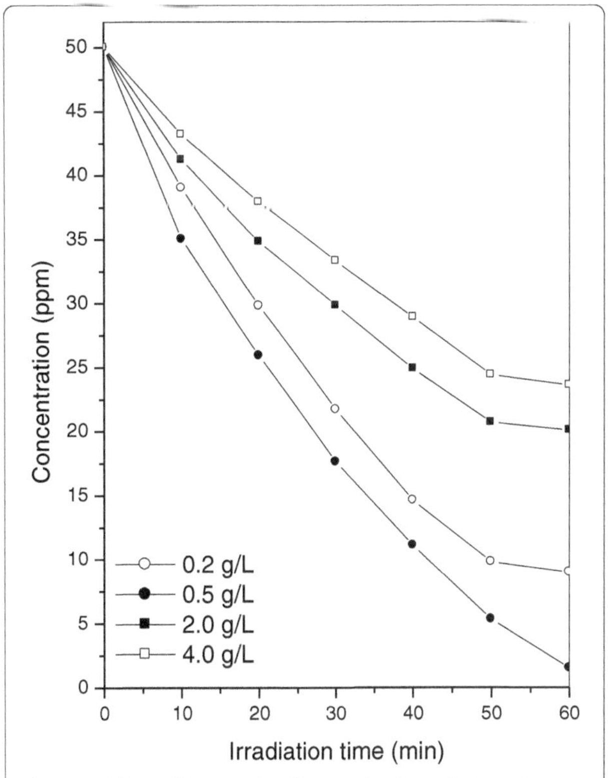

Figure 6 Effect of catalyst loading on the degradation of Rhodamine B. Having an initial concentration of 0.5 g/L over Cr_2O_3/SnO_2photocatalyst calcined at 400°C.

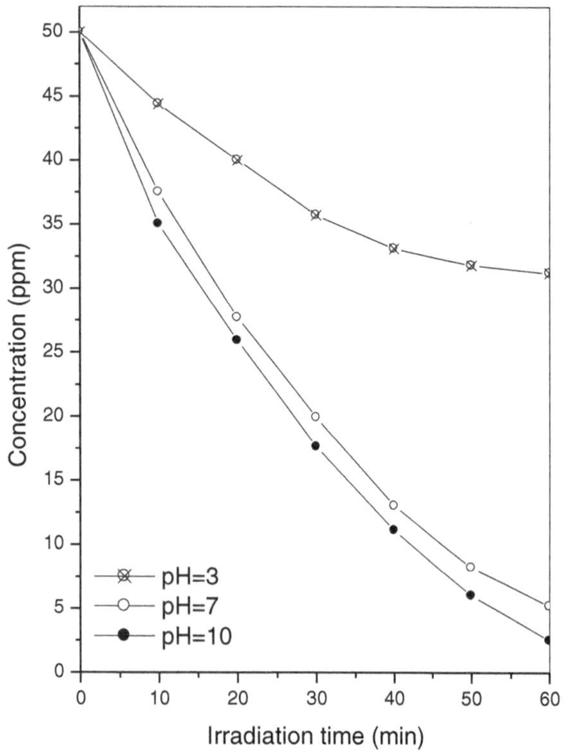

Figure 7 Effect of pH value on the degradation of Rhodamine B. Having an initial concentration of 0.5 g/L over Cr_2O_3/SnO_2 photocatalyst calcined at 400°C.

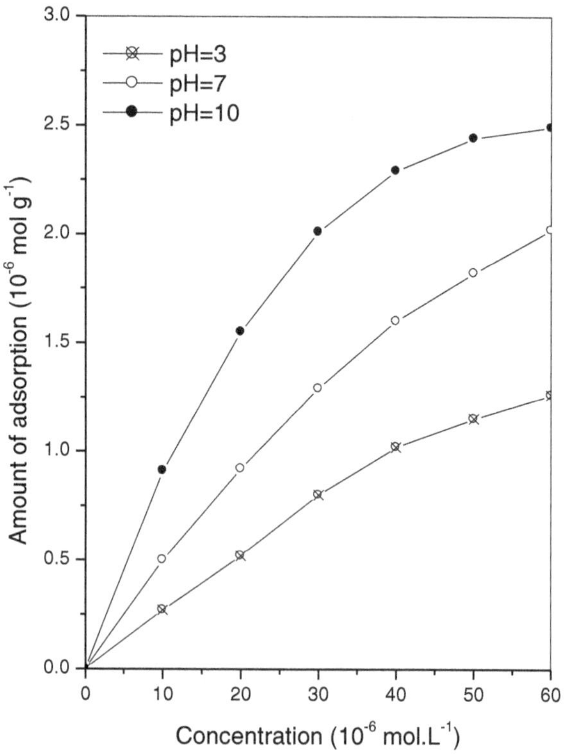

Figure 8 pH-dependent adsorption of Rhodamine B on Cr_2O_3/SnO_2 photocatalyst calcined at 400°C.

Conclusions

The novel visible-light-activated Cr_2O_3/SnO_2 nanocomposite photocatalyst was prepared by coprecipitation method. The characteristic patterns of XRD, BET, TEM, and UV–vis DRS displayed that the sample calcined at 400°C for 2 h (the molar ratio of Cr to Sn is 1:2) has better crystallization, smaller crystal size, and stronger response to visible light. The Cr_2O_3/SnO_2 photocatalyst showed remarkable photocatalytic activity compared with the standard P25 photocatalyst. Rhodamine B (98.0%) can be degraded in 60 min under illumination of the visible light (λ > 400 nm). At high catalyst loadings, penetration of the light inside the reaction medium was reduced because of the light scattering and shielding effect by catalyst particles. In addition, basic pH level of suspension was found to be beneficial for photocatalytic degradation.

Methods

Preparation of photocatalysts

The reagent grade chemicals used in preparing the samples, $CrCl_3{\cdot}6H_2O$ and $SnCl_4{\cdot}5H_2O$ were used as the starting materials. $CrCl_3{\cdot}6H_2O$ and $SnCl_4{\cdot}5H_2O$ were mixed (the molar ratio of Cr to Sn is 1:2) and dissolved in minimum amount of ethanol. The cationic surfactant cetyltrimethylammonium bromide (5%; 20 mL) in ethanol was dropped into the solution. The system was kept under constant stirring and sustaining the pH of 7 by simultaneous addition of ammonium hydroxide to form the precipitate. The precipitate was filtered and washed with deionized water until no Cl^- was found in the filtrates. Then, the wet powder was dried at about 100°C in air to form the precursor of the Cr_2O_3/SnO_2 photocatalyst. Finally, the precursors were calcined for 2 h at different temperatures in air to prepare the photocatalyst powders.

Characterization of photocatalysts

To determine the crystallite sizes and identities of the Cr_2O_3/SnO_2 nanocomposite photocatalysts, XRD analysis was carried out at room temperature using a model D8 Bruker AXS (Madison, WI, USA)with monochromatic Cu radiation (40 kV and 30 mA), over the 2θ collection range of 20° to 80°. The shapes of the samples were tested using transmission electron microscopy; FEI, Tecnai F30 HRTEM, FEG (FEI, Hillsboro, OR, USA) operated at 300 kV. The BET surface areas were determined using a Micromeritics ASAP 2010 N_2 adsorption apparatus (Norcross, GA, USA). XPS was used to study the chemical composition of the sample. The monochromatic X-ray beams of Al K_α ($h\nu$ = 1486.6 eV) and Mg K_α ($h\nu$ = 1253.6 eV) radiations were used as the excitation source. A hemispherical sector analyzer and multi channel detectors were used to detect the ejected photoelectrons as a function of their kinetic energies. XPS spectra were recorded at pass energy of 50 eV, 5-mm slit width, and a take-off angle of 55°. The spectrometer was calibrated by determining the binding energy values of Au $4f_{7/4}$ (84.0 eV), Ag $3d_{5/2}$ (368.4 eV), and Cu $2P_{3/2}$ (932.6 eV) levels using spectrograde materials. The instrumental resolution under these conditions was 1.6 eV full-width at half-maximum for Au $4f_{7/4}$ level. The Cls (285 eV) and Au $4f_{7/4}$ (84.0 eV) were used as internal standards whenever needed.

UV–vis DRS were recorded in air at room temperature in the wavelength range of 200 to 800 nm using a PE LAMBDA35 spectrophotometer (PerkinElmer, Waltham, MA, USA).

Rhodamine B adsorption experiment

All batch equilibrium experiments were conducted in the dark. The study of Rhodamine B adsorption has been performed at room temperature. In each test, 1.5 g of Cr_2O_3/SnO_2 sample calcined at 400°C were added to 50 mL of 50 mg/L Rhodamine B solution. The equilibrium concentration was determined using centrifugation and filtration, through a Millipore filter of the suspension. The amounts of Rhodamine B adsorbed were calculated as follows:

$$n(\text{ads}) = V\Delta C$$

where n(ads) was the number of moles adsorbed; ΔC was the difference between the initial concentration, C_0 and equilibrium concentration, C_c; and V was the volume (50 mL).

Photocatalytic activity measurements

Photocatalytic degradation of Rhodamine B in aqueous solution (0.5 g/L) was carried out using a Quartz reactor under visible-light irradiation (tungsten lamp, 500 W). Air was bubbled into the solution throughout the entire experiment. A cutoff filter was placed outside the Quartz jacket to completely remove all wavelengths less than 400 nm to ensure irradiation with visible light (λ > 400 nm). About 0.5 g of photocatalyst was immersed into a 50-mL aqueous Rhodamine B. Prior to irradiation, the suspensions were magnetically stirred in the dark for 30 min to ensure establishment of an adsorption-desorption equilibrium among the photocatalyst, Rhodamine B, and atmospheric oxygen. At given irradiation time intervals, 10 mL of the suspensions are collected, and then filtered through a Millipore filter to separate the photocatalyst particles. The changes in Rhodamine B concentration were analyzed by UV-visible spectroscopy.

Competing interests

The authors declare that they have no competing interests.

Authors' contributions

RB carried out the photocatalytic degradation study, SP carried out XRD characterization of samples, BP synthesized and calcined the samples, GM carried out XPS analysis, and AG drafted the manuscript and rechecked the whole manuscript. All authors read and approved the final manuscript.

Acknowledgements

This work is financed by the University Grants Commission, New Delhi, India (Grant no. 47–2028/11).

Author details

[1]D.B.F. Dayanad College of Arts and Science, Solapur, Maharashtra 413002, India. [2]Department of Physics, Sant Gadge Baba Amravati University, Amravati, Maharashtra 444602, India. [3]Department of Chemistry, Shri Anand College of Science, Pathardi, Ahmednagar, Maharashtra 414102, India.

References

1. Hoffmann, MR, Martin, ST, Choi, W, Bahnemann, DW: Environmental applications of semiconductor photocatalysis. Chem. Rev. **95**, 69–96 (1995)
2. Khan, SU, Al-shahry, M, Lngler Jr, WB: Efficient photochemical water splitting by a chemically modified n-TiO_2. Science **297**, 2243–2245 (2002)
3. Zhang, T, Oyama, T, Horikoshi, S, Zhao, J, Serpone, N, Hidaka, H: Photocatalytic decomposition of the sodium dedecylbenzene sulfonate surfactant in aqueous titania suspensions exposed to highly concentrated solar radiation and effects of additives. Appl. Catal. B: Environ. **42**, 13–24 (2003)
4. San, N, Hatipoglu, A, Kocturk, G, Cinar, Z: Photocatalytic degradation of 4-nitrophenol in aqueous TiO_2 suspensions, theoretical prediction of the intermediates. J. Photochem. Photobiol. A: Chem. **146**, 189–197 (2002)
5. Chen, S, Cao, G: Study on the photocatalytic reduction of dichromate and photocatalytic oxidation of dichlorvos. Chemosphere **60**, 1308–1315 (2005)
6. Cun, W, Jincai, Z, Xinming, W, Bixian, M, Guoying, S, Ping'an, P, Jiamo, F: Preparation, characterization and photocatalytic activity of nano-sized ZnO/SnO_2 coupled photocatalysts. Appl. Catal. B: Environ. **39**, 269–279 (2002)
7. Khan, R, Kim, TJ: Preparation and application of visible-light-responsive Ni-doped and SnO_2-coupled TiO_2 nanocomposite photocatalysts. J. Hazardous Mater. **163**, 1179–1184 (2009)
8. Guzman, C, del Angel, G, Gomez, F, Galindo-Hernandez, F, Angeles-Chavez, C: Degradation of the herbicide 2,4-dichlorophenoxyacetic acid over Au/TiO_2-CeO_2 photocatalysts: effect of the CeO_2 content on the photoactivity. Catal. Today **166**, 146–151 (2011)
9. Kanai, N, Nuida, T, Ueta, K, Hashimoto, K, Watanabe, T, Ohsaki, H: Photocatalytic efficiency of TiO_2/SnO_2 thin film stacks prepared by DC magnetron sputtering. Vaccum **74**, 723–727 (2004)
10. Wang, C, Shi, H, Li, Y: Synthesis and characterization of natural zeolite supported Cr-doped TiO_2 photocatalysts. Appl. Surf. Sci. **258**, 4328–4333 (2012)
11. Hamadanian, M, Jabbari, V, Gravand, A: Dependence of energy conversion efficiency of dye-sensitized solar cells on the annealing temperature of TiO_2 nanoparticles. Mater. Sci. Semicond. Process. **15**, 371–379 (2012)
12. Weckhuysen, B, Wachs, I, Schoonheydt, R: Surface chemistry and spectroscopy of chromium in inorganic oxides. Chem. Rev. **96**, 3327–3350 (1996)
13. Radecka, M, Zakrzewska, K, Wierzbicka, M, Gorzkowska, A, Komornicki, S: Study of the TiO_2-Cr_2O_3 system for photoelectrolytic decomposition of water. Solid State Ion. **157**, 379–386 (2003)
14. Yim, SD: Nam, I-S: Characteristics of chromium oxides supported on TiO_2 and Al_2O_3 for the decomposition of perchloroethylene. J. Catal. **221**, 601–611 (2004)
15. Hakuli, A, Harlin, ME, Backman, LB, Krause, AOI: Dehydrogenation of *i*-butane on Cr_2O_3/SiO_2 catalysts. J. Catal. **184**, 349–356 (1999)

11

Separation of the defect-free Fe_3O_4-Au core/shell fraction from magnetite-gold composite nanoparticles by an acid wash treatment

Hojatollah Montazeri[1], Amir Amani[1], Hamid Reza Shahverdi[2], Emad al din Haratifar[2] and Ahmad Reza Shahverdi[3,4*]

Abstract

Some undesired nanoparticles, such as malformed structures or incomplete core/shell Fe_3O_4-Au nanocomposites, might be formed when Fe_3O_4-Au core/shell nanocomposites are being synthesized. These impurities should be separated before any applications are performed. In this investigation, magnetic cores (Fe_3O_4) were synthesized using a conventional fabrication method involving coprecipitation of Fe^{2+} and Fe^{3+}. Carboxyl-capped magnetite-gold composite nanoparticles, measuring ≤50 nm, were then synthesized using a plant extract (*Eucalyptus camaldulensis*). The prepared carboxyl-capped magnetite-gold composite nanoparticles were further subjected to acid treatment for 18 h and characterized with different instrumentation methods. The results of the investigation showed that acid treatment can be applied successfully to separate defect-free Fe_3O_4-Au core/shell nanoparticles from different types of Fe_3O_4-Au nanocomposites, without any considerable changes in their physiochemical properties.

Keywords: Magnetite nanocomposite, Fe_3O_4-Au core/shell, Acid washing, Biosynthesis, Separation

Background

Magnetic nanoparticles (MNPs) have been known as one of the nanostructures that provide the widest uses in biomedicine [1-5]. For example, MNPs that are composed of magnetite (Fe_3O_4) or maghemite (γ-Fe_2O_3) have unique thermal, chemical, and magnetic properties that make these particles particularly well suited for medical applications [5,6]. However, producing nanoparticles (NPs) with high stability and biocompatibility presents one of the greatest challenges. To address this, MNPs should be covered with an external quite inert shell, in order to protect the magnetic core against chemical changes. In addition, the surfaces of these particles should be functionalized with organic molecules, to allow them to bind chemically to other biomolecules such as DNA, proteins, amino acids, etc. [7-9]. Gold is a chemically quite inert element, and no single acid dissolves it. This metal is resistant to chemical reactions that take place in biological fluids and is a very useful element for providing a coating layer to protect MNPs [10,11]. Also, gold metal does not decrease the magnetic properties of Fe_3O_4 MNPs and, consequently, will not limit the applications in which these particles are used [12,13].

At present, a number of different chemical and biological methods have been reported for fabricating magnetite Fe_3O_4-Au composite NPs [4,6,9]. These nanocomposites contain both defect-free Fe_3O_4 core@Au shell MNPs and other types of Fe_3O_4-Au nanocomposites such as malformed structures or incomplete core/shell Fe_3O_4-Au nanocomposites (Figure 1). As mentioned above, the importance of the external inert shell in protecting the magnetic core is well established, and different reports have been published on the synthesis of Fe_3O_4 core@Au shell MNPs [4,13]. However, in these studies, no further treatments such as acid washing have been applied for separating the defect-free Fe_3O_4 core@Au MNPs from other types of Fe_3O_4-Au nanocomposites. An intact gold layer on the Fe_3O_4 nucleus can protect the core against acid wash treatment. In contrast, Fe_3O_4 minerals in malformed structures or other structures with incomplete inert shell will be dissolved with exposure to acid. In the current study, carboxyl-capped

* Correspondence: shahverd@tums.ac.ir
[3]Biotechnology Research Center, Faculty of Pharmacy, Tehran University of Medical Sciences, Tehran 1417614411, Iran
[4]Pharmaceutical Sciences Research Center, School of Pharmacy, Tehran University of Medical Sciences, Tehran 417614411, Iran
Full list of author information is available at the end of the article

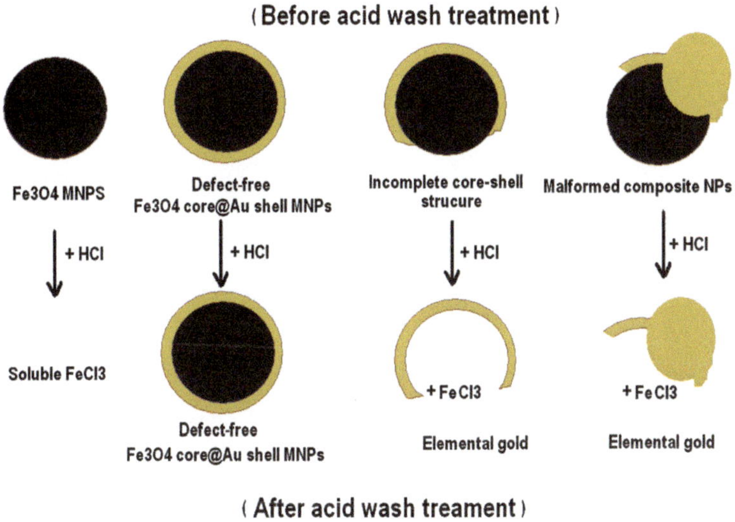

Figure 1 The effect of acid wash treatment on the different structure types of magnetite nanoparticles.

Fe_3O_4-Au composite MNPs were prepared by a combined chemical and biological reducing method [14]. Hydrochloric acid (HCl) was used to dissolve all types of Fe_3O_4-Au composite MNPs, except for Fe_3O_4@Au core/shell MNPs. In the next step, the remaining Fe_3O_4 MNPs with defect-free gold shell, which has been designated in this report as 'Fe_3O_4 core@Au shell MNPs,' were separated from the reaction mixture with a magnet and fully characterized using different instrumentation methods.

Results and discussion

This study reports a simple acid wash treatment for separating defect-free Fe_3O_4 core@Au shell MNPs from other types of Fe_3O_4-Au composite NPs (i.e., malformed structures and incomplete core/shell MNPs) (Figure 1). First, crude magnetite Fe_3O_4-Au nanocomposites were synthesized using the method described above, and their shapes and sizes were studied by transmission electron microscopy (TEM). Figure 2 shows two representative TEM images recorded from the drop-coated film of the Fe_3O_4-Au nanocomposites that were fabricated by a recently described semi-biosynthesis method [14]. Two dark and gray particles with sizes of ≤50 nm can be observed in the TEM images that have been illustrated in Figure 2a,b.

It should be noted that Fe_3O_4 magnetite NPs are seen as a gray color because elemental gold (Au) has a higher electron density than Fe_3O_4 and, consequently, more electrons are transmitted in bright field imaging [15]. The analysis of the gray particles by energy-dispersive spectroscopy (EDS) confirmed the presence of Fe and O element signals (Figure 2c). The peaks shown in Figure 2c are attributed to Fe and O elements, and this confirmed the existence of iron-oxide NPs (copper and carbon peaks existed due to the grid used for TEM imaging). The EDS pattern of the dark particles indicated the presence of gold, iron, and oxygen, as illustrated in Figure 2d. The EDS experiment was repeated to analyze different other dark NPs, and the obtained EDS spectra confirmed that the other dark NPs were also made up of both elemental gold and Fe_3O_4 magnetite ingredients (Figure 3a,b). In addition, further analysis of some gray particles (Figure 3a) by EDS showed a gold element signal with lower density, confirming the presence of other types of Fe_3O_4-Au nanocomposites, especially malformed Fe_3O_4-Au composite NPs (Figure 3c).

In the next step, a simple dissolution method using hydrochloric acid (1 N) was performed to separate acid-stable magnetite particles, coated with an intact gold layer, from those containing bare Fe_3O_4 MNPs, malformed structures or incomplete core/shell nanocomposites (Figure 1). It should be noted that defect-free Fe_3O_4 core@Au shell MNPs are stable against acid and can easily be separated after the acid wash process by applying an external magnetic field. In contrast, during acid wash treatment, malformed structures or incomplete Fe_3O_4@Au core/shell MNPs are decomposed to gold metal and soluble ferric iron (Figure 1). In addition, bare Fe_3O_4 MNPs are also completely dissolved in HCl solution (1 N) (Figure 1). Figure 4a shows a test tube containing Fe_3O_4-Au nanocomposites that has been mixed with HCl and incubated for 18 h. Also, Figure 4b,c demonstrates two tubes containing acid-stable NPs before and after separation by a magnet.

The upper illustration in Figure 5 shows different TEM images of the acid-stable NPs after washing with distilled water. As shown in this illustration, dark NPs with sizes of less than 50 nm can be observed in TEM images, indicating different magnetite NPs with higher electron density.

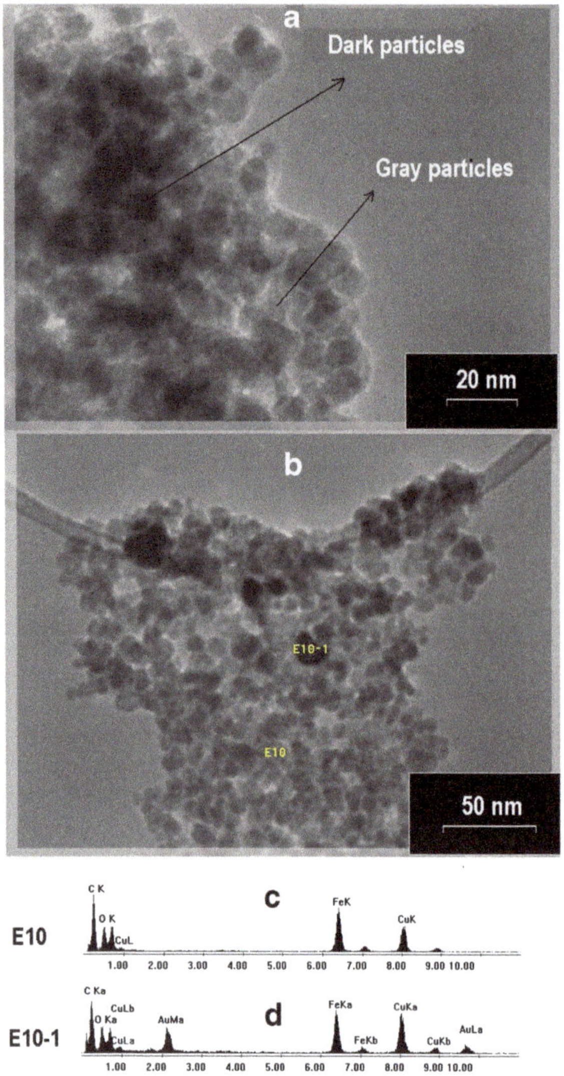

Figure 2 Transmission electron microscopic images. Of both magnetic-gold composite nanoparticles (dark particles) and bare Fe_3O_4magnetite nanoparticles (gray particles), recorded by different magnification (**a** and **b**). Lower illustrations show energy-dispersive spectroscopy patterns of these gray (**c**) or black (**d**) nanoparticles.

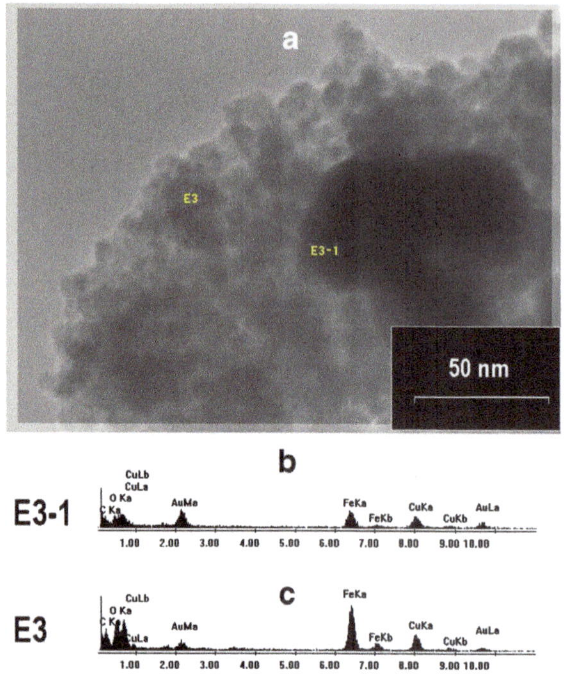

Figure 3 Another transmission electron micrograph of magnetic-gold composite nanoparticles, prepared by the semi-biosynthetic method (a). Lower illustration shows energy-dispersive spectroscopy patterns of a core/shell Fe_3O_4-Au composite nanoparticle (**b**) and other types of Fe_3O_4-Au nanocomposites, especially malformed Fe_3O_4-Au composite nanoparticles (**c**).

The chemical composition of these NPs was studied using the X-ray diffraction (XRD) method and is depicted in the lower illustration in Figure 5. The Fe_3O_4 composite MNPs were observed after the acid leaching process and verified the connection between MNPs and elemental gold (also lower illustration in Figure 5). Some negligible impurities may be remained after acid leaching, so advanced characterization methods such as high resolution TEM or elemental mapping is necessary to differentiate between the core/shell structure and other impurities. The magnetite properties of prepared Fe_3O_4-Au composite MNPs before and after acid wash treatment with HCl (1 N) were also investigated, using the vibrating sample magnetometer (VSM) method. The magnetic hysteresis curves of bare (a)

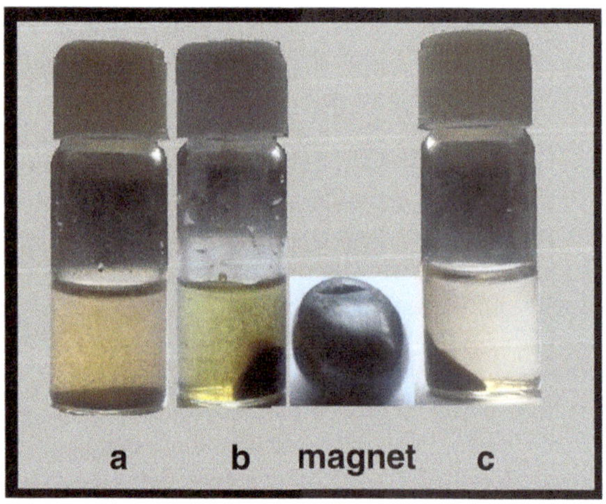

Figure 4 The images of different test tubes that contain Fe_3O_4-Au composite MNPs. In exposure with 1 N hydrochloric acid (**a**) and after separation with external magnetite field (**b**). Tube (**c**) shows the defect-free core/shell fraction of Fe_3O_4-Au nanocomposites separated by a magnet and washed three times with distilled water.

Fe_3O_4 MNPs, (b) Fe_3O_4-Au composites MNPs, and (c) Fe_3O_4 core@Au shell MNPs are demonstrated in Figure 6 and show magnetite saturation conditions for all magnetite nanomaterials. No considerable change was observed in superparamagnetism properties of this magnetite material (Fe_3O_4) either before or after the acid washing process with HCl (1 N) for 18 h (Figure 6b,c). Furthermore, we have tested HCl solutions with higher normality of 2, 3, 4, and 5 N for the dissolution of non-core/shell composite NPs, but in all experiments, non-magnetic properties were observed from the remaining nanomaterials (data not shown).

The existence of a carboxyl group on the surface of the prepared magnetite-gold nanocomposites fabricated by a semi-biosynthetic with ethanol extract of *Eucalyptus camaldulensis* has previously been demonstrated by the Fourier transform infrared spectroscopy (FTIR) method [10] and confirmed in this investigation (Figure 7a). IR peaks, centered at 3,430 and 1,720 cm^{-1}, confirmed the presence of a carboxyl group on the surface of the semi-biosynthesized Fe_3O_4-Au nanocomposites (Figure 7a). Furthermore, this carboxyl group can also be detected on the surface of Fe_3O_4-Au core/shell composites MNPs after acid wash treatment (Figure 7b) showing that carbonyl groups have not been chemically changed or degraded during acid treatment of Fe_3O_4-Au composite MNPs.

However, it should be noted that the magnetite properties of Fe_3O_4-Au composite MNPs was not optimized during this study, and further optimization process should be performed on the magnetization of the Fe_3O_4-Au composite MNPs which makes them better candidate for biomedical applications. Also, no assay was carried out to determine the biocompatibility of the above Fe_3O_4-Au core/shell composites MNPs and merit further investigation.

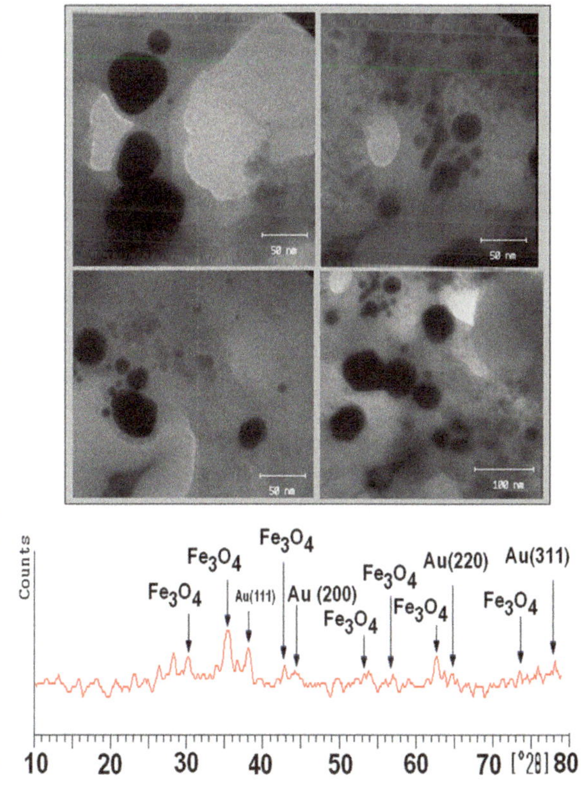

Figure 5 Representative transmission electron microscopic images. Of Fe_3O_4-Au composite nanoparticles treated with hydrochloric acid (1 N) for 18 h at room temperature and separated by an external magnetic field (upper illustration). Lower illustration shows X-ray diffraction pattern of these nanoparticles verifying the connection of magnetite nanoparticles with elemental gold.

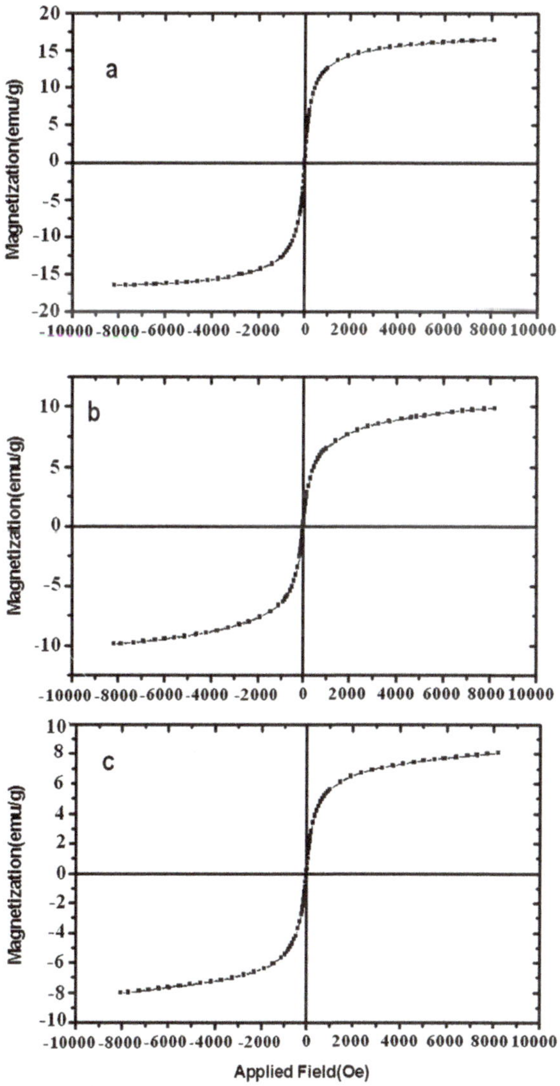

Figure 6 Magnetite properties of fabricated nanoparticles. Measured by vibrating sample magnetometer; bare Fe_3O_4 nanoparticles (**a**), Fe_3O_4-Au nanocomposites before (**b**) and after (**c**) acid wash treatment.

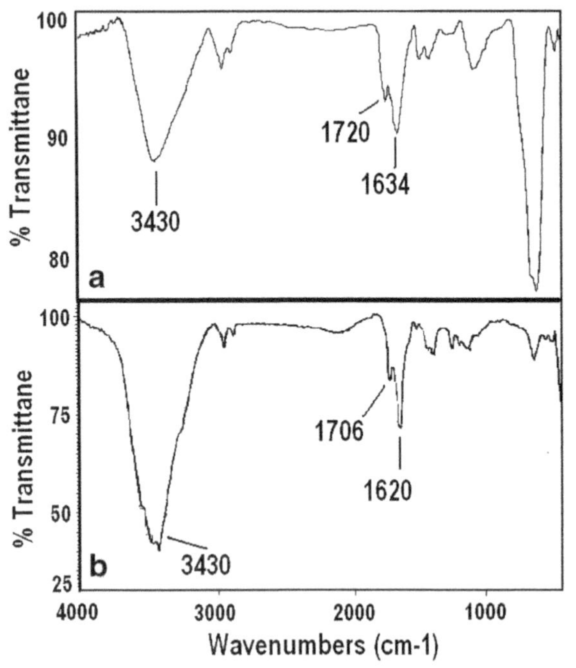

Figure 7 Infrared spectra of prepared Fe_3O_4-Au composite nanoparticles. Before (**a**) and after (**b**) acid wash treatment by HCL (1 N).

Conclusions

Different chemical or biological methods have been reported for synthesizing Fe_3O_4 MNPs and Fe_3O_4 core@Au shell MNPs [16-20]. However, when Fe_3O_4-Au core/shell nanocomposites are synthesized, some undesired NPs, such as malformed structures or incomplete core/shell Fe_3O_4-Au nanocomposites, may be formed. These types of MNPs are categorized as impurities and need to be removed from the final product. In the current study, an acid wash treatment has been developed to remove all undesired nanomaterials (Figure 1). The physiochemical properties of isolated Fe_3O_4-Au composite MNPs before and after acid wash treatment were investigated by different characterization methods, such as TEM, EDS, XRD, VSM, and FTIR. The results show that the acid wash process did not change the surface chemistry and magnetite properties of Fe3O4 core@Au shell MNPs and can be used to reduce impurities and obtain the defect-free core/shell fraction of Fe_3O_4-Au composite MNPS.

Methods

Semi-biosynthesis of magnetite Fe_3O_4-Au nanocomposites

Carboxyl-capped magnetite Fe_3O_4-Au nanocomposites were prepared at room temperature, using the previously described methods [14,21]. For this purpose, $FeCl_3{\cdot}6H_2O$ (0.487 g) and $FeSO_4{\cdot}7H_2O$ (0.25 g) were precisely weighted and dissolved in 200 ml of deionized water.

The mixture was then treated through a drop-by-drop addition of 10 ml NaOH solution (0.15 M), with mixing (200 rpm), under an argon atmosphere. The final pH of the reaction mixture was adjusted to pH12, with an excess volume of NaOH (0.15 M). The reaction mixture was then further incubated for 30 min at 80°C. The resulting black precipitate was separated by centrifugation (8,000×*g*). The separated MNPs were washed four times with deionized water. The prepared Fe_3O_4 MNPs were separated from the solution with a magnet. The reduction of Au^{3+} onto the surface of Fe_3O_4 MNPs was carried out using an ethanol extract prepared from *E. camaldulensis* leaves [14].

The *E. camaldulensis* leaves were air-dried at room temperature and then pulverized (50 g). The ethanol extract was prepared by macerating the powder (250 g) for 72 h with three changes of the solvent (1,000 ml) at room temperature. The combined solvent extracts were evaporated to yield a brownish or greenish viscous residue. A stock solution (10 mg/ml) of the plant extract was prepared in ethanol for further experiments and reserved in the refrigerator, at 4°C. Chloroauric acid was purchased from Merck, Darmstadt, Germany. MNPs (20 mg) were dispersed in 30 ml of aqueous $HAuCl_4$ (chloroauric acid) solution (1 mM) and mixed with 1 ml of an ethanol extract of *E. camaldulensis* (10 mg/ml). The resulting mixture was allowed to stand for 45 min at room temperature, and then the prepared magnetite Fe_3O_4-Au nanocomposites were separated from the colloid by a magnet. Finally, the NPs were washed three times with distilled water and subjected to the next experiments.

Acid wash treatment

HCl solution (1 N) was used to remove bare Fe_3O_4 MNPS or other malformed structures from magnetite Fe_3O_4-Au nanocomposites (Figure 1). Prepared composite NPs (100 mg) were mixed with HCl solution (600 ml, 1 N) and incubated for 18 h at room temperature. Then, the remaining composite NPs were separated from the solution by an external magnetic field. These separated NPs were washed three times with deionized water and subjected to further characterization.

Characterizations

All samples were characterized by transmission electron microscopy (model EM 208 Philips, Amsterdam, The Netherlands), EDS, XDS (Philips X'Pert Pro), and VSM. The surface chemistry of the composite MNPs before and after acid treatment was also studied by the FTIR(Nicolet Magna 550, Madison, WI, USA).

Competing interests

The authors declare that they have no competing interests.

Authors' contributions

ARS is the leader of the group and the main writer of this study's manuscript. HM and AA both participated in synthesis and characterization of nanoparticles. Also, AA and HS commented on EDS and XRD experiments. All authors read and approved the final manuscript.

Acknowledgements

This work has been supported by the Pharmaceutical Sciences Research Center, School of Pharmacy, Tehran University of Medical Sciences, Tehran, Iran. Also, the authors kindly thank the team of Scribendi Company for the language editing and proofreading assistance provided for this article.

Author details

[1]Department of Medical Biotechnology, School of Advanced Medical Technologies, Tehran University of Medical Sciences, Tehran 1417614411, Iran. [2]Department of Materials Engineering, University of Tarbiat Modares, 14115-143, Tehran, Iran. [3]Biotechnology Research Center, Faculty of Pharmacy, Tehran University of Medical Sciences, Tehran 1417614411, Iran. [4]Pharmaceutical Sciences Research Center, School of Pharmacy, Tehran University of Medical Sciences, Tehran 417614411, Iran.

References

1. Corchero, JL, Villaverde, A: Biomedical applications of distally controlled magnetic nanoparticles. Trends Biotechnol. **27**, 468–476 (2009)
2. Jain, TK, Richey, J, Strand, M, Leslie-Pelecky, DL, Flask, CA, Labhasetwar, V: Magnetic nanoparticles with dual functional properties: drug delivery and magnetic resonance imaging. Biomaterials **29**, 4012–4021 (2008)
3. Chertoka, B, Moffatb, BA, Davida, AE, Yua, F, Bergemannc, C, Rossb, BD, Yang, VC: Iron oxide nanoparticles as a drug delivery vehicle for MRI monitored magnetic targeting of brain tumors. Biomaterials **29**, 487–496 (2008)
4. Mahmoudi, M, Milani, AS, Stroeve, P: Synthesis, surface architecture and biological response of superparamagnetic iron oxide nanoparticles for application in drug delivery: a review. Int. J. Biomed. Nanosci. Nanotechnol. **1**, 164–201 (2010)
5. Mahmoudi, M, Simchi, A, Imani, M: Recent advances in surface engineering of superparamagnetic iron oxide nanoparticles for biomedical applications. J. Iran. Chem. Soc. **7**, S1–S27 (2010)
6. Lu, Q, Yao, K, Xi, D, Liu, Z, Luo, X, Ning, Q: Synthesis and characterization of composite nanoparticles comprised of gold shell and magnetic core/cores. J. Magn. Magn. Mater. **301**, 44–49 (2006)
7. Tsuji, M, Nishio, M, Jiang, P, Miyamae, N, Lim, S, Matsumoto, K, Ueyama, D, Tang, X: Role of chloride ions in the formation of Au@Ag core-shell nanocrystal structures by using a microwave-polyol method. Colloid. Surface. A. **317**, 247 (2008)
8. Rosenholm, JM, Zhang, J, Sun, W, Gu, H: Large-pore mesoporous silica-coated magnetite core-shell nanocomposites and their relevance for biomedical applications. Micropor. Mesopor. Mat. **145**(1–3), 14–20 (2011)
9. Maltas, E, Ozmen, M, Cingilli Vural, H, Yildiz, S, Ersoz, M: Immobilization of albumin on magnetite nanoparticles. Mater. Lett. **65**(23–24), 3499–3501 (2011)
10. Sun, Q, Reddy, BV, Marquez, M, Jena, P, Gonzalez, C, Wang, Q: Theoretical study on gold-coated iron oxide nanostructure: Magnetism and bioselectivity for amino acids. J. Phys. Chem. C **111**, 4159–4163 (2007)
11. Cho, SJ, Idrobo, JC, Olamit, J, Liu, K, Browning, ND, Kauzlarich, SM: Growth mechanisms and oxidation resistance of gold-coated iron nanoparticles. Chem. Mater. **17**, 3181–3186 (2005)
12. Liu, XG, Geng, DY, Liang, JM, Zhang, ZD: Magnetic stability of Al2O3-coated FCC-Co nanocapsules. J. Alloy Compound. **465**, 8–14 (2008)
13. Tartaj, P, Morales, MP, Veintemillas-Verdaguer, S, González-Carreño, T, Serna, CJ: The preparation of magnetic nanoparticles for applications in biomedicine. J. Phys. D: Appl. Phys. **36**, 182–197 (2003)
14. Haratifar, E, Shahverdi, HR, Shakibaie, M, MollazadehMoghaddam, K, Amini, M, Montazeri, H, Shahverdi, AR: Semi-biosynthesis of magnetite-gold composite nanoparticles using an ethanol extract of *Eucalyptus camaldulensis* and study of the surface chemistry. J. Nanomater. **2009**, 5 (2009)
15. Yan, A, Liu, X, Qiu, G, Wu, H, Yi, R, Zhang, N, Xu, J: Solvothermal, synthesis and characterization of size-controlled Fe3O4 nanoparticles. J. Alloy Compound. **458**, 487–491 (2008)
16. Li, Y, Xu, G, Zhu, YL, Ma, XL, Cheng, HM: SnO2/In2O3 one-dimensional nano-core-shell structures: synthesis, characterization and photoluminescence properties. Solid State Commun. **142**, 441–444 (2007)
17. Lu, H, Salabas, EL, Schuth, F: Magnetic nanoparticles: synthesis, protection, functionalization, and application. Angew. Chem. **46**, 1222–1244 (2007)
18. Qiua, JD, Penga, HP, Lianga, RP, Xiab, XH: Facile preparation of magnetic core–shell Fe3O4@Au nanoparticle/myoglobin biofilm for direct Biosensors and Bioelectronics. Biosense. Bioelectron. **25**, 1447–1453 (2010)
19. Pradhan, A, Jones, RC, Caruntu, D, O'Connor, CJ, Tarr, MA: Gold–magnetite nanocomposite materials formed via sonochemical methods. Ultrason. Sonochem. **15**, 891–897 (2008)
20. Lima, J, Tiltona, RD, Eggemanc, A, Majetichc, SA: Design and synthesis of plasmonic magnetic nanoparticles. J. Magn. Magn. Mater. **311**, 78–83 (2007)
21. Martínez-Mera, I, Espinosa-Pesqueira, ME, Pérez-Hernández, R, Arenas-Alatorre, J: Synthesis of magnetite (Fe_3O_4) nanoparticles without surfactants at room temperature. Mater. Lett. **61**, 4447–4451 (2007)

12

In situ oxidative polymerization of aniline in the presence of manganese dioxide and preparation of polyaniline/MnO_2 nanocomposite

Sajjad Sedaghat* and Fariba Golbaz

Abstract

The method used in this paper is *in situ* process for the synthesis of polyaniline (PANI)/manganese dioxide. Polymerization of aniline was carried out in a one-pot reaction vessel, and MnO_2 particles were influenced by the PANI pores to yield a PANI/MnO_2 nanocomposite. Morphology analysis showed soft and flat surfaces of the PANI/MnO_2 nanocomposite by scanning electron microscopy, and Fourier transform infrared spectroscopy revealed polyaniline being intercalated between manganese oxide layers. Findings showed that the MnO_2 particles should be doped on the network of polyaniline; therefore, the morphology of the product is in the same line with the results.

Keywords: Polyaniline; Nanocomposite; MnO_2; *In situ* polymerization

Background

Between metal oxides, MnO_2 shows a great potential as an alternative material because it is cheap. It is available in abundance and environmentally friendly. Manganese oxides have long been known as materials of technological importance for catalytic and electrochemical applications. However, in comparison with another oxidation, MnO_2 exhibits a much lower electrochemical capacitance [1-5]. In order to improve the performance of MnO_2, we can prepare nanocomposites of MnO_2 with conducting polymers [6,7]. Among these polymers, polyaniline (PANI)/MnO_2 nanocomposites have more applications because the capacitance of these nanocomposites was greatly improved compared with that of pure MnO_2 due to the high conductivity of PANI and the correlation effect between PANI and MnO_2 [8,9]. Synthesis and studies of nanocomposites of PANI and MnO_2 have attracted much attention due to low cost and environment friendliness of the active materials and also due to a good coating layer to restrain MnO_2 from dissolution in acidic environment [10]. It is reported that the nanocomposite obtained through intercalation of PANI into layers of MnO_2 showed an enhanced specific capacitance due to synergistic effects [11]. Polyaniline is one of the most promising conducting polymers due to its ease of preparation and good environmental stability; it is soluble in various solvents and exhibits a number of interesting properties such as chemical sensitivity and electrochromism [12-14]. The chemistry of PANI allows for several routes of synthesis and processing. It is easy to synthesize via simple chemical or electrochemical oxidation methods of aniline. PANI has a wide range of applications such as anticorrosion coating, batteries, and sensor [15,16]. *In situ* chemical polymerization techniques of PANI followed by dipping the partially oxidized PANI in a suspension of the metal oxide is reported for the synthesis of nanocomposites. It is known that MnO_2 can initiate the polymerization of aniline and simultaneously is reduced into soluble Mn^{2+} ions [17,18].

* Correspondence: sajjadsedaghat@yahoo.com
Department of Chemistry, Shahre-Qods Branch, Islamic Azad University, Tehran, Iran

Results and discussion

SEM measurements

Figure 1 shows the scanning electron microscopy (SEM) image of pure polyaniline. As can be seen from this picture, the pores of PANI are formed and the smooth surface is determined. It shows that the blank PANI is constructed with rodlike aggregates. PANI chains are prepared in acidic media and exist in the form of

Figure 1 SEM image of pure PANI.

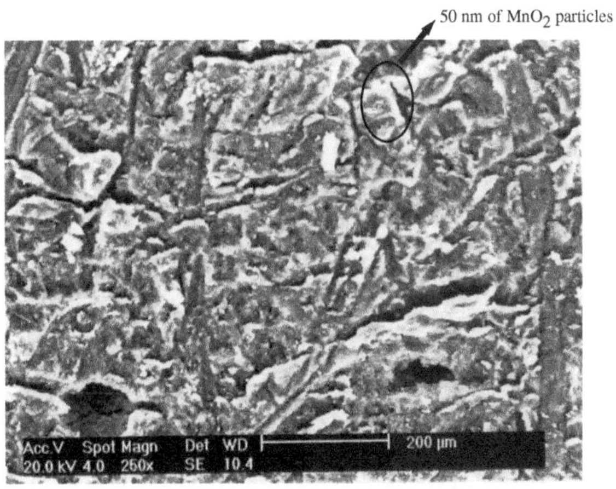

Figure 2 SEM image of PANI/MnO_2 nanocomposite.

polycations. The strong interactions between charged PANI chains [19] facilitate the formation of nanorods.

Findings showed that PANI/MnO_2 nanocomposite comprises platelike particles, which are about 50 nm in thickness (Figure 2). The morphology of the product after polyaniline being intercalated between manganese oxide layers is maintained. In addition, the incorporation of MnO_2 nanoparticles would effectively avoid the entanglement of PANI chains due to repulsion between charged MnO_2 nanoparticles in an acidic media. The SEM image of PANI/MnO_2 nanocomposite reveals that the nanocomposite is composed of many nanoparticles which are connected to each other resulting in a porous structure.

FT-IR characterization

Structural information of the synthesized PANI/MnO_2 nanocomposite was identified by Fourier transform infrared (FT-IR) spectroscopy. Figure 3 shows the FT-IR spectra of PANI. It can be seen in the spectra the range of 1,600 to 1,000 cm^{-1} which comes from the vibrational and rotational bands of the functional groups in PANI. The bands at 1,568 and 1,495 cm^{-1} can be assigned to C=C stretching of quinoid and benzenoid rings; the C-N stretchings for the benzenoid unit appear at 1,297 and 1,239 cm^{-1}, respectively; the peak at 1,144 cm^{-1} can be attributed to the in-plane bending vibration of aromatic C-H. The weak band at 3,320 cm^{-1} is attributed to N-H stretching.

The FT-IR spectra obtained from PANI/MnO_2 nanocomposite showed characteristic peaks of PANI (Figure 4). For example, vibrational peaks centered at 1,565 and 1,468 cm^{-1} are attributed to the stretching peaks of quinoid and benzenoid deformations of PANI. We also observed C-N stretching peaks at 1,238 and 1,300 cm^{-1} and C-H bending of protonated PANI at 1,124 cm^{-1}. The band in the regions over 400 cm^{-1} can be assigned to Mn-O stretching vibrations [20]. FT-IR spectra of the PANI/MnO_2 nanocomposite are similar to those of PANI, but the bands' characteristic of polymer backbone at 1,500 and 1,600 $cm^{-1}$are shifted to higher values after annealing, indicating deprotonation. The peak at 1,195 cm^{-1} is suppressed after annealing to a greater extent for PANI compared to that for the

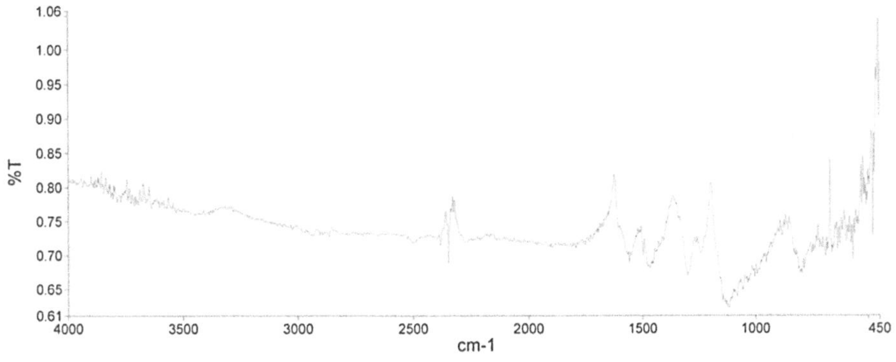

Figure 3 FT-IR spectra of pure PANI.

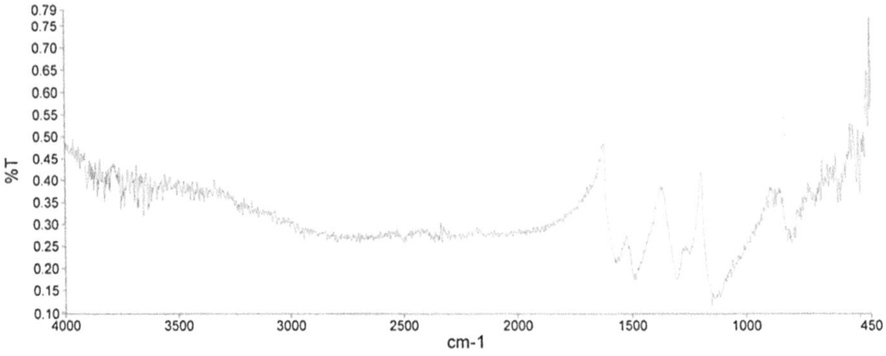

Figure 4 FT-IR spectra of PANI/MnO_2 nanocomposite.

nanocomposites, indicating a higher extent of deprotonation in pure PANI compared to nanocomposites. The results of FT-IR spectra confirm the presence of both components in the nanocomposite.

Conclusions

In this work, *in situ* polymerization of aniline in the presence of MnO_2 has been studied. Polyaniline intercalated layered manganese oxide nanocomposite shows great enhancement of morphology than the pristine PANI, which indicates that intercalation of PANI into the layered MnO_2. From the SEM and FT-IR results, we obtained that the nanocomposite is synthesized and that the information are in the same line with each other. We expect that this method can be used to produce other polymer/metal oxide nanocomposites.

Methods

Materials

Aniline was purified by distillation; hydrochloric acid solution (1 M) and manganese oxide were all purchased from the Merck (Darmstadt, Germany) and used as received.

Instrumentation

The morphology of the samples was studied using a scanning electron microscope XI-300 model Philips (Amsterdam, The Netherlands). Also, FT-IR spectra were recorded by PerkinElmer Spectrum 100 (Branford, CT, USA). KBr pellets were used to prepare samples, with a sweeping range of 400 to 4,000 cm^{-1} at a 4-cm^{-1} resolution.

Synthesis of PANI/MnO_2 nanocomposite

First, 30 ml of 1 M HCl solution was prepared, and 0.5 g MnO_2 and 0.46-ml monomer of aniline were added to the solution simultaneously. As soon as they were added, it obtained a dark color. After 30 min of sonication in an ice bath, the temperature of the solution decreased to 0°C until polymerization was stopped. After filtering, the precipitates were washed with deionized water and ethanol (96%). The product was then dried at 60°C in an oven for 1 h.

Competing interests

Both authors declare that they have no competing interests.

Acknowledgments

The authors thank Islamic Azad University, Shahre-Qods Branch for their support. References.

References

1. Devaraj, S, Munichandraiah, N: J. Phys. Chem. C. **112**, 4406 (2008)
2. Toupin, M, Brousse, T, Belanger, D: Chem. Mater. **14**, 3946 (2002)
3. Xu, MW, Kong, LB, Zhou, W, Li, HL: J. Phys. Chem. C. **111**, 19141 (2007)
4. Athouel, L, Moser, F, Dugas, R, Crosnier, O, Belanger, D, Brousse, T: J. Phys. Chem. **C112**, 7270 (2008)
5. Omomo, Y, Sasaki, T, Wang, LZ, Watanabe, M: J. Am. Chem. Soc. **125**, 3568 (2003)
6. Wang, YG, Wu, W, Cheng, L, He, P, Wang, CX, Xia, YY: Adv. Mater. **20**, 2166 (2008)
7. Zang, JF, Bao, SJ, Li, CM, Bian, HJ, Bao, XQL, Sun, CQ, Guo, J, Lian, K: J. Phys. Chem. **C112**, 14843 (2008)
8. Zhang, X, Ji, LY, Zhang, SC, Yang, WS: J. Power. Sources **173**, 1017 (2007)
9. Liu, FJ: J. Power. Sources **182**, 383 (2008)
10. Yuan, C, Su, L, Gao, B, Zhang, X: Electrochim. Acta **53**, 7039–7047 (2008)
11. Zhang, X, Ji, L, Zhang, S, Yang, W, Power, J: Sources **173**, 1017–1023 (2007)
12. Petty, MC, Bryce, MR, Bloor, D: An Introduction to Molecular Electronics. Edward Arnold, London (1995)
13. Scrosati, B: Applications of Electroactive polymers. Chapman and Hall, London (1993)
14. Alcacer, L: Conducting polymers- Special Applications. Reidel, Dordrecht (1989)
15. Huang, J: Pure Appl. Chem. **78**, 15 (2006)
16. Kulkarni, M, Viswanath, A, Aiyer, R, Khanna, P: J. Polym. Sci., Part B: Polym. Phys. **43**, 2161 (2005)
17. Pan, LJ, Pu, L, Shi, Y, Song, SY, Zhang, ZXUR, Zhang, YD: Adv. Mater. **19**, 461 (2007)
18. Gemeay, AH, EL-Sharkawy, RG, Mansour, IA, Zaki, AB: J. Colloid Interface Sci. **308**, 385 (2007)
19. Pan, LJ, Pu, L, Shi, Y, Sun, T, Zhang, R, Zheng, YO: Adv. Funct. Mater. **16**, 1279–1288 (2006)
20. Wang, YG, Wu, W, Cheng, L, He, P, Wang, CX, Xia, YY: Adv. Mater. **20**, 2166 (2008)

Magnetic properties of Cr-substituted Co-ferrite nanoparticles synthesized by citrate-gel autocombustion method

M Raghasudha[1], D Ravinder[2*] and P Veerasomaiah[3]

Abstract

A series of Co-Cr nanoferrites having the chemical formula $CoCr_xFe_{2-x}O_4$ (x = 0.0, 0.1, 0.3, 0.5, 0.7, 0.9, and 1.0) were synthesized by citrate-gel autocombustion method at a very low temperature (180°C). The X-ray diffraction analysis of as-synthesized powders and sintered powders has confirmed the formation of single-phase cubic spinel structure. The average particle size of the synthesized ferrites was 6 to 12 nm. Magnetic susceptibility measurements using Faraday magnetic susceptibility balance showed the paramagnetic nature of the ferrites. Magnetic properties of Co-Cr nanoferrites were measured using a vibrating sample magnetometer at room temperature in the applied field of 15 kOe. The saturation magnetization decreased from 33.84 to 13.83 emu/g with increase in Cr^{3+} concentrations, indicating the fact that the lesser magnetic Cr^{3+} ions substitute Fe^{3+} ions in the octahedral sublattice of the ferrite. With improvement in the magnetic properties, the synthesized nanoferrites become soft magnetic materials. Such materials are useful in transformer and motor cores to minimize the energy dissipation with the alternating fields associated with AC electrical applications. The coercivity of pure $CoFe_2O_4$ was larger than that of the Cr-doped cobalt ferrites.

Keywords: Co-Cr nanoferrites; Citrate-gel method; Magnetic susceptibility; Saturation magnetization; Vibrating sample magnetometer; Coercivity

Background

Nanoparticles are very important and have a distinct property, that is, they exhibit larger surface-area-to-volume ratio. This increase in surface-area-to-volume ratio leads to the change of the properties of nanoparticles better than that of the bulk particles [1]. Transition metal oxide nanoparticles represent a broad class of materials that have been investigated extensively due to their interesting catalytic, electronic, and magnetic properties relative to those of the bulk counterparts, and the wide scope of their potential applications [2]. Among these materials, ferrites have attracted immense attention of the scientific community because of their novel properties and technological applications especially when the size of the particles approaches to nanometer scale [3]. As magnetic materials, nano-sized ferrites cannot be replaced by any other magnetic material because they are relatively inexpensive, stable, and have a wide range of technological applications [4]. The spinel ferrites have remarkable magnetic and electrical properties. Among them, $CoFe_2O_4$ is interesting because of its perfect chemical properties, thermal stability, high electrical resistivity, and excellent magnetic properties [5]. Nanocrystalline $CoFe_2O_4$ with such properties have potential applications in high frequency devices, memory cores, recording media, and in biomedical field [6].

The properties of ferrite materials are sensitive to the grain size and also strongly influenced by the distribution of metallic ions among crystallographic crystal lattice sites. These in turn are sensitive to the method used to prepare those materials [7]. Various methods used to prepare nano-sized ferrites are co-precipitation [8], micro-emulsion procedures [9], microwave plasma [10], mechanical milling [11], and sol-gel autocombustion method [12]. The sol–gel auto combustion method in

* Correspondence: ravindergupta28@rediffmail.com
[2]Department of Physics, Nizam College, Basheerbagh Osmania University, Hyderabad 500001 Andhra Pradesh, India
Full list of author information is available at the end of the article

particular is one of the most useful and attractive techniques for the synthesis of nano-sized ferrite materials. This is because of its advantages such as good stoichiometric control and the production of ultrafine particles with a narrow size distribution in a relatively short processing time at a very low temperature.

Substitution of other metals for Fe in $CoFe_2O_4$ has been proposed as a method to tailor the magnetic and magneto-elastic properties for sensor applications [13]. Chromium ions (Cr^{3+}) with antiferromagnetic nature are known for achieving good control over magnetic parameters in developing technologically important materials. The substitution of Cr^{3+} ions for Fe^{3+} ions will alter magnetic properties marked by similar to that of nonmagnetic substitution.

The present work reports the synthesis of nano-sized chromium-substituted cobalt ferrites by citrate-gel autocombustion method and consequent changes on their magnetic properties. Co-Cr nanoferrites were synthesized with particle size ranging from 6 to 12 nm, which will have a great effect on its magnetic properties. Magnetic properties of these ferrites were reported here.

Results and discussions

X-ray diffraction studies

X-ray diffraction (XRD) was performed on the as-synthesized powders as well as on the powders calcined at 500°C, and the XRD patterns of all the samples were shown in Figure 1a,b. Obtained XRD patterns and crystalline phases were identified by comparison with reference data from the ICSD card no. 22–1086 for cobalt ferrites ($CoFe_2O_4$) and have been indexed. It confirms the formation of a homogeneous well-defined single-phase cubic spinel structure without any impurity peak belonging to the space group Fd3m (ICSD reference). The strongest reflection comes from the (311) plane that indicates the spinel phase.

From the XRD patterns of the as-synthesized powders, it is clear that the as-burnt powder is also in single phase with a spinel structure which indicates that the ferrite can be directly formed after the autocombustion of the gel without heat treatment. The broad peaks in the XRD patterns indicate a fine particle nature of the particles.

Figure 1a,b clearly shows that the positions of the reflection peaks for as-burnt powders are almost identical to the corresponding peaks for the calcined material. This implies that the basic structure of the nanoparticles is essentially same as that of the bulk material. The average particle sizes of the as-synthesized Co-Cr ferrites and calcined Co-Cr ferrites for different compositions were calculated from Scherrer formula [14] using the maximum intensity peak (311) and were shown in Table 1.

From the table, it is clear that nano-sized Co-Cr ferrite powders can be directly synthesized by Citrate-gel auto combustion method. After calcination at 500°C for 4 h, the reflection peaks of the samples become sharper and narrower, indicating the improvement of crystallinity. Comparing Figure 1a,b it was found that there were no differences between the same compositional samples, except for the relative intensity. Similar behavior was reported in by Toksha et al. [15].

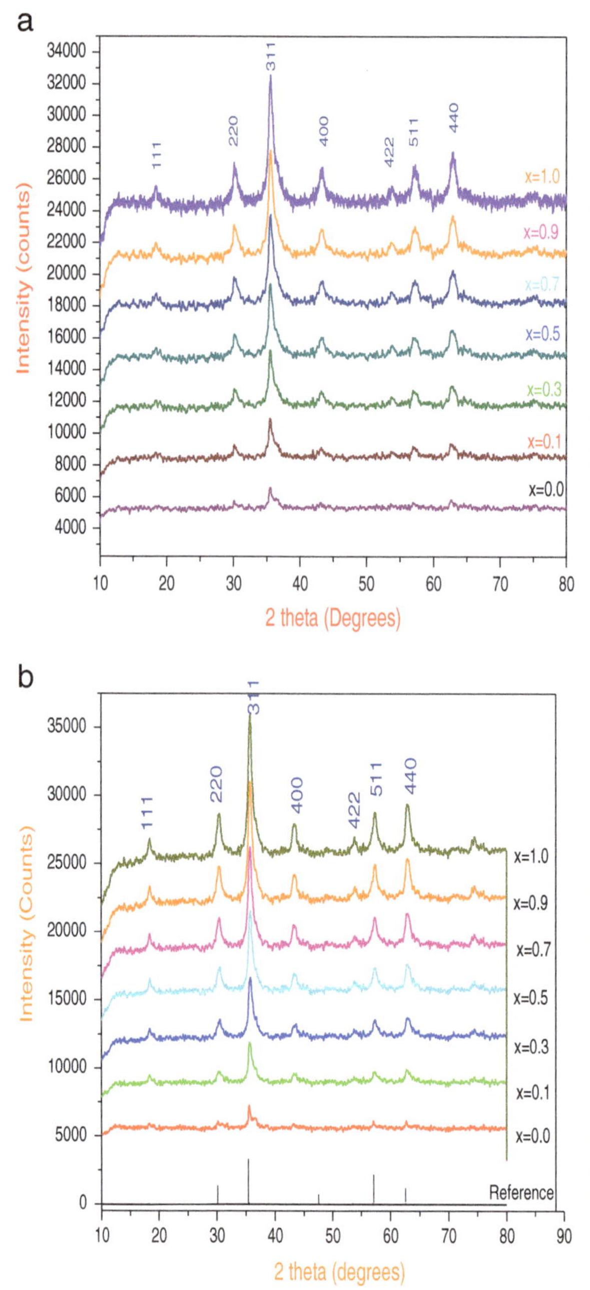

Figure 1 XRD patterns of the Co-Cr ferrites. (a) XRD patterns of the as-synthesized Co-Cr ferrites. **(b)** XRD patterns of the Co-Cr ferrites calcined at 500°C.

Table 1 Crystallite size of as-synthesized and heat-treated $CoCr_xFe_{2-x}O_4$ nanoferrites

Composition	Crystallite size of as-synthesized samples (nm)	Crystallite size of heat-treated samples (nm)
$CoFe_2O_4$	16.5	6.5
$CoCr_{0.1}Fe_{1.9}O_4$	9.95	12.14
$CoCr_{0.3}Fe_{1.7}O_4$	9.29	12.79
$CoCr_{0.5}Fe_{1.5}O_4$	9.39	11.86
$CoCr_{0.7}Fe_{1.3}O_4$	9.06	10.55
$CoCr_{0.9}Fe_{1.1}O_4$	10.01	10.69
$CoCrFeO_4$	10.98	12.32

x = 0.0, 0.1, 0.3, 0.5, 0.7, 0.9, and 1.0.

Magnetic susceptibility using faradays balance

The magnetic susceptibility is a dimensionless proportionality constant that indicates the degree of magnetization of a material in response to an applied magnetic field. Magnetic susceptibility of heat-treated Co-Cr nanoferrites at room temperature is measured using very sensitive instrument known as a Faraday magnetic susceptibility millibalance. In the Faraday balance, the sample is placed in between the magnetic poles, and the force depends only on the total mass of the material present. The force is measured as a weight change, using a magnetic balance. Working of the Faraday balance is clearly described by Carlin [16]. To calibrate the field gradient (the force experienced by a standard sample), mercury tetra thiocyanato cobaltate was measured which is known to have a gram susceptibility of 16.44 × 10^{-6} cgs units at 20°C [17]. The diamagnetic materials are weakly repelled by an external magnetic field, resulting in a negative reading. Paramagnetic materials are attracted to an external magnetic field and give a positive reading.

The gram magnetic susceptibility (χ_g) of synthesized samples may be calculated from the following equation.

$$\chi_g = \left(\chi_g\right)_s \left(\frac{w_s}{\Delta w_s}\right)\left(\frac{\Delta w_c}{w_c}\right),$$

where χ_g is the gram susceptibility of compound,

$(\chi_g)_s$ is gram susceptibility of 16.44 × 10^{-6} cgs units at 20°C, and

W_s is the weight of the standard sample in the absence of magnetic field,

W_c denotes the weight of the compound in the absence of magnetic field,

Δw_s the change in weight of the standard sample after the applied magnetic field, and

Δw_c the change in weight of the compound after the applied magnetic field.

The molar magnetic susceptibility (χ_m) values were then calculated from the gram magnetic susceptibility values using the following equation and were tabulated in Table 2.

$$\chi_m = \chi_g(Molecularweight)$$

The effective magnetic moment of the samples can be calculated from the gram magnetic susceptibility using the following equation and were tabulated in Table 2.

$$\mu_{eff} = 2.83\left(\chi_m.T\right)^{1/2BM}$$

where μ_{eff} stands for effective magnetic moment in Bohr magnetons (BM),

χ_m is the molar magnetic susceptibility, and

T is the absolute temperature.

It is reported that the χ_m value ranges between 10^{-5} and 10^{-6} for the paramagnetic substances and between -10^{-5} and -10^{-6} for the diamagnetic substances.

From the Table 2 it is clear that the molar susceptibility values of Co-Cr ferrites of various compositions are of the order of 10^{-6}, indicating the paramagnetic nature of Co-Cr ferrite system. Further, the effective magnetic moment calculated for various samples shows that μ_{eff} decreases with the increase in the concentration of dopant Cr^{3+}. This indicates that $CoFe_2O_4$ is super paramagnetic in nature. Its behavior has changed to paramagnetic by Cr substitution (due to antiferromagnetic nature of Cr). This behavior was confirmed by studying hysteresis loops from VSM measurements.

Magnetic properties using VSM

The magnetic measurements of various compositions of heat-treated Co-Cr nanoferrites were made using vibrating sample magnetometer at room temperature in 15 kOe. Figure 2 shows the magnetic hysteresis loops for all Co-Cr ferrite samples (calcined at 500°C) obtained from VSM measurements. Hysteresis loop gives the relation between the magnetization (M) and the applied field (H).

Magnetic parameters extracted from the hysteresis loops are saturation magnetization (M_s; maximum value of magnetization), remanent magnetization (M_r;

Table 2 Molar susceptibility (χ_m) and effective magnetic moment (μ_{eff}) of $CoCr_xFe_{2-x}O_4$

Composition	χ_m (cgs units)	μ_{eff} (BM)
$CoFe_2O_4$	1576368.35 × 10^{-6}	62.053
$CoCr_{0.1}Fe_{1.9}O_4$	718362.59 × 10^{-6}	41.889
$CoCr_{0.3}Fe_{1.7}O_4$	647697.77 × 10^{-6}	39.775
$CoCr_{0.5}Fe_{1.5}O_4$	606994.92 × 10^{-6}	38.505
$CoCr_{0.7}Fe_{1.3}O_4$	625344.65 × 10^{-6}	39.082
$CoCr_{0.9}Fe_{1.1}O_4$	592982.68 × 10^{-6}	38.058
$CoCrFeO_4$	675140.33 × 10^{-6}	40.607

x = 0.0, 0.1, 0.3, 0.5, 0.7, 0.9, and 1.0.

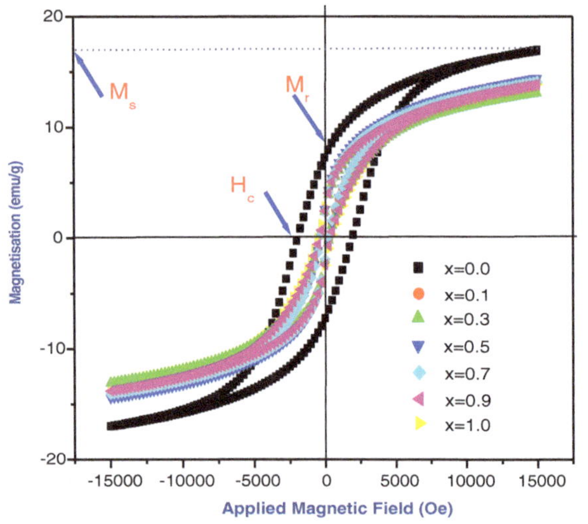

Figure 2 Magnetic hysteresis loops for all Co-Cr ferrite samples.

magnetization at zero field), and coercivity (H_C; magnetic field required to reduce the magnetization of that material to zero after the magnetization of the sample has been driven to saturation). These magnetic parameters are used to characterize the magnetic properties of materials.

The magnetic parameters of all the samples of Co-Cr ferrites were calculated from the individual *M-H* loops and were tabulated in Table 3. The *M-H* loops of $CoFe_2O_4$ and $CoCrFeO_4$ were shown in Figure 3.

From Table 3, it is clear that M_s at room temperature decreased from 33.8448 to 13.8347 emu/g with the increase in Cr^{3+} ion concentration in the Co-Cr ferrites. This decrease in M_s value may be due to the fact that the Fe^{3+} (magnetic moment 5 μB) are replaced by lesser magnetic Cr^{3+} ions (magnetic moment 3 μB) in the octahedral (B) sites of the ferrite sublattice. As the concentration of the Cr^{3+} ion increases, it decreases the $Fe^{3+}(B)/Fe^{2+}(A)$ ratio, that is, A-B super exchange interaction decreases [18].

H_c which is a measure of the magnetic field strength required for overcoming anisotropy to flip the magnetic moments is clearly affected by the chromium substitution. It can be seen from Table 3 that the value of H_c decreased from 1,954.66 Oe to 298.57 kOe up to the ferrite sample with Cr composition x = 0.7 and then increased for x = 0.9 and 1.0. The decrease in coercivity with increase in Cr^{3+} concentrations may be due to the decrease in anisotropy field which in turn decreases the domain wall energy [19]. The increase in coercivity for ferrites with composition x = 0.9 and 1.0 might be due to the increase in magnetic crystalline anisotropy. The coercivity of the nanoferrites has a contribution from their finite size namely surface anisotropy [20]. Normally, for a given composition of ferrite, when the crystal size is less, the coercivity will be more as the surface becomes much more dominant [21]. Higher Coercivity of $CoFe_2O_4$ in the present case may also be related to the particle size effect.

From the values of H_c and M_s, anisotropy constant (*K*) can be calculated [22] and tabulated in Table 4. It is seen that the value of *K* decreases with increase of Cr^{3+} concentrations.

The magnetic moment per formula unit in Bohr magneton (μ_B) was calculated [23] and tabulated in Table 4. Magnetic moment values were found to decrease with increase in Cr^{3+} concentration which is attributed to greater occupancy of Cr^{3+} at B sites.

From all these results, it is clear that the increase of Cr concentration decreases the magnetization, and the material is being converted into a soft magnetic material. Hysteresis curves also show that $CoFe_2O_4$ has large area inside the hysteresis loop, where as $CoCrFeO_4$ has a smaller area inside the hysteresis loop. This indicates the fact that the increase in Cr substitution has made the material magnetically soft.

The present study confirms that soft magnetic Co-Cr nanoferrite materials that were synthesized can be easily magnetized and demagnetized. Such materials are desirable for transformer and motor cores to minimize the energy dissipation with the alternating fields associated with AC electrical applications.

Table 3 Magnetic parameters from Hysteresis loops

Composition	Saturation magnetization (emu/g)	Remanent magnetization (emu/g)	Coercivity (kOe)	Remanence ratio (M_r/M_s)
$CoFe_2O_4$	33.8448	14.540	1,954.66	0.4296
$CoCr_{0.1}Fe_{1.9}O_4$	15.3200	2.912	381.03	0.1900
$CoCr_{0.3}Fe_{1.7}O_4$	14.9600	2.805	355.04	0.1875
$CoCr_{0.5}Fe_{1.5}O_4$	14.4232	2.617	319.25	0.1814
$CoCr_{0.7}Fe_{1.3}O_4$	14.0734	2.391	298.57	0.1698
$CoCr_{0.9}Fe_{1.1}O_4$	13.8651	2.423	389.73	0.1747
$CoCrFeO_4$	13.8347	2.784	500.22	0.2012

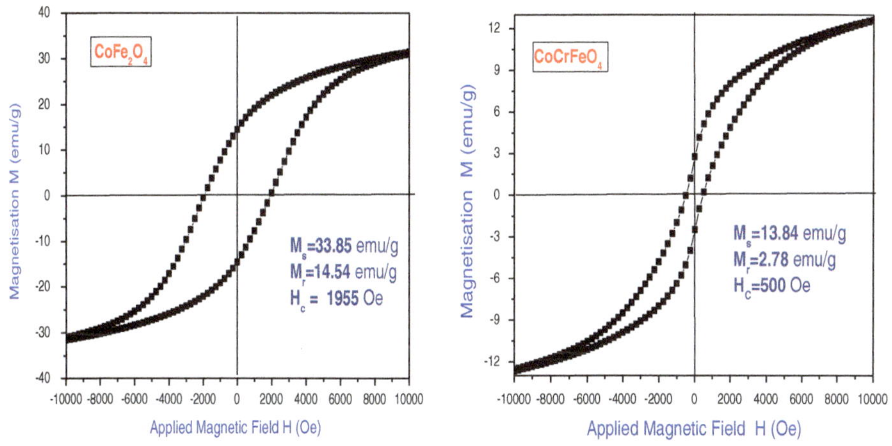

Figure 3 *M-H* **curves for $CoCr_xFe_{2-x}O_4$ (x = 0.0 and 1.0) nano particles synthesized by Citrate-gel technique.**

Conclusions

Based on the discussion above, we have drawn the following conclusions:

- A series of $CoCr_xFe_{2-x}O_4$ (x = 0.0, 0.1, 0.3, 0.5, 0.7, 0.9, and 1.0) nanoparticles have been successfully synthesized by citrate-gel autocombustion technique at a low sintering temperature.
- The XRD pattern confirms the cubic spinel structure for all compositions of both as-synthesized powders and heat-treated ferrite powders.
- The crystallite size of the synthesized Co-Cr ferrites was from 6 to 12 nm which is a novelty of this paper.
- Magnetic susceptibility measurements indicated the formation of super paramagnetic nano-sized ferrites and with increase in Cr composition superparamagnetic behavior has changed to paramagnetic behavior (soft magnetic materials).
- The incorporation of Cr^{+3} ions in Co ferrites resulted in decrease of saturation magnetization, coercivity, and magnetic moment because the replacement of Fe^{+3} by Cr^{+3} ions weakens the sublattice interaction and lowers the magnetic moments of the unit cells.
- Magnetic measurements from VSM confirmed the formation of magnetically soft materials.
- These characteristics of ferrites are desirable for their utility in transformer and motor cores to minimize the energy dissipation with the alternating fields associated with AC electrical applications.

Table 4 Anisotropy constant and Magnetic moment values for various Co-Cr ferrites

Composition	Anisotropy constant (erg/Oe)	Magnetic moment (BM)
$CoFe_2O_4$	67,505.2	1.4339
$CoCr_{0.1}Fe_{1.9}O_4$	5,956.97	0.6425
$CoCr_{0.3}Fe_{1.7}O_4$	5,419.79	0.6253
$CoCr_{0.5}Fe_{1.5}O_4$	4,698.58	0.6009
$CoCr_{0.7}Fe_{1.3}O_4$	4,287.65	0.5844
$CoCr_{0.9}Fe_{1.1}O_4$	5,513.89	0.5738
$CoCrFeO_4$	7,061.63	0.5716

Methods

Synthesis

The starting materials for the synthesis of $CoCr_xFe_{2-x}O_4$ (x = 0.0, 0.1, 0.3, 0.5, 0.7, 0.9 and 1.0) using citrate-gel autocombustion method were cobalt nitrate, ferric nitrate, chromium nitrate, citric acid, ammonia (99% pure; S D Fine-Chem Limited, Mumbai, India). The synthesis of Co-Cr nanoferrites by citrate-gel autocombustion technique was clearly reported in our earlier publication [14]. The synthesized powders were ground in agate mortar and pestle and calcined in a muffle furnace at 500°C for 4 h to obtain spinel phase.

Characterization

The structural characterization of the as-synthesized and heat-treated powders was carried out by Philips X-ray diffractometer (model 3710; Koninklijke Philips N.V., The Netherlands) using Cu $K\alpha$ radiation (λ = 1.5405 Å) at room temperature by continuous scanning in the range of 2 to 85 $\theta°$ to analyze the phase and crystallite size.

The magnetic susceptibility of the different compositions of Co-Cr ferrites at room temperature was studied by Faraday magnetic susceptibility millibalance (model-7550).

The magnetic properties of synthesized Co-Cr nanoferrites were studied using vibrating sample magnetometer

(VSM) (Lakeshore 665, Lake Shore Cryotronics, Inc., Westerville, USA) at room temperature in the applied field of 15 kOe.

Competing interests
The authors declare that they have no competing interests.

Authors' contributions
All the authors were involved with the whole research work presented here. The author MR has synthesized the nano particles and performed various experiments. DR has helped in interpreting the data and communicating the article to the journal. PV has provided required facilities for measurements and discussed the obtained data. All the authors read and approved the final manuscript.

Authors' information
M. Ragashuda, M.Sc., M. Phil. Associate Professor. Department of Chemistry, Jaya Prakash Narayan College of Engineering, Mahabubnagar, Andhrapradesh, India. Prof. D. Ravinder, Ph.D. Professor and Head, Department of Physics, Nizam College, Basheerbagh, Osmania University, Hyderabad, Andhrapradesh, India. Prof. P. Veerasomaiah, Professor of Chemistry, Osmania University, Hyderabad, Andhrapradesh, India.

Acknowledgments
MRS is thankful to chairman K.S. Ravi Kumar, Jayaprakash Narayan College of Engineering for his support in establishing the lab for the synthesis of the samples. DR is grateful to Prof. T.L.N. Swamy, principal of Nizam College, for his encouragement to carry out this research work. The authors are thankful to Prof. C. Gyana Kumari, head of the Department of Chemistry, Osmania University, Hyderabad for her encouragement in carrying out research activities. The authors are also thankful to Prof. M. Vithal, Department of Chemistry, Osmania University for his support in the characterization of the samples.

Author details
[1]Department of Chemistry, Jayaprakash Narayan College of Engineering, Mahabubnagar 509001 Andhra Pradesh, India. [2]Department of Physics, Nizam College, Basheerbagh Osmania University, Hyderabad 500001 Andhra Pradesh, India. [3]Department of Chemistry, Osmania University, Hyderabad 500007 Andhra Pradesh, India.

References
1. Vanaja, M, Gnanajobitha, G, Paulkumar, K, Rajeshkumar, S, Malarkodi, C, Annadurai, G: Phytosynthesis of silver nanoparticles by *Cissus quadrangularis*: influence of physicochemical factors. J. Nanostruct. Chem. **3**, 17 (2013)
2. Farhadi, S, Pourzare, K, Sadeghinejad, S: Simple preparation of ferromagnetic Co_3O_4 nanoparticles by thermal dissociation of the $[Co^{II}(NH3)_6](NO_3)_2$ complex at low temperature. J. Nanostruct. Chem. **3**, 16 (2013)
3. Subhash, C, Srivastava, BK, Anjali, K: Magnetic behaviour of nano-particles of $Ni_{0.5}Co_{0.4}Fe_2O_4$ prepared using two different routes. Indian J. Pur. Appl. Phy **42**, 366–367 (2004)
4. Costa, MM, GFM, PJ, Sombra, ASB: Dielectric and impedance properties, studies of the lead doped (PbO)-CO_2Y type hexa ferrite ($Ba_2CO_2Fe_{12}O_{22}$ (CO_2Y)). Int. J. Mater. Chem. Phy. **123**, 35–39 (2010)
5. Muzquiz-Ramos, EM, Cortes-Hernandez, DA, Herrera-Romero, OA, Escobedo-Bocardo, J.C.: Preparation and properties of $CoFe_2O_4$ synthesized by the modified citrate-gel method. Mater Sci Forum **644**, 39–42 (2010)
6. Pervaiz, E, Gul, IH: Enhancement of electrical properties due to Cr^{3+} substitution in Co-ferrite nanoparticles synthesized by two chemical techniques. J. Magn. Magn. Mater. **324**, 3696–3703 (2012)
7. Bahout, MM, Bertrand, S, Pena, O: Synthesis and characterization of $Zn_{1-x}Ni_xFe_2O_4$ spinels prepared by a citrate precursor. J. Solid. State. Chem. **178**, 1080–1086 (2005)
8. Kim, YI, Kim, D, Lee, CS: Synthesis and characterization of $CoFe_2O_4$ magnetic nanoparticles prepared by temperature controlled coprecipitation method. Phys B: Condensed Matter **337**, 42–51 (2003)
9. Liu, C, Zou, B, Rondinone, AJ, Zhang, ZJ: Chemical control of superparamagnetic properties of magnesium and cobalt spinel ferrite nanoparticles through atomic level magnetic couplings. Chem. Soc. **122**, 6263–6267 (2000)
10. Chau, JLH, Hsu, MK, Kao, CC: Microwave plasma synthesis of Co and SiC-coated Co nanopowders. Mater. Lett. **60**, 947–951 (2006)
11. Deng, HM, Ding, J, Shi, Y., Liu, XY, Wang, J: Ultrafine zinc oxide powders prepared by precipitation/mechanical milling. J. Mater. Sci. **36**, 3273–3276 (2001)
12. Pradeep, A, Priya Darshini, P, Chandra Sekharan, G: Production of single phase nano size $NiFe_2O_4$ particles using sol–gel auto combustion route by optimizing the preparation conditions. Mater. Chem. Phys. **112**, 572–576 (2008)
13. Zheng, H, Wang, J, Lofland, SE, Ma, Z, Mohaddes-Ardabili, L, Zhao, T., Salamanca-Riba, L, Shinde, SR, Ogale, SB, Bai, F., Viehland, D, Jia, YG, Schlom, D, Wutting, M, Roytburd, A, Ramesh, R: Multiferroic $BaTiO_3$-$CoFe_2O_4$ nanostructures. Science **303**, 661–663 (2004)
14. Raghasudha, M, Ravinder, D, Veerasomaiah, P: Characterization of chromium substituted cobalt nano ferrites synthesized by citrate-gel auto combustion method. Adv. Mat. Phy. Chem. **3**, 89–96 (2013)
15. Toksha, BG, Sagar, E, Shisath, Mane, ML, Patange, SM, Jadhav, SS, Jadhav, KM: Autocombustion high-temperature synthesis, structural, and magnetic properties of $CoCr_xFe_{2-x}O_4$ ($0 \leq x \leq 1.0$). J. Phys. Chem. **115**, 20905–20912 (2011)
16. Carlin, RL: Magnetochemistry. Springer, Germany (1986)
17. Figgis, BN, Nyholm, RS: A convenient solid for calibration of gouy magnetic susceptibility apparatus. J. Chem. Soc. **4**, 4190–4216 (1958)
18. Gul, IH, Maqsood, A: Structural, magnetic and electrical properties of cobalt ferrites prepared by the sol–gel route. J. Alloy. Comp. **465**, 227–231 (2008)
19. Gul, IH, Abbasi, A, Amin, F, Anis-ur-Rehman, M, Maqsood, A: Structural, magnetic and electrical properties of $Co_{1-x}Zn_xFe_2O_4$ synthesized by co-precipitation method. J. Magn. Magn. Mater. **311**, 494 (2007)
20. Shirsath, SE, Jadhav, SS, Toksha, BG, Patange, SM, Jadhav, KM: Influence of Ce^{4+} ions on the structural and magnetic properties of $NiFe_2O_4$. J. Appl. Phy. **110**, 013914 (2011)
21. Batoo, KM, Kumar, S, Lee, CG, Alimuddin: Study of dielectric and ac impedance properties of Ti doped Mn ferrites. Curr Appl Phy. **9**, 1397–1406 (2009)
22. Mane, DR, Birajdar, DD, Shirsath, SE, Telugu, RA, Kadam, RH: Structural and magnetic characterizations of Mn-Ni-Zn ferrite nanoparticles. Phys. Status. Solidi. A. **207**, 2355–2363 (2010)
23. Singhal, S, Chanda, K: Cation distribution and magnetic properties in chromium-substituted nickel ferrites prepared using aerosol route. J. Solid. State. Chem. **180**, 296–300 (2007)

14

Effect of mercaptoethanol and Na_2S dropwise addition rate on zinc sulfide semiconductor nanocrystals: synthesis and characterization

Abbas Rahdar

Abstract

Zinc sulfide (ZnS) semiconductor nanocrystals with mercaptoethanol (ME) as a stabilizer (capping agent) were synthesized by coprecipitation method in room temperature using the solution of zinc chloride ($ZnCl_2$) and sodium sulfide (Na_2S) as starting material. The effect of ME and Na_2S dropwise addition rate on the preparation of these samples was measured using UV–vis absorption and X-ray diffraction (XRD). The ultraviolet–visible (UV–vis) absorption and XRD of the prepared ZnS nanoparticles show increase of band gap and decrease of particle size with decrease in ME and Na_2S dropwise addition rate to the reaction medium. This behavior is related to the size quantization effect due to the small size of the particles. The photoluminescence emission peak positions exhibit obvious blue shift from 510 to 455 nm. The particle sizes were obtained from transmission electron microscopy images.

Keywords: ZnS; Co-precipitation; Mercaptoethanol; Capping agent; Optical band gap; XRD

Background

Nanocrystalline semiconductor particles have attracted considerable interest in recent years because of their novel properties, such as large surface-to-volume ratio and the three-dimensional confinement of the electrons [1-7]. Interest in semiconductor nanoparticles is justified by the fact that their fundamental physical and chemical properties can be very different from those of the bulk materials. Their reduced dimensions enable one to reduce the size of electronic circuitry. They are expected to have higher quantum efficiencies due to increased oscillator strengths as a result of quantum confinement. A threshold for the occurrence of quantum effects is given by the value for the Bohr radius of exciton in the bulk material [8]. As a representative of wide band gap II-VI semiconductor nanocrystals, zinc sulfide (ZnS) has been synthesized by different methods [9-14]. ZnS has a band gap of 3.66 eV at the temperature of 300 K. This corresponds to ultraviolet (UV) radiation for optical interband transitions. It is an important inorganic material for a variety of applications including photoconductors, solar cells, field effect transistors, sensors, transducers, optical coatings, and light-emitting materials [15-18]. It has been investigated extensively because of its potential optical applications [19-26]. The systematic exploration of growing inorganic nanocolloids is an increasingly popular research subject. Physical methods such as X-ray or electron beam lithography [27], which are commonly used for the production of low-dimensional solids, have inherent resolution limits that restrict their use at the nanometer scale. On the other hand, the colloid chemistry route offers a simple means to synthesize such particles with good control of the size and, more importantly, the size distribution. There have been extensive reports in the past few years [28,29] demonstrating that by changing the reaction conditions (i.e., the concentration of the starting materials, the nature of the solvents, and the suitable capping/stabilizing agents), it is possible to synthesize a variety of nanoparticles with different sizes. In this paper, we report the synthesis of ZnS quantum dots by the chemical coprecipitation method at room temperature and in aqueous solutions with zinc chloride ($ZnCl_2$), sodium sulfide (Na_2S) as starting materials, and mercaptoethanol (ME) as a capping agent for controlling particles size and double-distilled water as dispersing solvent. The effect of ME and Na_2S dropwise addition rate to the reaction

Correspondence: a.rahdar@uoz.ac.ir
Department of Physics, Faculty of Science, University of Zabol, Zabol, Iran

medium on the preparation of these samples has been studied.

Results and discussion

Figure 1 shows the X-ray diffraction (XRD) pattern of the ZnS nanoparticles with different dropwise addition rates at room temperature. Three different peaks are obtained at 2θ values of 29.0°, 48.1°, and 57.0° for the ZnS nanoparticles, which indicate that the samples have face centered cubic structure and the peaks correspond to diffraction at (111), (220), and (311) planes, respectively. Peaks related to high angles have submerged in the background due to the large line broadening, which is attributed to the nanoscale size of the particles.

The peak broadening at the lower angle is more meaningful for the calculation of particle size. Using the (111) reflection from the XRD pattern, the mean crystallite size (D) of nanoparticles was also estimated using the Scherrer formula [30] as follows:

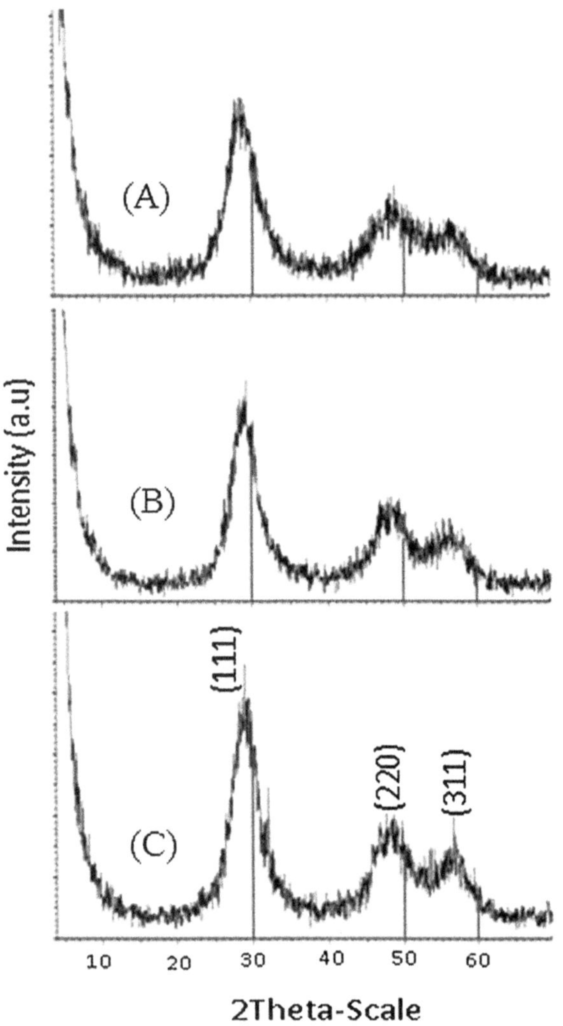

Figure 1 XRD patterns of ZnS nanoparticles at different dropwise addition rates. (A) 2 to 3 s, **(B)** 4 to 5 s, and **(C)** 6 to 7 s

$$D = \frac{0.9\lambda}{B\cos\theta}, \quad (1)$$

where λ, B, and θ are the X-ray wavelength of the radiation used (Kα(Cu) = 0.154056 nm), the full width at half maximum of the diffraction peak, and the Bragg diffraction angle (with 2θ ranging from 10° to 100°), respectively. The lattice parameter has been computed to be 6.29 Å, which is close to the standard value (5.42 Å) [31].

The optical absorption spectra of the nanoparticles were measured using a USB-2000 UV–VIS spectrophotometer (Ocean Optics, Dunedin, FL, USA). The absorption edge is observed in the range of 290 to 320 nm. The band gap of nanoparticles was also calculated using the formula as follows:

$$E_g = h\nu = \frac{hc}{\lambda} \quad (2)$$

where h, c, λ, and E_g are the Planck's constant, the speed of light in vacuum, the optical absorption wavelength (range of 290 to 320 nm), and the band gap, respectively. This blue shift of the absorption edges for different nanocrystals is related to the size decrease of the particles and is attributed to the quantum confinement effect of nanoparticles. A threshold for the occurrence of quantum effects is given by the value for the Bohr radius of exciton in the bulk material. The absorption spectrum of B sample nanoparticles synthesized at 4 to 5 rates is

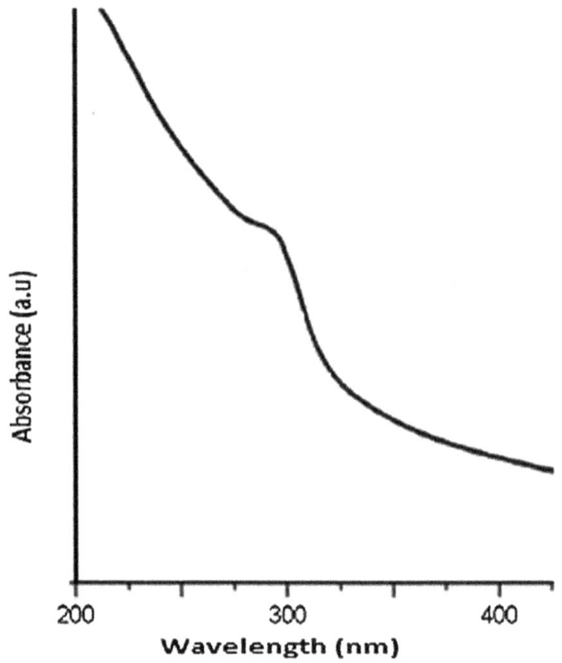

Figure 2 Absorption spectra of B sample ZnS nanoparticle.

shown in Figure 2. It shows an excitonic absorption peak at 300 nm (4.13 eV). There is a clear blue shift of band gap with respect to bulk ZnS, since band gap of bulk ZnS with cubic structure is 3.6 eV (345 nm). This is due to the modification of valence and conduction bands by quantum confinement effects. Even though the nanocrystallites exhibit bulk-like crystal structure, they are too small to have bulk-like electronic wave functions. Applying the confinement effects, the optical band gap energy of nanocrystallites is given by Brus equation [32] as follows:

$$E_g(\text{nano}) = E_g(\text{bulk}) + \frac{\pi^2\hbar}{2R^2}\left[\frac{1}{m_e}+\frac{1}{m_h}\right] - \frac{1.8e^2}{\varepsilon R}, \quad (3)$$

where E_g(bulk) in eV is the band gap energy of bulk, m_e and m_h are electron and hole effective masses, and R in nanometers is the particle size. For cubic ZnS, E_g = 3.6 eV, m_e = 0.34 m_o, and m_h = 0.23 m_o; m_o is the free electron rest mass and ε = 8.76 is the permittivity of the sample [33]. Taking the band gap of synthesized nanoparticles as 4.13 eV with substitution in Equation 4 gives the particle optical size as 2.69 nm. Table 1 shows the calculated size of ZnS nanoparticles from Scherrer formula and Brus equation. The band gap of a particle can be tuned by changing the size of nanocrystals below Bohr radius of exciton in the bulk material. Table 1 shows the variation of the optical band gaps, particle sizes, and absorption edges at different dropwise addition rates.

These semiconductor nanocrystals were then characterized using transmission electron microscopy (TEM, JEOL JEM3010; JEOL Ltd., Tokyo, Japan). Figures 3 and 4 show the TEM image of the ZnS semiconductor nanocrystals. According to the TEM images, the agglomeration occurred in the ZnS semiconductor nanocrystals. The agglomeration is due to van der Waals force between particles [34]. The TEM image shows clearly that the particles are not spherical.

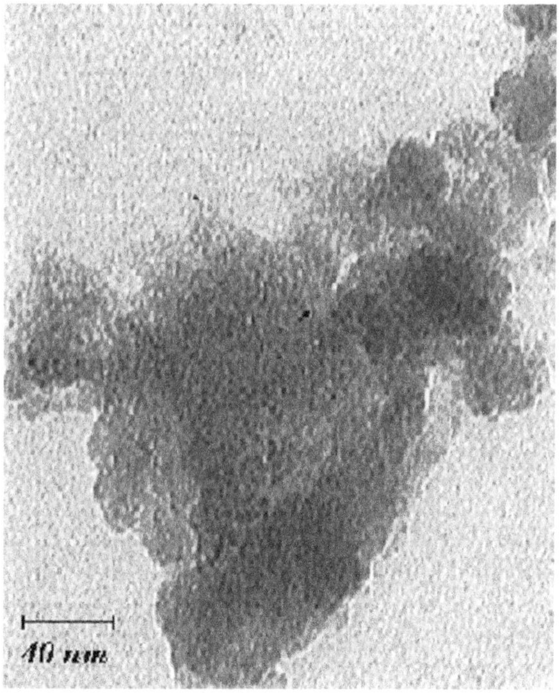

Figure 3 TEM image of ZnS semiconductor nanocrystals at the dropwise addition rate of B (6 to 7 s).

Conclusions

The nanoparticles of zinc sulfide have been successfully synthesized using a simple chemical method. It is also observed that the decrease in size of the particle and the increase of band gap occur with the decrease in ME and Na_2S dropwise addition rates to the reaction medium. The XRD analysis shows that the samples prepared are in a cubical phase. The broad peak of the XRD pattern indicates the nanocrystalline behavior of the particles. The absorption peak appears at around 290 to 320 nm, which is fairly blue-shifted from the absorption edge of the bulk (345 nm).

The solid-state theory based on the delocalized electron and hole within the confined volume can explain the blue-shifted optical absorption spectra [35,36].

Experimental procedure

First, 0.01 M of $ZnCl_2$, Na_2S, and ME as capping agent was dissolved in double-distilled water for 30 min at room temperature, and the obtained solution was magnetically stirred for 30 min at room temperature. Afterwards, the capping agent solution was added drop by drop at different rates of 2 to 3, 4 to 5, and 6 to 7 per second to the $ZnCl_2$ solution under vigorous stirring in N_2 atmosphere. After 10 min, the solution of Na_2S was poured drop by drop at different rates of 2 to 3, 4 to 5, and 6 to 7 per second to the solution containing $ZnCl_2$ and ME under vigorous stirring in N_2 atmosphere.

Table 1 Absorption edge and optical band gap of the ZnS nanoparticles with different dropwise addition rate

Sample	Dropwise addition rate (s)	Absorption edge (nm)	Optical size (nm)	Average crystallite size (nm)	Optical band gap (eV)
A	2 to 3	290	2.40	2.38	4.27
B	4 to 5	300	2.69	2.60	4.13
C	6 to 7	320	2.92	2.82	3.87

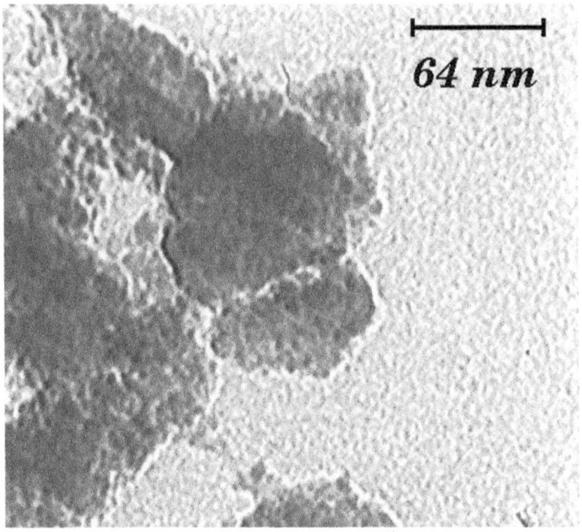

Figure 4 TEM image of ZnS semiconductor nanocrystals at the dropwise addition rate of C (6 to 7 s).

Nanoparticles of ZnS were fabricated by chemical reaction by Ying Wang [37] as follows:

$$ZnCl_2 + Na_2S \overset{ME}{\rightarrow} ZnS + 2NaCl \tag{4}$$

The nanoparticles were collected by centrifugation at 2,000 rpm for 15 min. In the final step, the obtained precipitate was filtered and dried at room temperature to remove water, organic cappings, and other by products formed during the reaction process. After sufficient drying, the precipitate was crushed to fine powder using mortar and pestle. The three samples, called A (2 to 3 s), B (4 to 5 s), and C (6 to 7 s) with different dropwise addition rates, were obtained.

Competing interests
The author declare that he has no competing interests.

Acknowledgments
The author would like to thank Dr. M. Ali-Ahmad, Mrs. M. Asudeh, Mr. H. Rahdar, and Mrs. Heidari Mokarrar for their support and assistance with this project.

References
1. Henglein, A: Small-particle research: physico-chemical properties of extremely small colloidal metal and semiconductor particles. Chem. Rev. **89**, 1861–1873 (1989)
2. Steigerwald, ML, Brus, LE: Semiconductor crystallites: a class of large molecules. Acc. Chem. Res. **23**, 183–188 (1990)
3. Bawendi, MG, Steigerwald, ML, Brus, LE: The quantum-mechanics of larger semiconductor clusters (quantum dots). Annu. Rev. Phys. Chem. **41**, 477–496 (1990)
4. Wang, Y, Herron, N: Nanometer-sized semiconductor clusters: materials synthesis, quantum size effects, and photo physical properties. J. Phys. Chem. **95**, 525–532 (1991)
5. Weller, H: Quantized semiconductor particles: a novel state of matter for materials science. Adv. Mater. **5**, 88–95 (1993)
6. Alivisatos, AP: Semiconductor clusters, nano crystals, and quantum dots. Science **271**, 933–937 (1996)
7. Eychmüller, A: Structure and photo physics of semi conductor nano crystals. J. Phys. Chem. B **104**, 6514–6528 (2000)
8. van Dijken, A., Janssen, AH, Smitsmans, MHP, Vanmaekelbergh, D, Meijerink, A: Size-selective photo etching of nano crystalline semiconductor particles. Chem. Mater. **10**, 3513–3522 (1998)
9. Xu, JF, Ji, W, Lin, JY, Tang, SH, Du, YW: Preparation of ZnS nanoparticles by ultrasonic radiation method. Appl Phys A **66**, 639–641 (1998)
10. Dinsmore, AD, Hsu, DS, Qadri, SB, Cross, JO, Kennedy, TA, Gray, HF, Ratna, BR: Structure and luminescence of annealed nanoparticles of ZnS:Mn. J. Appl. Phys. **88**, 4985 (2000)
11. Que, W, Zhou, Y, Lam, YL, Chan, YC, Kam, CH, Liu, B, Gan, LM, Chew, CH, Xu, GQ, Chua, SJ, Xu, SJ, Mendis, FVC: Structural and luminescence properties of nanostructured ZnS:Mn. Appl. Phys. Lett. **73**, 2727–2729 (1998)
12. Huang, J, Yang, Y, Xue, S, Yang, B, Liu, S, Shen, J: All-inorganic light emitting device based on ZnO nanoparticles. Appl. Phys. Lett. **70**, 2335–2337 (1997)
13. Shao, LX, Chang, Q, Hwang, HL: Zinc sulfide thin films deposited by RF reactive sputtering for photovoltaic applications. Appl. Surf. Sci. **212**, 305–310 (2003)
14. Habib Ullah, M, Il, K, Ha, CS: pH selective synthesis of ZnS nanocrystals and their growth and photoluminescence. Mater. Lett. **61**, 4267–4271 (2007)
15. Chestnoy, N, Hull, R, Brus, LE: Higher excited electronic states in clusters of ZnSe, CdSe, and ZnS: spin-orbit, vibronic, and relaxation phenomena. J. Chem. Phys. **85**, 2237–2242 (1996)
16. Lee, SM, Hwang, CS: Synthesis of a white-light-emitting ZnSe:Mn nanocrystal via thermal decomposition reaction of organometallic precursors. Bull Korean Chem Soc **34**, 321–324 (2013)
17. Brus, LE: Electron–electron and electron-hole interactions in small semiconductor crystallites: The size dependence of the lowest excited electronic state. J. Chem. Phys. **80**, 4403 (1984)
18. Wang, Y, Herron, N: Nanometer-sized semiconductor clusters: materials synthesis, quantum size effects, and photo physical properties. J. Phys. Chem. **91**, 257 (1987)
19. Awschalom, DD, Kikkawa, JM: Electron spin and optical coherence in semiconductors. Phys. Today **52**, 33 (1999)
20. Wang, X, Ding, Y, Summers, CJ, Wang, ZL: Large-scale synthesis of six-nanometer-wide ZnO nanobelts. J. Phys. Chem. B **108**, 8773–8777 (2004)
21. Narayanaswamy, A, Xu, HF, Pradhan, N, Kim, M, Peng, X: Synthesis of urea capped ZnS nanoparticles. J. Am. Chem. Soc. **128**, 10310 (2006)
22. Khosravi, AA, Kundu, M, Jatwa, L, Deshpande, SK, Bhagwat, UA, Sastry, M, Kulkarni, SK: Green luminescence from copper doped zinc sulphide quantum particles. Appl. Phys. Lett. **67**, 2702–2704 (1995)
23. Li, Y, Meng, GW, Zhang, LD, Phillipp, F: Ordered semiconductor ZnO nanowire arrays and their photoluminescence properties. Appl. Phys. Lett. **76**, 2011–2013 (2000)
24. Zhang, YS, Wang, LS, Liu, XH, Yan, YJ, Chen, CQ, Zhu, J: Effect of surface stress on the stiffness of micro/nanocantilevers: nanowire elastic modulus measured by nano-scale tensile and vibrational techniques. J. Phys. Chem. B **109**, 13091 (2005)
25. Ibanga, EJ, Le Luyer, C, Mugnier, J: Zinc oxide waveguide produced by thermal oxidation of chemical bath deposited zinc sulphide thin films. Matter Chem Phys **80**, 490 (2003)
26. Archana, J, Navaneethan, M, Ponnusamy, S, Hayakawa, Y, Muthamizhchelvan, C: Optical, structural and surface morphological studies of bean-like triethylamine capped zinc selenide nanostructures. Mat Lett **63**, 1931 (2009)
27. Kumar, A, Biebuyck, HA, Whitesides, GM: Patterning self-assembled monolayers: applications in materials science. Langmuir **10**, 1498 (1994)
28. Vossmeyer, T, Katsikas, L, Giersig, M, Popovic, IG, Diesner, K, Chemseddine, A, Eychmuller, A, Weller, H: CdS nano clusters: synthesis, characterization, size dependent oscillator strength, temperature shift of the excitonic transition energy, and reversible absorbance shift. J. Phys. Chem. **98**, 7665 (1994)
29. Murray, CB, Norris, DB, Bawendi, MG: Synthesis and characterization of nearly monodisperse CdE (E = sulfur, selenium, tellurium) semiconductor nanocrystallites. J. Am. Chem. Soc. **115**, 8706 (1993)
30. Patterson, A.: The Scherrer formula for X-ray particle size determination. Phys. Rev. **56**, 978–982 (1939). doi:10.1103/PhysRev.56.978. http://link.aps.org/doi/10.1103/PhysRev.56.978

31. Sharma, R, Chandra, BP, Bisen, DP: Optical properties of ZnS:Mn nanoparticles prepared by chemical routs. Chalcgenide Letters **6**, 339–342 (2009)
32. Bawendi, MG, Steigerwald, ML, Brus, LE: The quantum mechanics of larger semiconductor cluster. Rev Phys Chem **41**, 477–496 (1990)
33. Landolt-Bornstein: Numerical data and functional relationships in science and technology, p. 168. Springer Verlag, Berlin (1987)
34. Hartley, PA, Parfitt, GD, Pollack, LB: The role of the van der Waals force in the agglomeration of powders containing submicron particles. Power Tech. **42**, 35–46 (1985)
35. Rema Devi, BS, Raveendran, R, Vaidyan, AV: Synthesis and characterization of Mn^{2+}-doped ZnS nanoparticles. Pramana-J. Phys. **68**, 679–687 (2007)
36. Rahdar, A: Effect of 2-mercaptoethanol as capping agent on ZnS nanoparticles: structural and optical characterization. J. Nano. Chem. **3**, 1–5 (2013)
37. Wang, Y, Herron, N: Quantum size effects on the exciton energy of CdS clusters. Physical Review B **42**, 7253–7255 (1990)

15

Trace detection of thiols with Fourier transforms cyclic voltammetry technique in flowing solutions

Mohammad Barshan-Tashnizi[1], Bahman Ebrahimi-Hoseinzadeh[1*], Parandis Daneshgar[2] and Parviz Norouzi[3]

Abstract

Electrochemical experiments have an increasing demand for determining the concentration of biological and environmental samples as simple, inexpensive, and rapid analytical methods. Most developments in electroanalytical chemistry in recent years have originated from advances in sensor design, chemical modification, and functionalization of electrodes for enhanced sensitivity and selectivity of electroanalysis. In this work, a novel method for the fast monitoring of a thiol compound in flow injection systems has been developed. The fast Fourier transform continuous cyclic voltammetry at carbon paste electrode in flowing solution system was used for the determination of a thiol compound. The suggested technique demonstrated some advantages over other reported methods. Picomolar detection limit that is fast enough for the determination of any thiol compound was achieved. The linearization equation was current = 24.601x + 3.7068, with the correlation coefficients of 0.9812. The obtained detection limit was calculated to be 14.80 × 10^{-14} mol thiocholine (thiol compound sample). This system offers the necessary accuracy, sensitivity, and selectivity to assay a thiol compound in various sensors.

Keywords: Thiol compound; Fourier transform; Flow injection; Carbon nanotube paste electrode

Background

Most developments in electroanalytical chemistry in recent years have originated from advances in sensor design, chemical modification, and functionalization of electrodes for enhanced sensitivity and selectivity of electroanalysis. Organophosphorous (OP) compounds are highly toxic that often causes respiratory paralysis and death and can irreversibly inhibit acetylcholinesterase (AChE) which is essential for the central nervous system. Therefore, the rapid, sensitive, selective, and reliable quantification of these compounds is highly essential at a lower cost. Chromatographical techniques have been extensively used for the sensitive detection of OP insecticides, but they require trained staff, complicated sample pretreatments, and are often not suitable for field conditions. In recent years, biosensors show remarkable advances for the detection of toxic compounds based on enzymatic reactions. A variety of enzymes such as organophosphorous hydrolase, alkaline phosphatase, ascorbate oxidase, tyrosinase, and acid phosphatase have been employed in the preparation of pesticide biosensors [1]. Based on the inhibition action of pesticides and insecticides on cholinesterases, AChE and butyrylcholinesterase have been widely used due to the stability and sensitivity of the enzymes [2]. This method generally uses either single enzyme [3] or bienzyme (AChE and choline oxidase) [4] systems by monitoring the electrochemical oxidation of thiocholine or *p*-aminophenol and hydrogen peroxide, respectively. Some of the reports on immobilized AChE-based single enzyme system apply voltages of +400 to +700 mV for the oxidation of thiocholine [5].

Traditional methods to detect thiocholine involve spectrophotometric techniques [6]. The Ellman reaction involves the reaction of 5,5′-Dithiobis(2-nitrobenzoic acid) and 5-thio-2-nitrobenzoate (DTNB) with thiocholine liberated from its esters by enzymatic hydrolysis so the yellow 5-thio-2-nitro-benzoate product is measured spectrophotometrically at 412 nm. The thiocholine also generates a highly fluorescent product with 7-diethylamino-3-(40-maleimidylphenyl)-4-methylcoumarin, which is monitored by microfluorometric assay to detect nanomoles of released thiocholine as thiol compound. In AChE-based biosensors, acetylthiocholine is commonly used as a substrate. The thiocholine produced

* Correspondence: bahman.ebrahimi@ut.ac.ir
[1]Department of Life Science Engineering (LSE), Faculty of New Sciences and Technologies (FNST), University of Tehran, Tehran 14174, Iran
Full list of author information is available at the end of the article

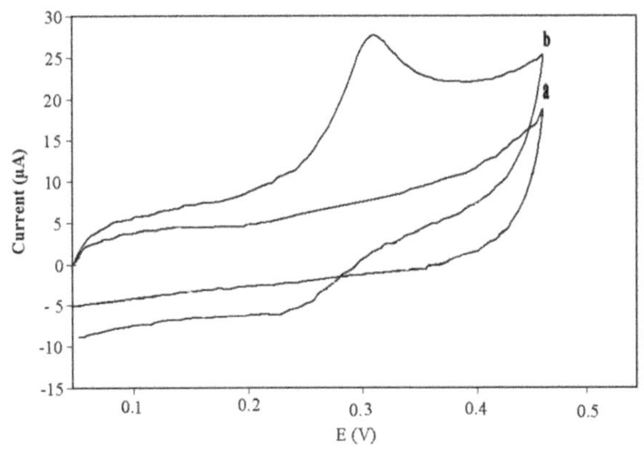

Figure 1 Cyclic voltammograms at 10 V.S^{-1} of 0.1 M PBS (pH 7.4) at a) PBS, b) TCh injection to the FFTCCV system.

during the catalytic reaction can be monitored using spectrometric, amperometric, or potentiometric methods. The enzyme activity is indirectly proportional to the pesticide concentration.

AChE can be immobilized on electrode surface using a variety of matrices such as cross-linked polymers, cross-linked bovine serum albumin [7], chitosan and cellulose [8], different support matrices such as nylon [9], controlled pore glass [10], magnetic particles [11], or the strong affinity linking with concanavalin A and metal ions [12].

Carbon paste electrode (CPE) is a special kind of heterogeneous carbon electrode consisting of mixture prepared from carbon powder and a suitable water-immiscible or nonconducting binder [13,14]. CPEs are widely applicable in both electrochemical studies and electroanalysis, thanks to their advantages such as ease and speed of preparation and of obtaining a new reproducible surface, very low background current (compared to solid graphite or noble metal electrodes), low cost, large potential window, and simple surface renewal process. The level of background currents cannot be exactly defined; it strongly depends on the composition of carbon paste as well as upon the type of measurement. For instance, in stripping voltammetry, carbon pastes containing Nujol or silicone oil exhibit background currents of about 10 to 500 nA. Whereas the background of tricresyl phosphate-based CPE may exceed the level of microampere due to unique interactions of the binder [15]. Besides the advantageous properties and characteristics listed before, the feasibility of incorporation of different substances during the paste preparation (which results in the so-called modified carbon paste electrode) allows the fabrication of electrodes with desired composition and, hence, with predetermined properties [16,17].

The modification of electrodes using redox modifiers is an interesting field in analytical chemistry. Redox modifiers are electroactive compounds that effectively shuttle electrons between the analyte and the electrode. One of the most important effects of any modifier is a reduction of the overpotential required for electrochemical reaction, which enhances the sensitivity and selectivity of the method. Carbon nanotubes (CNTs) are new kinds of carbon nanostructure materials possessing properties such as high electrical conductivity, high surface area, chemical stability, and significant mechanical strength. CNTs have demonstrated to be extremely useful for the development of new electrode materials. Their electrocatalytic properties have been widely demonstrated in connection with several compounds of clinical, biological, and environmental interest [18]. Also, the application of chemically modified electrodes in electroanalysis offers several advantages. They can lower the overpotential, increase the reaction rate and sensitivity, and improve selectivity [19].

Recently, stripping voltammetric methods were used in the determination of heavy metal ions and some organic compounds in flowing solutions with a parts-per-billion sensitivity range. Indeed, application of such techniques needs fast analyte accumulation and fast potential sweeping, which is not appropriate for large electrodes [20]. Nonetheless, this study aims to present a novel method for the rapid determination of thiocholine in ultratrace amounts.

Stripping voltammetric techniques have the advantages of being both rapid and economical in the determination of thiols in aqueous systems, with sensitivity for trace amounts. In fact, because of the passage of an analyte through the zone in front of the electrode in an electrochemical flow cell, the application of such techniques in flowing solutions requires fast accumulation of the analyte and fast potential sweeping, which is not appropriate in traditional voltammetry but is applicable in fast Fourier transform (FFT) voltammetry.

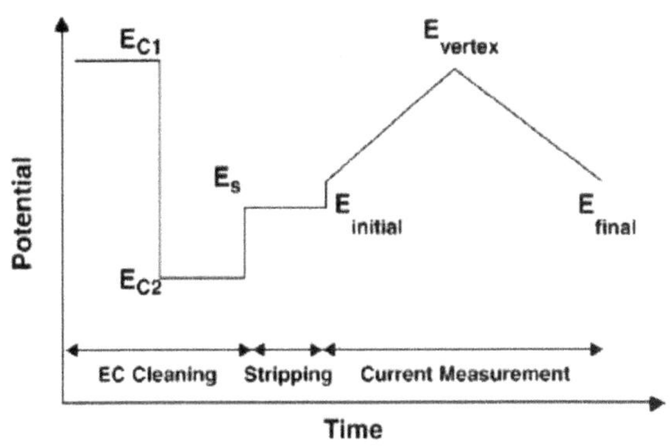

Figure 2 Diagram of the applied potential waveform.

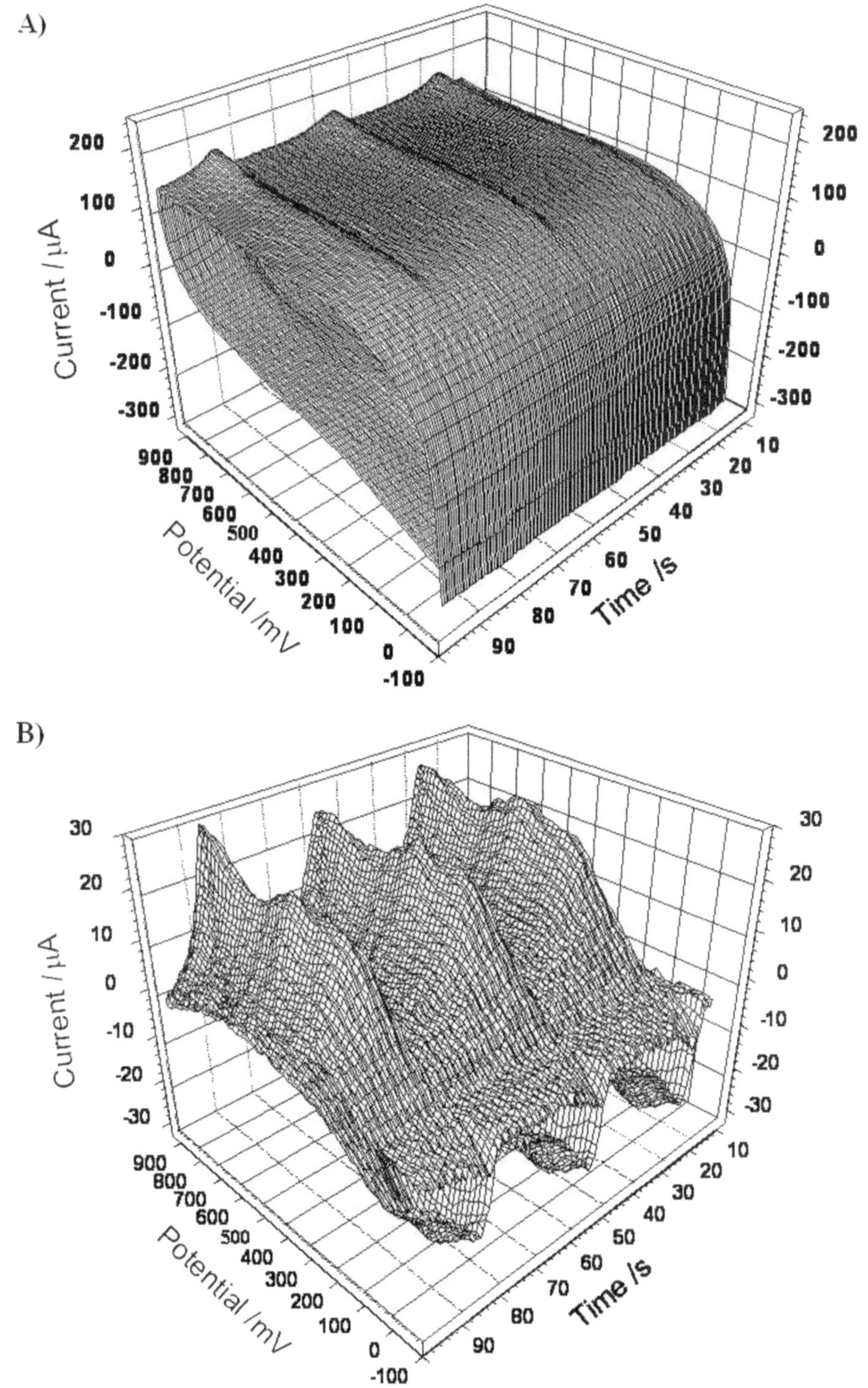

Figure 3 Cyclic voltammograms and Curves result (A) Cyclic voltammograms at carbon nanotube paste electrode recorded during a flow injection experiment. The eluent was phosphate buffer (0.1 M and pH 7.4), the flow rate was 3mL/min, and the sweep rate was 10 $V.s^{-1}$. Each scan was preceded by 100 ms (at 1600 mV) and 200 ms (at 300 mV) conditioning, respectively. The injected solution (50 μL) contained 1 mM thiocholine. **(B)** Curves result from the subtraction of a CVs average (in the absence of the analyte) from test of the CVs in **(A)**.

In this work, the proposed biosensor based on the immobilization of AChE on MWCNT-silicate sol composite showed high sensitivity for thiocholine which could be employed for flow injection analysis of OPs. The objective of this paper is the presentation of the Fourier transform continuous cyclic voltammetry (FFTCCV) technique for the determination of thiocholine through a flow injection system. The combination of biosensors with flow injection analysis makes it possible to control all the stages of the reagent additions, measure the enzyme activity, improve the sample throughput, and achieve the completely automated determination.

Results and discussion

Electrochemical behavior of system

The cyclic voltammograms of 1.0 mM acetylthiocholine (ATCh) are shown in Figure 1. Thiocoline (TCh) was injected to the flow system and showed an oxidation peak at +310 mV, while no detectable signals were observed at phosphate buffer saline (PBS) solution contain enzyme. In Figure 1A, it did not show any detectable response in the absence of TCh. Obviously, the peak came from the oxidation of thiocholine.

In this system, the stability of the background was the most essential problem. This was mainly due to the changes that occurred in the electrode surface during the oxidation and reduction of the thiocholine in each potential cycle. The best potential waveform, producing a stable background and an unchanged electrode surface, was the waveform displayed in Figure 2. Figure 3 presents an example of this waveform. Figure 3b shows the cyclic voltammetry (CV) sequences recorded during the flow analysis of 50 μL 1 mM thiocholine into the eluent solution, containing phosphate buffer (0.1 M and pH 7.4). The potential axis of this graph represents the potential applied to the working electrode during each sweep. The time axis represents the time passing between the beginning of the flow injection experiment and the beginning of a particular sweep. As can be observed, in the absence of thiocholine, the shape of the CV curves is typical for phosphate buffer (0.1 M and pH 7.4).

Figure 3a,b illustrates the absolute current changes in the CV curves, after the subtraction of background current. As can be seen, this way of presentation of the electrode response gives more details about the effect of adsorbed ion on currents of the CV. The curves show that current changes mainly take place at the potential regions of the oxidation and reduction of thiocholine. While the electrode-solution interface is exposed to thiocholine, the oxidation of TCh will be occurring. For the achievement of the detector maximum performance, the effect of the experimental parameters (such as the pH of the supporting electrolyte, and the sweep rate) were examined and optimized.

Sweeping rate effect

An acetylthiocholine concentration of 750 μM was used for the investigation of the sweeping rate effect. The detector responses to the potential sweep rates were recorded at different sweep rates in the range of 4 to 20 V s^{-1}. The result was shown in Figure 4, where it is clear that the detector exhibits the maximum sensitivity at the sweep rate value of 10 V s^{-1}. It should be mentioned that the sweep rate effects influence the detection performance in two ways. The first influence concerns the speed in data acquisition. The second influence concerns the kinetic factors of the thiocholine redox behavior. Also, the use of this detection method in conjunction with fast separation techniques such as capillary electrophoresis requires the employment of high sweep rates. From this point of view, it is important to check how the sensitivity of the method is affected by the sweep rate. In fact, to detect analyte on the electrode surface, high sweep rates must be employed to sense the correct concentration zone without any influence on kinetic of electron transferring. Nonetheless, the potential sweep rate is the determining factor which defines the sensitivity of the detection system, basically because of the kinetic factors and the instrumental limitations. In line with the resulting data for the examined thiocholine on the electrode, the height of the electrode response is almost constant within the experimental error for the sweep rates of 4 to 20 V s^{-1}.

Calibration curve and detection limit

Figure 5 presents the calibration graph for the thiocholine monitoring in phosphate buffer (0.1 M and pH 7.4). The experimental conditions were set at the optimum values to obtain the lowest detection limits. It should be noted that, similar to the classical stripping voltammetric method, in the calculation method, the analyte response is proportional to the redox current of TCh. The redox reaction of TCh in different concentration makes different currents.

The linearization equation was current = $24.601x + 3.7068$, with the correlation coefficients of 0.9812 (Figure 5).

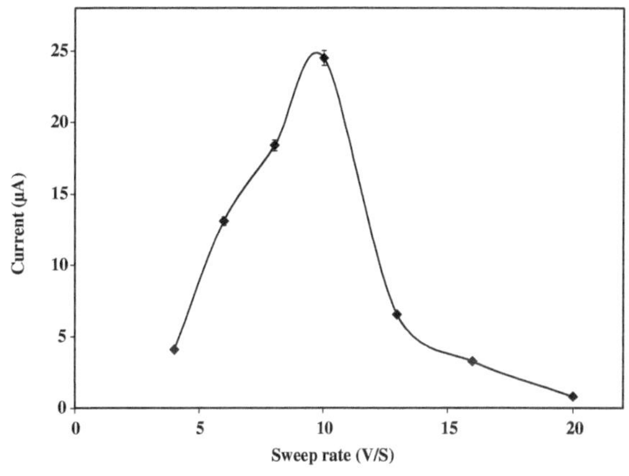

Figure 4 Sweep rate effect on the carbon paste electrode response to the injection of 1 mM thiocholine in phosphate buffer (0.1 M and pH 7.4).

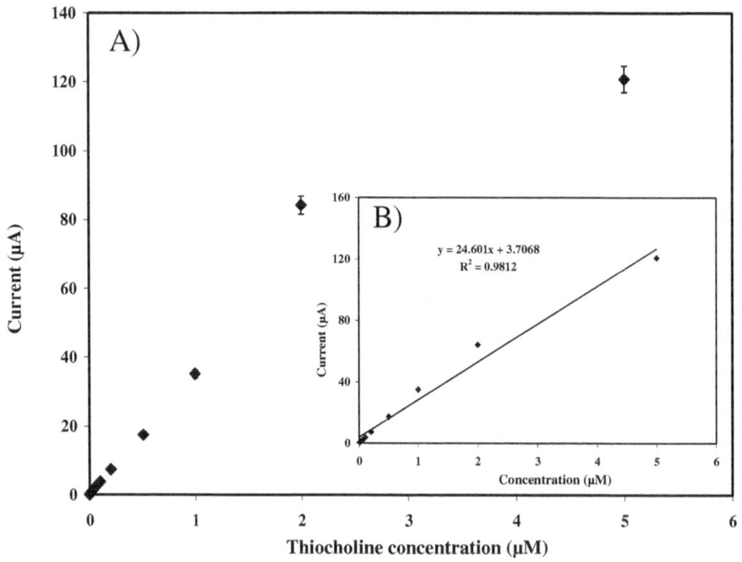

Figure 5 Calibration curves obtained for thiocholine on the carbon nanotube paste electrode in phosphate buffer (0.1 M and pH 7.4).

Measurements carried out for small analyte concentrations allow the estimation of the detection limit C_{DL}:

$$C_{\mathrm{DL}} = 3S_b/m \qquad (1)$$

where S_b is the standard deviation (2.43%) of the baseline around the flow injection peak and m is the slope of calibration curve [21]. To ensure the best S/N ratio, the measurements were carried out at high sweep rates. The obtained detection limit was calculated to be 2.96×10^{-9} M and based on 50 μL of thiocholine injection 14.80×10^{-14} mol thiocholine. The detection limit of the proposed method is about 100 times lower than that of the most sensitive reported (3×10^{-7} M) method [22].

Conclusions

In this work, it is demonstrated that the concentration of thiocholine in flowing solution can be measured indirectly via monitoring the current changes at oxidation and reduction of the carbon nanotube paste electrode surface. In this method, the S/N ratio is enhanced using the fast Fourier transform of the analyte and signal integration. Also, for improving the sensitivity, the method takes advantage of the adsorption of the analyte on the electrode, and the influence of adsorbed possible impurity in the eluent was removed by background subtraction. FFTCCV can be considered as a new sensitive, accurate, and fast method for the determination of drugs and some pesticides. However, in order to obtain better sensitivity for a specific target, experimental parameters should be optimized. Finally, such detection limit (in nanomolar level) makes the method suitable for bioanalysis. The obtained detection limit was calculated to be 2.96×10^{-9} M and based on 50 μL of thiocholine injection 14.80×10^{-14} mol thiocholine.

Methods

Reagents

Multiwall carbon nanotubes with 95% purity, length 1 to 10 μm, external diameter 37 nm, and wall no. 3 to15 were obtained from Research Institute of Petroleum Industry, Iran. DTNB with 99% purity was supplied from Merck Company, Darmstadt, Germany. Triton X-100 was obtained from Sigma-Aldrich Chemicals Co., St. Louis, MO, USA. Acetylthiocholine chloride with 99% purity was purchased from Fluka, Dorset, England, UK. Dimethylformamide was obtained from Sigma. Acetylcholinesterase (EC: 3.1.1.7, type V-S: electric eel source, 500 U/0.5 mg) was purchased from Sigma. Ceramic cylinders with average length 1.8 mm, external diameter 1.9 mm, and internal diameter 1 mm were obtained from Tailor China Shop, Iran. Phosphate buffers with various pH values were prepared according to manual.

Flow injection setup part

In the flow system (Figure 6A), a peristaltic pump to impel solutions was used. The various system elements were connected with a silicon tube. The equipment for flow injection analysis included a ten-roller peristaltic pump (UltrateckLabs Co., Iran) and a four-way injection valve (Supelco Rheodyne Model 5020, Sigma) with a 50-μL sample injection loop. The solutions were introduced into the sample loop by means of a plastic syringe. The electrochemical cell used in flow injection analysis is shown in Figure 6B. The volume of the cell was 100 μL. In

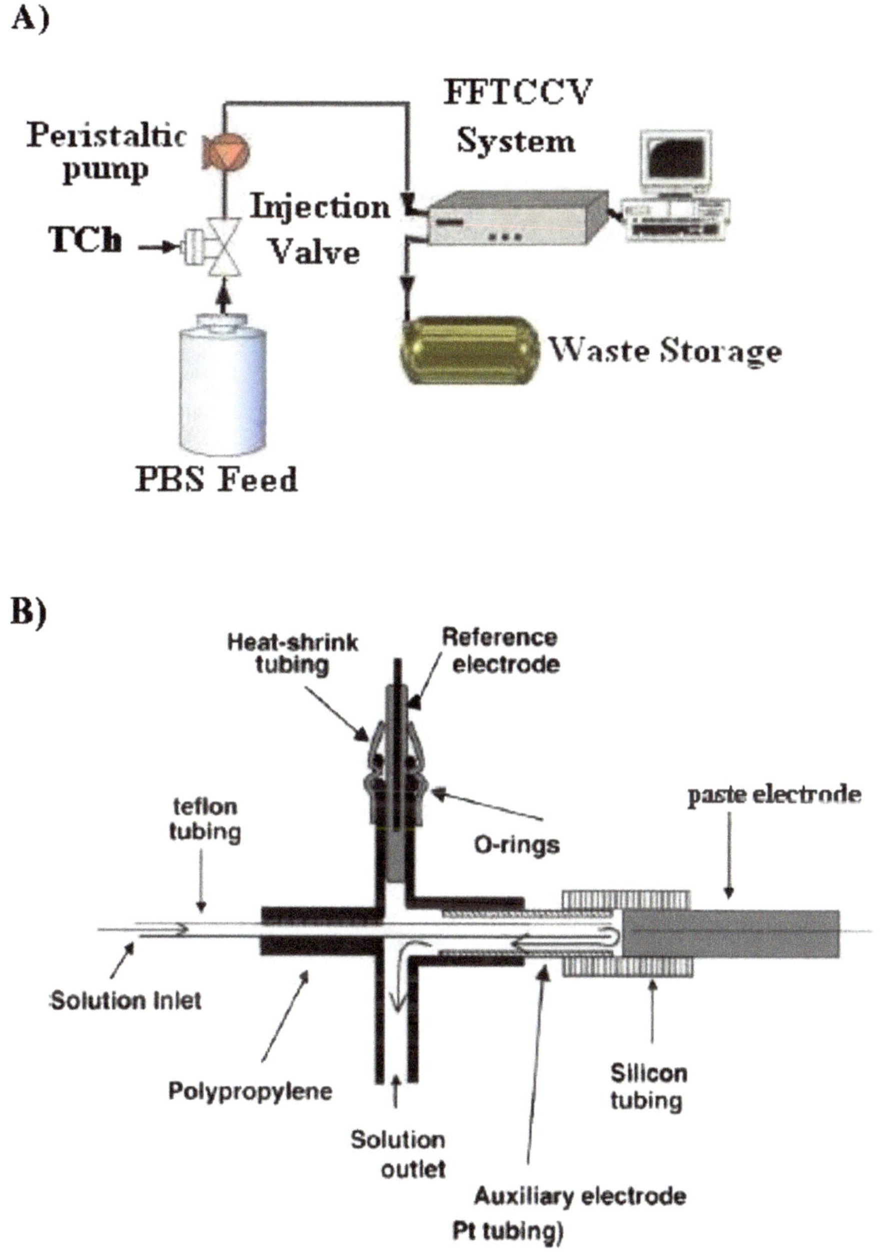

Figure 6 Schematic diagram of (A) flow-injection system; (B) the electrochemical cell.

all experiments described in this paper, the flow rate of eluent solution was 3 mL/min. Flow system feed was phosphate buffered (0.1 M and pH 7.4).

Electrode preparation

For the preparation of the carbon nanotube paste electrode, a sampler head tube (5 cm in length) with the hole at one end (2 mm in diameter) was served for filling with carbon nanotube paste as the electrode body. Electrical contact was provided by a copper piston within the center of the tube. Carbon nanotube paste was prepared by adding 4.5 g paraffin oil to 1.25 g carbon nanotube powder and mixed for 20 min until a uniformly wetted paste was obtained. The paste was then packed into the hole of the electrode holder. For every five to seven experiment runs, a small amount of carbon nanotube paste was cut, and the electrode surface was polished for 1 min using a fine paper. The electrode was washed with water before being placed in the cell. Additionally, an Ag(s)/ AgCl(s)/ KCl (aq, 1 M) reference electrode was used for all measurements, while the auxiliary electrode was made of a Pt wire (1 cm in length and 0.5 mm in diameter, Figure 6B).

Standard solutions for flow injection analysis

A TCh (as thiol compound) standard stock solution was prepared in the phosphate buffer (0.1 M and pH 7.4). This solution was prepared by the enzymatic reaction of the AChE and ATChCl solution for 30 min in the phosphate buffer. The final concentration of TCh was determined by spectrophotometry after reaction with DTNB, as described by Ellman [23,24]. Standard solutions of TCh and ATCh for studies were prepared by dilution of their stock solution with 0.1 M phosphate buffer.

Data acquisition and processing

A setup of a PC PIII Pentium 300 MHz microcomputer, equipped with a data acquisition board (PCL-818HG, Advantech. Co., Milpitas, CA, USA), and a custom-made potentiostat comprised the equipment needed for the data acquisition and data processing. The corresponding programs were developed in Delphi 6 program environment. In Figure 2, the applied waveform potential diagram during the cyclic voltammetric measurements is shown. The waveform potential consists of two sections: first, the electrode conditioning and, second, the measurements (the section of accumulation of analyte was eliminated in this work and was not applied by instrument). The electrode conditioning starts with the cleaning process, in which the electrode potential is held at a positive potential, Ec_1 (1,600 mV), at least for 100 ms. Then, for the reduction and desorption of any fouling component on the surface, it is held at the Ec_2 potential (–400 mV) for 100 ms. Finally, the current measurements are performed during potential ramp. Signal calculation in this method is established based on the integration of net current changes over the scanned potential range. It must be noted that in this case, the current changes (result of injected analyte) at the voltammograms can be caused by redox processes of TCh, which take place at the electrode surface. Indeed, in order to see the influence of the change in the current of oxidation and reductions peaks of the electrode surface, the scan rate must be in optimum which is in correlation with redox kinetic and data acquisition. The starting point for the electrochemical oxidation process of the carbon nanotube paste electrode surface is as follows:

$$2\ \text{thiocholine (red)} \rightarrow 2\text{thiocholine (ox)} + 2e + 2H^+ \quad (2)$$

Abbreviations

ATCh: Acetylthiocholine; CV: Cyclic voltammetry; FFTCCV: Fourier transform continuous cyclic voltammetry; TCh: Thiocoline; OP: Organophosphorous compounds.

Competing interests

All authors declare that they have no competing interests.

Authors' contributions

MBT and PD carried out the electrochemical studies. BEH and PN participated in the design of the study and performed the statistical analysis. All authors read and approved the final manuscript.

Author details

[1]Department of Life Science Engineering (LSE), Faculty of New Sciences and Technologies (FNST), University of Tehran, Tehran 14174, Iran. [2]Institute of Biochemistry and Biophysics, University of Tehran, Tehran 14174, Iran. [3]Center of Excellence in Electrochemistry, University of Tehran, Tehran 14174, Iran.

References

1. Trojanowicz, M: Determination of pesticides using electrochemical enzymatic biosensors. Electroanalysis **14**, 1311–1328 (2002)
2. Timur, S, Telefoncu, A: Acetylcholinesterase (AChE) electrodes based on gelatin and chitosan matrices for the pesticide detection. Artif. Cells Blood Substit. Immobil. Biotechnol. **32**, 427–442 (2004)
3. Joshi, KA, Tang, J, Haddon, R, Wang, J, Chen, W, Mulchandani, A: A disposable biosensor for organophosphorus nerve agents based on carbon nanotubes modified thick film strip electrode. Electroanalysis **17**, 54–58 (2005)
4. Lin, Y, Lu, F, Wang, J: Disposable carbon nanotube modified screen-printed biosensor for amperometric detection of organophosphorus pesticides and nerve agents. Electroanalysis **16**, 145–149 (2004)
5. Gogol, EV, Evtugyn, GA, Marty, JL, Budnikov, HC, Winter, VG: Amperometric biosensors based on nafion coated screen-printed electrodes for the determination of cholinesterase inhibitors. Talanta **53**, 379–389 (2000)
6. Cremisini, C, Di Sario, S, Mela, J, Pilloton, R, Palleschi, G: Evaluation of the use of free and immobilised acetylcholinesterase for paraoxon detection with an amperometric choline oxidase based biosensor. Anal. Chim. Acta **311**, 273–280 (1995)
7. Hendji, AMN, Jaffrezic-Renault, N, Martelet, C, Clechet, P, Shlu'ga, AA, Strikha, VI, Netchiporuk, LI, Soldatkin, AP, Wlodarski, WB: Sensitive detection of pesticides using a differential ISFET-based system with immobilized cholinesterases. Anal. Chim. Acta **281**, 3–11 (1993)
8. Bachmann, TT, Schmid, RD: A disposable multielectrode biosensor for rapid simultaneous detection of the insecticides paraoxon and carbofuran at high resolution. Anal. Chim. Acta **401**, 95–103 (1999)
9. Pariente, F, La Rosa, C, Galan, F, Hernández, L, Lorenzo, E: Enzyme support systems for biosensor applications based on gold-coated nylon meshes. Biosens. Bioelectron. **11**, 1115–1128 (1996)
10. Lee, H-S, Kim, YA, Cho, YA, Lee, YT: Oxidation of organophosphorus pesticides for the sensitive detection by a cholinesterase-based biosensor. Chemosphere **46**, 571–576 (2002)
11. Kindervater, R, Künnecke, W, Schmid, RD: Exchangeable immobilized enzyme reactor for enzyme inhibition tests in flow-injection analysis using a magnetic device. Determination of pesticides in drinking water. Anal. Chim. Acta **234**, 113–117 (1990)
12. Andreescu, S, Barthelmebs, L, Marty, J-L: Immobilization of acetylcholinesterase on screen-printed electrodes: comparative study between three immobilization methods and applications to the detection of organophosphorus insecticides. Anal. Chim. Acta **464**, 171–180 (2002)
13. Afkhami, A, Madrakian, T, Ghaedi, H, Khanmohammadi, H: Construction of a chemically modified electrode for the selective determination of nitrite and nitrate ions based on a new nanocomposite. Electrochim. Acta **66**, 255–264 (2012)
14. Mazloum-Ardakani, M, Sheikh-Mohseni, MA, Abdollahi-Alibeik, M, Benvidi, A: Application of nanosized MCM-41 to fabrication of a nanostructured electrochemical sensor for the simultaneous determination of levodopa and carbidopa. Analyst **137**, 1950–1955 (2012)
15. Švancara, I, Schachl, K: Testing of unmodified carbon paste electrodes. Chem Listy **93**, 490–499 (1999)
16. Mazloum-Ardakani, M, Beitollahi, H, Amini, MK, Mirjalili, B-F, Mirkhalaf, F: Simultaneous determination of epinephrine and uric acid at a gold electrode modified by a 2-(2,3-dihydroxy phenyl)-1, 3-dithiane self-assembled monolayer. J. Electroanal. Chem. **651**, 243–249 (2011)

17. Mazloum-Ardakani, M, Sheikh-Mohseni, MA, Abdollahi-Alibeik, M, Benvidi, A: Electrochemical sensor for simultaneous determination of norepinephrine, paracetamol and folic acid by a nanostructured mesoporous material. Chem **171–172**, 380–386 (2012)
18. Mazloum-Ardakani, M, Beitollahi, H, Ganjipour, B, Naeimi, H, Nejati, M: Electrochemical and catalytic investigations of dopamine and uric acid by modified carbon nanotube paste electrode. Bioelectrochem **75**, 1–8 (2009)
19. Goyal, RN, Bishnoi, S: A novel multi-walled carbon nanotube modified sensor for the selective determination of epinephrine in smokers. Electrochim. Acta **56**, 2717–2724 (2011)
20. Kissinger, PT, Heineman, WR: Laboratory techniques in electroanalytical chemistry. Marcel Dekker, New York (1996)
21. Norouzi, P, Ganjali, MR, Daneshgar, P, Alizadeh, T, Mohammadi, A: Development of fast Fourier transformation continuous cyclic voltammetry as a highly sensitive detection system for ultra trace monitoring of penicillin V. Anal. Biochem. **360**, 175–181 (2007)
22. Liu, G, Riechers, SL, Mellen, MC, Lin, Y: Sensitive electrochemical detection of enzymatically generated thiocholine at carbon nanotube modified glassy carbon electrode. Electrochem. Commun. **7**, 1163–1169 (2005)
23. Ellman, GL: A colorimetric method for determining low concentrations of mercaptans. Arch. Biochem. Biophys. **74**, 443–450 (1958)
24. Ellman, GL, Cotney, KD, Andress, V, Featherstone, RM: A new and repid colorimetric determination of acetycholinesterase activity. Biochem. Pharmacol. **7**, 88–95 (1961)

16

Anionic dye (acid green 25) adsorption from water by using polyaniline nanotubes salt/silica composite

Mohamad M Ayad[1,2*] and Ahmed Abu El-Nasr[2]

Abstract

Acid green 25 (AG) dye was adsorbed from water by using polyaniline nanotubes (PANI NT) salt/silica composite as adsorbent. Comparison between the removal of the dye by using PANI NT salt/silica composite and the conventional PANI salt/silica composite was investigated. This result was evidenced from the measurements of the UV-visible absorbance of AG dye solutions after different times. Experiments were conducted by varying parameters, namely initial concentration of the dye and contact time. The percentage of color removal decreased with increase in initial dye concentration. Adsorption equilibrium of color removal was reached after 60 min of contact time. Equilibrium data were fitted to Langmuir, Freundlich, and Tempkin isotherms, and their constants were determined. Using the linear correlation coefficients showed that the Langmuir isotherm best fits the AG adsorption data on PANI NT salt/silica composite. The experimental data were fitted into the following kinetic models: pseudo-first-order, pseudo-second-order, and the intraparticle diffusion model. It was observed that the pseudo-second-order kinetic model described the adsorption process better than any other kinetic models. The results obtained indicate that PANI NT salt/silica composite could be employed as a much more efficient adsorbent than the conventional PANI salt/silica composite for dye removal from water.

Keywords: Polyaniline nanotube salt/silica composite, Acid green 25, Adsorption isotherm, Kinetics

Background

Industrial effluents are one of the major causes of environmental pollution because effluents discharged from dyeing industries are highly colored with a large amount of suspended organic solid [1]. Untreated disposal of this colored water into the receiving water body either causes damage to aquatic life or to human beings by mutagenic and carcinogenic effect. As a matter of fact, the discharge of such effluents is worrying for both toxicological and environmental reasons [2]. Dyes can be classified as anionic (acid dyes), cationic (basic dyes), and non-ionic (disperse dyes) [3]. Acid dyes are organic sulfonic acids; the commercially available forms are usually sodium salts, which exhibit good water solubility. In sequence of their importance, acid dyes are mostly used with certain fiber types such as polyamide, wool, silk, modified acrylic, and polypropylene fibers as well as blends of the aforementioned fibers with other fibers such as cotton, rayon, polyester, regular acrylic [4], etc. Various techniques like precipitation, membrane filtration, coagulation, electrochemical, ion exchange, chemical oxidation, adsorption [5,6], etc. are used for the removal of dyes from wastewater. Adsorption is a procedure of choice for the removal of dyes from wastewater [7]. The major advantages of this technique are its low generation of residues and the possibility of its adsorbent being recycled and reused [8]. Several effective, selective, and cheaper adsorbent materials were developed such as waste orange peel [9], banana pith [10], rice husk [11], clay [12], neem leaf powder [13], powdered activated sludge [14], and activated carbon/charcoal [15]. Recently, conducting polymer was tested in the adsorption of dye effluent [16-18].

Composites which are formed by the incorporation of organic–inorganic materials are very fruitful enterprise as the merits of organic and inorganic moieties can be combined in the composites. Polyaniline (PANI) is one

* Correspondence: mohamad.ayad@ejust.edu.eg
[1]Department of Chemical and Petrochemical Engineering, Egypt-Japan University of Science and Technology, Borg El Arab, Alexandria 21934, Egypt
[2]Department of Chemistry, Faculty of Science, University of Tanta, Tanta 31111, Egypt

of the organic conducting polymers that can be used as an electrode material [19], in microelectronics [20], as an electrochromic device [21], in radiation shielding, and in recordable optical disks [22]. It was also reported to be utilized as an adsorbent for the adsorption of protein [23] and DNA [24]. Polyaniline is considered to be one of the most promising classes of polymers due to their well-behaved electrochemistry, easy protonation reversibility, excellent redox recyclability [25], good environmental stability [26], electrochromism [27], easy doping [28], and easy preparation. The properties of PANI can be improved by incorporating other functional materials, including oxides, polyacidic anions, etc. [29]. Among those inorganic moieties in the organic–inorganic composites, silica has received great attention because of its unique properties and wide applications [30]. Silica particles have been included in the composites through many ways, including surface polymerization of aniline adsorbed onto silica surfaces [31], *in situ* hydrolysis and condensation of tetraethoxysilane in PANI solutions or on solid PANI surfaces [32], etc. Although silica is insulating, some of the composites were in the same conductivity level as that of PANI [32], and some even displayed enhanced conductivity [32], providing potential applications of the composites in electrochemistry and related fields. Aniline polymerization in an aqueous acidic medium yields the most conductive form of PANI; the emeraldine salt (ES) was deprotonated by treatment with an alkali to polyaniline emeraldine base (EB) [33], as shown in Scheme 1. Recently, we have explored PANI coating on the electrode of the quartz crystal microbalance as chlorinated hydrocarbon and alcohol vapor [34-36] sensors. The diffusion and the adsorption kinetics of the adsorbed vapors into the polymer were studied.

Nanostructured PANI receives a great deal of attention due to its properties which differ from those of the coarse structures. The increasing surface area of PANI makes it an interesting material for sensor construction and gas storage [37]. Recently, PANI nanotubes (NT) were used for the reduction of silver nitrate to produce PANI-silver nanoparticle composites [38]. Also, the PANI NT base has been utilized as an adsorbent for the removal of cationic dyes such as MB from aqueous solution [39]. It has been concluded that the NTs provide a more efficient adsorbent than conventional PANI. The kinetic mechanism and the isotherm models were investigated. In the same context, the present paper was devoted to study the adsorption of the anionic acid green 25 dye (AG) as a model onto PANI NT salt/silica composite by using UV-visible spectroscopy. The difference between PANI NT salt/silica composite and the conventional PANI salt/silica composite in the adsorption of AG was studied.

Results and discussion

Scanning electron microscopy (SEM) was used to provide morphological information of both the PANI NT salt and PANI NT salt/silica composite, Figure 1a,b, respectively. It can be seen that the NTs are of 100 to 300 nm in diameter. The lengths of PANI NTs were 2 to 40 μm. The cavities in the NTs were easily visible (Figure 1a). The inner diameter differed substantially among the individual NTs, in the range of 20 to 250 nm, but it was relatively constant within each individual nanotube. Some nanorods with no cavities were also formed. These nanostructures obtained in the present study are similar to those reported in the bulk solution [40]. Figure 1b shows the SEM image of the PANI NT/silica composite with silica particles that appear as cluster spot. The PANI NT salt composite plays an important role in the enhancement of polymer surface area and helps in increasing the adsorption rate of polymer towards the dye.

Adsorption of AG onto the PANI NT salt/silica composite

Adsorption studies

The adsorption of AG onto the PANI NT salt, the conventional PANI, and their composites with silica composite was tested using UV-visible spectroscopy. Adding 0.05 g of different substrates such as PANI NT salt/silica composite, conventional PANI salt/silica composite, PANI NT salt, and conventional PANI salt to 18.6 mg L^{-1} AG

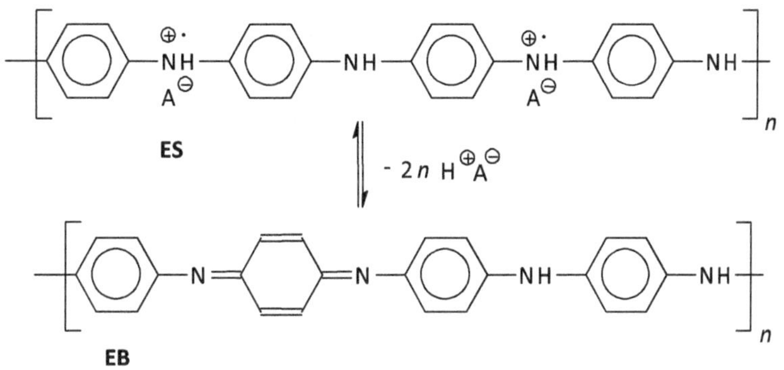

Scheme 1 Protonated (doped) polyaniline ES is deprotonated (dedoped) by treatment with an alkali to polyaniline EB.

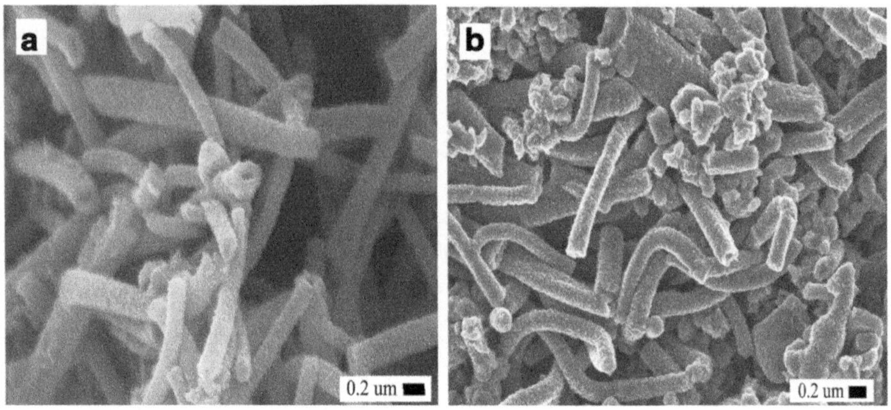

Figure 1 SEM images of PANI NT salt (a) and PANI NT salt/silica composite (b).

solution at the same conditions leads to a decrease in the absorbance of AG with increasing time (Figure 2a,b,c,d, respectively). The amount of dye adsorbed, Q_e (mg g^{-1}), onto per unit weight of PANI NT salt/silica composite, PANI NT salt, conventional PANI salt/silica composite, and conventional PANI salt at equilibrium is equal to 3.73, 3.56, 3.51, and 1.48 mg g^{-1}, respectively. Figure 3 shows the decrease in AG concentration at different times in the presence of PANI NT salt/silica composite. The amount of dye adsorbed into per unit weight of the polymer adsorbent, Q_e (mg g^{-1}), was calculated from the mass balance equation given by

$$Q_e = (C_0 - C_e)Vm^{-1} \tag{1}$$

where C_0 is the initial dye concentration in liquid phase (mg L^{-1}), C_e is the liquid-phase dye concentration at equilibrium (mg L^{-1}), V is the volume of the dye solution used (L), and m is the mass of the adsorbent used (g). It is observed that adsorption amount increases in the following order: PANI NT salt/silica composite > PANI NT salt > conventional PANI salt/silica composite > conventional PANI salt. The extent of AG dye adsorption onto the PANI NT salt/silica composite is highly accessible as compared to the other substrates. This result is explained by the higher surface area of the PANI NT salt/silica composite. The binding sites of the interactions available in the PANI NT salt/silica composite would be larger; hence, more interaction with the anionic dye AG occurred. The dye adsorption kinetics and isotherms of the PANI NT salt/silica composite will be studied in full details in the following sections.

Adsorption kinetics

In order to evaluate the kinetic mechanism which controls the process, the pseudo-first-order [41], pseudo-second-order [42], and intraparticle diffusion [43] models were tested, and the validity of the models were verified by the linear equation analysis log ($Q_e - Q_t$) vs. t, (t/Q_t) vs. t, and Q_t vs. $t^{1/2}$, respectively. Good correlation with the kinetic data explains the dye adsorption mechanism in the solid phase [41-43].

The first model is the pseudo-first-order rate equation:

$$\log(Q_e - Q_t) = \log Q_e - \frac{k_1}{2.303}t \tag{2}$$

where Q_e and Q_t (mg g^{-1}) refer to the amount of dye adsorbed at equilibrium and time t (min), respectively, and k_1 is the rate constant. Figure 4a shows the plot of the pseudo-first-order, and the parameters k_1, Q_e, and the correlation coefficient (R^2) values were determined (Table 1). The curve-fitting plots with correlation coefficient (R^2 = 0.9894) are shown. The second model, the pseudo-second-order reaction, is dependent on the amount of solute adsorbed on the surface of the adsorbent and the amount adsorbed at equilibrium.

The pseudo-second-order model can be represented in the following form:

$$t/Q_t = 1/k_2{Q_e}^2 + t/Q_e \tag{3}$$

where k_2 is the rate constant of the pseudo-second-order model (g mg^{-1} min^{-1}). Figure 4b shows the curve-fitting plot of Equation 3 ((t/Q_t) vs. t), and the parameter values of k_2, Q_e, and R^2 were determined (Table 1). The curve-fitting plots with excellent linearity (R^2 = 0.9995) confirm the applicability of the pseudo-second-order equation.

The third model is an intraparticle diffusion model. Weber and Morris [44] stated that if intraparticle diffusion is the rate-controlling factor, uptake of the adsorbate varies with the square root of time. The root time dependence was expressed by Equation 4:

$$Q_t = k_i t^{1/2} + C \tag{4}$$

where k_i is an intraparticle diffusion rate parameter. Figure 4c shows a plot of Q_t vs. $t^{1/2}$.

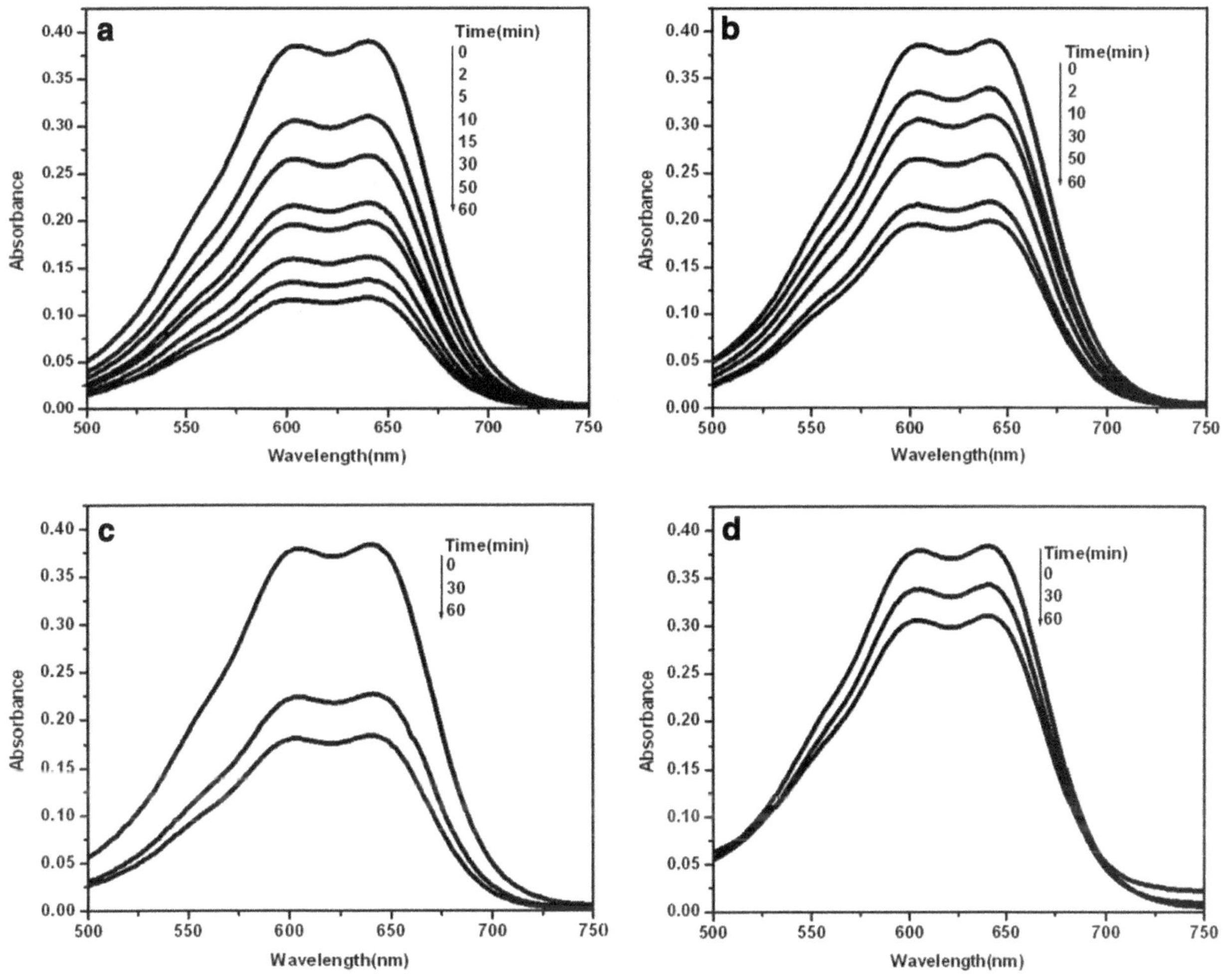

Figure 2 Time-resolved absorption spectra of AG dye adsorption (18.6 mg L^{-1}). With 0.05 g PANI NT salt/silica composite **(a)**, 0.05 g conventional PANI salt/silica composite **(b)**, 0.05 g PANI NT salt **(c)**, and 0.05 g conventional PANI salt **(d)** in the dark at 25°C.

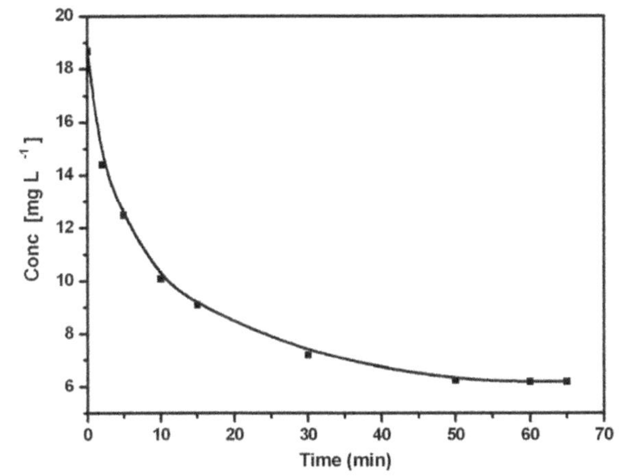

Figure 3 Concentration profile of AG dye with time in the presence of PANI NT salt/silica composite.

As shown in Figure 4c the external surface adsorption (stage 1) is the fastest and completed before 10 min, and then, the stage of intraparticle diffusion control (stage 2) is attained and continues from 10 to 60 min. The slope of the first linear portion (stage 2) characterizes the rate parameter corresponding to the intraparticle diffusion, whereas the intercept of this portion is proportional to the boundary layer thickness. The R^2 value for this diffusion model is 0.9851, and the other values were determined (Table 1). This indicates that the adsorption of AG onto PANI NT salt/silica composite can be followed by an intraparticle diffusion in about 60 min. However, the lines do not pass through the origin (the plots have intercept of 0.62 mg g^{-1}), indicating that intraparticle diffusion is involved in the adsorption process but it is not the only rate-limiting mechanism and that some other mechanisms also play an important role. Surface adsorption and intraparticle diffusion were likely to take place simultaneously.

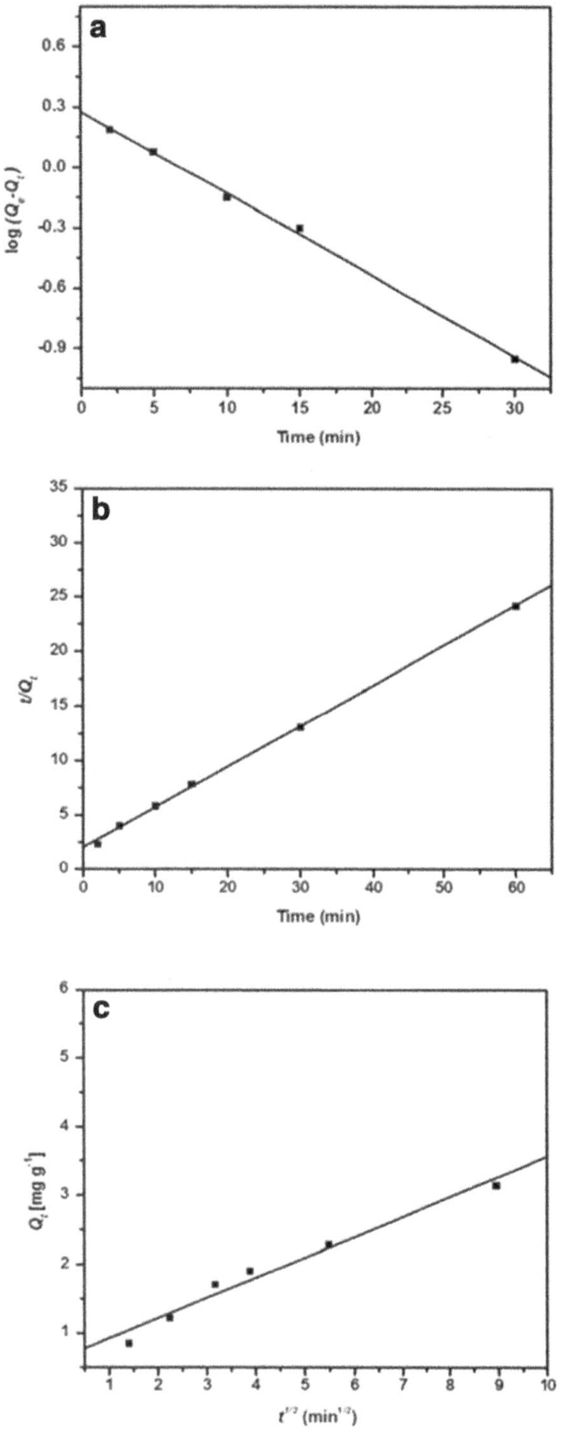

Figure 4 Kinetic plot for AG adsorption onto PANI NT salt/silica composite. **(a)** The pseudo-first-order, **(b)** pseudo-second-order, and **(c)** intraparticle diffusion kinetic plot for the adsorption of AG onto PANI NT salt/silica composite (dye concentration, 18.6 mg L^{-1}; adsorbent dose, 0.05 g; $T = 25°C$).

Adsorption isotherms

The most common adsorption models were used to fit the experimental data Langmuir [45], Freundlich [45], and Tempkin [45] isotherms. The first model is the Langmuir isotherm, which is given as

$$C_e/Q_e = 1/K_l + a_l C_e/K_l \tag{5}$$

where C_e is the equilibrium concentration of the adsorbate (mg L^{-1}), Q_e is the amount adsorbed per unit mass of adsorbent (mg g^{-1}), K_l (L mg^{-1}) is a constant related to the affinity between the adsorbent and the adsorbate, K_l/a_l is the theoretical monolayer saturation capacity Q_0. A linear plot was obtained when C_e/Q_e was plotted against C_e, as shown in Figure 5a. The correlation coefficient R^2 was calculated and equals 0.9961.

The second model is the Freundlich model, which is given by

$$\ln Q_e = \ln K_f + 1/n \ln C_e \tag{6}$$

where n is the Freundlich constant, and K_f ((L mg^{-1})1/n) is the constant correlated to the maximum adsorption capacity. Figure 5b shows that the plots of ln Q_e vs. ln C_e are linear with R^2 equals 0.9945.

The third model is the Tempkin isotherm, which is represented by the following equation:

$$Q_e = B \ln K_t + B \ln C_e \tag{7}$$

K_t is the equilibrium binding constant and corresponds to the maximum binding energy, and B is the constant related to the heat of adsorption. A linear plot was obtained when Q_e was plotted against ln C_e, as shown in Figure 5c. The R^2 was calculated and equals 0.9703.

The results obtained from the adsorption isotherms for the dye by the polymer are shown in Table 2. For the three studied systems, the Langmuir isotherm correlated better than the Freundlich and Tempkin isotherms. The Langmuir adsorption isotherm is the most widely used for the adsorption of a pollutant from a liquid solution, assuming that the adsorption takes place at specific homogeneous sites within the adsorbent [46].

The essential feature of the Langmuir isotherm can be expressed in terms of a dimensionless factor called

Table 1 Kinetic parameters for the adsorption of AG onto PANI NT salt/silica composite

Models	Model coefficients	R^2
Pseudo-first-order	$Q_e = 1.89$ mg g^{-1}	0.9894
	$k_1 = 0.092$ min^{-1}	
Pseudo-second-order	$Q_e = 2.7$ mg g^{-1}	0.9995
	$k_2 = 0.67$ g mg^{-1} min^{-1}	
Intraparticle diffusion	$k_i = 0.29$ mg g^{-1} min^{-1}	0.9851
	$t^{1/2} = 3.97$ min	
	$C = 0.62$ mg g^{-1}	

25°C; initial AG concentration = 18.6 mg L^{-1}.

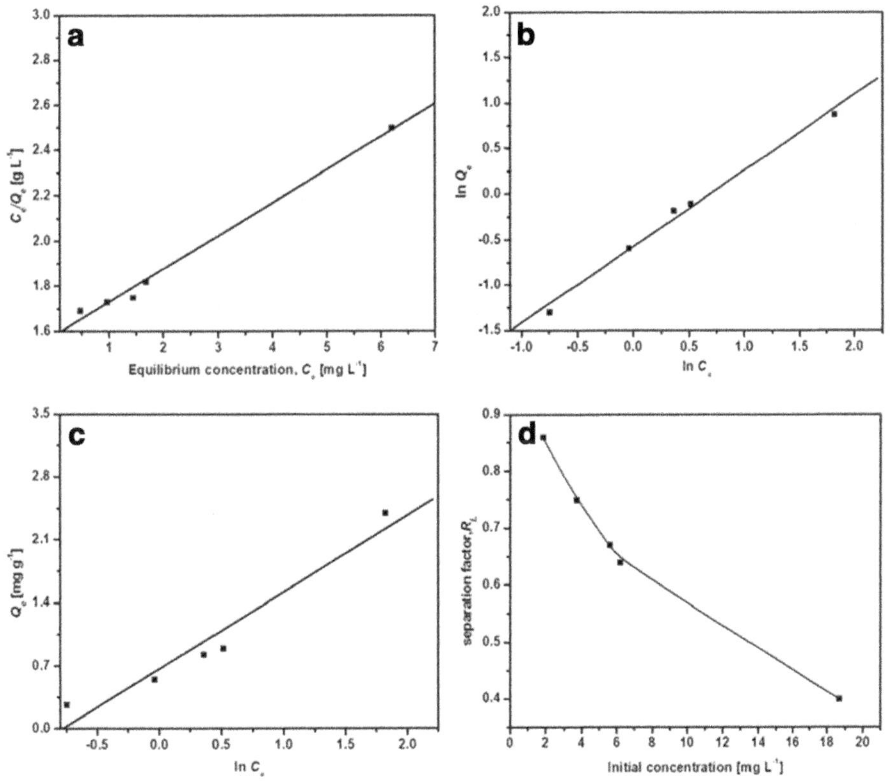

Figure 5 Adsorption isotherm models. (a) Langmuir isotherm model, **(b)** Freundlich isotherm model, **(c)** Tempkin isotherm model, and **(d)** separation factor vs. initial AG concentration.

separation factor (R_L), which is defined by the following equation [47]:

$$R_L = 1/(1 + a_l C_0) \tag{8}$$

where C_0 (mg L^{-1}) is the initial adsorbate concentration, and a_l (L mg^{-1}) is the Langmuir constant related to the energy of adsorption. The value of R_L indicates the shape of the isotherm to be either unfavorable ($R_L > 1$), linear ($R_L = 1$), favorable ($0 < R_L < 1$) or irreversible ($R_L = 0$). Figure 5d shows the variation of R_L with initial AG concentrations.

Table 2 Summary of the Langmuir, Freundlich, and Tempkin isotherm constants, separation factor and linear regression coefficients

Model	Parameters
Langmuir	$Q_0 = 6.896$ mg g^{-1}
	$K_L = 0.6$ L g^{-1}
	$a_L = 0.087$ L mg^{-1}
	$R_L = 0.38$
	$R^2 = 0.9961$
Freundlich	$K_f = 0.56$ L g^{-1}
	$n = 1.19$
	$R^2 = 0.9945$
Tempkin	$K_t = 2.18$ L/mg
	$B = 0.85$
	$R^2 = 0.9703$

The results indicate that R_L values were in the range of 0 to 1, indicating that the adsorption of AG onto PANI NT salt/silica composite is favorable. The observed rate constant $K_{obs,}$ is equal to 0.037 min^{-1}, and it was calculated from Figure 6 by using the following equation:

$$\ln\frac{A_0}{A} = Kt \tag{9}$$

where A_0 (mg L^{-1}) is the initial concentration of AG, A (mg L^{-1}) is the concentration of AG after different times, and t (min) is the time.

Effect of initial AG concentration

The experiment was conducted with different initial concentrations of AG in the presence of 0.05 g of PANI NT salt/silica composite for 60 min. When the initial concentration of the dye was 1.8 mg L^{-1}, the dye was completely absorbed in 10 min. At higher concentrations, the dye was not completely absorbed, indicating

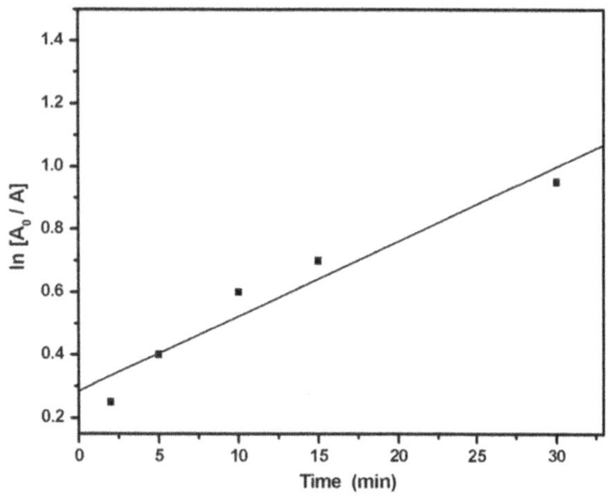

Figure 6 The observed rate constant $K_{obs.}$ Where AG concentration is 18.6 mg L^{-1} and the adsorbent dose is 0.05 g at 25°C.

that there is a saturation limit for the polymer, above which it does not remove the dye. The effects of initial concentration and time on the adsorption of AG by PANI NT salt/silica composite are described in Figure 7.

At lower concentrations, all AG present in the adsorption medium could interact with the binding sites on the surface of the adsorbent, so higher adsorption yields were obtained. At higher concentrations, lower adsorption yields were observed because of the saturation of the adsorption sites.

Conclusions

The adsorption of anionic sulfonated dye AG from aqueous solutions onto PANI NT salt/silica composite was studied with variations in AG concentration. The uptake of the dye was studied using UV-visible spectroscopy. It is concluded that the adsorption amount increases in the following order: PANI NT salt/silica composite > PANI NT salt > conventional PANI salt/silica composite > conventional PANI salt. This is due to the increase in the surface area of polymer nanotube substrates, and hence, the PANI NT salt/silica composite provides a more efficient adsorbent than the conventional PANI salt/silica composite substrates.

The kinetic models were calculated, and the data showed that the second-order kinetic model is the best linearity with the data more than any other kinetic model. Intraparticle diffusion was likely to take place. The data have been analyzed using the isotherm models, and the data showed that Langmuir is the best linearity with the data more than any other isotherm model.

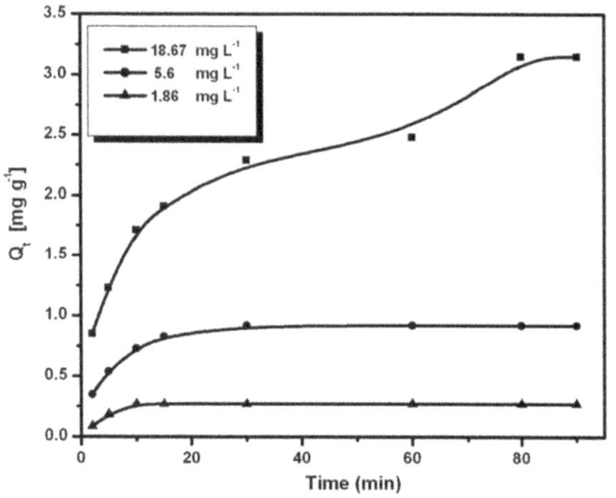

Figure 7 Effect of initial concentration and time on AG adsorption by PANI NT salt/silica composite.

Methods

Chemicals

Aniline (Adwic, Cairo, Egypt) was distilled twice under atmospheric pressure using zinc dust. Ammonium peroxodisulphate (APS) (MP Biomedicals, Santa Ana, CA, USA), acetic acid (Adwic), sulfuric acid (Adwic), silica (60 to 120 meshes with particle size of 0.13 to 0.25 mm (Aldrich, St. Louis, MO, USA), and acid green 25 (Aldrich) were used without further purification.

Preparation of PANI NT salt/silica composite and conventional PANI salt/silica

A solution 0.2 mol L^{-1} aniline and 0.25 mol L^{-1} APS in 250 ml of 0.5 mol L^{-1} acetic acid in the presence of 1.8 g silica according to Stjekal's procedure [40] for the preparation of PANI NT salt/silica composite was used. In the present investigation, the composite was prepared as in the bulk. On the following day, the precipitate was collected on a filter paper, rinsed with 0.5 mol L^{-1} acetic acid, and dried. The conventional PANI salt/silica composite was prepared in the bulk solution as described earlier for the PANI NT salt/silica composite, except that

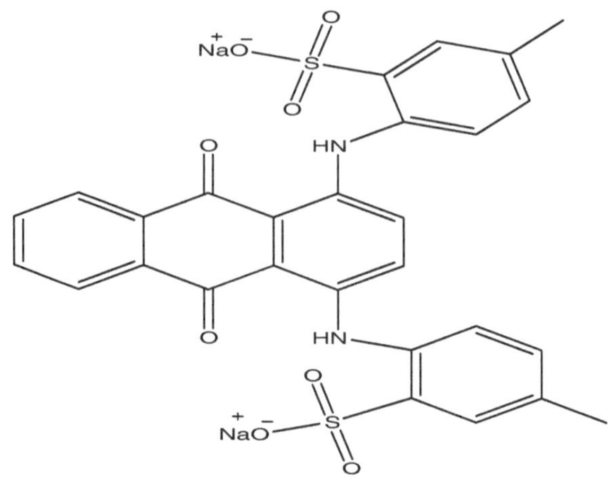

Scheme 2 Molecular structure of AG.

the acetic acid was changed to aqueous sulfuric acid of concentration 0.1 M.

Preparation of acid green 25 solution

The AG has a molecular formula of $C_{28}H_{20}N_2Na_2O_8S_2$ with a molecular weight of 622.57, and Scheme 2 shows the molecular structure of AG. It is a water-soluble dye that is green in color. A stock solution was prepared by dissolving an appropriate quantity of AG in a volume of distilled water. The working solutions were prepared by diluting the stock solution with distilled water to give the appropriate concentration of the working solutions.

Adsorption experiments and characterization

UV-visible absorption spectroscopy (Labomed, Inc, Culver City, CA, USA) was used for the determination of the initial and final concentrations of AG solutions by measuring absorbance at 642 nm. Ten-milliliter AG solutions of different concentrations were mixed with 0.05 g of PANI NT salt/silica composite or conventional PANI salt/silica composite and were stirred at 700 rpm by an electric stirrer for different times in the dark at room temperature. The mixtures of the polymer and dye were separated first by syringe then by a centrifuge (EBA 20, Hitachi, Tokyo, Japan). The morphology of the polymer NTs and their composite with silica were observed with a scanning electron microscope (Fesem Hitachi H-8100). The samples were dispersed in acetone and then transferred onto silicon wafer substrates. The latter was sputtered with a thin layer of Pt/Pd at an atomic ratio of 8:2.

Competing interests

The authors declare that they have no competing interests.

Authors' contributions

Mohamad M Ayad participated in the sequence alignment and drafted the manuscript and the design of the study and performed the statistical analysis. Ahmed Abu El-Nasr participated in some practical work. All authors read and approved the final manuscript.

Acknowledgements

Acknowledge to Mohamad M Ayad who contributed by his lab.
Also acknowledge to Omid Moradi who is the editor to significant revision of the manuscript.

References

1. Ozcan, AS, Ozcan, A: Adsorption of acid dyes from aqueous solutions onto acid-activated bentonite. J. Colloid Interf. Sci. **276**, 39–46 (2004)
2. Baskaralingam, P, Pulikesi, M, Elango, D, Ramamurthi, V, Sivanesan, S: Adsorption of acid dye onto organobentonite. J. Hazard Mater. **128**, 138–144 (2006)
3. Mall, ID, Srivastava, VC, Agarwal, NK: Removal of orange-G and methyl violet dyes by adsorption onto bagasse fly ash kinetic study and equilibrium isotherm analyses. Dyes Pigments. **69**, 210–223 (2006)
4. Attia, AA, Rashwan, WE, Khedr, SA: Capacity of activated carbon in the removal of acid dyes subsequent to its thermal treatment. Dyes Pigments. **69**, 128–136 (2006)
5. Stephenson, RJ, Sheldon, JB: Coagulation and precipitation of mechanical pueffluent.1. removal of carbon and turbidity. Water Res. **30**, 781–792 (1996)
6. Chiou, MS, Chuang, GS: Competitive adsorption of dye metanil yellow in acidic solutions on chemically cross-linked chitosan beads. Chemosphere. **62**, 731–740 (2006)
7. Forgacs, E, Tibor, C, Gyula, O: Removal of synthetic dyes from wastewaters. Environ. Int. **30**, 953–971 (2004)
8. Annadurai, G, Chellapandian, M, Krishnan, MRV: Adsorption of reactive Chitin. Environ. Monit. **59**, 111–119 (1999)
9. Namasivayam, C, Muniasamy, N, Gayatri, K, Rani, M, Ranganathan, K: Removal of dyes from aqueous solution by cellulosic waste orange peel. Bioresour. Technol. **57**, 37–43 (1996)
10. Namasivayam, C, Prabha, D, Kumutha, M: Removal of direct red and acid brilliant blue by adsorption onto banana pith. Bioresour. Technol. **64**, 77–79 (1998)
11. McKay, G, Ramprasad, G, Mowli, PP: Equilibrium studies for the adsorption of dyestuffs from aqueous solutions by low-cost materials. Water. Air. & Soil Pollution. **29**, 273–283 (1986)
12. Ramkrishna, KR, Viaraghavan, T: Dye removal using low cost adsorbents. Water Sci. Technol. **36**, 189–196 (1997)
13. Bhattacharya, KG, Sharma, A: Kinetics and thermodynamics of methylene blue adsorption on neem (Azadirachta indica) leaf powder. Dyes Pigm. **65**, 51–59 (2005)
14. Dogan, M, Alkan, M, Turkyilmaz, A, Ozdemir, Y: Kinetics and mechanisms of removal of methylene blue by adsorption on perlite. J. Hazard. Mater. **109**, 141–148 (2004)
15. Sharma, YC, Uma, SN, Upadhyay, GF: Adsorptive removal of a basic dye from water and wastewater by activated carbon. J. of Applied Science in Environmental Sanitation. **4**, 21–24 (2009)
16. Chowdhury, AN, Jesmeen, SR, Hossain, MM: Removal of dyes from water by conducting polymeric adsorbent. Polym. Adv. Technol. **15**, 633–638 (2004)
17. Debajyoti, M, Giridhar, M, Radhakrishnan, S, Satish, P: Adsorption and desorption kinetics of anionic dyes on doped polyaniline. J. Phys. Chem. B. **113**, 2293–2299 (2009)
18. Mahanta, D, Giridhar, S, Madras, R, Patil, S: Adsorption of sulfonated dyes by polyaniline emeraldine salt and its kinetics. J. Phys. Chem. B. **112**, 10153–10157 (2008)
19. MacDiarmid, AG, Chiang, J, Halpern, M, Wu, WS, Somasiri, NLD, Yaniger, SI: Polyaniline: Interconversion of Metallic and Insulating Forms. Mol. Cryst. Liq. Cryst. **121**, 173–180 (1985)
20. Lofton, EP, Thackeray, JW, Wrighton, MS: Amplification of electrical signals with molecule-based transistors: power amplification up to a kilohertz frequency and factors limiting higher frequency operation. J. Phys. Chem. **90**, 6080–6083 (1986)
21. Kitani, A, Yano, J, Sasaki, K: ECD materials for the three primary colors developed by polyanilines. J. Electroanal. Chem. **209**, 227–232 (1986)
22. Ayad, MM, Salahuddin, NA, Abou-Seif, AK, Alghaysh, MO: Chemical synthesis and characterization of aniline and o-anthranilic acid copolymer. Eur. Polym. J. **44**, 426–435 (2008)
23. Miksa, B, Slomkowski, S: Polypyrrole core/polyacrolein shell latex for protein immobilization. Colloid Polym. Sci. **273**, 47–52 (1995)
24. Saoudi, B, Jammul, N, Abel, ML, Chehimi, MM, Dodin, G.: DNA adsorption onto conducting polypyrrole. Synth. Met. **87**, 97–103 (1997)
25. Wu, MS, Wen, TC, Gopalan, A: In situ UV-visible spectroelectrochemical studies on the copolymerization of diphenylamine with anthranilic acid. Mater Chem Phys. **74**, 58–65 (2002)
26. Camalet, JL, Lacroix, JC, Aeiyach, S, Chane-ching, K, Lacaze, PC: Electrosynthesis of adherent polyaniline films on iron and mild steel in aqueous oxalic acid medium. Synth. Met. **93**, 133–142 (1998)
27. Salaneck, WR, Huang, WS, Lundstrom, I, MacDiarmid, AG: A two-dimensional surface state diagram for polyaniline. Synth. Met. **13**, 291–297 (1986)
28. Paul, EW, Ricco, AJ, Wrighton, WS: Resistance of polyaniline films as a function of electrochemical potential and the fabrication of polyaniline-based microelectronic devices. J. Phys. Chem. **89**, 1441–1447 (1985)
29. Liu, XX, Bian, LJ, Zhang, L, Zhang, LJ: Composite films of polyaniline and molybdenum oxide formed by electrocodeposition in aqueous media. J. Solid State Electrochem. **11**, 1279–1286 (2007)
30. Liu, P, Liu, WM, Xue, QJ: In situ chemical oxidative graft polymerization of aniline from silica nanoparticles. Mater Chem Phys. **87**, 109–113 (2004)
31. Feng, XM, Yang, G, Liu, YG, Hou, WH, Zhu, JJ: Electrodeposition of hybrid film of polyaniline/silica and its pseudocapacitive properties. J Appl Polym Sci. **101**, 2088–2094 (2006)

32. Wang, Y, Wang, X, Li, J, Mo, Z, Zhao, X, Jing, X, Wang, F: Conductive polyaniline/silica hybrids from Sol–Gel process. Adv Mater. **13**, 1582–1585 (2001)
33. Spinks, GM, Dominis, AJ, Wallace, GG: Comparison of Emeraldine Salt, Emeraldine Base, and Epoxy Coatings for Corrosion Protection of Steel During Immersion in a Saline Solution. CORROSION SCIENCE. **59**, 22–31 (2003)
34. Ayad, MM, El-Hefnawey, G, Torad, NL: Quartz crystal microbalance sensor coated with polyaniline emeraldine base for determination of chlorinated aliphatic hydrocarbon. Sens. Actuators B: Chem. **134**, 887–894 (2008)
35. Ayad, MM, El-Hefnawey, G, Torad, NL: A sensor of alcohol vapours based on thin polyaniline film and quartz crystal microbalance. J. Hazard. Mat. **168**, 85–88 (2009)
36. Ayad, MM, Torad, NL: Alcohol vapours sensor based on thin polyaniline salt film and quartz crystal microbalance. Talanta. **78**, 1280–1285 (2009)
37. Tagowska, M, Palys, B, Jackwska, K: Polyaniline nanotubes-anion effect on conformation and oxidation state of polyaniline studied by raman spectroscopy. Synth. Met. **142**, 223–229 (2004)
38. Ayad, MM, Prastomo, N, Matsuda, M, Stejskal, J: Sensing of silver ions by nanotubular polyaniline film deposited on quartz-crystal in a microbalance. Synth. Met. **160**, 42–46 (2010)
39. Ayad, MM, Abu, E-NA: Adsorption of cationic dye (methylene blue) from water using polyaniline nanotubes base. J. Phys. Chem. C. **114**, 14377–14383 (2010)
40. Konyushenko, EN, Stejskal, J, Sedenkova, I, Trchova, M, Sapurina, I, Cieslar, M: Polyaniline nanotubes: the conditions of formation. Polym. Int. **55**, 31–33 (2006)
41. Wu, F, Tseng, R, Juang: KineticModeling of liquid-phase adsorption of reactive dyes and metal ions on chitosan. Water Res. **35**, 613–618 (2001)
42. Chiou, MS, Li, HY: Equilibrium and kinetic modeling of adsorption of reactive dye on cross-linked chitosan beads. J. Hazard. Mat. **93**, 233–248 (2002)
43. Uzun, I: Kinetics of the adsorption of reactive dyes by chitosan. Dyes Pigm. **70**, 76–83 (2006)
44. Weber, WJ, Morris, JC: Kinetics of adsorption on carbon from solution. J. Sanit Eng. Div. ASCE. **89**, 31–59 (1963)
45. Langmuir, I: The adsorption of gases on plane surfaces of glass, mica and platinum. J. Am. Chem. Soc. **40**, 1361–1403 (1918)
46. Langmuir, I: The constitution and fundamental properties of solids and liquids. J. Am. Chem. Soc. **38**, 2221–2295 (1916)
47. Weber, TW, Chakravorti, RK: Pore and solid diffusion models for fixed-bed adsorbers. J. Am. Inst. Chem. Eng. **20**, 228–238 (1974)

Effect of 2-mercaptoethanol as capping agent on ZnS nanoparticles: structural and optical characterization

Abbas Rahdar

Abstract

In this work, we report the effect of a capping agent on the structural and optical properties of nanocrystalline ZnS particles, which have been synthesized by co-precipitation method. The structural properties of ZnS nanoparticles have been characterized by X-ray diffraction (XRD) analysis. The XRD patterns show a hexagonal structure in the nanoparticles. The mean crystallite size calculated from the XRD patterns has been found to be in the range of 1.80 to 2.45 nm with the increase in molar concentration of the capping agent. Absorption spectra have been obtained using a UV-vis spectrophotometer to find the optical direct band gap. We also found that optical band gap (E_g) increases with the increase in molar concentration of the capping agent. This behavior is related to size quantization effect due to the small size of the particles.

Keywords: ZnS nanoparticles, Chemical co-precipitation method, Capping agent, Optical band gap

Background

Semiconductor nanocrystals represent a class of materials that have hybrid molecular and bulk properties. They have attracted much attention over the past few years because of their novel properties which originate from quantum confinement effect [1-4]. Quantum confinement effect modifies the electronic structure of the nanocrystals when the sizes of the nanoparticles are comparable to that of the Bohr excitonic radius of those materials. When the particle radius falls below the excitonic Bohr radius, the band gap energy is widened, leading to a blueshift in the band gap, emission spectra, etc. On the other hand, the surface states will play a more important role in the nanoparticles, due to their large surface-to-volume ratio with a decrease in particle size (surface effects). In the case of semiconductor nanoparticles, radiative or nonradiative recombination of an exciton at the surface states becomes dominant in its optical properties with a decrease of particle size. Therefore, the decay of an exciton at the surface states will influence the qualities of the material for an optoelectronic device. These size-dependent optical properties have many potential applications in the areas of solar energy conversion, light-emitting devices, chemical/biological sensors, and photocatalysis [5-9].

Wide band gap II-VI semiconductors are expected to be the novel materials for the optoelectronic devices. ZnS, which is an important member of this family, has been extensively investigated as it has numerous applications to its credit. ZnS has been used widely as an important phosphor for photoluminescence, electroluminescence, and cathodoluminescence devices due to its better chemical stability compared to other chalcogenides such as ZnSe. In optoelectronics, it finds use as light-emitting diode, reflector, dielectric filter, and window material. The synthesis of ZnS remains a topic of interest for researchers, as new synthetic routes are being explored to get a single phase material via an economically and technically viable method. In the present paper, chemical method has been used to synthesize the ZnS nanoparticles. In the present paper, ZnS nanoparticles are synthesized by using mercaptoethanol as capping agent. This paper is organized as follows: The structural and optical properties of synthesized nanoparticles are presented in the 'Results and discussion' section in detail, followed by a 'Conclusion' section. In the 'Methods' section, we give a description of materials and synthesis method of ZnS nanoparticles with different molar concentrations of the capping agent.

Correspondence: a.rahdar@uoz.ac.ir
Department of Physics, Faculty of Science, University of Zabol, Zabol 98615538, Iran

Results and discussion

Structural characterization

The X-ray diffraction (XRD) patterns of the prepared ZnS nanoparticles with different molar concentrations of the capping agent ($HOCH_2CH_2SH$) are shown in Figure 1. All of the crystalline Bragg peaks in the XRD pattern ((111), (220), and (311) planes) are in good agreement with the diffraction data of a hexagonal structure with cell parameters a = 3.800 Å and c = 6.230 Å from the JCPDS card. Furthermore, the peak broadening in the XRD patterns clearly indicates the formation of ZnS nanocrystals with a very small size.

The peak broadening at a lower angle is more meaningful for the calculation of particle size. The mean crystallite size (D) of nanoparticles was also estimated using the Scherrer formula (Scherrer 1918) using (111) reflection from the XRD pattern as follows:

$$D = \frac{0.9\lambda}{B\cos\theta}, \quad (1)$$

where λ, B, and θ are the X-ray wavelength of the radiation used (K_α(Cu) = 0.154056 nm), the full width at half maximum of the diffraction peak, and the Bragg diffraction angle, respectively.

The values of mean crystallite size obtained from XRD for different molar concentrations of the capping agent are listed in Table 1.

Optical characterization

The absorption spectra of the different samples are shown in Figure 2. The absorption edge is observed in the range of 324 to 270 nm, which is blueshifted compared to bulk ZnS. As the molar concentration of the capping agent increases, the absorption edge shifts to a lower wavelength compared to ZnS (A). This blueshift of the absorption edges for different-sized nanocrystals is related to the size decrease of particles and is attributed to the quantum confinement limit reaching of nanoparticles. The quantum confinement effect is expected for semiconducting nanoparticles, and

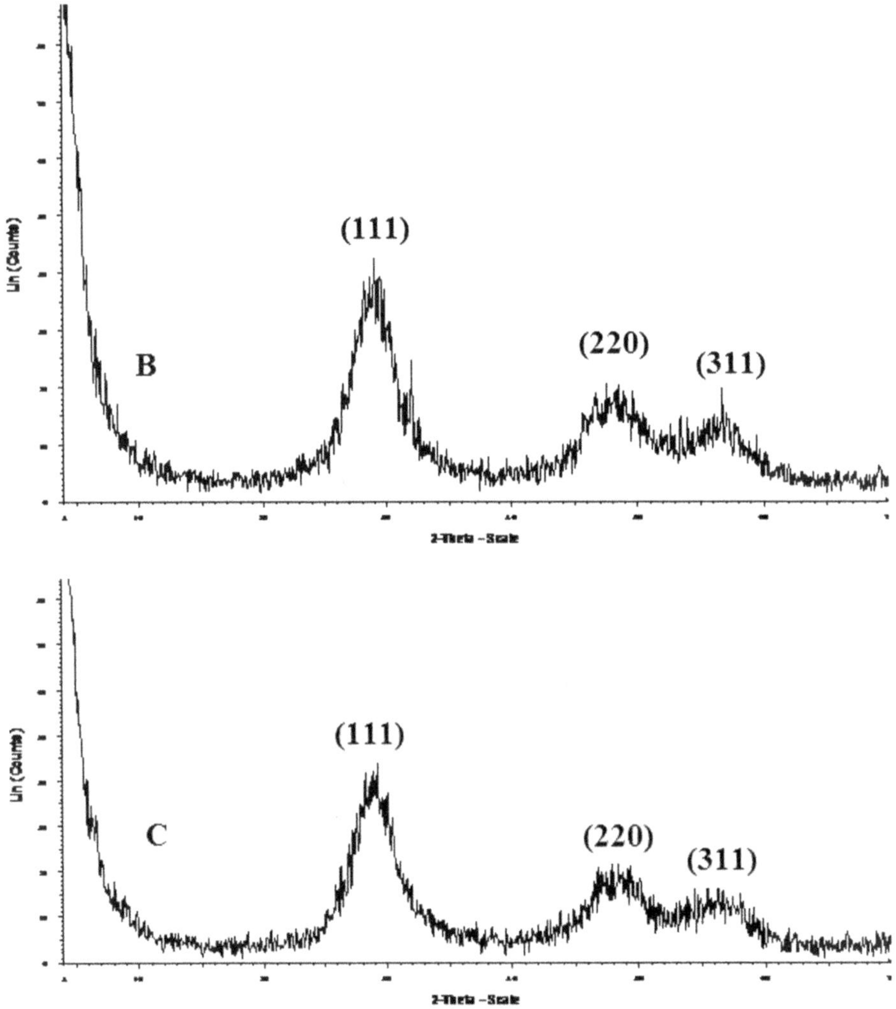

Figure 1 XRD patterns of ZnS nanoparticles. With different molar (M) concentrations of capping agent: 0.1 M (B) and 0.3 M (C).

Table 1 Mean crystallite size and optical band gap variation of ZnS nanoparticles

Sample	Molar concentration of capping agent	Mean crystallite size (nm)	Optical band gap E_g (eV)
ZnS (A)	0.0	2.45	3.83
ZnS (B)	0.1	2.38	4.27
ZnS (C)	0.3	2.22	4.49
ZnS (D)	0.5	1.80	4.59

the absorption edge will be shifted to a higher energy when the particle size decreases [10,11].

The obtained direct optical band gap values for different samples are shown in Table 1. It is necessary to mention that the optical direct band gap values of the ZnS nanoparticles were determined using Tauc's relation [12,13]:

$$\alpha h\nu = \alpha_0 \left(h\nu - E_g\right)^{1/2}, \quad (2)$$

where hv, α_o, and E_g are the photon energy, a constant, and optical band gap of the nanoparticles, respectively. Absorption coefficient (α) of the powders at different wavelengths can be calculated from the absorption spectra. Finally, the values of E_g were determined using extrapolations of the linear regions of the plot of $(\alpha h\nu)^2$ versus $h\nu$.

As seen in Table 1, the values of optical band gap 'E_g' increase with the increase in molar concentration of the capping agent, and therefore, the decrease in particle size as mentioned earlier is due to the quantum confinement effect. Also, the variation of optical band gap with particle size is shown in Figure 3. As illustrated in Figure 3, the optical band gap values of nanoparticles have changed from 3.83 to 4.59 eV by decreasing the particle sizes. The increase in band gap with increase in molar concentration of the capping agent is attributed to size quantization effect due to the small size of the particles [14].

These semiconductor nanocrystals were then subjected to transmission electron microscopy (TEM; JEOL JEM3010, JEOL Ltd., Tokyo, Japan) for characterization. Figure 4 shows the TEM image of ME-capped ZnS semiconductor nanocrystals.

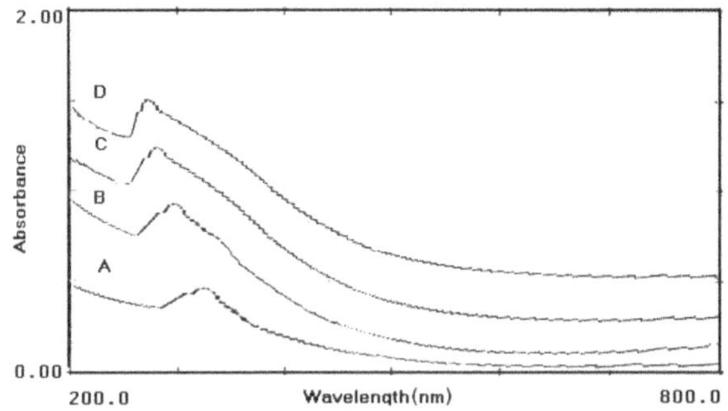

Figure 2 **Absorption spectra of different samples of ZnS nanoparticles.** 0 M (**A**), 0.1 M (**B**), 0.3 M (**C**), and 0.5 M (**D**).

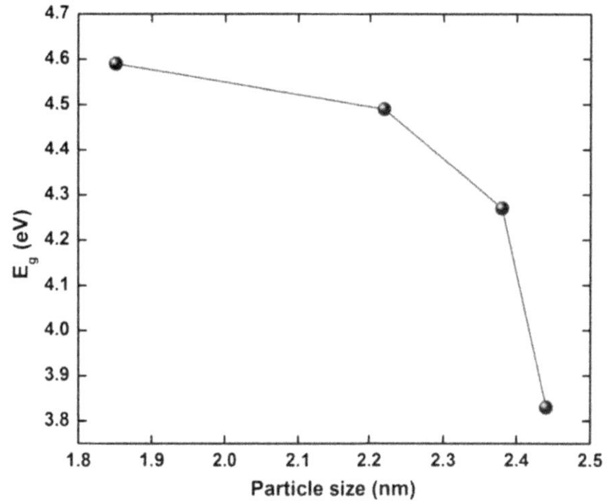

Figure 3 **The variation of optical band gap with particle size.**

Relation particle size with chemical reaction rate

Chemical reaction rate directly affects the time evolution of the number of nuclei, which determines both nucleation and growth process. First, the influence on nucleation is obvious: nucleation is faster when the chemical reaction is faster. Second, growth will be strongly influenced by the nuclei number which already formed at a given time. A great number of nucleation favor a fast autocatalytic growth, giving rise to a large number of small particles. Chemical reaction controls this kind of growth, in which the autocatalytic growth is faster when the chemical reaction is faster. However, in nanoparticle formation, there is another contribution to the growth molecules on the surface of a small particle which will tend to diffuse through the solution and add to the surface of a larger particle (growth by ripening). A slow chemical reaction

Figure 4 **TEM image of ME-capped ZnS semiconductor nanocrystals (0.1 M (B)).**

favors continuous nuclei, keeping always a certain number of nuclei in the system. As a result, growth by ripening can take place during the whole process. This fact explains the bigger particle size obtained from a slow reaction. One can conclude that a slow chemical reaction rate is associated with a more important ripening contribution to the growth. A high number of nuclei are still forming at this stage when the reaction is slow at the same time; some particles have already grown to the final value of size. This means that in this case (slow reaction rate), nucleation and growth take place simultaneously. This overlapping of nucleation and growth processes, which is more pronounced as the chemical reaction is slower, leads to larger nanoparticle sizes [15].

Rate of reaction depends on the molar concentration of the reactant solution and increases with the increase in molar concentration of the reactant solution. In the present study, the molar concentration of the capping agent varies from 0 to 0.5 M; the reaction rate is highest for the 0.5 M solution, and hence, the particle size obtained is smallest for the 0.5 M solution as compared to other materials in the series, which is in consistent with the above made argument.

Conclusions

It is possible to produce different-sized ZnS nanoparticles using a simple chemical method using different molar concentrations of the capping agent. XRD and optical band gap data have been obtained to confirm the nanosize of these materials. It is also observed that the particle size depends on the molar concentration of the capping agent. A decrease in formation rate of nanoparticles gives rise to a larger final particle size for all the studied synthesis conditions. As the particle size depends on the molar concentration of the doping agent, a decrease in the size of particle is observed with the increase of molar concentration of the capping agent. The mean crystallite size range of particles was between 1.85 and 2.44 nm, depending on molar concentration of the capping agent. The optical band gap values of ZnS nanoparticles have changed from 3.83 to 4.59 eV by increasing the molar concentration of the capping agent. These values exhibit a blueshift in E_g which is related to the size decrease of the particles and to the quantum confinement limit reaching of nanoparticles.

Considering these results, the chemical co-precipitation method using 2-mercaptoethanol as a capping agent is very efficient for the preparation of ZnS nanoparticles in order to control the particle size and also for modern optoelectronic technology and other industrial applications.

Methods

Materials

Zinc chloride ($ZnCl_2$) and sodium sulfide (Na_2S) as starting materials, 2-mercaptoethanol ($HOCH_2CH_2SH$) as

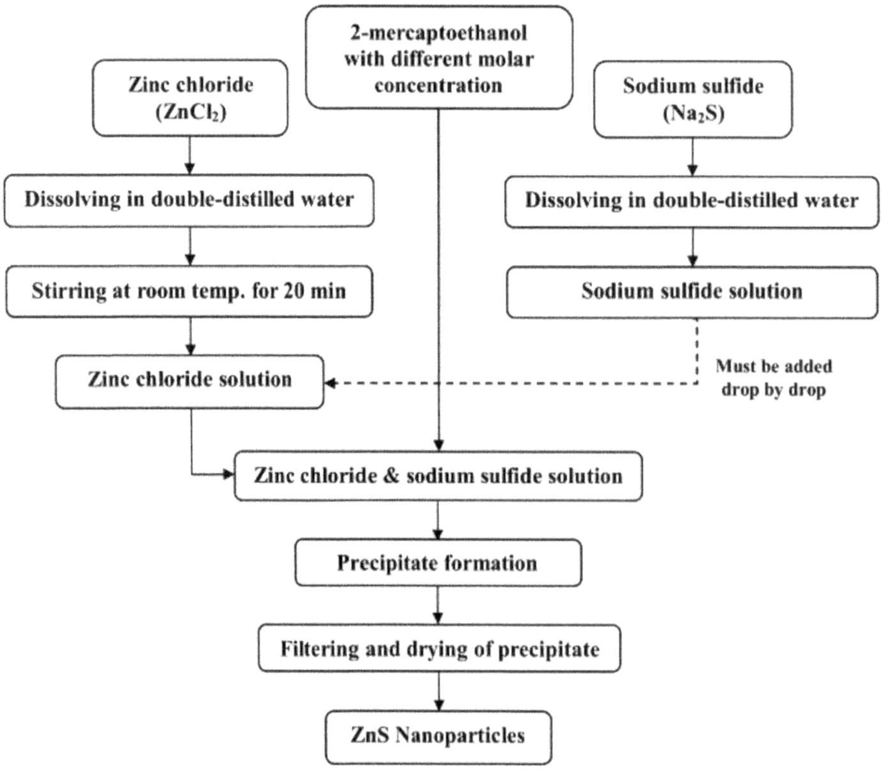

Figure 5 Flow chart of the preparation ZnS nanoparticles by chemical co-precipitation method.

a capping agent for the control particle size, and double-distilled water as dispersing solvent were used to prepare ZnS nanoparticles.

Synthesis of ZnS nanoparticles by co-precipitation method

The ZnS nanoparticles were synthesized using the chemical co-precipitation method as follows. First, $ZnCl_2$ was dissolved in double-distilled water (0.1 M), and then, the obtained molar solution was stirred for 20 min at room temperature to achieve complete dissolution. Sodium sulfide (Na_2S) was also dissolved in double-distilled water separately as per molar concentration. Afterwards, first, sodium sulfide solution was added drop by drop to the zinc chloride solution, and then, an appropriate amount of 2-mercaptoethanol as a capping agent was added to the reaction medium to control the particle size of ZnS. As a result of this, the white precipitate of the ZnS nanoparticles is formed slowly in the solution. In the final step, the obtained precipitate was filtered and dried at room temperature to remove both water and organic capping and other byproducts which formed during the reaction process. After sufficient drying, the precipitate was crushed to fine powder with the use of mortar and pestle. It is mentioned that the synthesis was done by passing nitrogen gas.

For the study of the effect of the capping agent on structural and optical properties of ZnS nanoparticles, different samples of nanoparticles have been obtained by changing the molar concentration of the capping agent, namely ZnS (A), ZnS (B), ZnS (C) and ZnS (D), as the amount of molar concentration of the capping agent used in the preparation is 0, 0.1, 0.3and 0.5 M, respectively. The synthesis process of ZnS nanoparticles by the co-precipitation method is summarized in a flow chart shown in Figure 5.

Powder characterization

The XRD patterns of ZnS nanoparticles were recorded using a Bruker system (Bruker Optik GmbH, Ettlingen, Germany) using CuK_α radiation (λ = 0.154056 nm) with 2θ ranging from 5° to 70°. The optical absorption spectra of nanoparticles were measured using a USB-2000 UV-vis spectrophotometer (Ocean Optics, Inc., Dunedin, FL, USA). Therefore, the obtained nanopowders have been suspended in glycerol using a magnetic stirrer, and their optical absorption spectra have been recorded at room temperature over the range of 200 to 800 nm to determine the optical band gap values.

Competing interests

The author has no competing interests.

Acknowledgment

The author would like to thank Mr. A. Rahdar, Mr. R. Hakimi-Pooya, Mrs. M. Asudeh, Mr. H. Rahdar, and Mrs. Heidari Mokarrar for their support and assistance with this project.

References

1. Rossetti, R, Ellison, JL, Gibson, JM, Brus, LE: Size effects in the excited electronic states of small colloidal CdS crystallites. J. Chem Phys. **80**, 4464 (1984)
2. Sang, W, Qian, Y, Min, J, Li, D, Wang, L, Shi, W, Liu, Y: Microstructural and optical properties of ZnS:Cu nanocrystals prepared by an ion complex transformation method. Solid State Commun. **121**, 475–478 (2002)
3. Prabhu, RR, Abdul Khadar, M: Characterization of chemically synthesized CdS nanoparticles. Pramana- J. Phys. **5**, 801 (2005)
4. Wageh, S, Ling, ZS, Xu-Rong, X: Growth and optical properties of colloidal ZnS nanoparticles. J. Cryst. Growth **255**, 332–337 (2003)
5. Rajeswar, K, Tacconi, NR, Chenthamarakshan, CR: Semiconductor-based composite materials: preparation, properties, and performance. Chem. Mater. **13**, 2765 (2001)
6. Alivisatos, AP: Semiconductor clusters, nanocrystals, and quantum dots. Science **271**, 933–937 (1996)
7. Anderson, MA, Gorer, S, Penner, RM: A hybrid electrochemical/chemical synthesis of supported, luminescent cadmium sulfide nanocrystals. J. Phys. Chem. B **101**, 5895 (1997)
8. Henshaw, G, Parkin, IP, Shaw, G: Convenient low-energy synthesis of metal sulfides and selenides; PbE, Ag_2E, ZnE, CdE (E = S, Se). Chem. Commun. **10**, 1095–1096 (1996)
9. Hirai, T, Bando, Y, Komasawa, I: Immobilization of CdS nanoparticles formed in reverse micelles onto alumina particles and their photocatalytic properties. J. Phys. Chem. B **106**, 8967–8970 (2002)
10. Zhu, H, Yang, D, Yu, G, Zhang, H, Yao, K: A simple hydrothermal route for synthesizing SnO_2 quantum dots. Nanotechnology **17**, 2386 (2006)
11. Sarhaddi, R, Shahtahmasebi, N, Rezaee Rokn-Abadi, M, Bagheri-Mohagheghi, MM: Effect of post-annealing temperature on nano-structure and energy band gap of indium tin oxide (ITO) nano-particles synthesized by polymerizing–complexing sol–gel method. Physica E **43**, 452–457 (2010)
12. Tauc, J: Optical Properties of Solids. Academic Press, New York (1966)
13. Liu, CM, Zu, XT, Wei, QM, Wang, LM: Fabrication and characterization of wire like SnO_2. J. Phys D: Appl. Phys. **39**, 2494 (2006)
14. Amaranatha Reddy, D, Divya, A, Murali, G, Vijayalakshmi, RP, Reddy, BK: Synthesis and optical properties of Cr doped ZnS nanoparticles capped by 2-mercaptoethanol. Physica B **406**, 1944–1949 (2011)
15. de Dios, M, Barroso, F, Tojo, C, Blanco, MC, Lopez-Quintela, MA: Effects of the reaction rate on the size control of nanoparticles synthesized in microemulsions. Colloids and Surfaces A: Physicochem. Eng. Aspects **83**, 270 (2005)

Novel design of combinational and sequential logical structures in quantum dot cellular automata

Bahniman Ghosh[1], Shoubhik Gupta[2], Smriti Kumari[2] and Akshaykumar Salimath[2*]

Abstract

Quantum dot cellular automata (QCA) is a emerging nanotechnology that promises smaller size and lower power consumption, with faster speed compared to the transistor-based technology. In this paper, we have proposed novel 8-into-3 bit simple encoder, 4-into-2 bit priority encoder, scan flip-flop, and pseudo-random bit sequence generator designs in QCA. These circuits are useful components for the design of many logical and functional circuits. Simulation results of the proposed QCA circuits are obtained by using the QCA designer tool. The correctness of the proposed circuits is hence confirmed.

Keywords: Quantum dot cellular automata, Priority encoder, Simple encoder, Scan flip-flop, Pseudo-random bit sequence generator

Background

Complementary metal-oxide-semiconductor (CMOS) technology is in rampant use in current-day semiconductor fabrication. But there is a paradox in CMOS that power consumed increases with increase in speed. But some applications require more speed and less power. New technologies like carbon nanofield effect transistor and quantum dot cellular automata (QCA) have the capability to provide more integration and speed along with less power consumption and high parallel processing [1,2]. An encoder is a digital circuit that has two *n* input lines and *n* output lines. Corresponding to the input value, binary codes are generated by output lines [3]. A priority encoder is an encoder circuit named so because it carries out the priority function, i.e., in case more than one inputs are equal to 1 and at the same time, precedence will be given to the input having the highest priority. Scan flip-flop contains a multiplexer (MUX) to select either a normal operation with data input or scan operation with scan input [4]. It has a control input to select either data or scan input. It is bigger than normal flip-flop (as MUX is included here).It adds nearly 20% to 30% of area per flip-flop. This paper deals with the QCA implementation of some of these combinational and sequential circuits.

* Correspondence: akshaykumarls@gmail.com
[2]Department of Electrical Engineering, Indian Institute of Technology, Kanpur, 208016, India
Full list of author information is available at the end of the article

QCA basics

QCA cells

The principle of quantum dot cellular automata was first proposed by Prof. CS Lent [1]. In order to implement a system that encodes information in the form of electron position, it becomes necessary to construct a vessel in which an electron can be trapped and 'counted' as present or absent.

As shown in Figure 1, a quantum dot does just this by establishing a region of low potential surrounded by a ring of high potential. In ordinary form, this technology is based on bi-stable QCA cell's interaction. The cell is constructed from four quantum dots and charged with two free electrons, which are able to tunnel between adjacent dots [5]. These electrons tend to occupy antipodal sites as a result of their mutual electrostatic repulsion. Thus, there exist two equivalent energetically minimal arrangements of the two electrons in the QCA cell, as shown in Figure 1. These two arrangements are denoted as cell polarization. By using cell polarization to

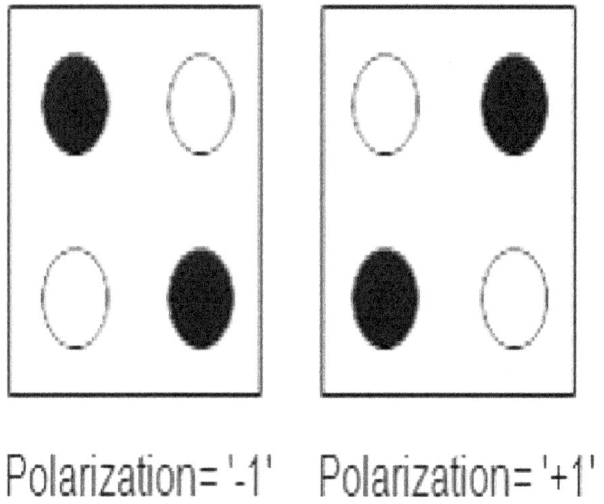

Figure 1 QCA cells.

represent logic '1' and to represent logic '0', binary information is encoded in the charge configuration of the QCA cell.

QCA wires

Other important component in QCA designing is the wire. In a QCA wire, the propagation of binary signal takes place from input to output due to the electrostatic interactions between cells. Since the polarization of each cell tends to align with that of its neighbors, a linear arrangement of standard cells is used to transmit binary information from one point to another [6,7]. In this wire, all free cells align in the same direction as the driving cell (input cell), so the information contained in the state of the input is transmitted down the wire. Besides, the separation between dots and cells is a key parameter giving Coulombic effect in QCA application in conventional form.

A QCA wire is shown in Figure 2. Moreover, in a QCA wire, all of the computational power is provided by the Coulomb interaction between cells in absence of electrical current flow between cells and hence no power dissipation. Similar to the binary wire, there is another type of wire that is used to transfer data from one place to another. This is called inversion chain. In an inversion chain all the QCA cells are 45° rotated from the normal QCA cells. Each cell of the chain inverts the signal due to Coulombic forces between the electrons. Hence, depending on the number of cells between input and output, the output can be same or the inverted value of the input [8]. For wire crossover, it uses a total of three layers for designing crossovers as shown in Figure 3. The multilayer crossover appears conceptually simple but there are questions about its realization, since it requires two overlapping active layers.

QCA clock zones

QCA circuits require a clock to function properly. QCA clocks change the tunneling barrier between quantum dot so as to facilitate the movement of electrons within a cell and allow them to change their configuration in a predefined manner [9]. There are two types of switching of tunneling barrier possible, which are adiabatic and abrupt. In adiabatic switching the tunneling barriers among the dots are raised slowly to bring the cell in a fixed polarization. Similarly, the barriers are lowered slowly to make electrons free to move among the quantum dots. Adiabatic switching is preferred over abrupt switching because it ensures that at every point of time during the operation, the circuit will remain in a ground (stable) state and will not move to an excited state permanently [10,11]. In this work we have used adiabatic switching in clocks. There are four QCA clocks, namely, clock 0, clock 1, clock 2, and clock 3 as shown in Figure 4. Each clock has four parts which are high level, falling edge, low level, and rising edge.

These phases are named *hold, release, relax,* and *switch,* respectively. Figure 4 shows the four phases of clock. The main function of QCA clocks is to ensure the proper transfer of the data from one place to another in a circuit [12]. During the *hold* state the tunneling barrier of the quantum dots is high; hence, electrons are highly localized and have a fixed polarization. During the release state, the tunneling barrier is decreased and electrons slowly become free to move. In the relax state electrons are completely free to tunnel among quantum dots and do not have any influence on the neighboring cells. In the switch state the tunneling

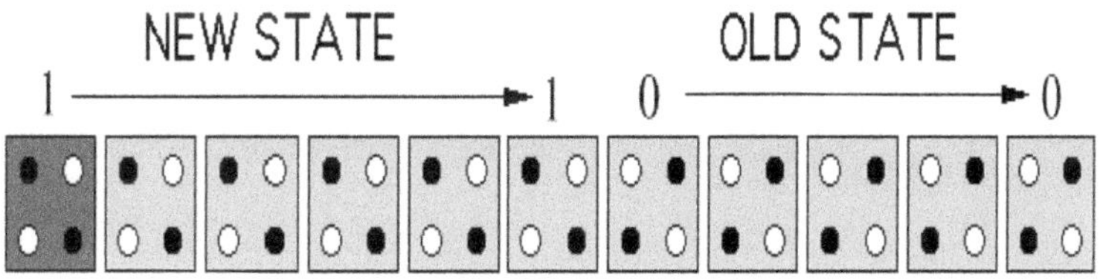

Figure 2 QCA wire.

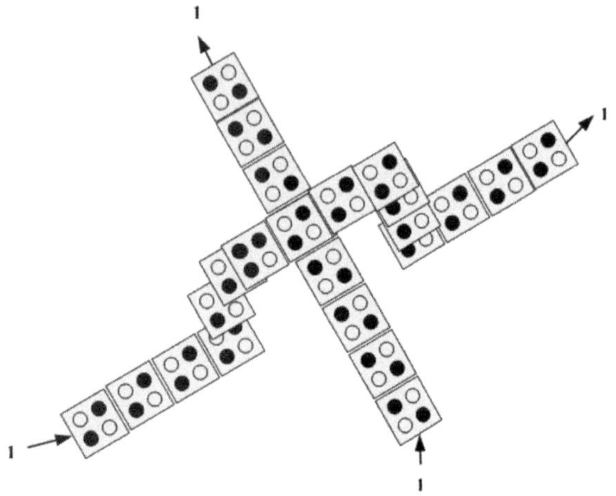

Figure 3 Multilayer crossover.

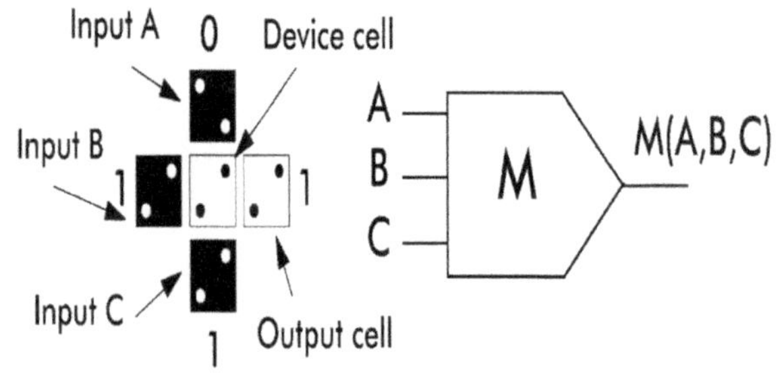

Figure 5 Majority gate.

barrier is slowly raised constraining the electrons to the most stable state [13,14].

Majority gate and inverters

Any QCA circuit can be efficiently built using only majority gates and inverters. Figure 5 shows a majority gate with three inputs, i.e., A, B, and C, and output M (A, B, and C). A majority gate performs following function: Maj (A, B, C) = AB + BC + CA. AND and OR gates can be constructed using the majority gate easily by assigning one of the inputs to the majority gate as 0 and 1, respectively [15]. Figure 6 shows the QCA layout of inverter.

Proposed designs

The 8-3bit simple encoder

Figure 7 shows the QCA design of 8:3 bit simple encoder where D0 to D7 are inputs and Y0, Y1, and Y2 are outputs, respectively. Figure 8 shows the simulation result of proposed circuit.

The 4–2 bit priority encoder

Priority encoders are used in a number of computer systems as well as other applications. For example, when several processes, modules, or units request a single hardware (or software) resource, a decision has to be made to allow a single request to use such a resource. The priority encoder implements a fixed selection function where the resource is granted to the request with the highest priority. The basic function of a 4-bit *priority encoder* is to convert the 4-bit input into a binary representation. If the input n is high, all lower inputs (n – 1... 0) are ignored. In Figure 9, D1, D2, D3, and D4 are four inputs respectively, and Y1, Y2, and V are outputs. Figure 10 shows the simulation result of the proposed circuit.

Scan flip-flop

Flip-flops and latches are fundamental building blocks of sequential digital circuits. Due to significant inter- and intra-die process variations, it is necessary to

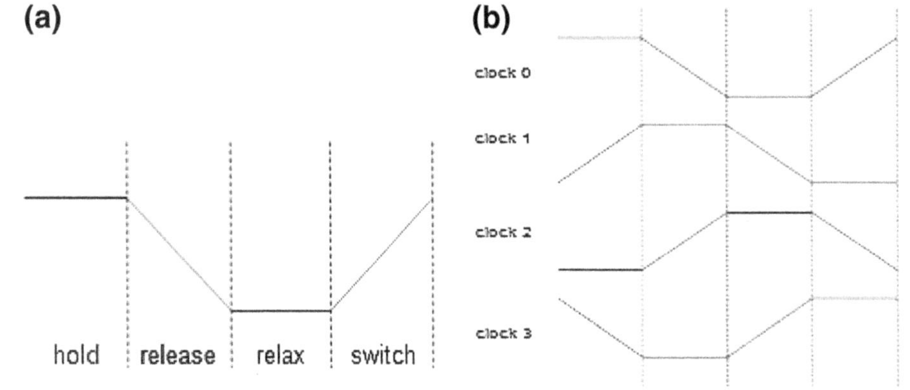

Figure 4 Clocking scheme.

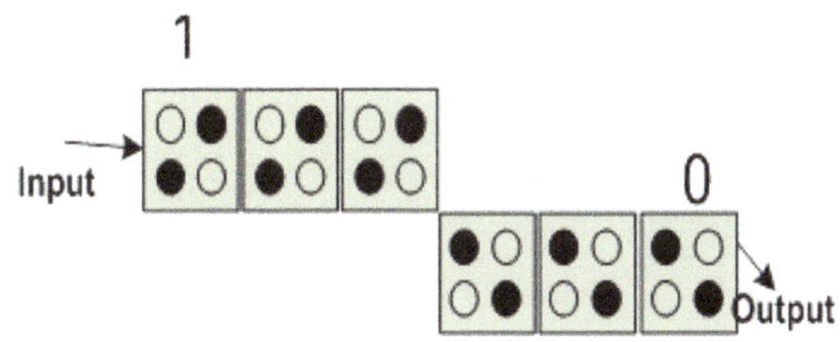

Figure 6 QCA inverter.

ensure correct logical and temporal functionality, so we need to carry out both functional as well as timing checks on the fabricated design [16]. Figure 11 shows the block diagram of a scan flip-flop, along with the D flip-flop it replaces. It consists of a multiplexer and a D flip-flop. A and B are the inputs of the multiplexer and S acts as a select signal. The S signal is low during normal operation of the scan flip-flop. In scan mode SE is high. Now the D input of the flip-flop is driven by the scan-in signal B. They are very similar to the flip-flops with an additional enable pin, enabling them to function both as an ordinary (operating mode) flip-flop as well as the scan flip-flop(testing mode). They are used during testing and are much preferred over built-in self-test units. Figure 12 shows the QCA layout of scan flip-flop where A and B are input to MUX and S is select pin. The output of MUX is input to D flip-flop and Q is the output. Figure 13 shows the simulation result of proposed design.

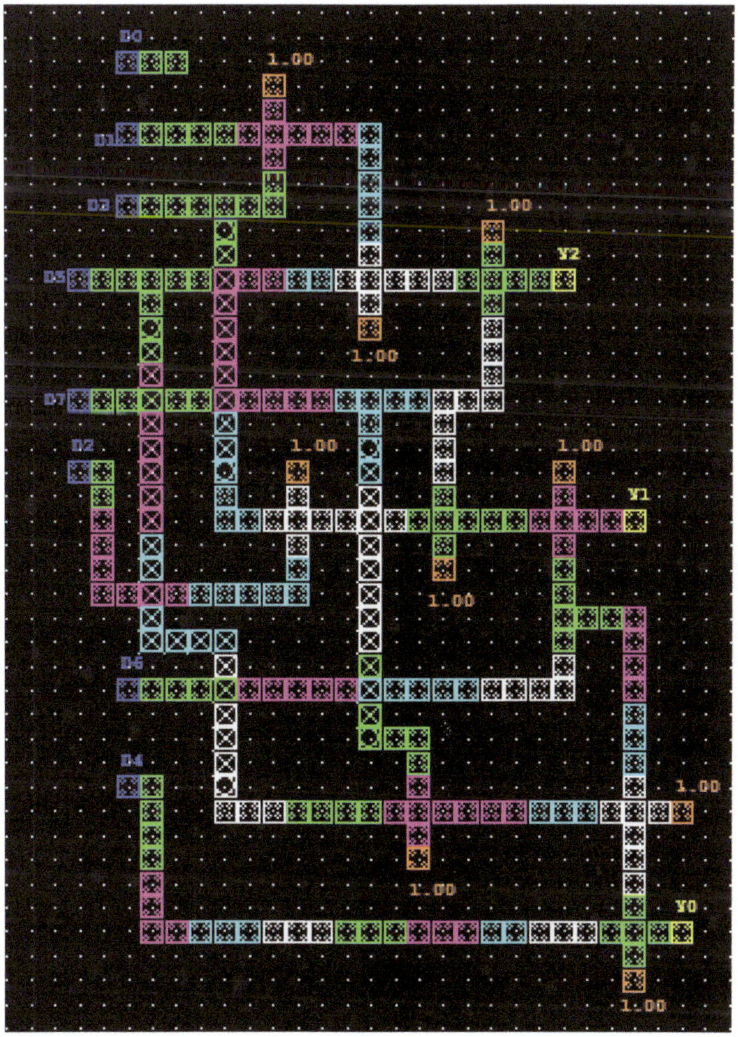

Figure 7 QCA layout of 8-3 bit simple encoder.

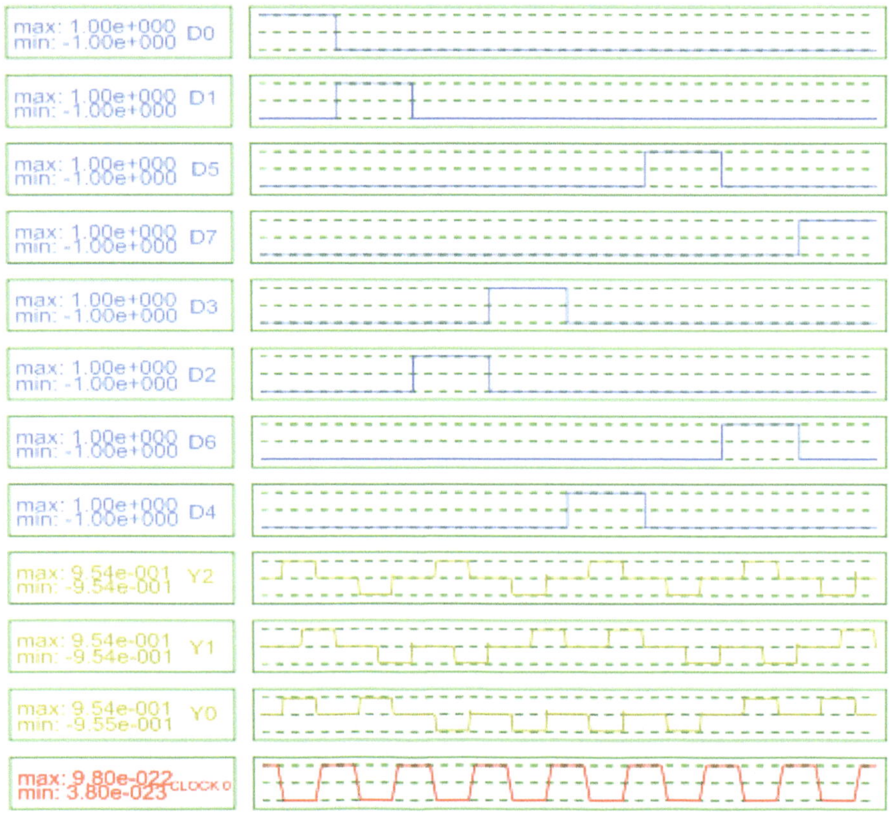

Figure 8 Simulation result of 8-3bit simple encoder.

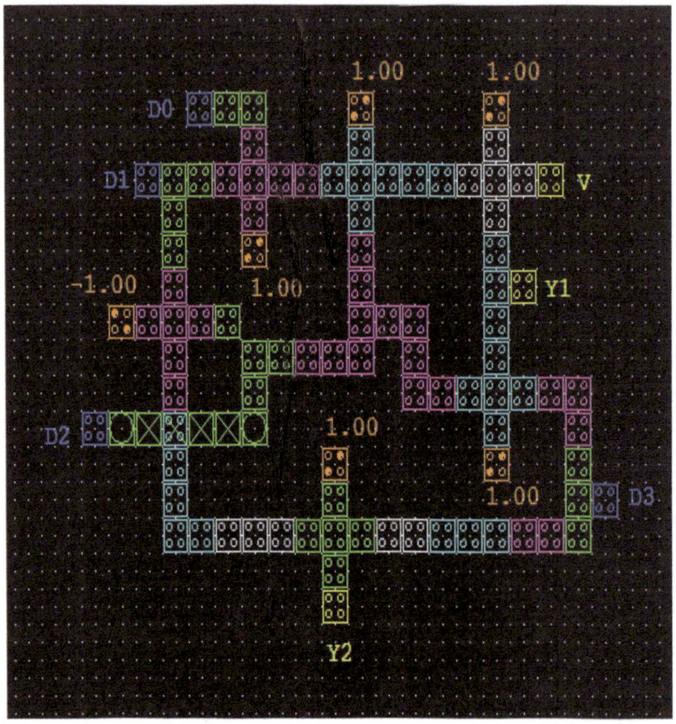

Figure 9 QCA layout of 4–2 bit priority encoder.

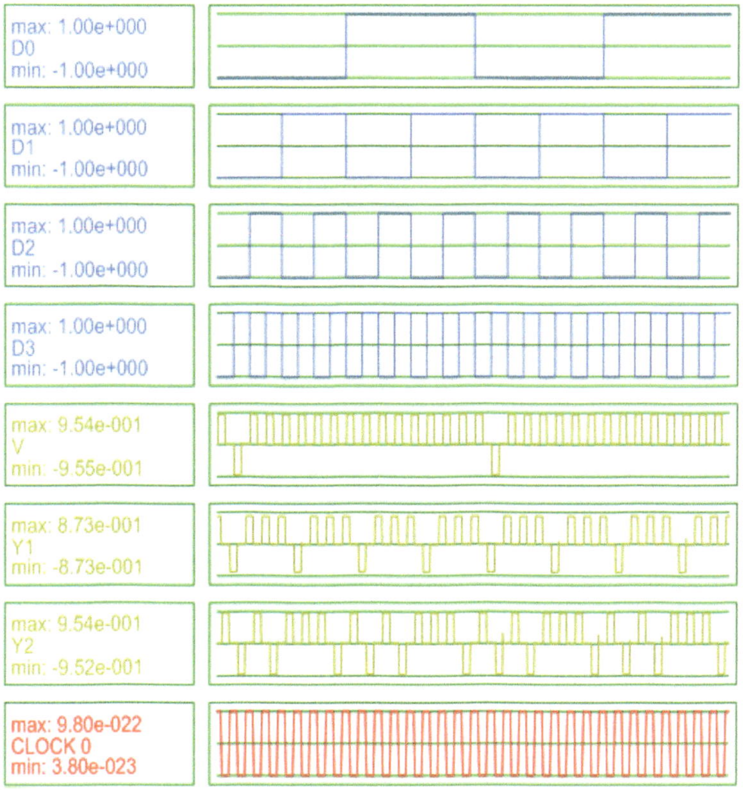

Figure 10 Simulation result of 4–2 bit priority encoder.

The 3-bit PRBS generator

An *N*-bit pseudo-random binary sequence (PRBS) is a binary sequence and random in a sense that the value of any element is independent of all other elements' values, similar to real random sequences, and pseudo because it starts to repeat itself after every *N* elements. PRBS can be used for supplying known binary sequences which are used as test signals for making bit error measurement.

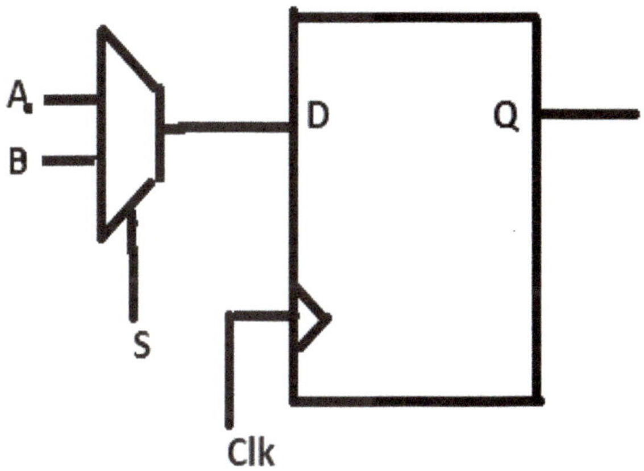

Figure 11 Schematic of scan flip-flop.

Figure 14 shows the QCA design of proposed PRBS generator.

Results and discussion

Circuits use lesser number of cells compared to traditional circuits and are more densely packed. These circuit designs have been simulated using QCA designer and the results are shown. Ninety QCA cells are used to design a scan flip-flop with area of 0.19 μm^2 whereas the 4–2 bit priority encoder, which contains 100 cells, has an area of 0.13 μm^2. Some of the larger and complex circuits such as 8–3 bit encoder have 281 cells with total area of 0.06 μm^2, and PRBS generator has 381 cells and area 0.61 μm^2. These and some more basic circuit designs can be clubbed together to create more complex circuit designs such as arithmetic logic unit or ALU, field programmable gate array or FPGA. In extension to this work, a fault-tolerant majority gate can be used and hence making the circuit more suitable for physical implementation.

Conclusions

This paper has demonstrated the design of a QCA 4:2 bit priority encoder and 8:3 bit simple encoder, scan flip-

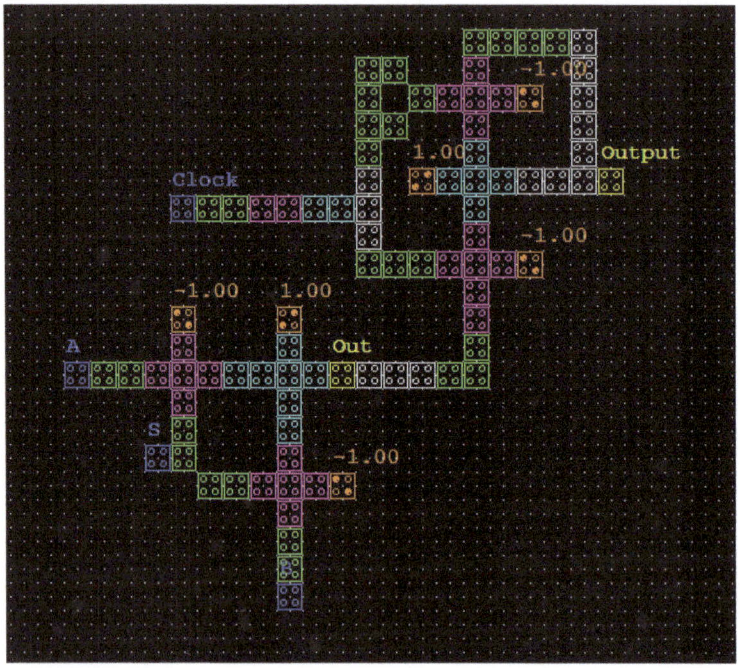

Figure 12 QCA layout of scan flip-flop.

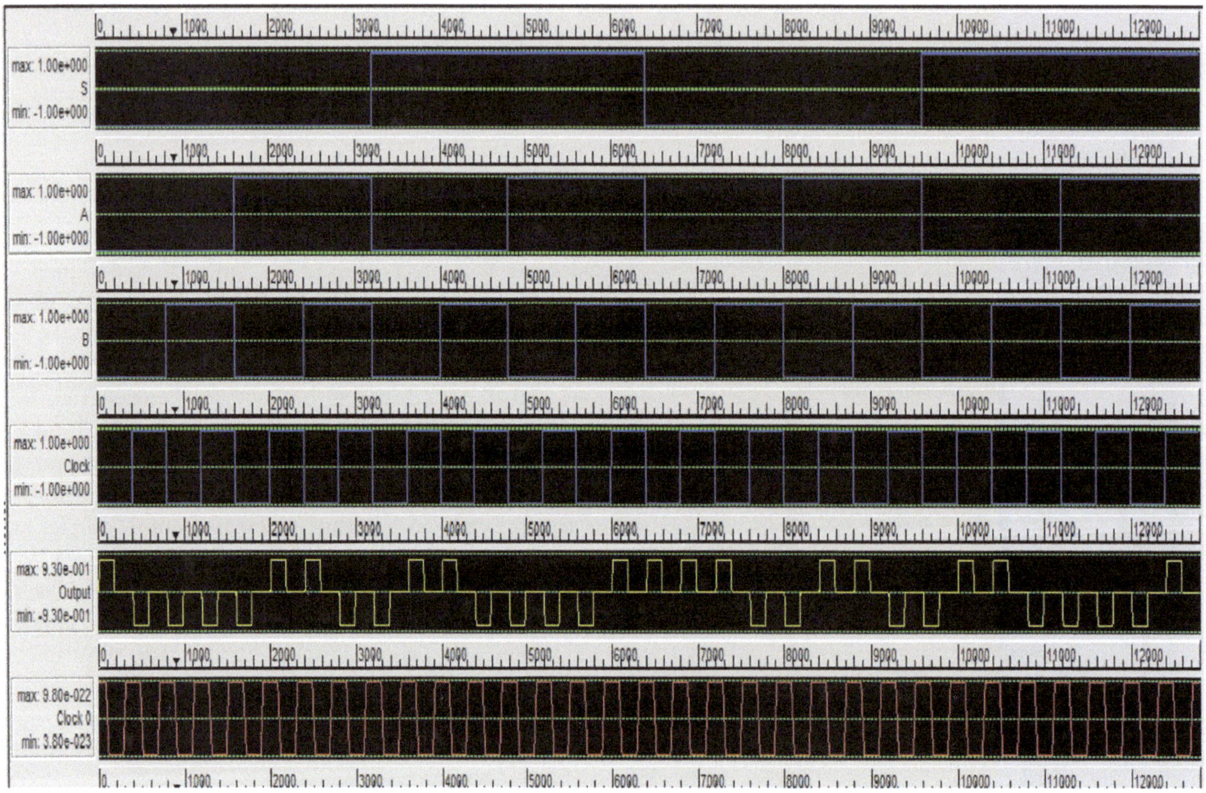

Figure 13 Simulation result of scan flip-flop.

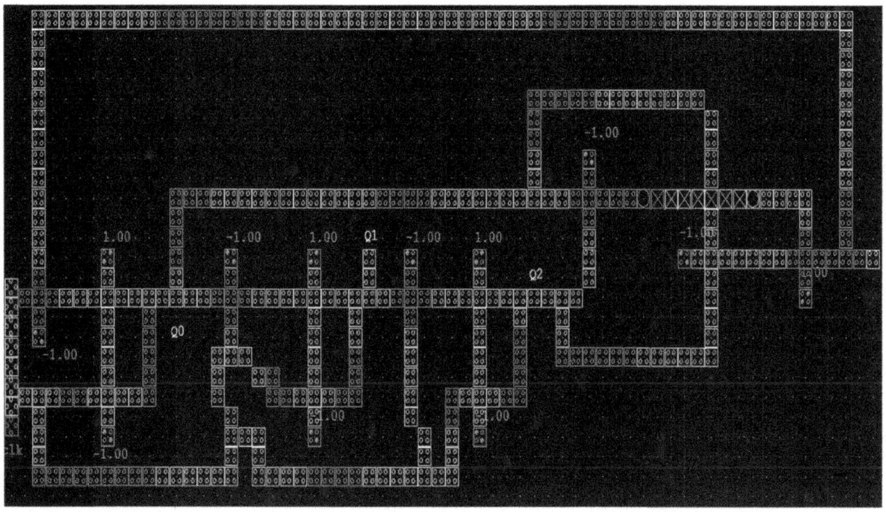

Figure 14 QCA layout of 3 bit PRBS.

flop, and PRBS. The proposed designs are significantly smaller than the same circuits using CMOS technology. The sizes of layouts are measured on the basis of size of QCA cells. All designs are carefully clocked and were functionally verified using QCA designer. Finally, in Table 1, designs are compared according to number of cells, area, and delay.

Methods

First of all, the logic behind any proposed circuit is deduced and then the circuit diagram is drawn at gate level. The gate level circuit is converted to QCA layout using majority gates, inverters, etc. as described in the above sections and then these designs are simulated in QCA designer which is the product of an ongoing research effort by the Walus Group at the University of British Columbia to create a design and simulation tool for QCA. The designer tool allows the designer to layout a QCA design and simulates it quickly. QCA designer has provided a new platform for developers; results from simulations, using this tool, have been published by many international groups [11-16]. Results obtained by this tool are then compared to theoretical values to verify the correctness of the circuit.

Table 1 Comparison of proposed designs

Design	Number of cells	Area (in μm^2)	Delay
8-3 bit simple encoder	281	0.22 × 0.30 = 0.06	2.25
4-2 bit priority encoder	100	0.32 × 0.42 = 0.13	1.25
Scan flip-flop	90	0.46 × 0.41 = 0.19	2
PRBS generator	381	1.28 × 0.478 = 0.61	4.5

Competing interest
The authors declare that they have no competing interests.

Authors' contributions
SK designed the logic of proposed circuits and SG simulated them on QCA designer, and both of them drafted the manuscript. BG verified the simulation result. AKS helped in literature study and in modifying the manuscript. All authors read and approved the final manuscript.

Authors' information
BG is a professor in the Department of EE, IIT Kanpur. His interest lies in spintronics and semiconductor device modeling. SG and SK are third year undergraduate students in the Department of EE, IIT Kanpur. AS is pursuing his Ph.D. degree under Prof. BG at the Department of EE, IIT Kanpur. His research interests include spintronics and device modeling.

Acknowledgment
We express our thanks to Mr. DiwakarAgarwal, alumnus of IIT Kanpur, for his help in understanding the basics of QCA technology and simulation.

Author details
[1]Microelectronics Research Center, University of Texas at Austin, 10100, Burnet Road, Bldg. 160, Austin, TX 78758, USA. [2]Department of Electrical Engineering, Indian Institute of Technology, Kanpur, 208016, India.

References

1. Wilson, M, Kannangara, K, Smith, G, Simmons, M, Raquse, B: Nanotechnology: basic science and emerging technologies. Chapman and Hall, London (2002)
2. Lent, CS, Tougaw, PD, Porod, W, Bernstein, GH: Quantum cellular automata. Nanotechnology **4**(1), 49–57 (1993)
3. Delgado-Frias, JG, Jabulani, N: A high-performance encoder with priority lookahead. IEEE Trans Circuits Syst **47**(9), 1390–1393 (2000)
4. Kianpour, M, Sabbaghi-Nadooshan, R: A novel modular decoder implementation in quantum cellular automata. In: IEEE International Conference on Nanoscience, Technology and Society Implications, pp. 1–5. C. V. Raman College, Orissa (8–10 December 2011)
5. Sara, H, Mohammad, T: Keivan N An efficient quantum-dot cellular automata full-adder. Sci Res Essays **7**(2), 177–189 (2012)
6. Orlov, AO, Amlani, I, Toth, G, Lent, CS, Bernstein, GH, Snider, GL: Experimental demonstration of a binary wire for quantum-dot cellular automata. Appl Phys Lett **74**(19), 2875–2877 (1999)

7. Kim, K, Wu, K, Karri, R: Towards designing robust QCA architectures in the presence of sneak noise paths. In: Proceedings of the Design, Automation and Test in Europe, Munich, 7–11 March 2005 vol 2, pp. 1214–1219. , Munich (2005)
8. Heumpil, C, Swartzlander, EE: Adder designs and analyses for quantum-dot cellular automata. IEEE Trans Nanotechnol **6**(3), 374–383 (2007)
9. Hema Sandhya, J, Mousumi, S, Pavan Kumar, J: Quantum dot cellular automata based effective design of combinational and sequential logical structures. World Acad Sci Eng Technol **60**, 671–675 (2011)
10. Walus, K, Jullien, G: Design tools for an emerging technology: quantum-dot cellular automata. Proc IEEE **94**(6), 1225–1244 (2006)
11. Walus, K, Jullien, G, Dimitrow, V: Computer arithmetic structures for quantum cellular automata. In: Conference Record of the Thirty-Seventh Asilomar Conference on Signals, Systems and Computers, pp. 1435–1439. Pacific Grove (9–12 November 2003)
12. Orlov, A, Islamshah, A, Kummamuru, RK, Ramasubramaniam, R, Toth, G, Lent, CS, Bernstein, GH, Snider, GL: Experimental demonstration of clocked single-electron switching in quantum-dot cellular automata. Appl Phys Lett **77**(2), 295–297 (2000)
13. MostafaRahimi, A, Kavehei, O, Navi, K: A novel design for quantum-dot cellular automata cells and full adders. J Appl Sci **7**, 3460–3648 (2012)
14. Walus, K, Schulhof, G, Jullien, GA: High level exploration of quantum-dot cellular automata (QCA). In: Conf. Rec. 38th Asilomar Conf. Signals, Systems and Computers, Pacific Grove, p. 3033. (7–10 November 2004)
15. Townsend, WJ, Abraham, JA: Complex gate implementations for quantum dot cellular automata. In: Proc. 4th IEEE Conference Nanotechnology, Pacific Grove, pp. 625–627. (16–19 August 2004)
16. Kumar, R, Bollapalli, KC, Garg, R, Soni, T, Khatri, SP: A robust pulsed flip-flop and its use in enhanced scan design. In: IEEE International Conference on Computer Design 2009, ICCD 2009, pp. 97–102. , Lake Tahoe (4–7 October 2009)

19

The effect of sulfuric acid, oxalic acid, and their combination on the size and regularity of the porous alumina by anodization

Alireza Keshavarz[1], Zohreh Parang[2*] and Ahmad Nasseri[1]

Abstract

Porous alumina template was prepared by constant voltage in three different electrolytes (0.45 M sulfuric acid, 0.3 M oxalic acid, and mixed solution of 0.45 M sulfuric and 0.3 M oxalic acids). In this paper, the effects of the electrolyte type on the morphology is studied, so the anodization is carried out in an oxalic acid, sulfuric acid, and a mixture of oxalic and sulfuric acids. The effect of these three different electrolytes was investigated by scanning electron microscopy. Results show that in sulfuric acid, the pore structure is smaller than the oxalic and a mixture of oxalic and sulfuric acid. Density of holes in sulfuric acid was higher than other electrolytes; however, arrangement of pores in oxalic acid was more regular than that of the others.

Keywords: Alumina, Anodization, Porous

Background

The anodization of aluminum, a method of synthesis of ordered porous structures, seems to be a cheaper, simpler, and less time-consuming option for industrial purposes. The advantages of this method led to its further attraction in recent decades. One common method for nanofabrication is the use of membranes with nanosized holes [1-7]. The conditions could be controlled to achieve desired pore length and diameter, and they can be used in many different ways to prepare a variety of nanostructured materials, magnetic metals and alloys, semiconductor alloys, heterostructures, superconductors, carbon nanotubes, etc. A lot of alternative methods have been developed for the formation of nanopore and nanodot array on various substrates including sol-gel methods [8,9], CVD based on the vapor-liquid-solid growth method [10], and template synthesis, in particular, using anodized aluminum oxide membranes [11-13]. The aim of the present study is to investigate the self-organized formation of ordered arrays of nanopores by a two-step anodizing process of aluminum. We investigated the effect of the type of electrolyte on the size and regularity of the pores of samples. For this purpose, the effects of three different electrolytes such as the solution of 0.45 M sulfuric acid, 0.3 M oxalic acid, and a mixture of the same volume of 0.45 M sulfuric and 0.3 M oxalic acid were studied.

Results and discussion

By the anodization process, fabrication of porous arrays of aluminum was successfully performed in three different electrolytes, and the results of surface morphology of the films were studied with a scanning electron microscope (SEM, Leo-440I, Leo Electron Microscopy, Cambridge, UK) and a field emission scanning electron microscope (FESEM, Hitachi-54160, Chiyoda-ku, Japan).

The situation during a steady-state pore growth must be considered to explain the effect of self-organization. By applying voltage between the anode electrode and cathode electrode on the aluminum surface, a layer of flat oxide surface is produced and proceeds with the dissolution of some oxide area (surface defects) within the electrolyte, and with the dissolution tendency of the oxide layer of primary hollow points, holes are created on the surface.

The electric field was focused in the hollow areas, and concentration of the electric field causes the formation

* Correspondence: zparang@yahoo.com
[2]Department of Physics, Shiraz branch, Islamic Azad University, Shiraz, Iran
Full list of author information is available at the end of the article

of pores. It can be seen that the hole growths are the same, parallel to each other, and perpendicular to the surface. This is because the kind of conflict in relation to growth of the oxide layer and its dissolution and corrosion, simultaneously in millions of dots or holes, comes into existence, creating a kind of balance.

Another quantity of anodic alumina in the morphology is interpore distance which is the interval between two consecutive holes. In fact, an electric field exists in all the holes that caused the equal interpore distances.

In general, the shape and diameter of holes can be controlled by the anodization voltage, anodization temperature, and acid type and concentration. Sulfuric acid can produce the smallest diameter holes at lower voltages [14]. In the anodization process of porous alumina membranes, there is equilibrium between the dissolution of field-enhanced

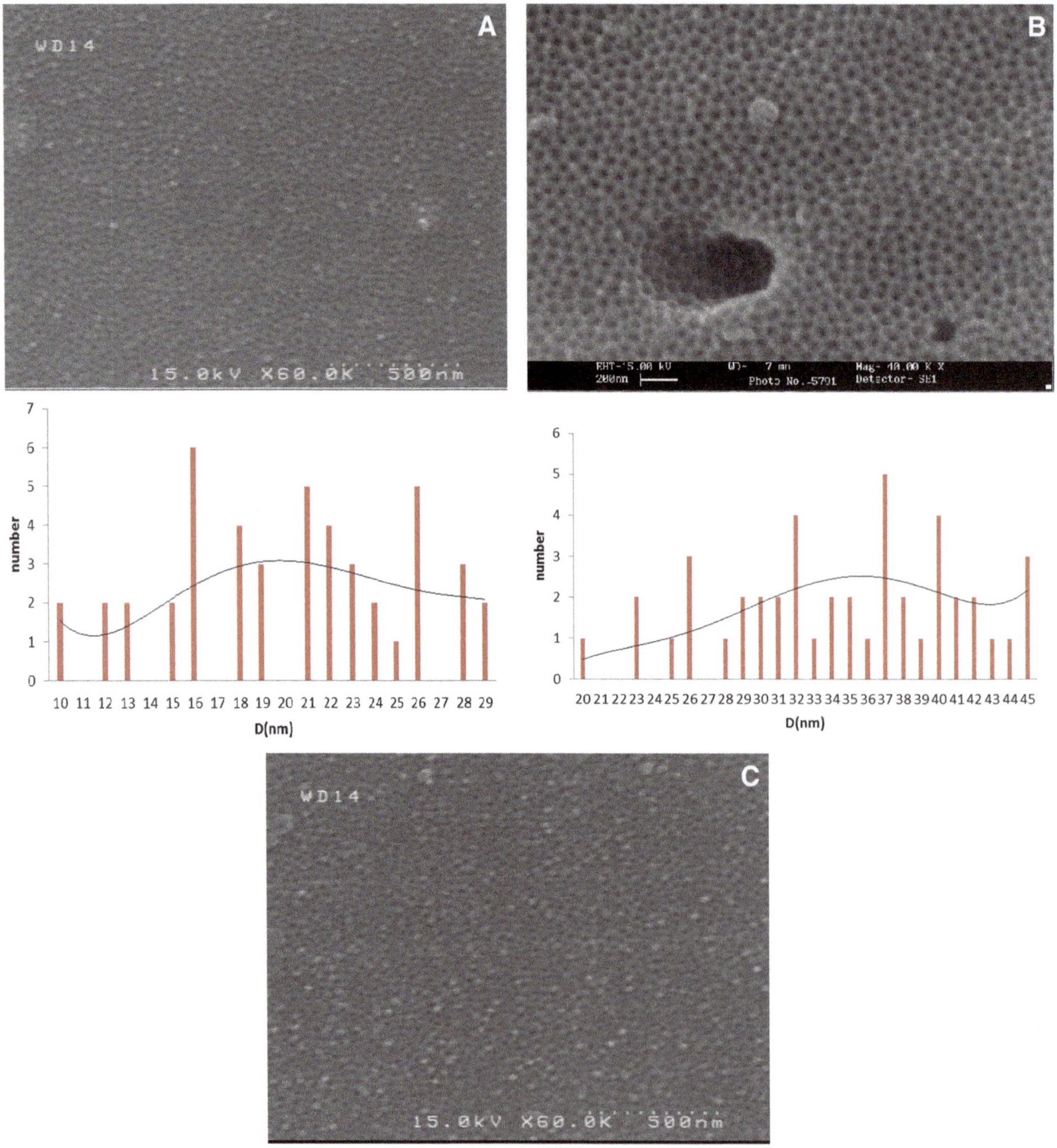

Figure 1 SEM image of anodization of Al and distribution of pores in 300-nm × 600-nm rectangle. (**A**) 0.45 M sulfuric acid, (**B**) FESEM image of anodization of Al in 0.3 M oxalic acid, and (**C**) a mixture of 0.3 M oxalic and 0.45 M sulfuric acid.

oxide in the interface electrolyte/oxide and oxide formation at the interface oxide/metal. This balance is critical for the formation of porous alumina because it causes the thickness of the barrier to remain constant throughout the anodization process [15].

SEM images in Figure 1 show that fabricated holes in 0.45 M sulfuric acid baths are very tiny so the density of the holes is high. The fabricated pores are arranged in a hexagonal structure.

As shown in Figure 1, pores are more regular in oxalic acid, but their sizes are bigger in comparison with other electrolytes. By considering a rectangle with 300-nm × 600-nm sides, the average diameter of pores was almost 34.5, 20.26, and 25 nm, respectively, for samples which were anodized in 0.3 M oxalic acid, 0.45 M sulfuric acid, and a mixed solution of 0.3 M oxalic and 0.45 M sulfuric acids (Table 1).

Variations of current with the applied anodizing potential are very important. So, the time evolutions of the current which pass through the electrolytes with different solutions in the first and second anodization processes were shown in the Figure 2.

In all of the curves, the current stabilized after enough time has passed (Figure 1A). Current during anodization is a key factor of the self-ordering process. So, variations of current with time are very important. When the rate of aluminum oxide dissolution occurring at the base of pores equals the rate of oxide formation at the metal-oxide interface, from that moment, a stable growth of the porous oxide layer on anodized aluminum starts.

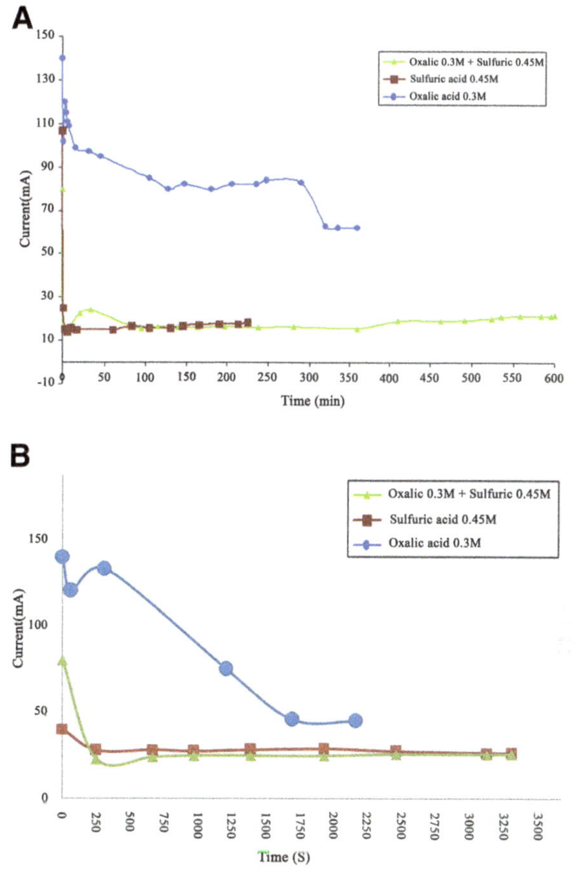

Figure 2 The current-time curve for first (A) and second (B) anodization processes of aluminum in different electrolytes.

Conclusions

By the anodization process, porous alumina was fabricated in different electrolytes. The pore diameter could be controlled by conditions of anodization such as temperature, applied voltage, annealing time, electrolyte, etc. In three different electrolytes which we investigated, the sizes of produced pores in sulfuric acid were smaller than the sizes of pores in the oxalic acid and a mixture of oxalic and sulfuric acids. Density of holes in sulfuric acid was higher than other electrolytes. Regularity of pores in oxalic acid was more than other electrolytes.

Table 1 The pore diameter in different electrolytes

Electrolyte	Average pore diameter (nm)
0.45 M Sulfuric acid	20.26
0.3 M Oxalic acid	34.5
Mixed solution of 0.45 M sulfuric and 0.3 M oxalic acids	25

Methods

An aluminum sheet with 2-mm thickness and 1-cm × 5-cm size was prepared. The sample was washed with acetone and distilled water in order to wash the impureness from the surface. The morphology of anodized alumina has direct relation to the annealing time, so the Al sheet was annealed at 450°C for 2 h in air [16]. In order to remove an oxide layer of the substrate, the samples were etched in a 1 M NaOH solution for 3 min. The Al sheet was electropolished in a mixture of perchloric acid and ethanol with a value ratio of 25:75 for 3 min to decrease the roughness of the substrate.

In the first process of anodization, the aluminum sheet was selected as anode and an aluminum electrode as cathode with a 2.5-cm separation. Then, a DC voltage of 40 V was applied to the electrodes. The solution was stirred by a magnetic stirrer during the anodization process. After that, the alumina film was removed by putting it in a mixture of 0.2 M of chromic acid and 0.5 M of phosphoric acid for 12 h. Then, the second step of anodization was performed under the same condition as the first step (for 1 h). This experiment was implemented at 17°C.

Competing interests
The authors declare that they have no competing interests.

Authors' contributions
ZP carried out the preparation of nanoporous alumina. ZP and AN participated in the study of pores' structure. AK, as the supervisor, participated in this work. AK, ZP, and AN carried out the manuscript preparation. All authors read and approved the final manuscript.

Authors' information
AK was born in Shiraz, Iran. He received his B.S. degree in Atomic Physics from Shiraz University in 1995, M.S. degree in Atomic and Molecular Physics from Kerman University in 1997, and the Ph.D. degrees in Physics (Laser and Optics) from Shiraz University in 2004. He has worked on much scientific research areas such as nonlinear optics and laser, nanophotonics and nanophysics. He joined the faculty of basic sciences in Shiraz University of Technology in 2004. He became an Associate Professor in 2012. At present, he is actively involved in optics and photonics and has M.S. and Ph.D. students in these fields. ZP was born in 1978, in Kuwait. She received her B.S. degree in Physics from Shiraz University, Iran, in 2001, and her M.S. degree from Islamic Azad University, Research and Science Branch, Tehran, Iran, in 2005, under the supervision of Dr. Dorranian and Dr. Ghorannevis. She received her Ph.D. degree in Physics (Nano) at the same university under the supervision of Dr. Alireza Keshavarz and Dr. Mohammad Elahi. She is now teaching Physics in Islamic Azad University, Shiraz Branch, Iran. She is interested in nanoscience especially bimetallic nanoparticles and nanostructures. AN was born in 1981, in Shiraz. He completed her B.S. in Applied Solid State Physics from Yasuj University, Iran in 2004 and M.S. in Atomic and Molecular Physics, Payam-e-noor University of Shiraz, Iran. He is an experimental officer at the nanophysics laboratory of Shiraz University of Technology, Iran.

Acknowledgment
This project was supported by the University of Technology of Shiraz.

Author details
[1]Department of Physics, Shiraz University of Technology, Shiraz, Iran.
[2]Department of Physics, Shiraz branch, Islamic Azad University, Shiraz, Iran.

References
1. Murray, CB, Kagan, CR, Bawendi, MG: Self-organization of CdSe nanocrystallites into three-dimensional quantum dot superlattices. Science **270**, 1335–1338 (1995)
2. Andres, RP, Bielefeld, JD, Henderson, JI, Janes, DB, Kolagunta, VR, Kubiak, CP, Mahoney, WJ, Osifchin, RG: Self-Assembly of a two-dimensional superlattice of molecularly linked metal clusters. Science **273**, 1690–1693 (1996)
3. Shi, J, Gider, S, Babcock, K, Awschalom, DD: Magnetic clusters in molecular beams, metals, and semiconductors. Science **271**, 937 (1996)
4. Billas, IML, Becker, JA, Chatelain, A, de Heer, WA: Magnetic momentsof iron clusters with 25 to 700 atoms and their dependence on temperature. Phys. Rev. Lett. **71**, 4067–4070 (1993)
5. White, RL, New, RMH, Pease, RFW: Magnetic domains: the analysis of magnetic microstructures. IEEE Trans. Magn. **33**, 990–995 (1997)
6. Chou, SY, Wei, MS, Krausse, PR, Fischer, PB: Single-domain magnetic pillar array of 35 nm diameter and 65 Gbits/in^2 density for ultrahigh density quantum magnetic storage. J. Appl. Phys **76**, 6673 (1994)
7. Ross, CA, Smith, HI, Savas, T, Schattenburgh, M, Farhoud, M, Hwang, M, Walsh, M, Abraham, MC, Ram, RJ: Fabrication of patterned media for high density magnetic storage. J. Vac. Sci. Technol. B **17**, 3168 (1999)
8. Yang, P, Deng, T, Zhao, D, Feng, P, Pine, D, Chmelka, BF, Whitesides, GM, Stucky, GD: Hierarchically ordered oxides. Science **282**, 2244–6 (1998)
9. Pavasupree, S, Suzuki, Y, Pivsa-Art, S, Yoshikawa, S: Synthesis and characterization of nanoporous, nanorods, nanowires metal oxides. Sci. Tech. Adv. Mater. **6**, 224–229 (2005)
10. Chik, H, Xu, JM: Nanometric superlattices: non-lithographic fabrication, materials, and prospects. Mater Sci Eng R **43**, 103 (2004)
11. Jain, K, Lakshmikumar, ST: Porous alumina template based nanodevices. IETE Technical Rev **19**, 293–306 (2002)
12. Peng, X, Chen, A: Electrochemical fabrication of novel nanostructures based on anodic alumina. Nanotechnology **15**, 743–748 (2004)
13. Piao, Y, Lim, H, Chang, JY, Lee, W-Y, Kim, H: Nanostructured materials prepared by use of ordered porous alumina. Electrochim. Acta **50**, 2997–3013 (2005)
14. Jenny, M: Nunn S, pang Y T: Electrochemical synthesis of ordered alumina nanowire arrays. J Solid State Electrochem. **7**, 344–347 (2005)
15. John, S, Balasubramanian, V, Shenoi, AB: Hard anodizing of aluminum and its Alloys at 30 to 35°C. Metal Finishing **83**, 23–26 (1985)
16. Almasi Kashi, M, Ramazani, A: The effects of temperature and concentration on the self-organized pore formation in anodic alumina. J. Phys. **38**, 2396–2399 (2005)

20

Rugby-ball-shaped $Na_{0.5}La_{0.5}MoO_4:Eu^{3+}$ 3D architectures: synthesis, characterization, and their luminescence behavior

Rajagopalan Krishnan, Jagannathan Thirumalai*, Ithrees Basha Shameem Banu and Anthuvan John Peter

Abstract

Three-dimensional self-assembled rugby-ball-shaped microarchitectures of tetragonal $Na_{0.5}La_{0.5}MoO_4:Eu^{3+}$ crystal structure were synthesized by facile hydrothermal method at 200°C for 8 h using ethylenediaminetetraacetic acid (EDTA) as surfactant. Structure and morphology of $Na_{0.5}La_{0.5}MoO_4:Eu^{3+}$ were investigated by X-ray diffraction and scanning electron microscopy. In the hydrothermal process, EDTA not only acts as a chelating reagent to facilitate the formation of $Na_{0.5}La_{0.5}MoO_4:Eu^{3+}$ but also acts as a surface capping agent to adhere to the newly created surface and to promote crystal splitting. The molar concentration of EDTA plays an important role and can effectively tune the size of the particles. Photoluminescence study reveals that hierarchical structures of $Na_{0.5}La_{0.5}MoO_4:Eu^{3+}$ show a strong emission in the red region under 395-nm UV excitation.

Keywords: Hydrothermal method, Scheelite tetragonal structure, FTIR, Scanning electron microscope

Background

Homogenous and self-aggregated three-dimensional (3D) superstructures with uniform size distribution have attracted interest nowadays because of their aberrant behavior that tailors the physical and chemical properties of materials resulting from momentous advancement that has been developed for the fabrication of a new class of micro- and nanostructured devices [1,2]. The peculiar size-dependent properties of 3D hierarchical networks have initiated worldwide intense research to material scientists in different fields of applications such as opto-electronics, catalysts, display devices, and fluorescent probes for biological staining [3,4]. Recently, the interests have been expanded into controlling the shape of micro/nanomaterials and also in understanding the correlations between the material shape and their properties. Sheaf-like orthorhombic $Gd_2(MoO_4)_3:Eu^{3+}$ nanostructures [5], ordered shuttle-like $NaLa(MoO_4)_2:Eu^{3+}$ nanorods composed of nanoparticles [6], self-assembled 3D flower-like $NaY(MoO_4)_2:Eu^{3+}$ microarchitectures [7], and 3D bowknot-like hierarchical microstructures $Y_2(WO_4)_3:Ln^{3+}$ [8] are the few examples of the shape-controlled synthesis of 3D networks. Therefore, the development of a reliable and convenient synthetic route that can control the shape of nanostructures under ambient conditions remained an important research topic of investigation.

It is well known that among the various solution-phase techniques, the hydrothermal route has been exclusively investigated as a promising alternative technique for the fabrication of three-dimensional hierarchical structures under mild synthesis and feasible environment [9]. The reason is that this method is fast, simple, economical, and environment friendly [10]. Recently, scheelite-type crystalline structures of molybdate or tungstate compounds, in which Mo^{6+} or W^{6+} ions are coordinated tetrahedrally, show promising applications in the field of opto-electronics and laser technology [11,12]. The compounds with the general formula $Na_{0.5}R_{0.5}MoO_4$ (where R = La^{3+}, Gd^{3+}) possess the tetragonal scheelite structure in which Na^+ and R^{3+} jointly occupied the dodecahedral positions and the Mo^{6+} ions are at the centers of the tetrahedral symmetry [13]. Recently, europium-activated molybdates are more attractive due to strong red light emission under 393- to 464-nm excitations [14], and Eu^{3+} has been extensively studied in rare-earth ions due to its unique spectral character that shows enriched emission in various host materials.

With this view, we have reported $Na_{0.5}La_{0.5}MoO_4:Eu^{3+}$ rugby-ball-shaped microarchitectures prepared by the

* Correspondence: thirumalaijg@gmail.com
Department of Physics, B.S. Abdur Rahman University, Vandalur, Chennai, Tamilnadu 600048, India

hydrothermal route at 200°C for 8 h by modulating the ethylenediaminetetraacetic acid (EDTA) molar concentrations. The structure and phase purity of the samples were analyzed using X-ray diffraction (XRD) patterns. The characteristic vibrational behaviors of elements were analyzed by Fourier transform infrared spectrometry (FTIR). The final product morphology was examined by scanning electron microscopy (SEM) analysis, and the room-temperature photoluminescence properties of $Na_{0.5}La_{0.5}MoO_4:Eu^{3+}$ were investigated.

Results and discussion

Structural and morphological investigation

Figure 1 shows the XRD pattern of $Na_{0.5}La_{0.5}MoO_4:Eu^{3+}$ prepared by the hydrothermal method by modulating molar concentrations of EDTA. All the peaks are good in agreement with the standard JCPDS card no. 79–2243 of $Na_{0.5}La_{0.5}MoO_4$. It is clearly seen from the XRD that the as-synthesized products are highly crystalline and they belong to tetragonal phase with scheelite structure of the space group *I41/a*.

Figure 2a,b,c,d,e,f indicates the low- and high-magnification SEM images of $Na_{0.5}La_{0.5}MoO_4:Eu^{3+}$ rugby-ball-shaped structures which were synthesized by modulating the amount of EDTA with a fixed Na^+, $[La^{3+}/Eu^{3+}]$. When the molar ratio of EDTA was 1 mM, microspheres with an average length of 2.3 μm and diameter of 1.20 μm (Figure 2a) were obtained. The SEM image of the sample synthesized with 1.25 mM EDTA shows rugby-ball-shaped microstructures with an average length of 2.1 μm and diameter of 1.1 μm (Figure 2b). When the amount of EDTA was increased to 1.50 mM, the size of the particle was further reduced, showing an average length of 2.0 μm and diameter of 1 μm (Figure 2c,d). At 1.75 mM concentration of EDTA, the size was further reduced to an average length of 1.8 μm and diameter of 850 nm (Figure 2e). Still when the amount of EDTA was increased to 2.0 mM, the size of the particle was further reduced to 800 nm in length and 350 nm in diameter (Figure 2f). The increase in EDTA concentration enhances the chelating ability of the EDTA molecule that will lead to the reduction of particle size. Hence, it was found that the amount of EDTA introduced to the reaction system had a great influence on the morphology and size distribution of the final products. It could be seen from the SEM images that even a small change in the molar concentration of EDTA shows an extreme impact on the size and morphology of the final product. As a representative result, $Na_{0.5}La_{0.5}MoO_4:Eu^{3+}$ prepared with 1.50 mM of EDTA was taken for FTIR and photoluminescence studies.

FTIR analysis

Figure 3 shows the FTIR transmittance spectra of the $Na_{0.5}La_{0.5}MoO_4:Eu^{3+}$ rugby-ball-shaped structure prepared with 1.50 mM of EDTA. It consists of a strong and broad absorption band starting from 700 to 850 cm^{-1}, and it was assigned to the stretching vibrations of O-Mo-O in the $MoO4^{2-}$ group [15]. This absorption band was assigned to the antisymmetric stretch F_2 (v_3) of the scheelite tetragonal crystalline structure. The weak absorption at 1,407 cm^{-1} is assigned to the CH_2 group, and the sharp absorption at 1,572 cm^{-1} is due to H-O-H bending vibration of the absorbed water from air. In addition,

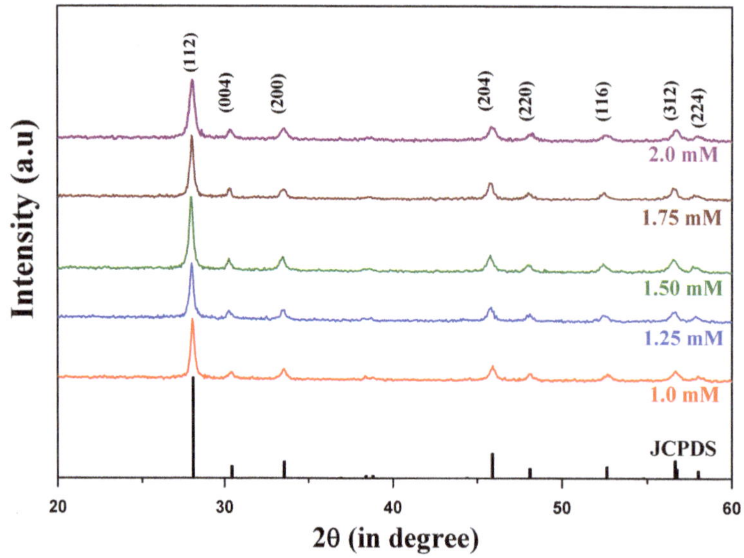

Figure 1 XRD pattern of $Na_{0.5}La_{0.5}MoO_4:Eu^{3+}$ prepared by hydrothermal method at 200°C for 8 h. With different EDTA concentrations: 1.0, 1.25, 1.50, 1.75, and 2.0 mM.

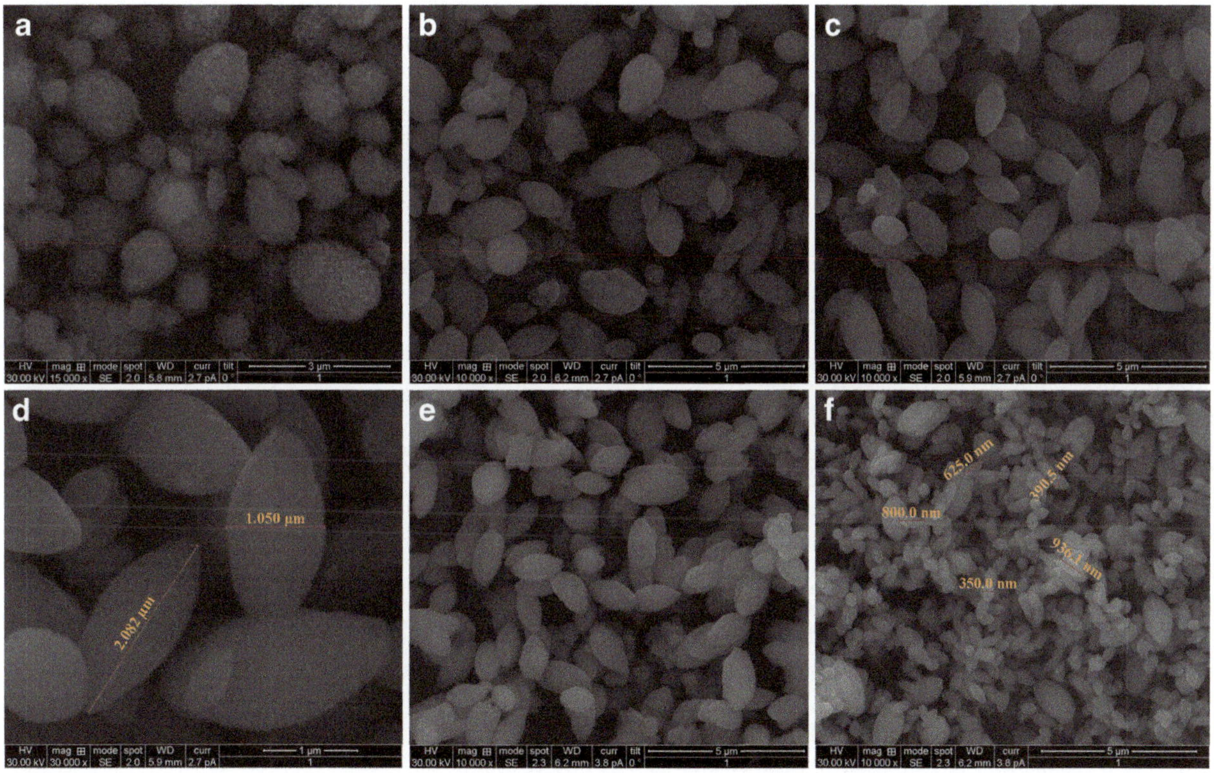

Figure 2 Low- and high-magnification SEM images of rugby-ball-shaped $Na_{0.5}La_{0.5}MoO_4$:Eu^{3+} 3D structures prepared by a hydrothermal route. At 200°C for 8 h with different EDTA concentrations: 1.0 (**a**), 1.25 (**b**), 1.50 (**c, d**), 1.75 (**e**), and 2.0 mM (**f**).

the wide bands corresponding to O-H stretch vibrations were observed around 3,358 cm^{-1}.

Photoluminescence studies

Figure 4 shows the room-temperature PL excitation and emission spectra of $Na_{0.5}La_{0.5}MoO_4$:Eu^{3+} rugby ball intra-configurational *f-f* transitions of Eu^{3+} that appeared in the wavelength region 350 to 450 nm. These characteristic configurations were assigned to the transition from ground state ($^7F_0$) to the Eu^{3+} upper excited states ($^5D_{4,3}$ and $L_{6,7}$), i.e., $^7F_0 \rightarrow {}^5D_4$ (362 nm), $^7F_0 \rightarrow {}^5L_7$ (382 nm), $^7F_0 \rightarrow {}^5L_6$ (395 nm), and $^7F_0 \rightarrow {}^5D_3$ (416 nm). While exciting with 395-nm UV irradiation, the emission spectra are dominated by the hypersensitive red

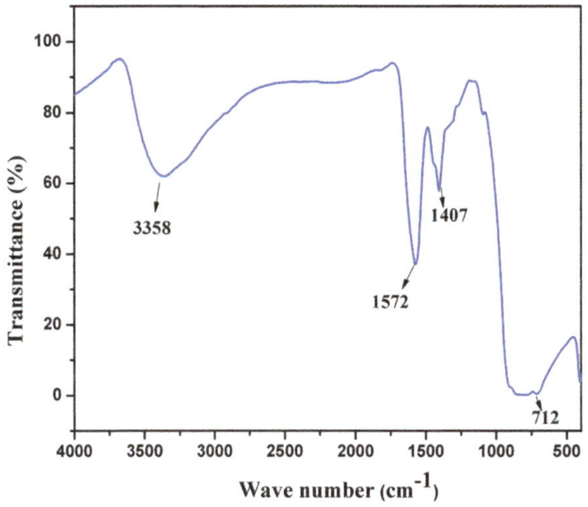

Figure 3 FTIR transmittance spectra of $Na_{0.5}La_{0.5}MoO_4$:Eu^{3+} rugby-ball-shaped microstructure.

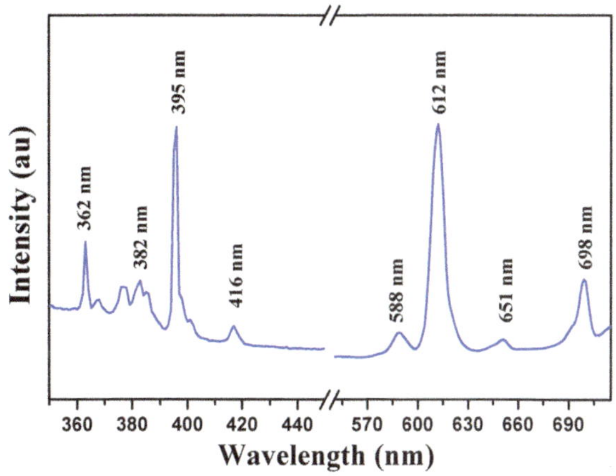

Figure 4 Photoluminescence excitation and emission spectra of $Na_{0.5}La_{0.5}MoO_4$:Eu^{3+} prepared with 1.50 mM EDTA concentration.

emission (612 nm), showing the transition $^5D_0 \rightarrow {}^7F_2$ (due to electric-dipole transition). The other transitions, $^5D_0 \rightarrow {}^7F_1$ (magnetic-dipole transition), $^5D_0 \rightarrow {}^7F_3$, and $^5D_0 \rightarrow {}^7F_4$, are relatively weak. The presence of strong luminescent intensity indicates the good crystalline nature of $Na_{0.5}La_{0.5}MoO_4{:}Eu^{3+}$ rugby-ball-shaped microstructures.

Conclusions

Self-assembled 3D superstructures of $Na_{0.5}La_{0.5}MoO_4{:}Eu^{3+}$ were successfully synthesized using hydrothermal route under mild synthesis condition with different molar concentrations of capping agent (EDTA). The hierarchical architectures of $Na_{0.5}La_{0.5}MoO_4{:}Eu^{3+}$ belong to the scheelite tetragonal structure with the space group *I41/a*. The FTIR results confirm the presence of stretching vibrations of metal-oxygen bands. The SEM images clearly indicate that the final product consists of rugby-ball-shaped microstructures with uniform size distribution. The molar ratio of EDTA introduced to the reaction system had a great influence on the morphology and size distribution of the particles. The photoluminescence study reveals that upon 395-nm UV excitation, the $Na_{0.5}La_{0.5}MoO_4{:}Eu^{3+}$ crystal structure shows strong emission in the red region at 615 nm. These results may be important for the synthesis of three-dimensional networks to fabricate new class of micro/nanostructures.

Methods

Synthesis of rugby-ball-shaped 3D hierarchical structures of $Na_{0.5}La_{0.5}MoO_4{:}Eu^{3+}$

All the chemicals were purchased from Sigma-Aldrich (St. Louis, MO, USA) with 99.99% purity. In the typical synthesis, the stoichiometric amount of Na_2MoO_4 was first dissolved in 30 mL of double-distilled water under vigorous stirring. Then, $LaCl_3$ and $EuCl_3$ were prepared by dissolving the corresponding rare-earth oxides (La_2O_3, Eu_2O_3) in diluted hydrochloric acid and stirred for 15 min. In order to remove the excess HCl, the solution was heated to 60°C to 80°C for 30 min. Then the separate rare-earth chloride solution was carefully mixed with the Na_2MoO_4 solution, and a white colloidal precipitate was obtained under vigorous stirring for 30 min. Finally, 1 to 2 mM of EDTA was dissolved in 20 mL of double-distilled water, and it is added to the resultant colloidal solution. Then the pH value of the final product was consequently adjusted to 7 by adding NaOH solution. After additional agitation for 30 min, the as-obtained white colloidal precipitate was transferred into a 100-mL Teflon autoclave, sealed in a stainless steel vessel, and heated at 200°C for 8 h. Subsequently, the autoclave was allowed to cool at room temperature, and the resultant solid products were centrifugally separated from the suspension, washed with distilled water and absolute ethanol several times, and dried at 60°C in air for 5 h.

Characterization

The XRD patterns were recorded on Philips X'Pert PRO system from PANalytical's diffractometer (Almelo, The Netherlands) with Cu Kα radiation (λ = 0.15406 nm) and a scanning rate of 0.05° s^{-1}. Morphology and elemental composition of the product were investigated by SEM (JEOL JSM-840, Akishima-shi, Japan). FTIR measurements were carried out in Nicolet 6700 FTIR (Waltham, MA, USA) equipped with a deuterated triglycine sulfate detector. Furthermore, the room-temperature photoluminescence properties of the sample were measured using a Horiba-Jobin Yvon nanolog spectrofluorometer (Kyoto, Japan).

Competing interests

The authors declare that they have no competing interests.

Authors' contributions

RK synthesized the materials and carried out all the characterization studies. JT participated in the sequence alignment, drafted the manuscript, conceived of the study, and participated in its design and coordination. SB and AJ participated in the sequence alignment. All authors read and approved the final manuscript.

Acknowledgements

The authors gratefully acknowledge Alagappa University, Karaikudi, and CUSAT (Saif), Cochin, for extending their instrumentation facilities for characterization.

References

1. Law, M, Sirbuly, DJ, Johnson, JC, Goldberger, J, Saykally, RJ, Yang, PD: Nanoribbon waveguides for subwavelength photonics integration. Science **305**, 1269–1273 (2004)
2. Duan, H, Wang, D, Sobal, NS, Giersig, M, Kurth, DG, M'ohwald, H: Magnetic colloidosomes derived from nanoparticle interfacial self-assembly. Nano Lett. **5**, 949–952 (2005)
3. Hu, JT, Ouyang, M, Yang, PD, Lieber, CM: Controlled growth and electrical properties of heterojunctions of carbon nanotubes and silicon nanowires. Nature **399**, 48–51 (1999)
4. Shen, J, Sun, LD, Yan, CH: Luminescent rare earth nanomaterials for bioprobe applications. Dalton Trans. **42**, 5687–5697 (2008)
5. Thirumalai, J, Krishnan, R, Shameem Banu, IB, Chandramohan, R: Controlled synthesis, formation mechanism and luminescence properties of novel 3-dimensional $Gd_2(MoO_4)_3{:}Eu^{3+}$ nanostructures. Mater Sci. Mater Electron. **24**, 253–2599 (2013)
6. Yang, M, You, H, Jia, Y, Qiao, H, Guo, N, Song, Y: Synthesis and luminescent properties of $NaLa(MoO_4)_2{:}Eu^{3+}$ shuttle-like nanorods composed of nanoparticles. Cryst. Eng. **13**, 4046–4052 (2011)
7. Huang, Y, Zhou, L, Yang, L, Tang, Z: Self-assembled 3D flower-like $NaY(MoO_4)_2{:}Eu^{3+}$ microarchitectures: hydrothermal synthesis, formation mechanism and luminescence properties. Opt. Mater. **33**, 777–782 (2011)
8. Huang, S, Zhang, X, Wang, L, Bai, L, Xu, J, Li, C, Yang, P: Controllable synthesis and tunable luminescence properties of $Y_2(WO_4)_3{:}Ln^{3+}$ (Ln = Eu, Yb/Er, Yb/Tm and Yb/Ho) 3D hierarchical architectures. Dalton Trans. **41**, 5634–5642 (2012)
9. Fang, YP, Xu, AW, You, LP, Song, RQ, Yu, JC, Zhang, HX, Li, Q, Liu, HQ: Hydrothermal synthesis of rare earth (Tb, Y) hydroxide and oxide nanotubes. Adv. Funct. Mater. **13**, 955–960 (2003)
10. Dhas, NA, Suslick, KS: Sonochemical preparation of hollow nanospheres and hollow nanocrystals. J. Am. Chem. Soc. **127**, 2368–2369 (2005)

11. Ryu, JH, Yoon, JW, Lim, CS, Oh, WC, Shim, KB: Microwave-assisted synthesis of $CaMoO_4$ nano-powders by a citrate complex method and its photoluminescence property. J. Alloy Compd. **390**, 245–249 (2005)
12. Gong, Q, Qian, X, Cao, H, Du, W, Ma, X, Mo, M: Novel shape evolution of BaMoO4 microcrystals. J. Phys. Chem. B **110**, 19295–19299 (2006)
13. Kuz'micheva, GM, Rybakov, VB, Panyutin, VL, Zharikov, EV, Subbotin, KA: Symmetry of (Na0.5R0.5) MO4 crystals (Ra = Gd, La; M = W, Mo). Russ. J. Inorg. Chem. **55**, 1148–1453 (2010)
14. Zheng Liang, W, Hong Bin, L, Liya, Z, Jing, W, Meng Lian, G, Qiang, S: NaEu0.96Sm0.04(MoO4)2 as a promising red-emitting phosphor for LED solid-state lighting prepared by the Pechini process. J. Lumin. **128**, 147–154 (2008)
15. Frost, RL, Cejka, J, Dickfos, MJ: Raman and infrared spectroscopic study of the molybdate containing uranyl mineral calcurmolite. J. Raman Spectrosc. **39**, 779–785 (2008)

21

Effect of trisodium citrate concentration on the particle growth of ZnS nanoparticles

Abdul Kareem Thottoli* and Anu Kaliani Achuthan Unni

Abstract

ZnS nanoparticles were prepared by simple wet chemical precipitation method to study the effect of sulfur and trisodium citrate concentration on the particle growth of ZnS nanoparticles. Particle sizes were seen varied with the concentration of the sulfur and trisodium citrate. Compressive strain was observed only for a particular concentration of sulfur, and that sample was chosen for trisodium citrate experiments. Interestingly, the compressive strain was seen varied with the concentration of trisodium citrate and seen maximum for 2 M concentration of trisodium citrate. Further, interband transition was observed in the case of smaller crystallites prepared in the trisodium citrate. It is clearly observed from the FTIR analysis that the trisodium citrate acts as a stabilizer and capped the particles from further growth.

Keywords: ZnS nanoparticles; Wet chemical preparation; Crystallites

Background

It is a well-known fact that the semiconductor nanoparticles exhibit novel optical, electronic, and physical properties than their bulk counterparts. When the size of the nanoparticles becomes less or comparable to the Bohr diameter of the exciton [1], various size quantization effects such as widening of band gap and formation of discrete orbitals come into the picture as explained by the Heisenberg uncertainty principle. Among the semiconducting nanoparticles, ZnS is one of the most important materials in the electronic/optoelectronic industry with prominent applications [2] in flat-panel displays [3], white light LEDs [4], electroluminescence devices [5], sensors [6], lasers [7], infrared windows [8], ultraviolet (UV) lasers [9], antireflection coating on solar cells [10], solar cells [11], and for investigating cellular interactions [12], for both *in vivo-in vitro* imaging applications [13], and memristor applications [14]. Basically, ZnS has negative conductivity [15,16] and has a small exciton Bohr radius of about 2.5 nm [17]. Particle size reduction has a tremendous effect on the properties of the ZnS such as blueshift in the optical absorption spectrum, increased luminescence, enhanced oscillator strength, nonlinear optical effects, geometrical structure, chemical bonds, ionization potential, mechanical strength, melting point, etc. Furthermore, ZnS relates more popular ZnO in terms of atomic structure and chemical properties. Certain properties of ZnS are unique and advantageous compared to that of ZnO. ZnS has a larger band gap than that of ZnO (approximately 3.4 eV [17]), and therefore, it is more suitable for visible UV light-based devices such as sensors/photodetectors. However, nano-sized ZnS has not been investigated in much detail compared to ZnO nanomaterials.

This paper discusses the preparation of zinc sulfide nanocrystals by simple wet chemical precipitation method [18-21] and studies the effect of the concentration of sulfur and trisodium citrate on nanoparticle formation since trisodium citrate acts as surfactant and capping agent that can modify the surface of ZnS particles and prevents the growth of the particles.

Results and discussion

ZnS nanocrystalline powder synthesis by varying sulfur concentration

The prepared ZnS powder samples were analyzed by powder X-ray diffraction technique, scanning electron microscopy (SEM), energy dispersive X-ray (EDX), and UV-visible absorption spectrophotometer. The X-ray diffractogram of the prepared samples well correlates with the American Society for Testing and Materials

* Correspondence: abdulkareem.t@gmail.com
PG & Research Department of Physics, Kongunadu Arts and Science College, GN Mills PO, Coimbatore, Tamilnadu 641 029, India

(ASTM) card no. 77–2100 of cubic structure, and the diffraction peaks from the planes (111), (220), and (311) were noted in the diffractograms shown in Figure 1. All of the X-ray diffractogram peaks and the peak broadening confirmed that the prepared sample contains pure cubic ZnS nanocrystallites. Their crystallite size and the strain were obtained from the Williomson-Hall plot [22]. The crystallite size, strain, lattice parameters, and the volume of the crystals of each sample are shown in Table 1, and it is seen that the prepared ZnS powder consists of nanoparticles. Table 1 clearly shows the compressive strain for the samples $ZnS_{0.1}$ and $ZnS_{0.05}$, a property of nanoparticles.

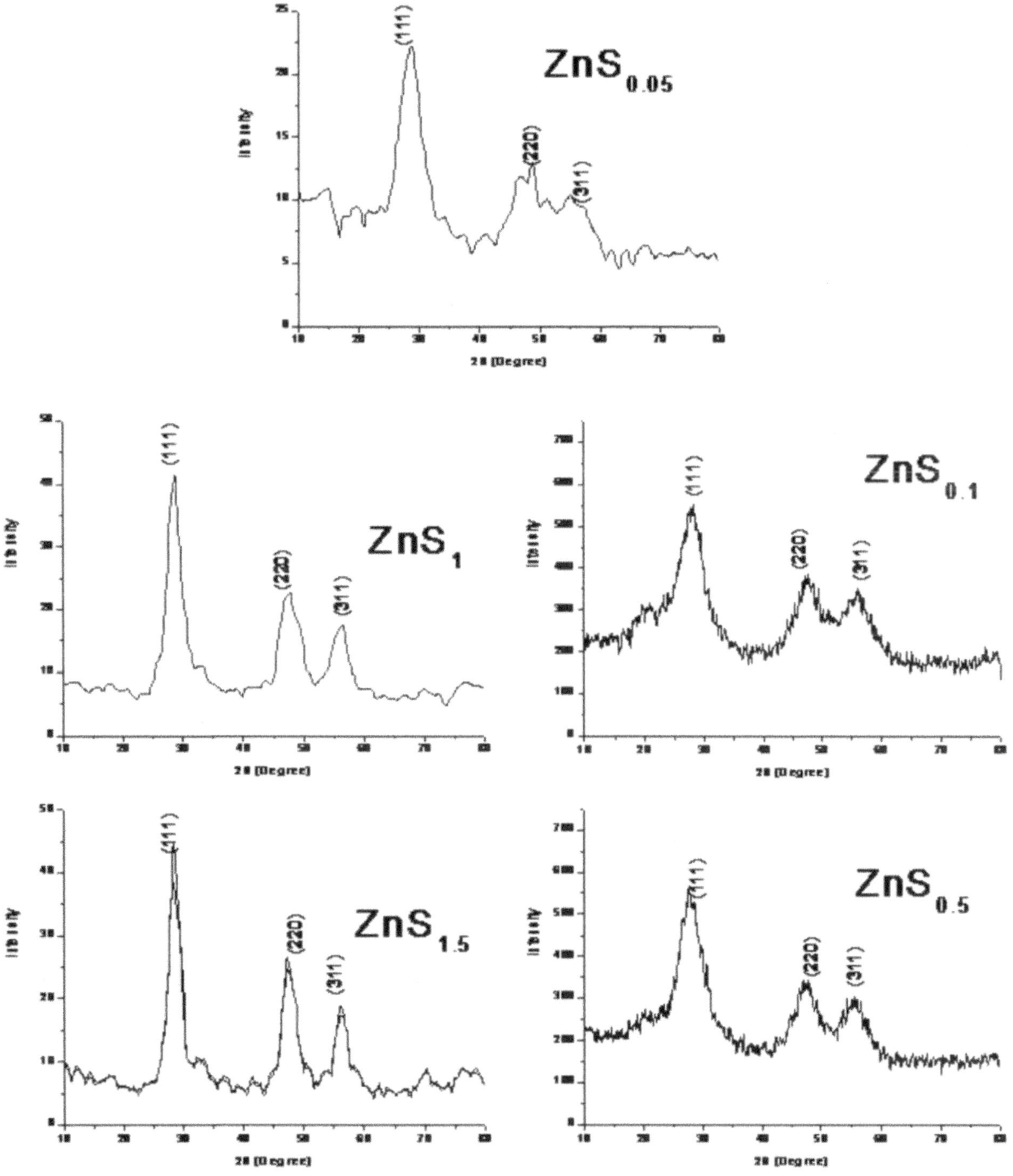

Figure 1 X-ray diffractogram of ZnS samples prepared for various sulfur molarity.

Table 1 Lattice parameters and volume of crystals

Samples	Lattice parameters (ASTM values are shown in bracket) (Å)			Average crystallite size (nm)	Average strain	Volume (10^{-28} m^3) (ASTM values are shown in bracket)
	a	b	c			
$ZnS_{1.5}$	5.4159 (5.4146)	-	-	6.91	0.0279	1.5905 (1.5875)
ZnS_1	5.4106 (5.4146)	-	-	5.35	0.0441	1.5839 (1.5875)
$ZnS_{0.5}$	5.5138 (5.4146)	-	-	2.91	0.0265	1.6763 (1.5875)
$ZnS_{0.1}$	5.4289 (5.4146)	-	-	2.43	−0.0167	1.6001 (1.5875)
$ZnS_{0.05}$	5.3949 (5.4146)	-	-	2.90	−0.0351	1.5702 (1.5875)

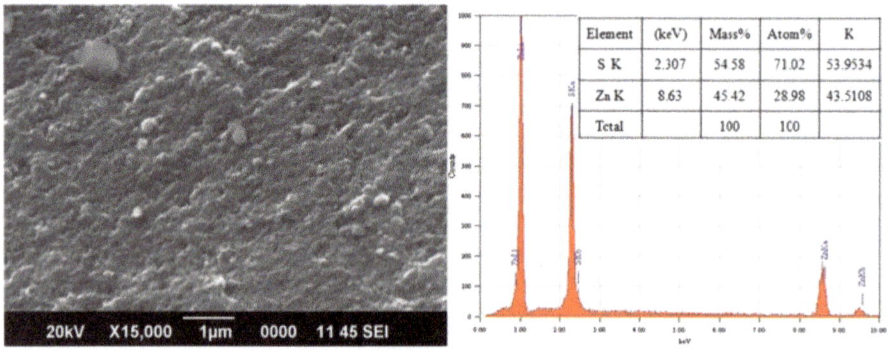

Figure 2 SEM micrograph and EDX of ZnS prepared with 1 M Na_2S solution (ZnS_1).

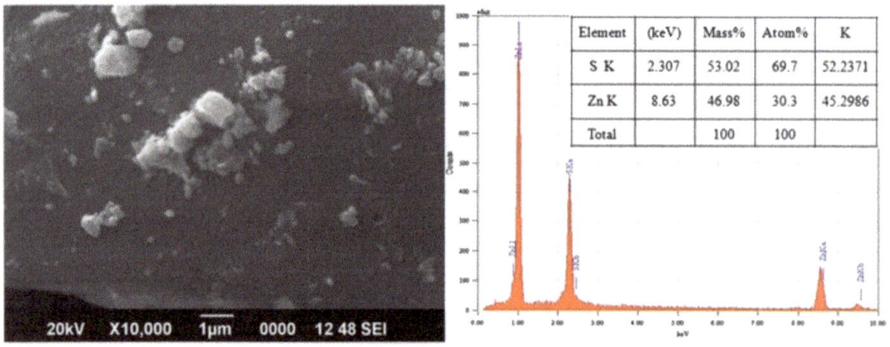

Figure 3 SEM micrograph and EDX of ZnS prepared with 0.5 M Na_2S solution ($ZnS_{0.5}$).

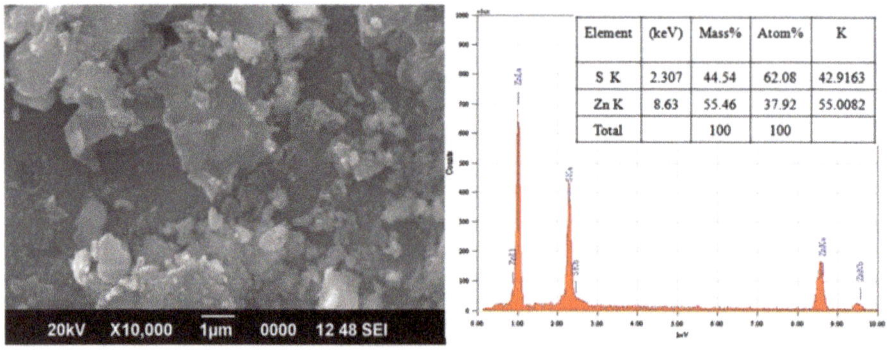

Figure 4 SEM micrograph and EDS of ZnS prepared with 0.1 M Na_2S solution ($ZnS_{0.1}$).

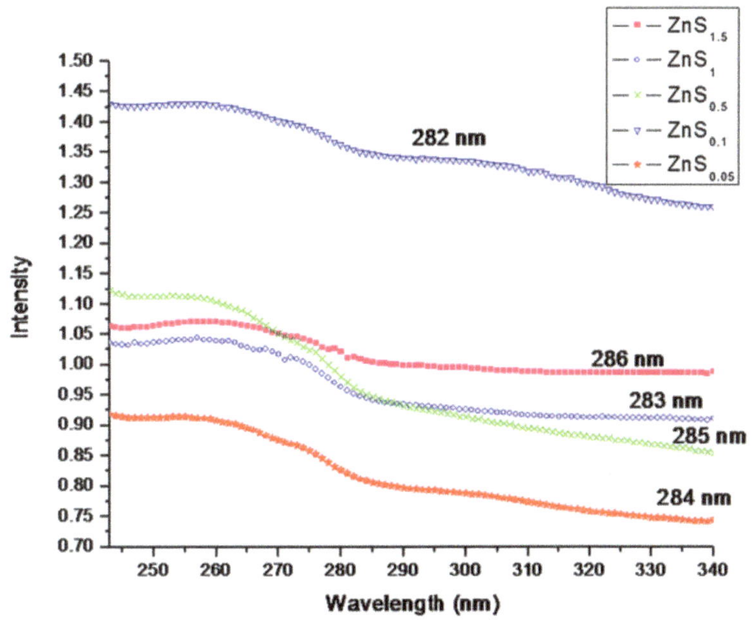

Figure 5 Absorption spectrograph of ZnS prepared in various concentrations of Na_2S solution.

EDX analysis of smaller crystallite samples were performed to find their elemental composition, and the results are shown in Figures 2, 3, and 4, where the elemental data are shown in the inset tables of Figures 2, 3, and 4. These data confirm that the prepared powder consist of only ZnS compound as seen from the X-ray diffractograms of those samples (Figure 1).

To obtain the visual image of the nanoparticles, these nanocrystalline powder samples were analyzed by SEM, and the micrographs are shown in Figures 2, 3, and 4. All of the SEM micrographs reveal that the nanograins are agglomerated due to the drying process, and the sizes of the particles are much larger than the values found from the XRD data. The discrepancy associated with the size of ZnS nanoparticles found from XRD and SEM may be because of the fact that SEM shows the lateral dimension of the particles whereas XRD gives the regularity in the atomic arrangement. Perhaps the lateral dimension is of the order of a few tens of nanometers, but the individual nanoparticle with regular atomic lattice extends only up to a few nanometers.

The study of optical absorption is important to understand the behavior of semiconductor nanocrystals since the fundamental property of semiconductors is the band gap that is the energy separation between the filled valence band and the empty conduction band. The optical excitation of electrons across the band gap is strongly allowed that produces an abrupt increase in absorption at the wavelength corresponding to the band gap energy. This feature of the optical spectrum is known as the optical absorption edge. High-quality ZnS structures have good ultraviolet optical properties at room temperature, and Figure 5 shows the optical absorption spectra of the prepared ZnS nanoparticles which is in the range of 800 to 200 nm. The absorption edge was obtained at shorter wavelengths compared to the absorption edge of the bulk materials due to the increase in band gap as a result of the quantum confinement effect as well as the discrete energy bands. Strong confinement is observed when the crystallite or particle size is smaller than the Bohr exciton radius of ZnS (approximately 2.5 nm [23]) as a result of this strong confinement absorption peaks which appear in the absorption spectra due to subband transitions, so that in this case, band gap can be found from the maximum of this absorption peak. However, when the crystallite size increases above the Bohr radius, the peak structure disappears and attains the normal interband transition spectrum [24] having an absorption edge at a lower wavelength than the bulk

Table 2 Crystallite size and band gap

Samples	Crystallite size (nm)	Band gap from Brus equation (1) (eV)	Band gap from absorption spectra (eV)	Absorption edge (nm)
$ZnS_{1.5}$	6.91	3.75	4.33	286
ZnS_1	5.35	3.83	4.38	283
$ZnS_{0.5}$	2.91	4.37	4.35	285
$ZnS_{0.1}$	2.43	4.71	4.40	282
$ZnS_{0.05}$	2.90	4.37	4.37	284

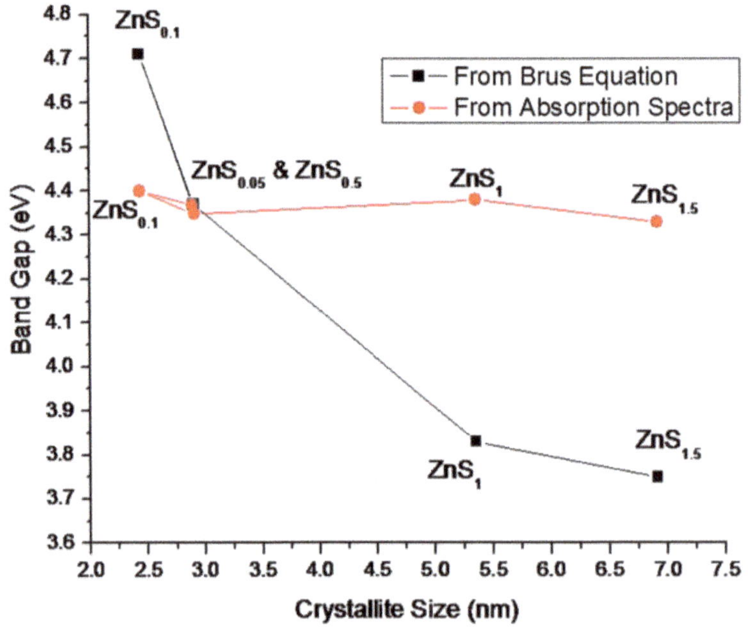

Figure 6 Crystallite size vs band gap.

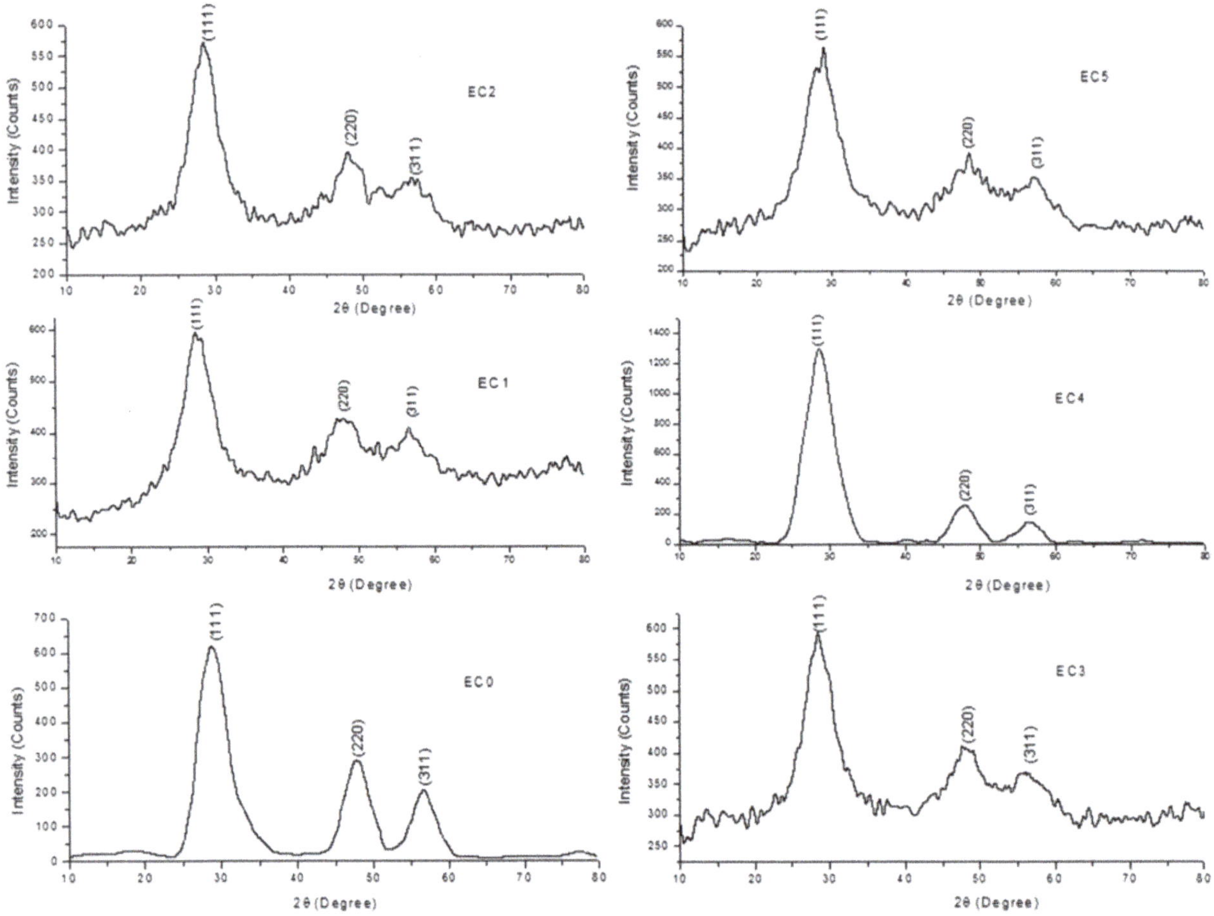

Figure 7 X-ray diffractogram of the samples of various citrate concentrations.

2θ (Degree)		(hkl)	Interplanar spacing 'd' (Å)	
Experimental	ASTM 77-2100		Experimental	ASTM 77-2100
28.66	28.53	(111)	3.1122	3.1261
47.66	47.454	(220)	1.9065	1.9143
56.55	56.307	(311)	1.6261	1.6325

Figure 8 SEM image and XRD information of ZnS prepared without any surfactant (C0).

2θ (Degree)		(hkl)	Interplanar spacing 'd' (Å)	
Experimental	ASTM 77-2100		Experimental	ASTM 77-2100
28.5708	28.53	(111)	3.1217	3.1261
48.3629	47.454	(220)	1.8804	1.9143
56.7	56.307	(311)	1.6221	1.6325

Figure 9 SEM image and XRD information of ZnS prepared in 0.5 M concentration of trisodium citrate (C1).

2θ (Degree)		(hkl)	Interplanar spacing 'd' (Å)	
Experimental	ASTM 77-2100		Experimental	ASTM 77-2100
28.7	28.53	(111)	3.1080	3.1261
47.83	47.454	(220)	1.9001	1.9143
56.78	56.307	(311)	1.6200	1.6325

Figure 10 SEM image and XRD information of ZnS prepared in 1 M concentration of trisodium citrate (C2).

2θ (Degree)		(hkl)	Interplanar spacing 'd' (Å)	
Experimental	ASTM 77-2100		Experimental	ASTM 77-2100
28.5071	28.53	(111)	3.1285	3.1261
47.5088	47.454	(220)	1.9122	1.9143
56.400	56.307	(311)	1.6300	1.6325

Figure 11 SEM image and XRD information of ZnS prepared in 1.5 M concentration of trisodium citrate (C3).

2θ (Degree)		(hkl)	Interplanar spacing 'd' (Å)	
Experimental	ASTM 77-2100		Experimental	ASTM 77-2100
28.54	28.53	(111)	3.1250	3.1261
47.989	47.454	(220)	1.8942	1.9143
56.459	56.307	(311)	1.6285	1.6325

Figure 12 SEM image and XRD information of ZnS prepared in 2 M concentration of trisodium citrate (C4).

2θ (Degree)		(hkl)	Interplanar spacing 'd' (Å)	
Experimental	ASTM 77-2100		Experimental	ASTM 77-2100
28.541	28.53	(111)	3.1249	3.1261
47.508	47.454	(220)	1.9123	1.9143
56.459	56.307	(311)	1.6285	1.6325

Figure 13 SEM image and XRD information of ZnS prepared in 3 M concentration of trisodium citrate (C5).

Table 3 Lattice parameters, crystallite size, strain, and volume of crystals

Samples	Lattice parameters (ASTM value in brackets) (Å)			Average crystallite size (nm)	Average strain	Volume (10^{-28} m^3) (ASTM value in brackets)
	a	b	c			
C0	5.3921 (5.4146)	-	-	2.94	−0.0136	1.5677 (1.5874)
C1	5.3686 (5.4146)	-	-	1.74	−0.0247	1.5474 (1.5874)
C2	5.3769 (5.4146)	-	-	1.73	−0.0872	1.5545 (1.5874)
C3	5.4113 (5.4146)	-	-	2.25	−0.0341	1.5846 (1.5874)
C4	5.3906 (5.4146)	-	-	1.46	−0.0119	1.5664 (1.5874)
C5	5.4075 (5.4146)	-	-	1.95	−0.0531	1.5812 (1.5874)

materials. In the bulk material or in thin films, the band gap can be found from the plot between $(\alpha h\nu)^2$ vs $h\nu$, whereas in the nanomaterials of sizes above Bohr radius, the band gap is obtained from the absorption edge, but when the nanomaterial size is less than Bohr exciton radius of the materials, it is obtained from the absorption peak.

When comparing the absorption spectrogram of each of the samples, it is well clear that the ZnS nanocrystalline powder prepared with 0.1 M of Na_2S has the highest absorption intensity. Highest intensity of absorption means that there is a higher number of particles to absorb as well as this sample has the highest capacity to absorb the incident light. Each of the samples has the absorption around 280 nm (4.4 eV) which was blueshifted from 336 nm (3.7 eV) of the bulk ZnS [25]. The blueshifted absorption also confirms that the results obtained from the X-ray diffractograms and the SEM micrographs, that the powder samples are made up of nanocrystals, since the reduction in the size of particles leads to widening of the band gap and this widening is observable in the optical absorption spectra, where a blueshift is observed with the reduction in particle size. Brus [26] has proposed a theoretical model relating the effective band gap of a material with the particle size. The simple model predicting the variation of exciton energy with particle size is based on the effective mass approximation. The lowest excited state of the crystallite is assumed to be the ground state of an electron hole pair. The ground state energy of an exciton or the increase in effective band gap as a function of crystalline size is estimated as

$$E_g(r) = E_{g,\text{bulk}} + \frac{h^2}{8r^2}\left(\frac{1}{m_e}+\frac{1}{m_h}\right) - 1.8\frac{e^2}{4\pi\varepsilon_r\varepsilon_0 r} \tag{1}$$

where E_g (r) is the band gap of the nanoparticle, $E_{g,bulk}$ is the bulk semiconductor band gap (approximately 3.68 eV for Cubic ZnS [25], approximately 3.9 eV for Wurtzite ZnS [25]), h is Planck's constant (6.6260×10^{-34} m^2 kg/s or 4.1356×10^{-15} eV · s), ε_r is the relative dielectric constant of the material (8.6 [25]), ε_o is the permittivity of free space, r is the radius of the particle, and m_e and m_h are the effective masses of the electron (approximately $0.3m_e$ for ZnS [27]) and hole (approximately $0.6m_e$ for ZnS [27]), respectively. Using the average crystallite size obtained from X-ray diffractograms, the band gap for each sample is calculated in Table 2. It is clear from Table 2 and Figure 6 that the band gap increases as the crystallite size decreases, and it is found that the sample $ZnS_{0.1}$ possesses the smallest crystallites and has high absorption as well as higher band gap when compared with other samples, so that the stoichiometry of the $ZnS_{0.1}$ was selected for further experiments.

Effect of trisodium citrate concentration

Here, trisodium citrate acts as a surfactant, and it allows easy spreading and lowering the surface tension of the solvent. Further, it acts as a microreactor for hosting the reaction and a steric stabilizer to inhibit the aggregation. X-ray diffractogram of the samples was compared with the standard ASTM cards of ZnS and ZnO, and it is found

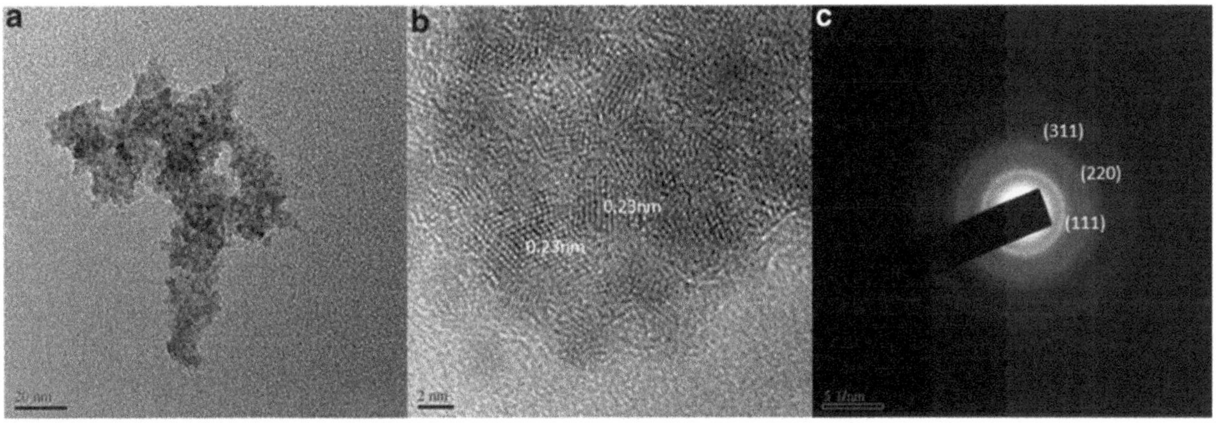

Figure 14 (a) TEM, (b) HRTEM, and (c) SAED images of ZnS prepared in 2 M concentration of citrate (C4).

Table 4 Comparison of interplanar spacing of sample C4 from different analyses

Sample	ASTM 77-2100 (nm)	XRD (nm)	SAED (nm)	HRTEM (nm)	Planes
C4	0.3126	0.3125	0.30	0.23	(111)
	0.1914	0.1894	0.18	-	(220)
	0.1632	0.1628	0.15	-	(311)

Table 5 Crystallite size and band gap of the ZnS prepared in 2 M trisodium citrate

Crystallite size (nm)	Band gap from Brus equation (1) (eV)	Band gap from absorption spectra (eV)
1.46	6.82	4.38

that all the diffractograms correlate with the ASTM card of ZnS 77–2100 (cubic). Their X-ray diffractograms are shown in Figure 7, and their broadened peaks reveal that the sample consisted of nano-sized crystallites. Interplanar spacing of the crystals was calculated and compared with ASTM data in the table given with Figures 8, 9, 10, 11, 12, and 13. Average crystallite sizes and strain were obtained from Williamson-Hall plot [22], and other crystal information is given in Table 3. It is seen from Table 3 that the ZnS crystals prepared in trisodium citrate have smaller crystal size when compared to the sample without this surfactant and formed various sizes of crystallites depending on the concentrations of the surfactants. Further, all of the samples show compressive strain, a property of nanomaterials. From all these experiments, ZnS prepared in 2 M of citrate has the lowest size of crystallites which is evident from Table 3. SEM micrographs of the samples shown in Figures 8, 9, 10, 11, 12, and 13 and the images reveal that well-dispersed particles of nanocrystallites have been formed as observed from the X-ray diffractogram. From this analysis, it is obvious that the ZnS particle growth can be controlled by the concentration of the surfactant sodium citrate.

Reported works of John et al. [28] and Yuan-yuan [20] on ZnS nanostructures without any surfactants show larger particles (12 and 35 nm, respectively) when compared to

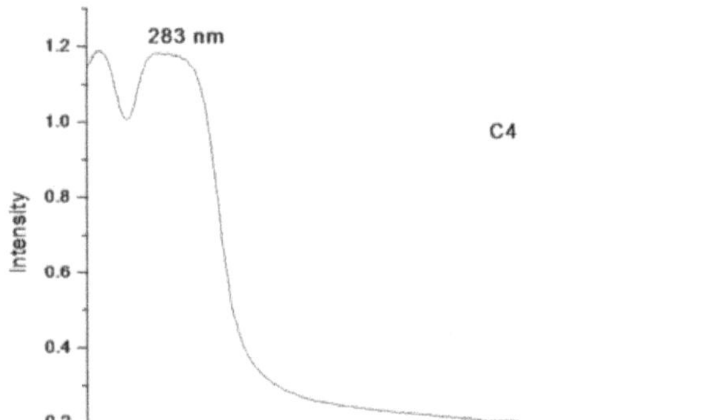

Figure 15 Absorption spectra of the ZnS prepared in 2 M trisodium citrate.

the work explained in this paper (1.46 to 2.25 nm). This comparison on particle size with the previous works [20,28] also showed that surfactant sodium citrate has influence on the crystallization of ZnS nanocrystals.

Further, the sample prepared in 2 M of trisodium citrate was analyzed by transmission electron microscopy (TEM) by dispersing the powder in ethanol. Figure 14 shows the TEM, high-resolution transmission electron microscopy (HRTEM), and selected area electron diffraction (SAED) images of this sample, and it is explicitly shown that the particles are well below 10 nm. The polycrystalline diffraction rings of the SAED images are labeled to the corresponding planes according to the ASTM card of ZnS 77–2100 (cubic). The interplanar spacing of 0.23 nm found from HRTEM matches with the (111) planes which is smaller than the XRD data since the powder was dispersed in ethanol for TEM analysis that disintegrate the agglomerated particles and that further compresses the crystallites. The interplanar spacing (d) obtained from different analysis are compared in Table 4.

Optical absorption spectrum of the ZnS prepared in 2 M citrate is shown in Figure 15, and the appearance of the peak shows interband transitions due to the strong quantum confinement effect. The band gap of the sample is calculated and shown in Table 5 with the crystallite size. It is seen that the band gap calculated from the Brus equation is larger than the band gap obtained from the absorption spectra, which shows that the Brus equation cannot be expected to be quantitatively correct for very small particles [29], since the absorption in the case of small particles having size smaller than their exciton Bohr radius is due to the interband transition. Further, the agglomeration of the nanoparticles also has an effect in their absorption spectra.

Fourier transform infrared spectroscopy (FTIR) analysis was performed after dispersing the powder in ethanol to find the presence of trisodium citrate in the final product that helps to differentiate the properties of the pure ZnS nanocrystals with the ZnS nanocrystals with trisodium citrate. The FTIR spectra of the trisodium citrate, ZnS prepared without trisodium citrate, and ZnS prepared with citrate (sample C4) are shown in Figure 16, and the absorption bands are assigned to the respective vibrations as shown in Table 6. The FTIR spectra show that ZnS is covered with the surfactant trisodium citrate.

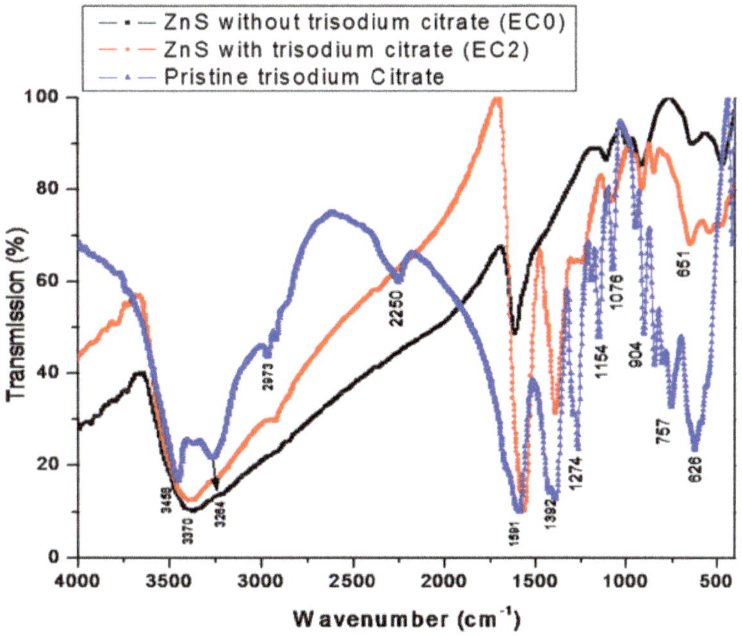

Figure 16 FTIR spectra of Trisodium citrate and ZnS with and without trisodium citrate.

Conclusions

The preparation of zinc sulfide nanocrystals by simple wet chemical precipitation method is discussed here. It is found that the concentration of sulfur and trisodium citrate has tremendous effect on ZnS nanoparticle formation. The particle size was seen to depend on the concentration of trisodium citrate, and the dependence of the optical behavior on Bohr exciton radius of the ZnS nanoparticles was observed.

Methods

ZnS nanocrystalline powder synthesis by varying sulfur concentration

It is essential to find a suitable reagent concentration and stoichiometry for ZnS nanoparticles preparation, and therefore, it was decided to fix the concentration of Zn precursor ($ZnCl_2$) and the surfactant trisodium citrate and to vary the concentration of the sulfur precursor (Na_2S). These studies revealed the suitable Zn precursor and sulfur precursor concentrations for processing ZnS nanocrystallites.

ZnS samples were prepared by varying the sulfur concentration from 0.05 to 1.5 M, and the samples were named as $ZnS_{0.05}$, $ZnS_{0.1}$, $ZnS_{0.5}$, ZnS_1, and $ZnS_{1.5}$. For this experiment, 50-ml stock solution of 1 M $ZnCl_2$ and 0.5 M trisodium citrate was prepared in deionized (DI) water and concentration of Na_2S was varied as shown in Table 7. Trisodium citrate solution (0.5 M) was added to each of the Na_2S solution and stirred well for 1 hr at room temperature. Then, each of the 10.5-ml Na_2S trisodium citrate solution was added dropwise using a burette to the 10-ml $ZnCl_2$ solution while continuing the stirring (200 rpm) at room temperature. It was found that the color of the solution changed from a transparent

Table 6 FTIR peak assignments of trisodium citrate and ZnS with and without trisodium citrate

Citrate				ZnS without citrate (C0)	Peak assignment	ZnS with citrate (C4)	Peak assignments
Peaks	Peak assignment	Peaks	Peak assignment				
3458	O-H stretching	1193	C-O stretching	3370	O-H stretching	3399	O-H stretching
3264	O-H stretching	1154	C-O stretching	1615	C-C	1563	C=O stretching
2973	C-H bending	1076	C-O stretching	1115	C-O stretching	1394	C-H bending
2921	C-H bending	949	C-O stretching	1002	C-O stretching	1293	C-O asymmetrical stretching
2250	C=O stretching	904	C-H bending	920	C-H bending	1077	C-O asymmetrical stretching
1591	C=O stretching	845	C-H bending	642	O-H bending	913	C-H bending
1392	C-H bending	757	C-H bending	-	-	849	C-H bending
1274	C-O stretching	626	C-H deformation	-	-	651	C-H bending

Table 7 Concentration of Na_2S for finding suitable stochiometry

Samples	Number of moles	Molarity (M)	Color of the product
$ZnS_{1.5}$	0.015	1.5	light yellow
ZnS_1	0.01	1	light sandal
$ZnS_{0.5}$	0.005	0.5	light white
$ZnS_{0.1}$	0.001	0.1	pure white
$ZnS_{0.05}$	0.0005	0.05	light brown

solution to milky; then, a white precipitate was formed. The excess water with the dissolved by-product NaCl ($ZnCl_2 + Na_2S \rightarrow ZnS + 2NaCl$) was decanted, and the precipitate was washed in ethanol and then washed four times with deionized water. Since, ethanol can avoid the new formation of ZnS nanocrystals from undissolved Na_2S and $ZnCl_2$ ($ZnCl_2$ partially dissolves in ethanol). After washing four times in DI water, the precipitate was again washed in ethanol and the excess ethanol was decanted, and then the precipitate was dried at 60°C in air since ethanol evaporates at around 50°C. After evaporation of the ethanol, a fine powder of ZnS was obtained and the color of the final powder was seen different for each sample as shown in Table 7.

Effect of trisodium citrate concentration

The effect of the concentrations of trisodium citrate on ZnS particle growth was studied by varying trisodium citrate concentration. Here, $ZnCl_2$ and Na_2S concentrations were fixed similar to the concentration of the sample $ZnS_{0.1}$ in the 'ZnS nanocrystalline powder synthesis by varying sulfur concentration' section based on the results discussed in the 'Results and discussion' section. Stock solutions of 1 M $ZnCl_2$ and 0.1 M Na_2S were prepared as in the 'ZnS nanocrystalline powder synthesis by varying sulfur concentration' section, and then the solutions of trisodium citrate for various concentrations were prepared as shown in Table 8.

Each concentration of trisodium citrate solution was made under magnetic stirring for 10 min to make it a clear solution, and then each of the trisodium citrate solution was mixed with 10 ml Na_2S solution. The stirring continued for 1 h until the solution became light blue, and then the Na_2S-trisodium citrate solution was added drop by drop using a burette to 10 ml $ZnCl_2$ solution while maintaining the stirring (about 200 rpm). It was seen that the solution became milky for each drop of Na_2S-trisodium citrate, and after mixing all the Na_2S-trisodium citrate solution, a white precipitate was obtained. Then, stirring was stopped to let the white precipitate settle at the bottom of the flask. The excess liquid was decanted, and then the settled white powder was washed in ethanol and then washed four times using DI water as explained in the 'ZnS nanocrystalline powder synthesis by varying sulfur concentration' section. After washing again with ethanol, the precipitate was taken into a heater which was maintained at about 60°C in air, and then finally, a white fine powder was obtained.

Table 8 Variation of citrate concentration for ZnS nanopowder synthesis

Samples	Molarity (M)
C0	0
C1	0.5
C2	1
C3	1.5
C4	2
C5	3

Competing interests
The authors declare that they have no competing interests.

Authors' contributions
AKT carried out the experiments, examined the results, and drafted the manuscript. AKAU coordinated the project and discussed the results. Both authors read and approved the final manuscript.

Acknowledgements
The authors thank STIC, Cochin University of Science and Technology, Cochin, Kerala and SAIF, North Eastern Hill University, Shillong, Meghalaya for XRD, SEM, FTIR, and TEM.

References
1. Suchita, K, Gosavi, SW, Urban, J, Kulkarni, SK: Nanoshell particles: synthesis, properties and applications. Curr. Sci. **91**, 8 (2006)
2. Chen, Z-G, Zou, J, Wang, D-W, Yin, L-C, Liu, G, Liu, Q, Sun, C-H, Yao, X, Li, F, Yuan, X-L, Sekiguchi, T, Lu, GQ, Cheng, H-M: Field emission and cathodoluminescence of ZnS hexagonal pyramids of zinc blende structured single crystals. Adv. Funct. Mater. **19**, 484–490 (2009)
3. Sakshi, S, Dhoble, SJ, Singh, P, Ramrakhiani, M: Fabrication of ZnS:Cu/PVA nanocomposite electroluminescence devices for flat panel displays. Adv. Materials Lett. **4**(2), 169-173 (2013)
4. Nizamoglu, S, Demir, HV: Excitation resolved color conversion of CdSe/ZnS core/shell quantum dot solids for hybrid white light emitting diodes. J. Appl. Phys. **105**, 083112 (2009)
5. Muller, GO, Mach, R, Selle, B, Ohnishi, H: Efficient ZnS-like alkaline earth sulfide electroluminescence. J. Crystal Growth **101**, 999–1003 (1990)
6. Wang, X, Xie, Z, Huang, H, Liu, Z, Chen, D, Shen, G: Gas sensors, thermistor and photodetector based on ZnS nanowires. J. Mater. Chem. **22**, 6845–6850 (2012). doi:10.1039/C2JM16523F
7. Sorokina, IT, Sorokin, E, Mirov, S, Fedorov, V, Badikov, V, Panyutin, V, Schaffers, KI: Broadly tunable compact continuous-wave Cr^{2+}:ZnS laser. Opt. Lett. **27**(12), 1040–1042 (2002)
8. Sumitomo Electric Industries, Ltd: Environmentally-resistant ZnS lens for far-infrared cameras (new products and techniques). Sei Tech. Rev. **71**, 113–115 (2010)
9. Onodera, C, Masaaki, Y, Tadayoshi, S, Tsunemasa, T: Threshold current density in ZnS/MgBeZnS quantum well ultraviolet lasers. Optical Rev. **17**(3), 159–160 (2010)
10. Gangopadhyay, U, Kyunghea, K, Dhungel, SK, Mangalaraj, D, Park, JH, Yi, J: Application of CBD zinc sulfide (ZnS) film to low cost antireflection coating on large area industrial silicon solar cell. Trans. Electrical and Electronic Materials **5**, 1 (2004)

11. Bredol, M, Matras, K, Szatkowski, A, Sanetra, J, Prodi-Schwa, A: P3HT/ZnS: A new hybrid bulk heterojunction photovoltaic system with very high open circuit voltage. Solar Energy Mater. Solar Cells **93**, 662–666 (2009)
12. Hsieh, M-F, Li, JK-J, Lin, JC-A, Huang, S-H, Sperling, RA, Parak, WJ, Chang, WH: Tracking of cellular uptake of hydrophilic CdSe/ZnS quantum dots/hydroxyapatite composites nanoparticles in MC3T3-E1 osteoblast cells. J. Nanosci. Nanotechnol. **8**, 12 (2008)
13. Hussain, S, Won, N, Nam, J, Bang, J, Chung, H, Kim, S: One-pot fabrication of high-quality InP/ZnS (core/shell) quantum dots and their application to cellular imaging. Chem. Phys. Chem. **10**, 1466–1470 (2009)
14. Bhadra, R, Singh, VN, Mehta, BR, Datta, P: Studies on some aspects of ZnS nanocrystals for possible applications in electronics. Chalcogenide Lett. **6**(5), 189–196 (2009)
15. Lan, C, Gon, J, Jiang, Y, Ding, Q: Fabrication of ZnS/SnO nanowire/nanosheet hierarchical nanoheterostructure and its photoluminescence properties. Cryst. Eng, Comm (2012). doi:10.1039/C2CE26053K
16. Shaban, SM, Saeed, NM, Raad, MS, Shaban, SM, Saeed, NM, AL-Haddad, RMS: Fabrication and study zinc sulfide schottky barrier detectors. Indian J. Sci. Technol. **4**, 4 (2011)
17. Coleman, VA, Jagadish, C: Basic properties and applications of ZnO. In: Jagadish, C, Pearton, S (eds.) Zinc Oxide Bulk, Thin Films and Nanostructures: Processing, Properties, and Applications, pp. 1–20. Elsevier, Oxford (2006)
18. Warada, HC, Ghosh, SC, Hemtanon, B, Thanachayanont, C, Dutt, J: Luminescent nanoparticles of Mn doped ZnS passivated with sodium hexametaphosphate. Sci. Technol. Adv. Mater. **6**, 296–301 (2005)
19. Dong, D-Q, Lan, LI, Zhang, X-S, Xu, HAN, Hai-Ping, AN: Effect of precursor molar ration of S^{2-}/Zn^{2+} on particles size and photoluminescence of ZnS: Mn^{2+} nanocrystals. Chin. Phys. Lett. **24**(9), 2661 (2007)
20. Yuan-yuan, SHE, Juan, YANG, Ke-qiang, QIU: Synthesis of ZnS nanoparticles by solid–liquid chemical reaction with ZnO and Na_2S under ultrasonic. Trans. Nonferrous Met. Soc. China **20**, s211–s215 (2010)
21. Zhao, Z, Geng, F, Cong, H, Bai, J, Cheng, H-M: A simple solution route to controlled synthesis of ZnS submicrospheres, nanosheets and nanorods. Nanotechnology **17**, 4731–4735 (2006). doi:10.1088/0957-4484/17/18/031
22. Williamson, GK, Hall, WH: X-ray line broadening from filed aluminium and wolfram. Acta Metall. **1**, 22 (1953). doi:10.1016/0001-6160(53)90006-6
23. Bhattacharjee, B, Ganguli, D, Iakoubovskii, K, Stesmans, A, Chaudhuri, S: Synthesis and characterization of sol–gel derived ZnS: Mn^{2+} nanocrystallites embedded in a silica matrix. Bull. Mater. Sci. **25**(3), 175–180 (2002)
24. Sahu, SN, Nanda, KK: Nanostructure semiconductors:physics and applications. PINSA **67**(A, 1), 103–130 (2001)
25. Wang, J, Isshiki, M: Wide-bandgap II–VI semiconductors: growth and properties. In: Kasap, S, Capper, P (eds.) Springer Handbook of Electronic and Photonic Materials, pp. 325–342. Springer, New York (2007)
26. Brus, LE: A simple model for the ionization potential, electron affinity and aqueous redox potentials of small semiconductor crystallites. J. Chem. Phys. **79**, 5566–5571 (1983)
27. Üzar, N, Arikan, MÇ: Synthesis and investigation of optical properties of ZnS nanostructures. Bull. Mater. Sci. **34**(2), 287–292 (2011)
28. John, R, Sasi Florence, S: Optical, structural and morphological studies of bean-like ZnS nanostructures by aqueous chemical method. Chalcogenide Lett. **7**(4), 269–273 (2010)
29. Prabhu, RR, Abdul Khadar, M: Characterization of chemically synthesized CdS nanoparticles. Pramana J. Phys. **65**(5), 801–807 (2005)

One-pot and chemoselective synthesis of bis(4-hydroxycoumarin) derivatives catalyzed by nano silica chloride

Ramin Karimian[1*], Farideh Piri[2], Ali Asghar Safari[3] and Seyed Javad Davarpanah[1]

Abstract

Nano silica chloride (nano SiO_2Cl) has been found to be an efficient, chemoselective and recyclable catalyst for facile and simple condensation of 4-hydroxycoumarin with aromatic and heteroaromatic aldehydes into bis(4-hydroxycoumarin)methanes in dry CH_2Cl_2. The products were obtained in high to excellent yields.

Keywords: 4-hydroxycoumarin; Knoevenagel condensation; Michael addition; bis(4-hydroxycoumarin)methane; Chemoselective; Recyclable; Nano silica chloride

Background

4-Hydroxycoumarin derivatives are of interest because of their widespread biological activities [1,2]. These compounds are used as anticoagulant and sustaining agents [3-5]. They have also been reported as antibiotics [6] and antitumor drugs [7]. Recently, for the preparation of biscoumarins by the reaction of 4- hydroxycoumarin and various aldehydes (one-pot Knoevenagel condensation and Michael addition), a variety of Lewis acid catalysts [8-12], phase transfer catalysts [13-18], microwave reactions [19-22], and molecular iodine were utilized. Although these methods may be effective, some of them have relatively long reaction times and unsatisfactory yields. This finding prompted us towards further investigation in search for a cheap catalyst, which will carry out the synthesis of biscoumarins under simpler experimental set-up and easy workup. Thus, in continuation of our ongoing research for the development of simple and efficient methods for the synthesis of various compounds [23,24], herein we wish to report a simple, economic, and efficient one-pot method for the synthesis of bis(4-hydroxycoumarin)methanes in CH_2Cl_2 using nano silica chloride as the catalyst. One such modified nano silica gel is nano silica chloride (nano SiO_2Cl). Our studies have shown that it was easily prepared by the addition of thionyl chloride to the nano silica gel. Nano silica chloride was reported to be an efficient catalyst for the synthesis of many organic compounds such as amorphous silica chloride [25-27]. It must be noted that the preparation of nano silica chloride is simple, clean, and without workup procedure. Nano silica chloride can be a useful, cheap, and chemoselective catalyst for the synthesis of bis(4-hydroxycoumarin)methanes, which was prepared by the readily available material and can also be easily removed from the reaction mixture.

* Correspondence: karimian.r@gmail.com
[1]Applied Biotechnology Research Center, Baqiyatallah University of Medical Sciences, Tehran, Iran
Full list of author information is available at the end of the article

Results and discussion

First, the effective catalytic route for the synthesis of bis (4-hydroxycoumarin)methanes is described. Amorphous silicon dioxide (75 mg), amorphous silica chloride (75 mg), nano silicon dioxide (75 mg), and nano silica chloride (75 mg) catalyzed the synthesis of bis(4-hydroxycoumarin) methanes. In order to show the applicability and efficiency of this method, our results have been compared with those of some of the same catalysts on the synthesis of bis (4-hydroxycoumarin)methanes [23,24]. As you can see in Table 1, nano silica chloride is superior to the other catalysts.

In the continuation of our study, the reaction of 4-hydroxycoumarin with 4-chlorobenzaldehyde (2j) was chosen as a model reaction in the presence of nano SiO_2Cl (55 mg) in dry dichloromethane at 40°C. The corresponding product (3j) was obtained after 8 h (71%) (Scheme 1; Table 2, entry 1). In a search for an even higher yield, we varied the amount of nano silica chloride;

Table 1 Comparison of results for the synthesis of 3,3′-((4-chlorophenyl)methylene)bis(4-hydroxycoumarin) catalyzed by some silicon catalysts at same conditions

Entry	Catalyst	Time (h)	Yield (%)
1	SiO_2	12	88
2	Nano SiO_2	5.5	92
3	SiO_2Cl	9	87
4	Nano SiO_2Cl	3	92

Table 2 Optimization studies using 4-hydroxycoumarin (1) and 4-chlorobenzaldehyde (2j)

Entry	SiO_2Cl (mg)	Temperature (°C)	Time (h)	Yield[a] (%)
1	55	40	8	71
2	65	40	5.5	85
3	75	40	3	92
4	75	room temperature	12	69
5	75	30	9.5	73
6	75	35	5.5	83

[a]Reaction performed with aldehyde 2j (0.5 mmol), 4-hydroxycoumarin (1 mmol), and product 3j as determined by GC based on the amount of 4-hydroxycoumarin.

gratifyingly, a 92% yield of 3j was obtained when 75 mg was used (Table 2, entry 3). Finally, the reaction was performed at different temperatures. When the reaction was carried out at 25°C, 30°C, and 35°C, the product was also obtained in low yields comparable to the reaction performed at 40°C (Table 2, entries 4 to 6 vs. entry 3). Furthermore, we kept the catalyst concentration constant, and a number of different solvents such as toluene, THF, Et_2O, n-hexane, CH_3CN, and CH_2Cl_2 were also investigated. We found that dichloromethane was better suited as solvent for this purpose.

Thus, the best condition for the reaction was the use of nano silica chloride (75 mg), 4-hydroxycoumarin (1.0 mmol), and aldehyde (0.5 mmol) in dry CH_2Cl_2 (5 ml) at 40°C under air atmosphere (Scheme 2). This result prompted us to investigate the scope and the generality of this new protocol for various aromatic and heteroaromatic aldehydes under optimized conditions (Table 3).

After the optimized reaction condition, several aldehydes were examined. The results of their reaction with 4-hydroxycoumarin are summarized in Table 3. Aromatic aldehydes both with electron-withdrawing and electron-donating groups underwent smooth transformation to the corresponding biscoumarins, without the formation of any side products, in high to excellent yields. However, synthesis could not be achieved in the absence of the catalyst.

Using the same method, the catalytic system is capable of selective condensation of bisaldehydes, as a terephthaldehyde, to the corresponding 4-(bis(4-hydroxy-2-oxo-2H-chromen-3-yl)methyl)benzaldehyde by controlling the molar ratio of 4-hydroxycoumarin (Scheme 3). The results showed that 83% (3v) would be achieved in the presence of two equivalents of 4-hydroxycoumarin, whereas a treatment using four equivalents of 4-hydroxycoumarin with terephthaldialdehyde led to the occurrence of the corresponding 3,3′,3″,3‴-(1,4-phenylenebis(methanetriyl)) tetrakis(4-hydroxy-2H-chromen-2-one) (three times).

This method is also a very good way for the chemoselective conversion of aryl aldehydes in the presence of aliphatic ketones and aliphatic aldehydes. For example, when a 1:1 mixture of 4-chlorobenzaldehyde and cyclohexanone was allowed to react with 4-hydroxycoumarin in the presence of nano SiO_2Cl, it was found that only 4-chlorophenyl-3,3′-bis(4-hydroxycoumarin)methane was obtained; there is no corresponding product of cyclohexanone. Also, in an equimolar mixture of aryl aldehyde and aliphatic aldehyde, it was found that the aryl aldehydes were chemoselectively converted to the corresponding bis(4-hydroxycoumarin) methane, but the aliphatic ones were converted slightly (Scheme 4). The product was obtained in high yield without the formation of any side reactions.

The formation of compound **3** can be explained by the mechanism that is presented in Scheme 5. The reaction

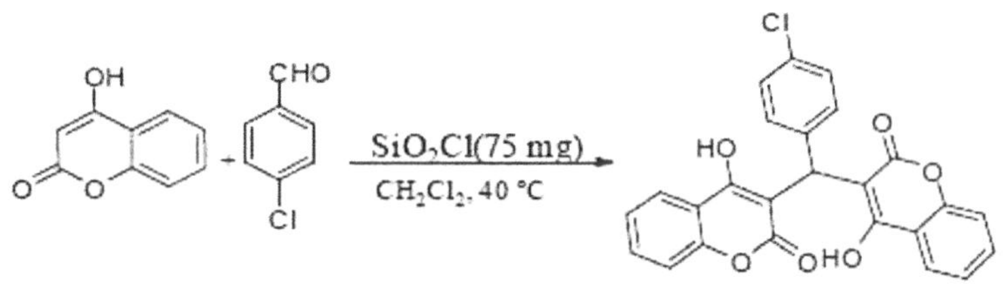

1 2j 3j

Scheme 1 The reaction of 4-hydroxycoumarin with 4-chlorobenzaldehyde (2j) and the corresponding product (3j).

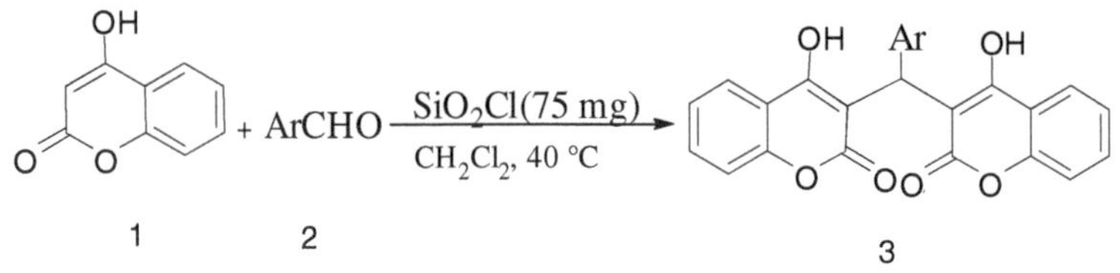

Scheme 2 Nano silica chloride-catalyzed synthesis of bis(4-hydroxycoumarin)methanes.

sequence is one-pot Knoevenagel condensation, Michael addition, and tautomerization. 4-Hydroxycoumarin adds easily to electron-poor alkenes in the Michael addition fashion. Thus, cascade reactions of addition, elimination, and addition could be achieved, but the intermediate alkenes **12** were not available even in 1:1 experiments according to the present reaction conditions [22]. The electron-poor alkenes **12** are useful intermediates for Michael additions.

Inspection of the SEM and TEM images of a sample catalyst from this reaction indicates the involvement of silica chloride nanoparticles with a size distribution of 15 to 35 nm (average ≈25 nm; Figures 1 and 2). After the reaction, nano silica chloride can be easily separated (by centrifuge and filtration) and reused without a decrease in its activity. For example, the reaction of 4-hydroxycoumarin with 4-chlorobenzaldehyde afforded the corresponding bis(4-hydroxycoumarin)methanes in 95%, 95%, and 91% isolated yield over three cycles. When we carried out the reaction in CH_2Cl_2, the reaction proceeded very slowly to give moderate yields.

Table 3 Nano silica chloride catalyzed synthesis of bis(4-hydroxycoumarin)methanes

Entry	ArCHO	Product	Time (h)	Yield[a,b](%)	Melting point (°C)
1	PhCHO	3a	2.5	85	231-234
2	2-OH PhCHO	3b	2	88	254-256
3	2-OMe PhCHO	3c	2.5	82	236-238
4	2-Cl PhCHO	3d	3	93	224-226
5	2-Br PhCHO	3e	3	95	256-258
6	2-NO_2 PhCHO	3v	3.5	91	104-106
7	3-NO_2 PhCHO	3f	3.5	90	234-236
8	4-NO_2 PhCHO	3g	3.5	85	232-234
9	4-CF_3 PhCHO	3h	3.5	71	258-260
10	4-F PhCHO	3i	6	71	212-214
11	4-Cl PhCHO	3j	3	92	254-256
12	4-Br PhCHO	3k	3.5	91	266-268
13	4-Me PhCHO	3l	2.5	90	266-268
14	4-OMe PhCHO	3m	2	78	246-248
15	4-NMe_2 PhCHO	3n	1	95	222-224
16	4-OH PhCHO	3o	2.5	90	222-224
17	3,4-$(OMe)_2$ PhCHO	3p	2	88	263-265
18	2,6-Cl_2 PhCHO	3q	3.5	78	178-180
19	Furfural	3r	2.5	70	200-202
20	2-Thenaldehyde	3s	3	75	210-212
21	2-Formylpyrrole	3t	2	75	161-163
22	3-Formylindole	3u	5.5	68	238-240

[a]Yields refer to crude product (isolated product).
[b]All products were identified by comparing their physical and spectral data with those of authentic samples.

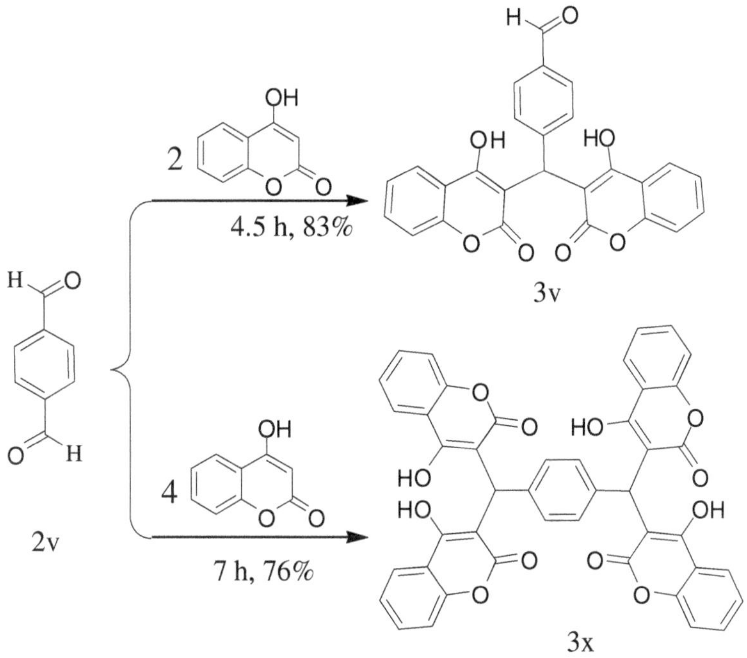

Scheme 3 Selective condensation of dialdehyde with 4-hydroxycoumarin.

Conclusion

A facile and highly efficient method for the condensation of 4-hydroxycoumarin with aromatic and heteroaromatic aldehydes has been described using nano silica chloride as a recyclable and chemoselective catalyst. The products were obtained in high to excellent yields after stirring for a few hours with operational simplicity.

Methods

Preparation of nano silica chloride

Nano SiO_2Cl powder was prepared according to the procedure reported [23,24]. To an oven-dried (120°C) nano silica gel (5 g) in a round-bottomed flask (250 mL) equipped with a condenser and a drying tube, thionyl chloride (toxic and should be used with caution) (100 mL) was added, and the mixture was refluxed for 1 week. The excess thionyl chloride was distilled off. The resulting white-grayish powder was flame-dried and stored in a tightly capped bottle.

Typical experimental procedure

To a mixture of 4-hydroxycoumarin (1 mmol) and aldehydes (0.5 mmol) in CH_2Cl_2 (5 mL), nano silica chloride (75 mg) was added at 40°C. The mixture was stirred for a specified period (Table 3). The progress of the reaction was monitored by thin layer chromatography (TLC).

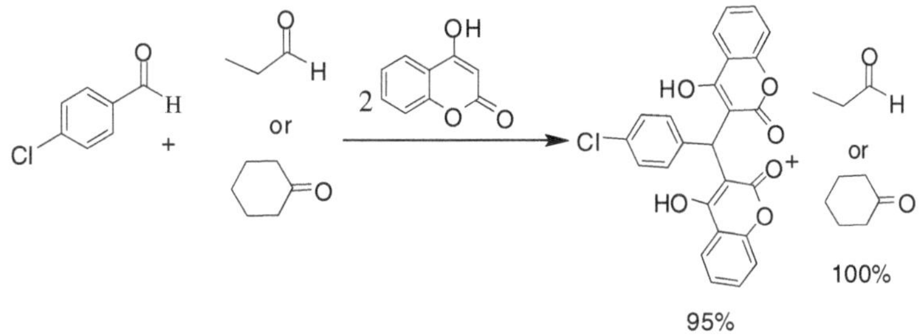

Scheme 4 Chemoselectivity of aryl aldehyde reacting with 4-hydroxycoumarin in the presence of ketone and aliphatic aldehyde.

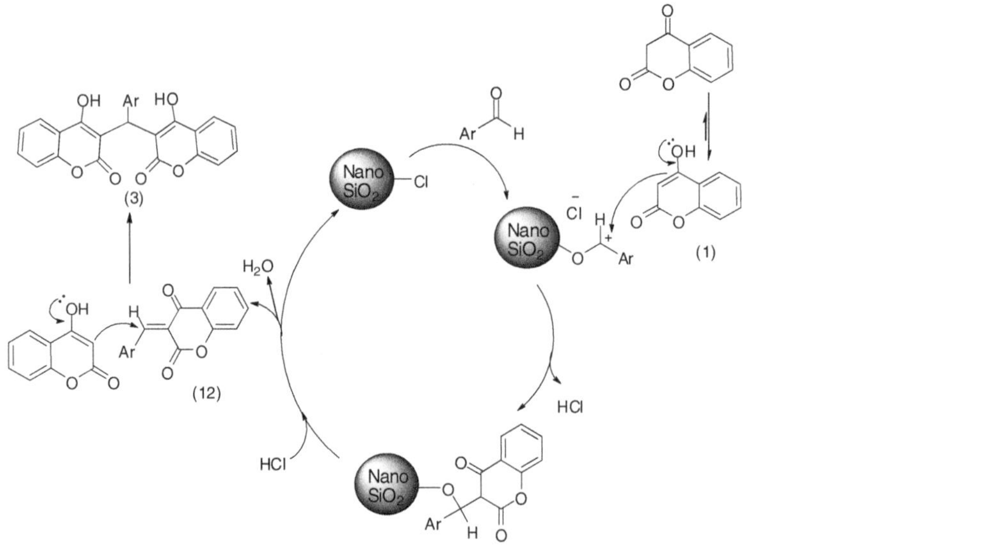

Scheme 5 Mechanism of synthesis of bis(4-hydroxycoumarin)methane.

After the complete conversion of the starting material, as indicated by TLC, the reaction mixture was filtered and concentrated under reduced pressure. The obtained product was filtered and recrystallized from ethanol to get the pure product. The products were characterized according to their ^1H nuclear magnetic resonance (NMR), IR, and melting point data. The representative spectral (^1H NMR and IR) data of bis(4-hydroxycoumarin)methane derivatives 3a to 3u are given below.

Compound 3a. White solid; ^1H NMR (250 MHz, $CDCl_3$): δ 6.18 (s, 1H), 6.82 to 7.56 (m, 13H), 10.72 (brs, 2OH); IR (KBr disk): 3034, 1652, 1608, and 754.

Compound 3g. Yellow solid; ^1H NMR (250 MHz, $CDCl_3$): δ 6.56 (s, 1H), 7.13 to 8.38 (m, 12H), 11.22 (brs, 2OH); IR (KBr disk): 3033, 1652, 1615, 1529, 1347, and 761.

Compound 3m. White solid; ^1H NMR (250 MHz, $CDCl_3$): δ 3.78 (s, 3H), 6.38 (s, 1H), 7.1 to 8.31 (m, 12H), 11.49 (brs, 2OH); IR (KBr disk): 3388, 3026, 1666, 1605, 1255, 1050, and 768.

Compound 3r. Black solid; ^1H NMR (250 MHz, $CDCl_3$): δ 6.09 (s, 1H), 6.35 to 6.55 (m, 3H), 7.23 to 8.38 (m, 8H), 12.02 (brs, 2OH); IR (KBr disk): 3038, 1668, 1615, and 767.

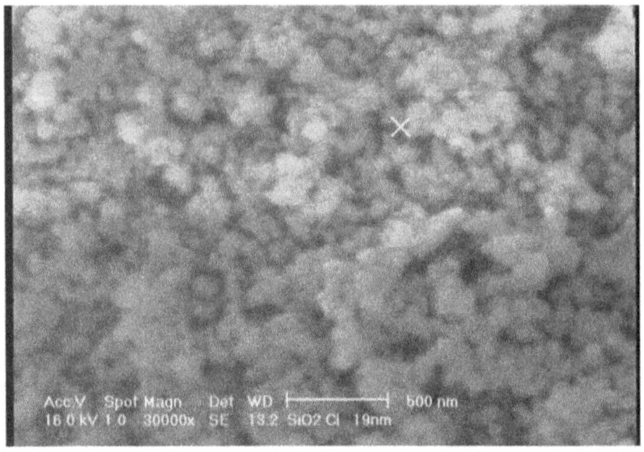

Figure 1 **SEM image of silica chloride nanoparticle.**

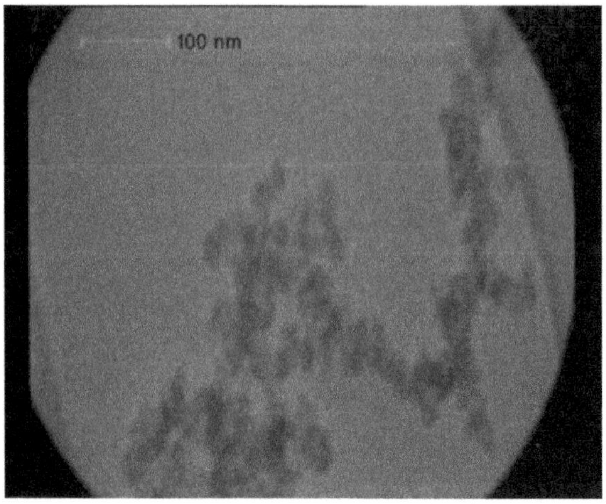

Figure 2 **TEM image of silica chloride nanoparticle.**

Competing interest

The authors declare that they have no competing interests.

Authors' contributions

RK and AAS have contributed to all experiments and has been supervised by FP who supervised the project. SJD has participated in the statistical analysis and in the preparation of the manuscript. All authors read and approved the final manuscript.

Authors' information

RK is an assistant professor of Organic Chemistry in Applied Biotechnology Research Center at Baqiyatallah University of Medical Sciences. FP is an assistant professor of Organic Chemistry at University of Zanjan. AAS is an assistant professor of Organic Chemistry at Islamic Azad University of Zanjan. SJD is an assistant professor of Biology in Applied Biotechnology Research Center at Baqiyatallah University of Medical Sciences.

Acknowledgments

This research work is financially supported by Baqiyatallah University of Medical Sciences and University of Zanjan.

Author details

[1]Applied Biotechnology Research Center, Baqiyatallah University of Medical Sciences, Tehran, Iran. [2]Department of Chemistry, Faculty of Science, Zanjan University, P.O. Box 45195-313, Zanjan, Iran. [3]Department of Chemistry, Faculty of Sciences, Zanjan Branch, Islamic Azad University, Zanjan, Iran.

References

1. Okenne, R, Thomes, RD: Coumarins: Biology Application and Modes of Action. Wiley, Chichester (1997)
2. Kostava, I, Manolov, I, Nicolova, I, Konstantonov, S, Karaivanova, M: New lanthanide complexes of 4 methyl 7 hydroxycoumarin and their pharmacological activity. Eur. J. Med. Chem. **36**, 339–347 (2001)
3. Chohan, ZH, Shaikh, AU, Rauf, A, Supuran, CT: Antibacterial, antifungal and cytotoxic properties of novel N-substituted sulfonamides from 4-hydroxycoumarin. J. Enzym. Inhib. Med. Chem. **21**, 741–748 (2006)
4. Zhao, H, Neamati, N, Hong, H, Mazumdar, A, Wang, S, Sunder, S, Milne, GWA, Pommier, Y, Burke, TR: Coumarin-based inhibitors of HIV integrase. J. Med. Chem. **40**, 242–249 (1997)
5. Stahmann, A, Ikawa, M, Link, KP: US Patent, 2427578, Chem. Abstr. **1948**(42), P603h (1947)
6. Murray, RDH, Mendez, J, Brown, SA: Natural coumarins. Wiley, Chichester (1982)
7. Hinman, JW, Hoeksema, H, Caron, EL, Jackson, WG: The partial structure of novobiocin (streptonivicin). J. Am. Chem. Soc. **78**, 1072–1074 (1956)
8. Sangshettia, JN, Nagnnath, D, Kokarea, B, Shinde, DB: Water mediated efficient one-pot synthesis of bis-(4-hydroxycoumarin)methanes. Green Chemistry Letters and Reviews **2**, 233–235 (2009)
9. Chen, YL, Wang, TC, Tzeng, CC, Chang, NC: Geiparvarin analogues: synthesis and anticancer evaluation of α-methylidene-γ-butyrolactone-bearing coumarins. Helv. Chim. Acta. **82**, 191–197 (1999)
10. Yadav, JS, Reddy, BVS, Basak, AK, Visali, B, Narsaiah, AV, Nagaiah, K: Phosphane-catalyzed Knoevenagel condensation: a facile synthesis of α-cyanoacrylates and α-cyanoacrylonitriles. Eur. J. Org. Chem. **3**, 546–551 (2004)
11. Narsaiah, AV, Nagaiah, K: An efficient Knoevenagel condensation catalyzed by $LaCl_3{\cdot}7H_2O$ in heterogeneous medium. Synth. Commun. **33**, 3825–3832 (2003)
12. Lehnert, W: Verbesserte variante der knoevenagel-kondensation mit $TiCl_4$/THF/pyridin(I). Alkylidenund arylidenmalonester bei 0–25°C. Tetrahedron Lett. **11**, 4723–4724 (1970)
13. Karimi-Jaberi, Z, Nazarifar, MR, Pooladian, B: Tris(hydrogensulfato) boron as a solid heterogeneous catalyst for the rapid synthesis of α, α′-benzylidene bis (4-hydroxycoumarin) derivatives. Chin. Chem. Lett. **23**, 781–784 (2012)
14. Mehrabi, H, Abusaidi, H: Synthesis of biscoumarin and 3,4-dihydropyrano[c] chromene derivatives catalysed by sodium dodecyl sulfate (SDS) in neat water. J. Iran. Chem. Soc. **7**, 890–894 (2010)
15. Jang, DO: Hypophosphorous acid mediated dehalogenation in water. Tetrahedron Lett. **37**, 5367–5368 (1996)
16. Yorimitsu, H, Shinokubo, H, Oshima, K: Radical cyclization reaction using a combination of phosphinic acid and a base in aqueous ethanol. Chem. Lett. **2**, 104–105 (2000)
17. Kita, Y, Nambu, H, Ramesh, NG, Anilkumar, G, Matsugi, M: A novel and efficient methodology for the C–C bond forming radical cyclization of hydrophobic substrates in water. Org. Lett. **3**, 1157–1160 (2001)
18. Khan, TA, Tripoli, R, Crawford, JJ, Martin, CG, Murphy, JA: Diethylphosphine oxide (DEPO): high-yielding and facile preparation of indolones in water. Org. Lett. **5**, 2971–2974 (2003)
19. Lang, S, Corr, M, Muir, N, Khan, TA, Schonebeck, F, Murphy, JA, Payne, AH, Williams, AC: First organophosphorus radical-mediated cyclisations to afford medium-sized rings: eight-membered lactones and seven- and eight-membered lactams. Tetrahedron Lett. **46**, 4027–4030 (2005)
20. Cravotto, G, Nano, GM, Palmisano, G, Tagliapietra, S: The reactivity of 4-hydroxycoumarin under heterogeneous high-intensity sonochemical conditions. Synthesis **8**, 1286–1291 (2003)
21. Deb, ML, Bhuyan, PJ: Uncatalysed Knoevenagel condensation in aqueous medium at room temperature. Tetrahedron Lett. **46**, 6453–6457 (2005)
22. Bansal, KM, Mothsra, P, Saxen, S, Somvanshi, RK, Dey, S, Singh, TP: Molecular iodine: a versatile catalyst for the synthesis of bis(4-hydroxycoumarin) methanes in water. J. Mol. Catal. A: Chem. **268**, 76–81 (2007)
23. Karimian, R, Piri, F, Karimi, B, Moghimi, A: Silica chloride nano particle catalyzed synthesis of 2,2′-(arylmethylene)bis(5,5-dimethylcyclohexane-1,3-dione) derivatives. Croat. Chem. Acta **84**, 111–115 (2011)
24. Karimian, R, Piri, F, Karimi, B, Moghimi, A: Silica chloride nano particle catalyzed ring opening of epoxides by aromatic amines. Chin. J. Chem. **29**, 2421–2427 (2011)
25. Firouzabadi, H, Iranpoor, N, Hazarkhani, H, Karimi, B: Reactions of silica chloride (SiO_2Cl)/DMSO, a heterogeneous system for the facile regeneration of carbonyl compounds from thioacetals and ring-expansion annelation of cyclic thioacetals. J. Org. Chem. **67**, 2572–2576 (2002)
26. Kamitori, Y, Hojo, M, Masuda, R, Kimura, T, Yoshida, T: Selective protection of carbonyl compounds. Silica gel treated with thionyl chloride as an effective catalyst for thioacetalization. J. Org. Chem. **51**, 1427–1431 (1986)
27. Sathe, M, Gupta, AK, Kaushik, MP: An efficient method for the esterification of phosphonic and phosphoric acids using silica chloride. Tetrahedron Lett. **47**, 3107–3109 (2006)

Comparison between antibacterial activity of some natural dyes and silver nanoparticles

Mohammad Mirjalili and Mina Abbasipour*

Abstract

Curcuma longa rhizome (turmeric), walnut, and henna are medicinal plants which are traditionally used for fiber colorations. In this study, the viscose rayon fabrics were dyed with different mordant at a variety of conditions. The dyed fabric and treated fabric with Ag nanoparticles were evaluated for antibacterial activity against pathogenic strain of Gram-negative (*Escherichia coli*) bacteria. Durability of antibacterial activity to laundering was also investigated. The results indicated that treated fabrics with these natural dyes had excellent antibacterial activity as well as Ag nanoparticles before and after wash.

Keywords: Natural dyes, Ag nanoparticles, Antibacterial activity, Durability

Background

Nowadays, the demand for antibacterial fabric and health care has increased; so, finding a method to get antibacterial textile is a challenge. Fibers, especially natural fibers (wool and cotton), provide basic requirements such as moisture and nutrients for bacterial growth and multiplication. This often leads to objectionable odor, dermal infection, product deterioration, and other related diseases [1-4]. A variety of materials like quaternary ammonium salt, metal oxide nanoparticles, and nanocomposites are reported as antibacterial agents [1-3]. Using synthetic non-biodegradable chemical compounds for these approaches cause environmental and health concerns. Although natural dyes are known for a long time for dyeing as well as medicinal applications, the structures and protective properties have been recognized only in the recent past [4,5]. Many of the plants used for dye extraction are classified as medicinal, and some of these have recently been shown to possess remarkable antimicrobial activity [6]. Several sources of plant dyes rich in naphthoquinons such as lawsone from henna, juglone from walnut, and lapachol from alkanet are reported to exhibit antibacterial and antifungal activities [7-10].

Curcuma longa, also known as 'turmeric', is used as a coloring agent, and has medicinal properties. Various sesquiterpenes and curcuminoids have been isolated from the rhizome of *C. longa*, attributing to its biological characteristics such as antioxidant, anti-inflammatory, wound healing, anticancer, antiproliferative, antifungal, and antibacterial properties [11,12].

In this study, viscose fabric was dyed with walnut, turmeric, and henna, and its antibacterial properties were those treated with Ag nanoparticles.

Results and discussion

Antimicrobial activity of natural dyes

Table 1 shows the antimicrobial activity of turmeric, walnut, and henna in different conditions. The dyed fabric with turmeric powder under alkaline condition created a greater effect in reducing bacteria concentration as compared to that under acidic condition. As results show, the dyed fabrics with these natural dyes show high bacterial reduction activity.

Table 2 shows the reduction of bacterial concentration of Ag nanoparticles. The fabric treated with silver nanoparticles performed very high activity with reduction of bacteria against *E. coli.*

As Table 3 shows, the dyed samples with turmeric, walnut shell, and henna in different conditions have good washing fastness.

Figure 1 shows a summary of the antibacterial activities of the samples dyed with turmeric, henna, and walnut and the sample treated with Ag nanoparticles after 20 washings, respectively. As it is evident, the dyed

* Correspondence: mina.abbasipour@gmail.com
Department of Textile Engineering, Yazd Branch, Islamic Azad University, Yazd, Iran

Table 1 Antimicrobial evaluation against *E. coli* in different conditions

Condition of dyeing	Mordant	% R (turmeric)	% R (henna)	% R (walnut shell)
Neuter	-	91.25	97.65	99.56
Acetic acid	-	99.3	99.1	99.6
NaOH	-	97.65	98.41	99.12
Acetic acid	Potash alum	97.41	99.96	99.64
NaOH	Potash alum	97.52	98	99.3
Acetic acid	Potassium bichromate	99.75	99.89	99.92
NaOH	Potassium bichromate	99.82	99.98	99.4

samples withstand severe laundering which indicate good durability.

Conclusion

Viscose fabric was dyed with three different natural dyes (walnut, turmeric, and henna) under different conditions. The antibacterial activity of dyed samples against *E. coli* were compared with treated samples containing silver nanoparticles. The results indicated that treated fabrics with these natural dyes had excellent antibacterial activity as well as those treated with Ag nanoparticles before and after wash. The dyed fabric with turmeric powder under alkaline condition created a greater effect in reducing bacteria concentration as compared to that under acidic condition. The dyed fabric with potassium bichromate mordant had good reduction of bacteria concentration in alkaline and acidic condition. Dyed sample with turmeric under neutral condition did not have good antibacterial activity.

Methods

Materials

In this study, the viscose rayon fabric, Ag nanoparticles, walnut, turmeric, and henna powders were used throughout experiments. Alum and potassium bichromate were purchased from Merk Co., Ltd (Whitehouse Station, NJ, USA). Silver nanoparticles (average size of <150 nm) were supplied from Sigma Aldrich Co., Ltd (St. Louis, MO, USA).

The method of dyeing

Viscose fabrics were dyed by exhaustion method in different conditions (acidic, alkaline, and neuter) as shown in Figure 2. In the dye preparation, 3 g of turmeric and henna powder were dissolved in 40 ml of distilled water, and temperature was gradually raised to 90°C over 30 min and maintained at 90°C for 5 min. In using walnut shell as the dye, 5 g of walnut were added to 50 ml of distilled water and temperature was maintained at 90°C for 40 min. Viscose fabrics were dyed with 2% (owf) of dye. Alum and potassium bichromate were used as mordant. The material-to-liquor ratio was 1:20 and temperature was raised to 90°C over 30 min and

Table 2 Antibacterial activity of Ag nanoparticles against *E. coli*

Sample	% R (*E. coli*)
Nanosilver-treated fabric	100

Table 3 Evaluation of washing fastness of treated samples

Condition of dyeing	Washing fastness	
	Color change	Staining on cotton
Turmeric		
Neuter	3	2
Acetic acid	5	3.5
NaOH	4.5	4.5
Acetic acid + Potash alum	3	3
NaOH + Potash alum	3	4
Acetic acid + Potassium bichroate	4.5	4.5
NaOH + Potassium bichroate	4.5	4.5
Henna		
Neuter	3	3
Acetic acid	3	3
NaOH	4.5	5
Acetic Acid + Potash alum	4.5	5
NaOH + Potash alum	4.5	5
Acetic acid + Potassium bichroate	4.5	5
NaOH + Potassium bichroate	4.5	5
Walnut shell		
Neuter	4	5
Acetic acid	4.5	5
NaOH	4.5	5
Acetic acid + Potash alum	5	4.5
NaOH + Potash alum	5	4.5
Acetic acid + Potassium bichroate	5	4.5
NaOH + Potassium bichroate	5	4.5

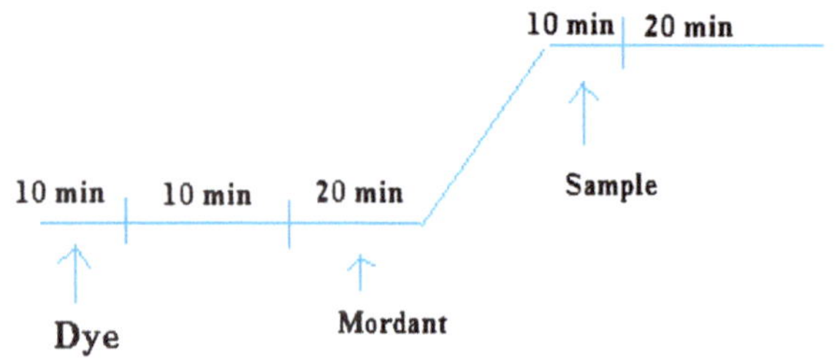

Figure 1 Dyeing profile for different natural dyes.

maintained at 90°C for 30 min. Dyed fabric was rinsed with cold water and dried.

Treated fabric with Ag nanoparticles

The fabric was dipped in the 5% Ag colloidal solution; dried at 80°C and cured at 160°C for 4 min.

Wash fastness test

The treated samples were washed according to the ISO 105 C10 (2006) to determine the color change and the antimicrobial effect of fabrics after laundering. A gray scale was selected for evaluating washing fastness.

Antibacterial test

The antimicrobial activity was quantitatively evaluated against *Escherichia coli* (ATCC 6538), a Gram-negative organism, according to the AATCC 100 test method. The fabric samples with 4.8 ± 0.1 cm in diameter were placed in a 250-ml glass jar with screw cap and absorbed 1.0 ± 0.1 ml of bacterial inoculum. After incubation over contact periods of 24 h, 100 ml of sterilized distilled water was added into the jar and shacked vigorously for 1 min. The solution was then serially diluted to 10^1, 10^2, 10^3, and 10^4. The diluted solution was plated on a nutrient agar and incubated for 24 h at 37°C ± 2°C. Colonies of bacteria recovered on the agar plate were counted,

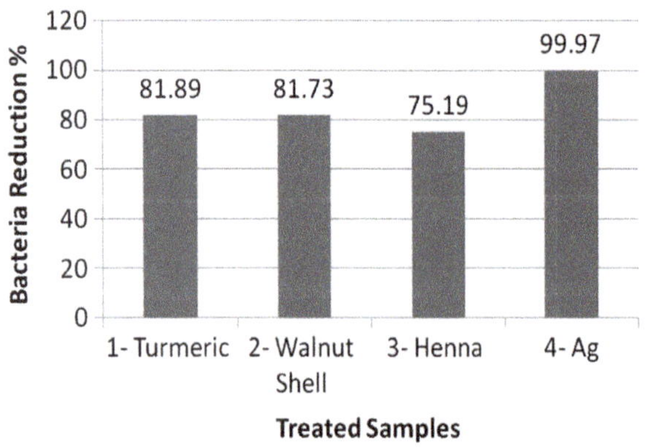

Figure 2 Comparison of bacteria reduction of dyed sample.
1- Turmeric (NaOH/Potassium Bichromate), 2- Walnut Shell (NaOH/ Potassium Bichromate), 3- Henna (NaOH/Potassium Bichromate) and 4- Ag Nanoparticles after home laundry 20 times.

and the percent reduction of bacteria (R) was calculated by the following equation:

$$R(\%) = (B–A) \times \frac{100}{B}, \qquad (1)$$

where A is the number of bacterial colonies from treated specimen after inoculation over 24 h of contact period and B is the number of bacterial colonies from untreated control specimen after inoculation at 0 contact time.

Competing interests

The authors declare that they have no competing interests.

Authors' contributions

MM participated in the design of the study and performed the statistical analysis. MA conceived of the study and participated in its design and coordination. Both authors read and approved the final manuscript.

Acknowledgements

The authors want to express their thanks to Prof. Omid Moradi.

References

1. Han, S, Yang, Y: Antimicrobial activity of wool fabric treated with curcumin. Dyes and Pigments **64**, 157–161 (2005)
2. Ali, S, Hussai, T, Nawaz, R: Optimization of alkaline extraction of natural dye from henna leaves and its dyeing on cotton by exhaust method. J. Cleaner Prod. **17**, 61–66 (2009)
3. Sarkar, RK, De, P, Chauhan, PD: Bacteria-resist finishes on textiles using natural herbal extracts. Indian Journal of Fiber and Textile Research **28**, 322–331 (2003)
4. Othmers, K: Encyclopedia of chemical technology. Interscience **7**, 7–81 (1967)
5. Hussein, SAM, Barakat, HH, Merfort, I, Nawwar, MAM: Tannins from the leaves of *Punica granatum*. Photochemistry **45**, 23–819 (1997)
6. Gerson, H: Fungi toxicity of 1,4-napthoquinones to *Candida albicans* and *Trichophyton mentagrophytes*. Can. J. Microbiol. **21**, 197–205 (1975)
7. Schuerch, AR: Wehrli, W: beta-Lapachone, an inhibitor of oncornavirus reverse transcriptase and eukaryotic DNA polymerase-alpha inhibitory effect, thiol dependence and specificity. Eur. J. Biochem. **84**, 197–205 (1978)
8. Wagner, H, Kreher, B, Lotter, H, Hamburger, MO, Cordell, GA: Structure determination of new isomeric naphthol [2,3,-b] furan- 4,9-diones from Tabebuia avellanedae by the selective INEPT technique. Helv. Chim. Acta **72**, 67–659 (1989)
9. Machado, TB, Pinto, AV, Pinto, M, Leal, IC, Silva, MG, Amaral, AC, Kuster, RM, Netto-dosSantos, KR: In vitro activity of Brazilian medicinal plants, naturally occurring naphthoquinones and their analogues, against methicillin-resistant Staphylococcus aureus. Int. J. Antimicrob. Agents **21**, 84–279 (2003)
10. Gulrajani, ML, Gupta, D: Wash fastness tests to test durability. In: NCUTE Workshop on Dyeing and Printing with Natural Dyes, pp. 3–5. IIT, Delhi (2001)
11. Barik, A, Mishra, B, Kunwar, A, Kadam, RM, Shen, L, Dutta, S, Padhye, S, Satpati, AK, Zhang, HY, Priyadarsini, KI: Comparative study of copper(II) ecurcumin complexes as superoxide dismutase mimics and free radical scavengers. Eur. J. Med. Chem. **42**, 431–439 (2007)
12. Singh, R, Jain, A, Panwar, S, Gupta, D, Khare, SK: Antimicrobial activity of some natural dyes. Dyes and Pigments **66**, 99–102 (2005)

Study of growth parameters on structural properties of TiO_2 nanowires

Abdollah MortezaAli[1] and Saeideh Ramezani Sani[2*]

Abstract

In this article, the effect of growth parameters on structural properties of TiO_2 nanowires, synthesized by thermal evaporation process, have been studied. The nanostructures are characterized by scanning electron microscopy and X-ray spectroscopy. Our synthesized TiO_2 nanowires were found, through X-ray diffraction measurements, to be crystalline rutile TiO_2. We studied the effect of growth parameters such as the deposition temperature, the system pressure, and carrier gas flow on the growth and structure of our synthesized nanowires. It was found that their growth was governed by vapor–liquid–solid mechanism because of using Au catalyst. In this mechanism, the Ti layer deposited on the substrate diffused in to the solid Au catalyst; then, a liquid alloy of Au-Ti was formed to serve as preferred sites for the growth of TiO_2 nanowires.

Keywords: TiO_2 nanowires, Growth parameters, Growth mechanism, Thermal evaporation

Background

Semiconductor nanowires are a fundamental importance to the study of size- and dimensionality-dependent chemical and physical phenomena [1]. Semiconducting nanowires have received considerable attention due to their unique physical properties and on dimensional structure. Among these nanostructure materials, TiO_2 nanostructures have emerged as one of the most promising materials because of their potential for gas sensors [2], optical devices [3,4], photocatalysis [5-8], and solar cells [9]. There are some reports to synthesize TiO_2 nanowires by chemical methods [10-13], but to date, there have been a few results on synthesizing of TiO_2 nanowires by thermal evaporation [14-19].

Understanding the growth mechanism is critical in controlling and designing nanostructures. Change of the morphology of nanostructure is a direct consequence of changing synthesis condition which gives them some new optical and electrical properties. Therefore, the technological importance of one-dimensional (1-D) nanostructure and the lack of information on their morphology underline the need for developing growth parameter studying. In the study of Wu et al. [16] in 2005, the influence of reaction on the temperature of synthesis at 850°C elucidated. They also considered the effect of the temperature of synthesis on the reaction time of 1 h on the structure of TiO_2 nanowires [18]. In 2006, Lee et al. [20] studied the controlled growth of TiO_2 nanowires on sapphire and silica by thermal evaporation. Amin et al. [21] in 2007 studied the growth of nanowires under different conditions of substrate, catalytic material, and chemical composition of Ti_xO_y precursor.

In this article, we report structural properties of TiO_2 nanowires, which have been synthesized by thermal evaporation. To explain the effect of growth parameters on the structure of nanowires, we studied the effect of several parameters such as deposition temperature, system pressure, and carrier gas flow on growth and structure of our synthesized nanowires. It was found that nanowires were in the crystalline rutile phase, and their growth was governed by vapor–liquid–solid (VLS) mechanism.

Results and discussion

The effect of deposition temperature

To investigate the effect of deposition temperature, similar substrates were placed at different distances from the source material downstream in a tube furnace, which resulted in different synthesis temperature due to the

* Correspondence: s_ramezanisani@yahoo.com
[2]Department of Mechanic, Islamic Azad University, Roudehen Branch, Roudehen, Iran
Full list of author information is available at the end of the article

natural temperature gradient along the furnace. The temperature of synthesis was varied among 750°C, 850°C, and 950°C for 1 h. The scanning electron microscopy (SEM) images of TiO_2 nanowires at different deposition temperatures have been represented in Figure 1. Figure 1a shows that no nanowire was obtained when the temperature was controlled at 750°C. Figure 1b (the deposition temperature is 850°C) shows the nanowires with the diameter of around 60 to 120 nm and length of about 500 to 2,000 nm. As the deposition temperature increases to 950°C, the nanowires become shorter and thicker with length of about 250 to 1,000 nm and diameters of 68 to 150 nm. It can be understood in the light of the fact that in the temperature of more than 850°C, the vapor of a material is very energetic and leads to a short length. It was found that a convenient temperature for controlling the diameter size of nanowires is about 850°C. Synthesis conditions and morphology characteristics of TiO_2 nanowires are listed in Table 1.

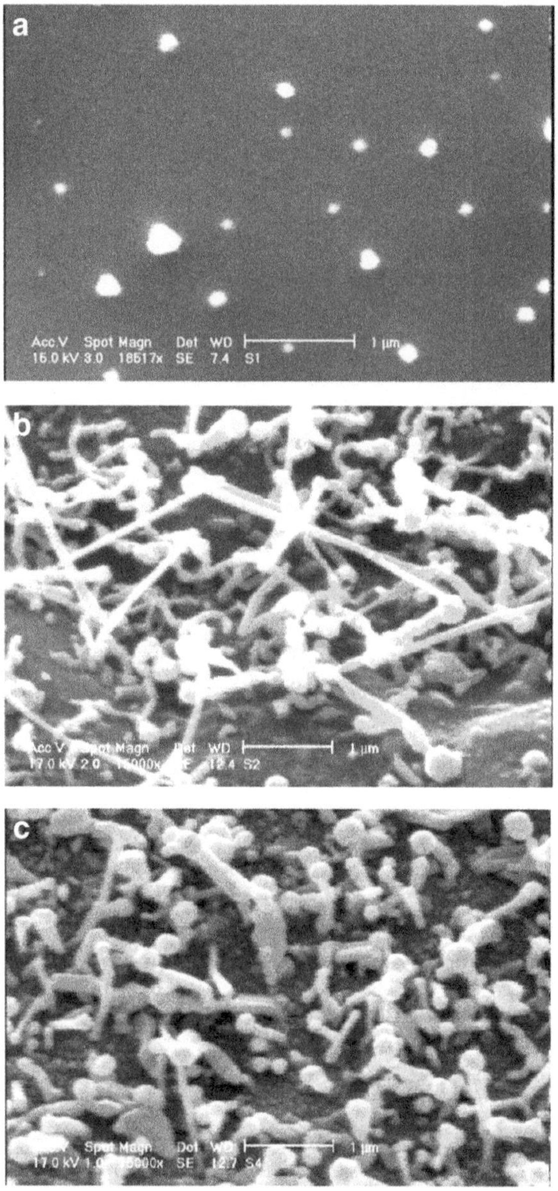

Figure 1 SEM images of as-synthesized TiO2 nanowires at different deposition temperatures. (**a**) Deposition temperature of 750°C where no nanowires was obtained. (**b**) Deposition temperature of 850°C and (**c**) 950°C. Scale bar is 1 μm; with the increase of the deposition temperature from 850°C to 950°C, the nanowires become shorter and thicker.

The effect of system pressure

To study the influence of system pressure on the crystal structure of nanowire, experiments were performed at different system pressures that were changed from 1,000 to 10 mbar by a rotary pump, while other parameters were maintained unchanged (Table 2). Figure 2 shows SEM images of TiO_2 nanowires at different system pressures. Figure 2a shows the nanowires with a diameter of around 60 to 120 nm. As the system pressure decreases, the nanowires become thinner so that in the system pressure of 60 mbar, the diameter of the nanowires is 60 to 100 nm (Figure 2b), and in the system pressure of 30 mbar, the diameter of nanowires decreases to 40 to 100 nm (Figure 2c). The advantage of using low pressure in the fabrication of 1-D nanostructures can be understood from the fact that lowering the system pressure increases the mean free path, and as a result, the gas phase transfer to deposition surface is facilitated. Also, a low pressure system is beneficial to establish the desired degree of supersaturation for 1-D nanostructure. Figure 2d shows the growth of nanowires at the system pressure of 10 mbar. The image shows that nanowires have irregular morphology. One possibility is that a loss in the absorption of oxygen on the surface causes a decrease in the growth of nanowires. It is noted that the thermal evaporation process is very sensitive to the concentration of oxygen in the growth system [22]. To achieve the high-yield growth of TiO_2 nanowires, a low concentration of O_2 is necessary. In our experiments, the low-concentration of O_2 is likely to come from the system leakage and the adsorption on surface. However, in the system pressure of 10 mbar, there is no enough leakage of air into the reaction system, and as a result, no enough O_2 is absorbed in the surface.

The effect of carrier gas flow

To understand the role of carrier gas flow on structure nanowires, new experiments involving deliberate introduction of Ar into the system were performed.

Table 1 Synthesis conditions and morphology characteristics of TiO_2 nanowires at different deposition temperature

Furnace center temperature (°C)	Carrier gas flow (sccm)	System pressure (mbar)	Deposition temperature (°C)	Reaction time (h)	Nanowire diameter (nm)	Nanowires length (nm)
1,050	-	1,000	750	1	-	-
1,050	-	1,000	850	1	60 to 120	500 to 2,000
1,050	-	1,000	950	1	68 to 150	250 to 1,000

Using the unchanged experimental parameters (furnace center temperature, reaction time, system pressure), the gas flow was varied from 50 to 150 sccm. Synthesis conditions of TiO_2 nanowires are listed in Table 3. The SEM images of the TiO_2 nanowires at different carrier gas flows have been represented in Figure 3. It was found that the yield of TiO_2 nanowires increases with the increase of gas flow. Figure 3a (gas flow is 50 sccm) shows the nanowires with a diameter of around 70 to 150 nm and length of about 400 to 1,000 nm. With the increase of the the carrier gas flow to 100 sccm, the nanowires become thinner and longer with a length of about 400 to 1,800 nm and diameter of 40 to 100 nm (Figure 3b). This is understandable from the fact that O_2 plays a key role in the growth of the TiO_2 nanowires. The increase of the concentration of oxygen causes a decrease of the growth of the nanowires. When oxygen concentration is high, in the VLS growth, Ti vapor reacts with oxygen before the formation of droplets of the alloy and prevents the creation of nanowires. Thus, low concentration of oxygen, which causes the increase of the growth of the nanowires, is necessary [19]. The increase of the gas flow from 50 to 100 makes a low-oxygen environment; therefore, more nanowires are produced. Figure 3c shows the growth of nanowires at a gas flow of 150 sccm; in this case, nanowires become shorter, thicker, and denser. The image shows that nanowires are rod-like and have irregular morphology. It was found that a convenient gas flow for controlling the diameter size of the nanowires is about 100 sccm [19].

Crystalline structure of the TiO_2 nanowires

To study the information regarding the crystallinity of the TiO_2 nanowires, X-ray diffraction (XRD) experiments were performed. All of the samples indicated that the TiO_2 nanowires are mainly composed in rutile phase. A typical XRD pattern of the sample is shown in Figure 4. It can be seen that TiO_2 nanowires are in rutile phase (JCPDS: 73–1765) with a dominate reflectance of the (101) plan. The unite cell constants of TiO_2 nanowires are $a = b = 4.58$ Å, and $c = 2.96$ Å, which matched the published value in the JCPDS database (# 73–1765, $a = 4.52$ Å, $c = 2.98$ Å). The Au diffraction peaks (JCPDS: 04–0784) were caused by the diffraction from the catalyst.

Growth mechanism

Understanding the growth mechanism is critical in controlling and designing nanostructures. One possible mechanism is the vapor-solid (VS) growth that TiO_2 nanowires form with no catalytic particles on their tips. In addition to the VS growth, the VLS growth mechanism might also protect the TiO_2 nanowires. According to the VLS growth mechanism, crystal growth was guided by a droplet of liquid alloy. VLS growth is also called catalyst growth. Since Au was used as a catalyst, the growth of the nanowires was mainly dominated by the VLS mechanism. The Ti buffer layer was partially oxidized into a TiO_2 layer by the O_2 gas of atmosphere at an elevated temperature. The Ti layer deposited on the silicon substrate was diffused into the solid Au catalyst; then, a liquid alloy of Au-Ti was formed. The interfacial region between the TiO_2 layer and the liquid Au catalyst formed by the buffer layer-assisted VLS mechanism was in a high-energy state; hence, it provided an active nucleation site for the TiO_2 nanowires. The Ti buffer was introduced onto the surface of the Si substrate to prevent any undesirable diffusion reaction between the Ti vapor source and the Si substrate and also to provide a good nucleation

Table 2 Synthesis conditions and morphology characteristics of TiO_2 nanowires at different system pressures

Furnace center temperature (°C)	Carrier gas flow (sccm)	System pressure (mbar)	Deposition temperature (°C)	Reaction time (h)	Nanowire diameter (nm)	Nanowire length (nm)
1,050	-	1,000	850	1	60 to 120	500 to 2,000
1,050	-	60	850	1	60 to 100	300 to 1,100
1,050	-	30	850	1	40 to100	300 to 2,500
1,050	-	10	850	1	irregular morphology	irregular morphology

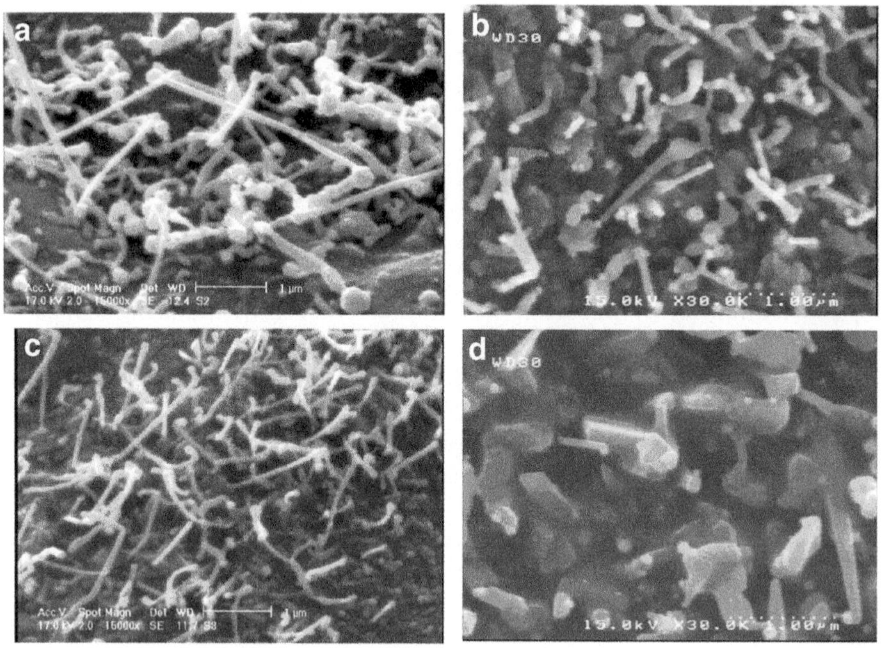

Figure 2 SEM images of as-synthesized TiO2 nanowires at different system pressures. System pressures of (**a**) 1,000, (**b**) 60, (**c**) 30, and (**d**) 10 mbar. Scale bar is 1. As the system pressure decreases, the nanowires become thinner.

site for the growth of the nanowires [19]. A typical SEM image and EDX of the TiO_2 nanowires are presented in Figure 5. Results show the nanowires with catalytic material on its tips (the Si signal comes from the substrate).

Conclusions

TiO_2 nanowires were grown by thermal evaporation on a silicon substrate coated with a catalytic Au-coated Ti layer. The effect of several parameters such as deposition temperature, system pressure, and carrier gas flow on growth and structural properties was investigated. It was found that a convenient temperature and gas flow for controlling the diameter size of nanowires are about 850°C and 100 sccm, respectively. The nanowires analyzed with XRD were found to be in rutile phase. The growth of the nanowires was dominantly governed by the VLS mechanism because of using Au as a catalyst.

Methods

TiO_2 nanowires were prepared using a thermal evaporation method similar to that described in our previous studies [19]. In general, the process of synthesis thermal evaporation technique is conducted in a tube furnace, where the source material is at a high-temperature zone and the substrate is in the low-temperature zone. Also, it consists of a rotary pump system and a gas supply. In a typical preparation procedure, a Ti buffer layer with thickness of 50 nm was deposited on the silicon substrate. Then, a gold thin film was deposited as a catalyst on the Ti layer. Then, the TiO_2 nanowires were grown by thermal evaporation in the low-temperature zone of a horizontal quartz furnace. There are several parameters such as deposition temperature, system pressure, carrier gas flow, and reaction temperature that can change the morphology of products. In this study, to explore the growth mechanism of TiO_2 nanowires, a series of experiments were carried out and investigated the effect

Table 3 Synthesis conditions and morphology characteristics of TiO_2 nanowires at different carrier gas flows

Furnace center temperature (°C)	Carrier gas flow (sccm)	System pressure (mbar)	Deposition temperature (°C)	Reaction time (h)	Nanowire diameter (nm)	Nanowire length (nm)
1,050	50	1,000	850	1	70 to 105	400 to 1,000
1,050	100	1,000	850	1	40 to 100	400 to 1,800
1,050	150	1,000	850	1	irregular morphology	irregular morphology

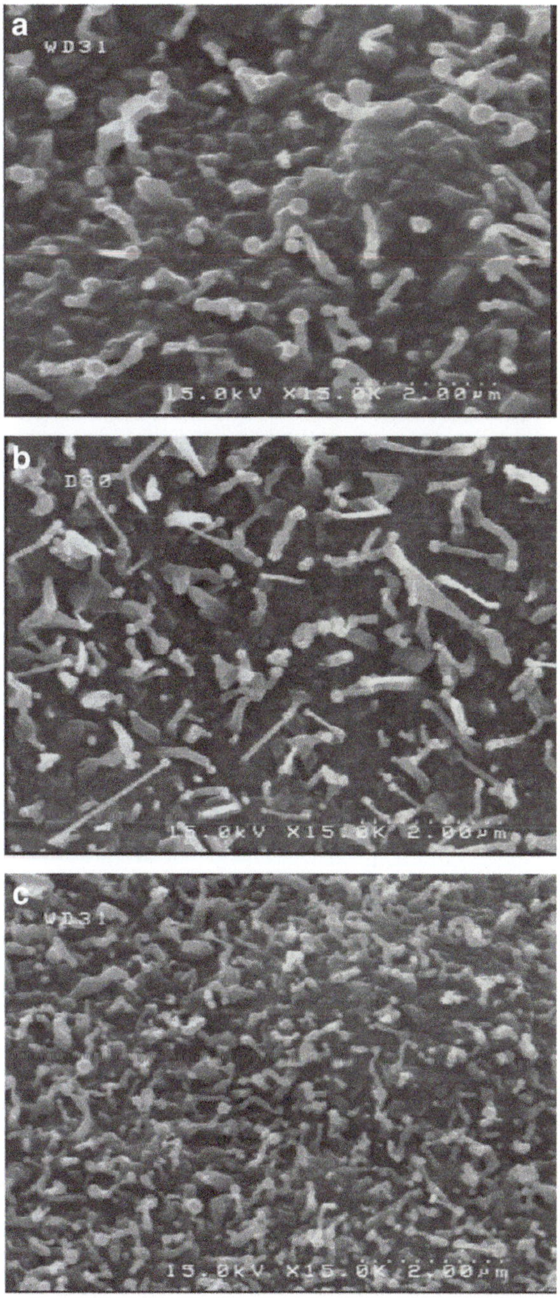

Figure 3 SEM images of as-synthesized TiO2 nanowires at different carrier gas flows. Gas flow at (**a**) 50, (**b**) 100, and (**c**) 150 sccm. Scale bar is 2 m. It was found that the yield of TiO2 nanowires increase with the increase of gas flow.

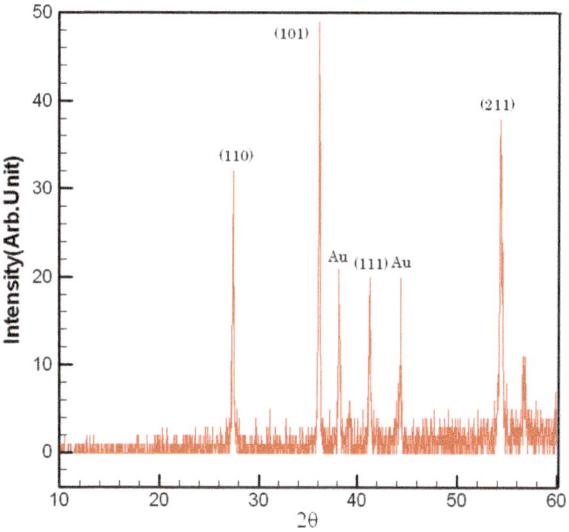

Figure 4 A typical XRD pattern of TiO_2 nanowires. It can be seen that TiO_2 nanowires are in rutile phase with a dominate reflectance of the (101) plan.

Figure 5 A SEM image of TiO_2 nanowires. With an EDX spectrum that displays the existence of Au catalysis on its tip.

of deposition temperature, system pressure, and carrier gas flow on the structural properties of TiO_2 nanowires.

The structure and morphology of the samples were respectively analyzed by XRD (JEOL JDX-8030, Akishima-shi, Japan) and SEM (XL130, Philips, Amsterdam, The Netherlands).

Competing interests
The authors declare that they have no competing interests.

Authors' contributions
SR carried out the thermal evaporation studies, synthesized TiO_2 nanowires, and wrote the manuscript for publication. AM participated in the idea and the design of the study. Both authors read and approved the final manuscript.

Acknowledgments
The authors would be like to thank Azzahra University for the kind support.

Author details
[1]Laboratory of Surface Physics, Department of Physics, Alzahra University, Tehran 19938, Iran. [2]Department of Mechanic, Islamic Azad University, Roudehen Branch, Roudehen, Iran.

References

1. Hu, J, Odom, TW, Lieber, CM: Chemistry and physics in one-dimension: synthesis and properties of nanowires and nanotubes. Acc. Chem. Res. **32**, 435 (1999)
2. Rothschild, A, Evakov, AY, Ashkenasy, N, Komen, Y: Surface photovoltage spectroscopy study of reduced and oxidized nano crystalline TiO_2 film. Surf. Sci. **419**, 532 (2003)
3. Wijnhoven, EGJ, Vos, WL: Preparation of photonic crystals made of air spheres in titania. Science **281**, 802 (1998)
4. Richel, A, Johnson, NP, McComb, D: Observation of Bragg reflection in photonic crystals synthesized from air spheres in a titania matrix. Appl. Phys. Lett. **76**, 1816 (2000)
5. Albu, SP, Ghicov, A, Macak, J, Hahn, MR, Schmuki, P: Self-organized, free-standing TiO_2 nanotube membrane for flow-through photocatalytic applications. Nano Lett. **7**, 862 (2007)
6. Armstrong, A, Armstrong, G, Canales, J, Garcia, R, Bruce, PG: Lithium-ion intercalation into TiO_2-B nanowires. Adv. Mater. **17**, 862 (2005)
7. Armstrong, A, Armstrong, G, Canales, J, Garcia, R, Bruce, PG: TiO_2 (B) nanotubes as negative electrodes for rechargeable lithium. Electrochem. Solid Lett **9**(A), 139 (2006)
8. Karimi, L, Zohoori, S: Superior photocatalytic degradation of azo dyes in aqueous solutions using TiO2/SrTiO3 nanocomposite. J. Nanostruct. Chem. **3**, 32 (2013)
9. Park, NG, Van de Lagemaat, J, Frank, AJ: Comparison of dye-sensitized rutile- and anatase-based TiO_2 solar cells. J. Phys. Chem. **104**, 8989 (2000)
10. Huang, MH, Wu, Y, Feick, H, Tran, N, Weber, E, Yang, P: Catalytic growth of zinc oxide nanowires by vapor. Transport. Adv. Mater. **13**, 69 (2001)
11. Dong, KY, Dong, YK: Novel approach to the fabrication of macroporous polymers and their use as a template for crystalline titania nanorings. Nano Lett. **3**, 207 (2003)
12. Du, GH, Chen, Q, Che, RC, Yuan, ZY, Pen, LM: Preparation and structure analysis of titanium oxide nanotubes. Appl. Phys. Lett. **79**, 3702 (2001)
13. Miao, L, Tanemura, S, Toh, S, Kaneko, K, Tanemura, M: Fabrication, characterization and Raman study of anatase-TiO_2 nanorods by a heating-sol–gel template process. J. Cryst. Growth **264**, 246 (2004)
14. Wu, J-M, Shin, H, Wu, W-T, Tseng, Y-K, Chen, I-C: Thermal evaporation growth and the luminescence property of TiO_2 nanowires. J. Cryst. Growth **281**, 384 (2005)
15. Wu, J-M, Wu, W-T, Shin, H: Characterization of single-crystalline TiO_2 nanowires grown by thermal evaporation. J. Electrochem. Soc. **152**, G613 (2005)
16. Wu, J, Shin, H, Wu, W: Growth of TiO_2 nanorods by two-step thermal evaporation. J. Vac. Sci. Technol. B **23**, 2122 (2005)
17. Zhang, H, Luo, X, Xu, J, Xiang, B, Yu, D: Synthesis of TiO_2/SiO_2 core/shell nanocable arrays. J. Phys. Chem. B **108**, 14866 (2004)
18. Wu, J-M, Shin, H, Wu, W-T: Formation and photoluminescence of single-crystalline rutile TiO_2 nanowires synthesized by thermal evaporation. Nanotechnology **17**, 105 (2006)
19. Ramezani Sani, S, Morteza Ali, A, Jafari, R: The effect of carrier gas flow on structural and optical properties of TiO_2 nanowires. Physica E **43**, 1809 (2011)
20. Lee, J-C, Park, K-S, Kim, T-G, Choi, H-J, Sung, Y-M: Controlled growth of high-quality TiO_2 nanowires on sapphire and silica. Nanotechnology **17**, 4317 (2006)
21. Seyed, SA, Alan, WN, Terry, TX: A facile approach to synthesize single-crystalline rutile TiO_2 one-dimensional nanostructures. Nanotechnology **18**, 445609 (2007)
22. Dai, ZR, Pan, ZW, Wang, ZL: Growth and structure evolution of novel tin oxide diskettes. J. Am. Chem. Soc. **124**, 8673 (2002)

25

Magnetic Fe_3O_4 nanoparticles as recovery catalyst for preparation of oximes under solvent-free condition

Behzad Zeynizadeh and Mostafa Karimkoshteh*

Abstract

Magnetically nano Fe_3O_4 efficiently catalyzes solvent-free conversion of various carbonyl compounds to the corresponding oximes with $NH_2OH{\cdot}HCl$. The reactions were carried out in oil bath (70°C to 80°C) at solvent-free conditions to afford oximes in high to excellent yields. This method offers some advantages in terms of clean reaction conditions, recyclable of catalyst, short reaction times, easy work-up procedure, and suppression of side products.

Keywords: Carbonyl compounds; Nano Fe_3O_4; Magnetic; $NH_2OH{\cdot}HCl$; Oxime

Background

Transformation of carbonyl compounds into oximes is one of the most important synthetic reactions in organic chemistry. Oximes are not only useful for protection, purification, and characterization of carbonyl compounds [1] but also they can serve as intermediates for conversion into nitro compounds [2], amides [3], nitriles [4], and amines [5]. In addition, all compounds that contain the oxime fragment, which represents a pharmacophore group, exhibit diverse biological activities. These derivatives are also recommended as insecticides, fungicides, pesticides, herbicides, protozoacides, and plant growth regulators [6]. A wide spectrum of biological activity was revealed for oximes of pyridine, indole, isatin, and pyrroleandquinoline series [7]. Oximes of α, β-unsaturated ketones are known as insecticides, vasorelaxants, and antimicrobial agents [6].

The literature review shows that many improvements have been carried out for the preparation of oximes. $NH_2OH{\cdot}HCl/K_2CO_3$ [8], DOWEX(R)50WX4 [9], ethylenediamine/oxone [10], ultrasound/Na_2SO_4 [11], Bi_2O_3 [12], heterogeneous polyoxometalates [13], phase transfer catalysis [14], and $TiO(acac)_2$ [15] are some of the reagents or protocols which have been reported for transformation of carbonyl compounds to oximes. Generally low yields, long reaction times, and limitation to use acid- or base-sensitive functionality in carbonyl compounds are the major limitations of these methods.

On the other hand, nanoparticles exhibit high catalytic activity and chemical selectivity under mild conditions [16]. The extremely small-size particles maximize the surface area which, when exposed to the reactant, allow more reactions to occur at the same time and thus speed up the process [17]. Thus, in the line of the outlined strategies, herein, we wish to report a green and practical method for solvent-free conversion of various carbonyl compounds to the corresponding oximes by $NH_2OH{\cdot}HCl$ catalyzed by magnetically nano Fe_3O_4 (Scheme 1).

Results and discussion

Magnetically nano Fe_3O_4 due to highly active and specific centers of ultra-size coupled with high magnetic specification has been found some useful applications in organic synthesis [18]. Continuation of our research program directed to the oximes reactions under mild and green reaction conditions [3] and lack of information to use magnetically nano Fe_3O_4 in preparation of oximes from carbonyl compounds by $NH_2OH{\cdot}HCl$ encouraged us to investigate the titled transformation.

* Correspondence: Karimkoshteh@gmail.com
Department of Chemistry, Faculty of Science, Urmia University, Urmia 57159-165, Iran

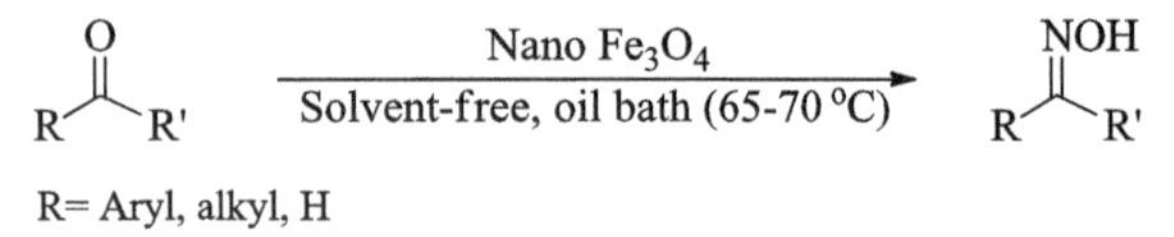

Scheme 1 A green and practical method for solvent-free conversion of various carbonyl compounds.

The promotion effect of nano Fe_3O_4 in the formation of oximes was studied by the reaction of benzaldehyde with $NH_2OH{\cdot}HCl$ under different reaction conditions. The effects of molar ratio of the reaction components and solvent were investigated in the typical experiment (Table 1). As it is seen, the combination system of hydroxylamine hydrochloride with nano Fe_3O_4 in a molar ratio of 2:1.5, respectively, in oil bath (70°C to 80°C) at solvent-free conditions showed a perfect efficiency for complete conversion of 1 mmol benzaldehyde to benzaldehydeoxime within 20 min(entry 6). This success encouraged us to further explore the capability of $NH_2OH{\cdot}HCl$/nano Fe_3O_4 system for oximation of different carbonyl compounds at the optimized reaction conditions. The summarized results in Tables 2 and 3 show that all reactions were carried out successfully in oil bath (70°C to 80°C) at solvent-free conditions, and the corresponding aldoximes or ketoximes were obtained in high to excellent yields within 15 min to 3 h. In addition, due to inherent less reactivity of ketones versus aldehydes, the oximation reactions took excess amounts of $NH_2OH{\cdot}HCl$ and nano Fe_3O_4 in longer reaction times (Table 3). α,β-Unsaturated aldehydes and ketones were successfully converted to the corresponding conjugated oximes with $NH_2OH{\cdot}HCl$/nano Fe_3O_4 system in high yields and without any side reaction (Table 2: entry 8 and Table 3: entry 9). We also observed that oximation of carbonyl compounds with two carbonyl moieties was carried out selectively. By using the controlled amounts of $NH_2OH{\cdot}HCl$ in an appropriate time, oximation of one or both of the carbonyl group took place efficiently (Table 3: entries 7 and 8).

A plausible mechanism for the nanoFe_3O_4 catalyzed was illustrated in Scheme 2. The nano Fe_3O_4 facilitates the oximation process through coordination of its Fe^{3+} with O of carbonyl.

In the next attempt, we turned our attention to recyclable of catalyst; the catalyst was easily separated from the reaction system and reused (after washing with distilled water/acetone and dried under vacuum) for three consecutive runs, and no obvious diminishing activity was observed (the conversions and selectivities were 100% and 100% for first run; 99% and 100% for second run; 91% and 95% for third run; 91% and 90% for fourth run).

Conclusions

In this paper, we have shown that magnetically nano Fe_3O_4 promoted oximation of various carbonyl compounds byhydroxylamine hydrochloride in oil bath (70°C to 80°C) under solvent-free conditions. Aldoximes and ketoximes were obtained in high yields within 15 min to 3 h. This method offers some advantages in term of clean reaction conditions, easy work-up procedure, short reaction times, and suppression of any side product. So we think that $NH_2OH{\cdot}HCl$/nano Fe_3O_4 system could be considered a new and useful addition to the present methodologies in this area.

Methods

General

All solvents and reagents were purchased from commercial sources with the best quality and they were used without further purification. Nano Fe_3O_4 are prepared in high purity according to the reported procedures in literature [19] (Figures 1, 2, and 3). ^1H, ^{13}C NMR, and IR spectra were recorded on 300 MHz Bruker Avance and Thermo Nicolet Nexus 670 FT-IR spectrometers (Bruker AXS, Inc., Madison, WI, USA). All products are known and were characterized by their spectral data. Yields refer to isolated pure products. Thin layer chromatography (TLC) was applied for monitoring of the reactions over silica gel 60 F_{254} aluminum sheet.

A typical procedure for solvent-free oximation of acetophenone to acetophenone-oxime with $NH_2OH{\cdot}HCl$/nano Fe_3O_4 system

A mixture of Acetophenone (0.120 g, 1 mmol) and nano Fe_3O_4 (0.462 g, 2 mmol) was ground in a mortar.

Table 1 Optimization experiment for oximation of benzaldehyde with $NH_2OH{\cdot}HCl$/nano Fe_3O_4 system under different reaction conditions (all reactions were carried out with 1 mmol of benzaldehyde)

Entry	Molar ratio[a]	Condition[b]	Time (min)	Convertion (%)
1	1:2:1.5	CH_3CH/reflux	120	30
2	1:2:1.5	EtOH/reflux	100	30
3	1:2:1.5	H_2O/reflux	120	60
4	1:2:1.5	THF/reflux	100	10
5	1:2:1	Solvent-free/oil bath	50	80
6	1:2:1.5	Solvent-free/oil bath	20	100
7	1:2:2	Solvent-free/oil bath	20	100

[a]Molar ratio as benzaldehyde/$NH_2OH{\cdot}HCl$/nano Fe_3O_4; [b]temperature of oil bath was (70° to 80°C).

Table 2 Oximation of aldehydes with $NH_2OH{\cdot}HCl$/nano Fe_3O_4 system (all reactions were carried out in oil bath (70°C to 80°C) under solvent-free conditions)

Entry	Substrate	Product	Molar ratio[a]	Time (min)	E:Z ratio[b]	Yield (%)[c]
1	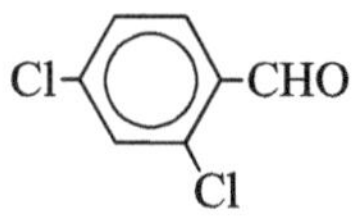	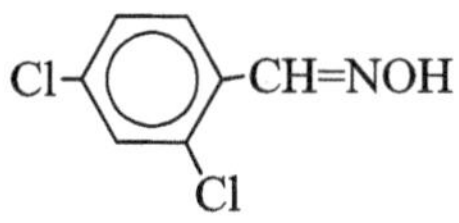	1:2:1.5	15	98:2	96
2	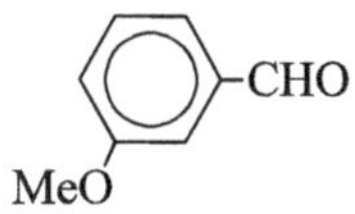	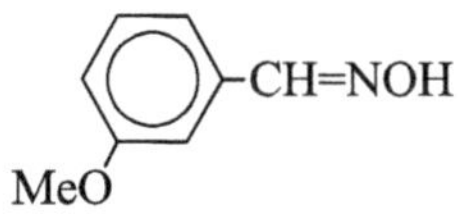	1:2:1.5	30	98:2	95
3	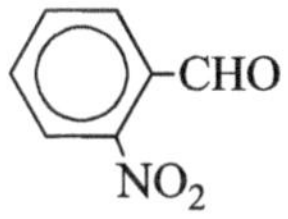	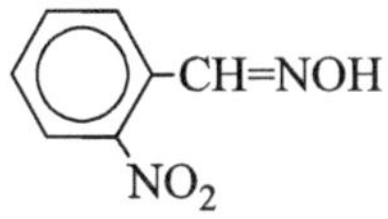	1:2:1.5	25	96:4	95
4	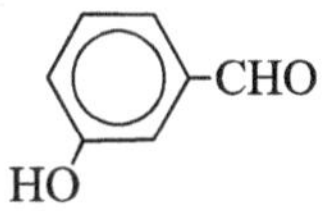	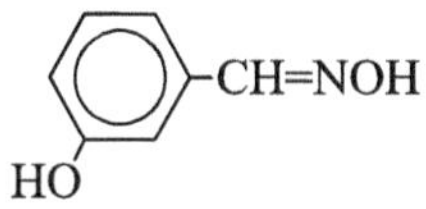	1:2:1.5	30	98:2	96
5	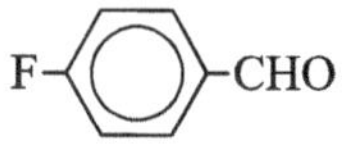	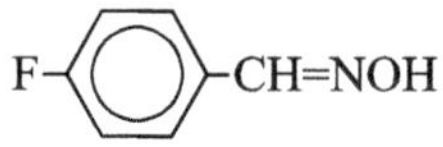	1:2:1.5	20	98:2	98
6	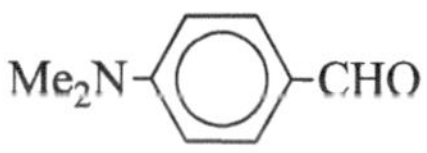	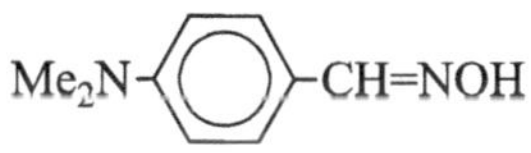	1:2:1.5	15	97:3	93
7	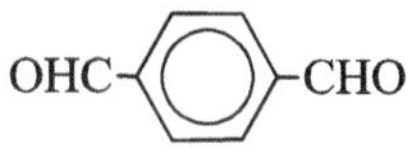	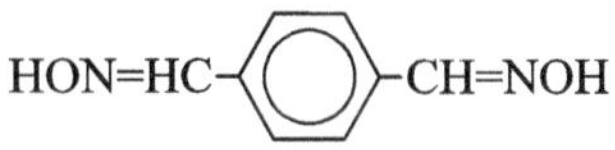	1:3:1.5	45	98:2	90
8	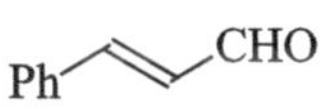	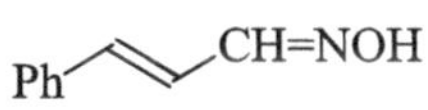	1:2:1.5	50	98:2	92
9	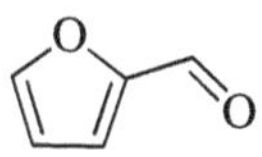	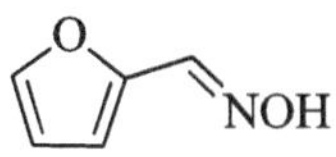	1:2:2	15	100	74

[a]Molar ratio as Subs./$NH_2OH{\cdot}HCl$/nano Fe_3O_4; [b]approximately determined form TLC; [c]yields refer to isolated pure products.

$NH_2OH{\cdot}HCl$ (0.2085 g, 3 mmol) was then added to the mortar, and the grinding of the reaction mixture was continued for a moment. The mixture was stirred magnetically in oil bath (70°C to 80°C) under solvent-free conditions for 60 min. TLC monitored the progress of the reaction (eluent; *n*-hexane/EtOAc: 5/3). After completion of the reaction, the mixture was washed with EtOAc (3 × 5 ml). Evaporation of the solvent affords the white solid Acetophenone Oxime in 90% yield (1215 g, Table 2: entry 1).white solid, m.p. 58°C to 60°C. IR (KBr, ν_{max}): 3,235, 3,084, 2,924, 1,496, 1,444, 1,370, 1,300, 1,079, 1,005, 924, 847, 773, 756, 688, and 651 cm^{-1}. ^1H NMR (300 MHz, $CDCl_3$, δ ppm): 8.86 (br s, 1H), 7.69 to 7.58 (m, 2H), 7.43-7.36 (m, 3H), 2.32 (s, 3H),^{13}C NMR (75 MHz, $CDCl_3$, δ ppm): 156.1, 136.6, 129.3, 128.6, 126.1, 12.3; MS (EI) *m*/*z* (% rel. intensity) 135 (M+, 86), 118 (14), 106 (22), 103 (26), 94 (32), 77 (100), 66 (16), and 51 (33).

Competing interests

Both authors declare that they have no competing interests.

Authors' contributions

BZ participated in the design of the study and performed the statistical analysis. MK carried out the nanoparticle studies and participated in its design and coordination. Both authors read and approved the final manuscript.

Table 3 Oximation of ketones with $NH_2OH·HCl$/nano Fe_3O_4 system (all reactions were carried out oil bath (70°C to 80°C) under solvent-free conditions)

Entry	Substance	Product	Molar ratio[a]	Time (h)	E:Z(%)[b]	Yield (%)[c]
1	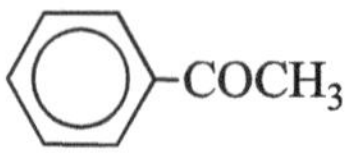	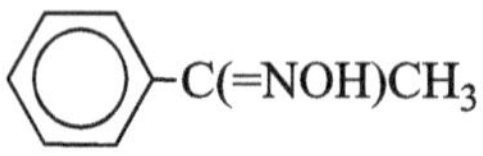	1:3:2	1	98:2	90
2	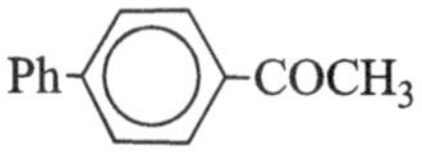	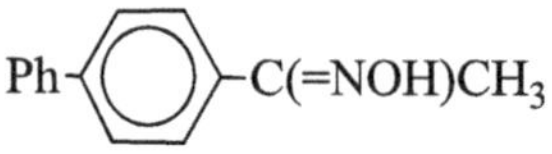	1:3:2	1.8	98:2	94
3	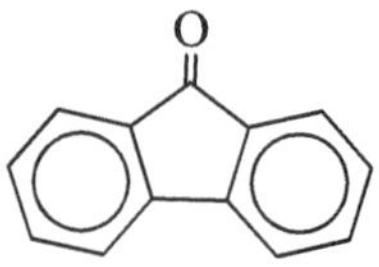	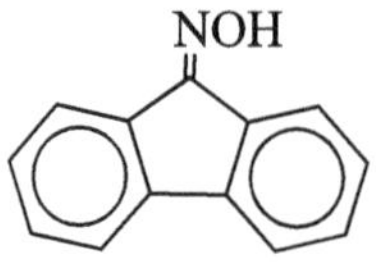	1:3:2	2	100:0	93
4	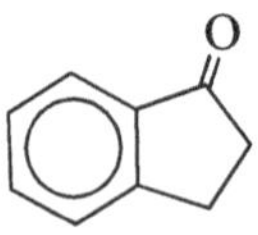	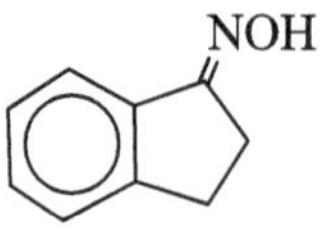	1:3:2	1.8	98:2	95
5	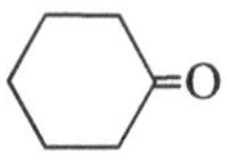	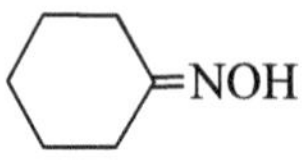	1:3:2	1.2	100:0	94
6	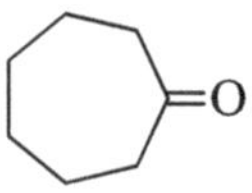	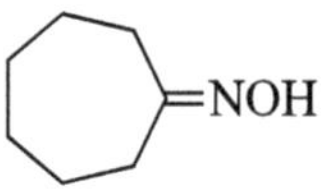	1:3:2	1	100:0	92
7	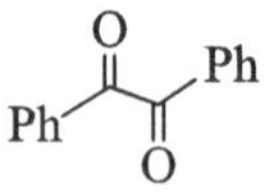	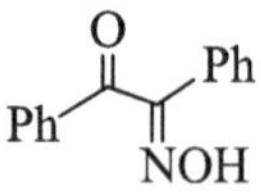	1:3:2	2.2	100:0	93
8	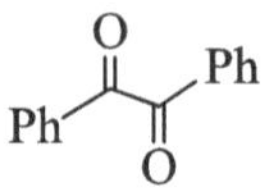	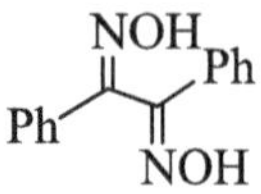	1:4:2	3	100:0	95
9	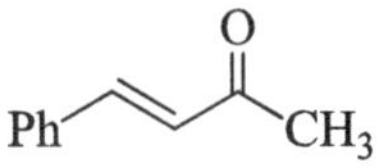	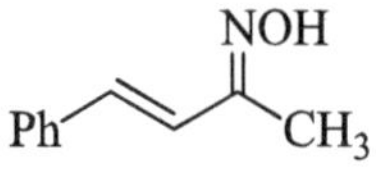	1:3:2	3	96:4	95

[a]Molar ratio as Subs./$NH_2OH·HCl$/nano Fe_3O_4; [b]approximately determined from TLC; [c]yields refer to isolated pure products.

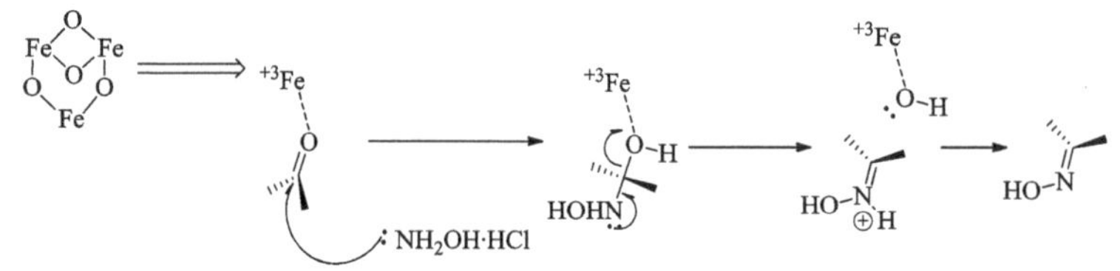

Scheme 2 A plausible mechanism for the nano Fe_3O_4 catalyzed.

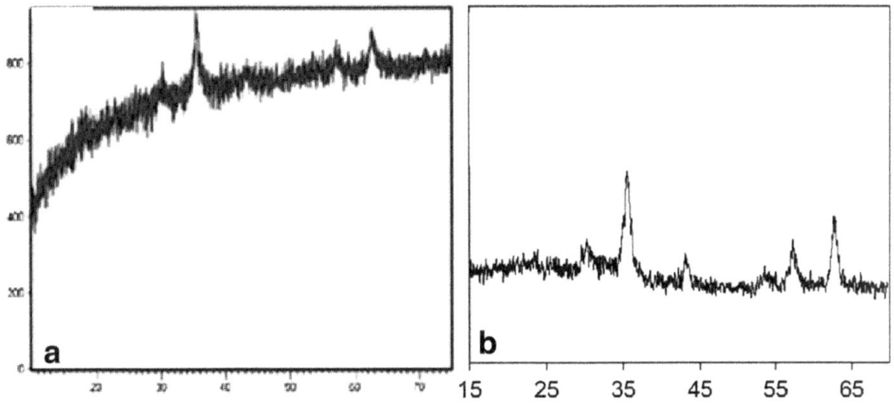

Figure 1 XRD pattern of obtained samples and XRD pattern of [19].

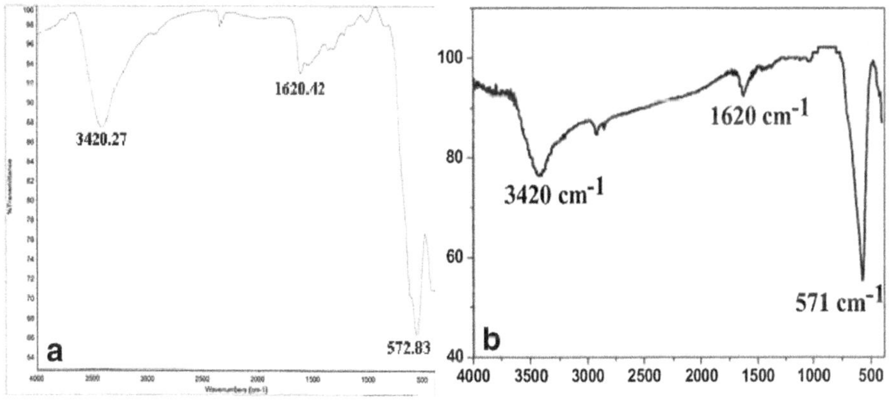

Figure 2 Infrared spectra of the samples (a) and (b) [19].

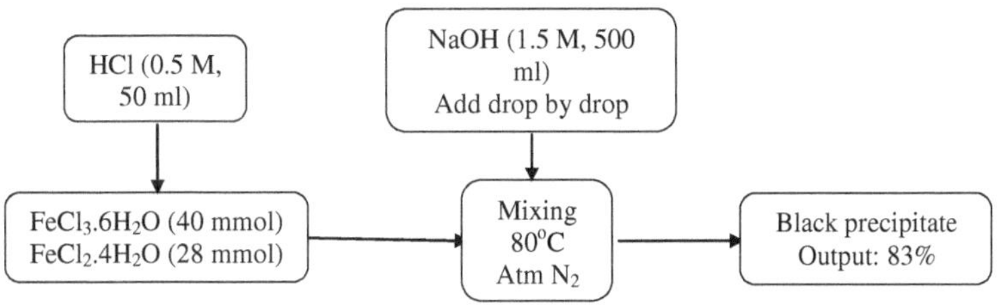

Figure 3 Process synthesis nano Fe_3O_4 [19].

Acknowledgments
The authors gratefully acknowledge the financial support of this work by the research council of Urmia University.

References

1. Mikhaleva, IA, Zaitsev, AB, Trofimov, BA.: Oximes as reagents. Russ Chem Rev **75**, 797–823 (2006)
2. Barnes, MW: An oxime to nitro conversion. a superior synthesis of secondary nitroparaffins. J Org Chem **41**, 733–735 (1976)
3. Bagheri, M, Karimkoshteh, M: Nano SiO_2/H_2SO_4 as catalyst for the Beckmann rearrangement anddeoximation of aldoximes. IJC **3**(1), 27–32 (2013)
4. Hegedüs, A, Cwik, A, Hell, Z, Horváth, Z, Esek, A, Uzsoki, M: Microwave-assisted conversion of oximes into nitriles in the presence of a zeolite. Green Chem **4**, 618–620 (2002)
5. Zeynizadeh, B, Kouhkan, M: A rapid and practical protocol for solvent-free reduction of oximes to amines with $NaBH_4/ZrCl_4/Al_2O_3$ system. Bull Korean Chem Soc **32**, 3448–3452 (2011)
6. Abele, E, Lukevics, E: Recent advances in the chemistry of oximes. Org Prep Proced Int **32**, 235–264 (2000)
7. Abele, E, Lukevics, E: Recent advances in the synthesis of heterocycles from oximes. Heterocycles **53**, 2285–2336 (2000)
8. Bo, RK, Gi, HS, Jeum, JK, Yong, JY: A development of rapid, practical and selective proces for preparation of Z-oximes. J Korean Chem Soc **57**, 295–299 (2013)
9. Davood, S, Behrooz, K, Seyran, E: Synthesis of oximes with $NH_2OH{\cdot}HCl$/DOWEX(R) 50W-X4 system. J Chin Chem Soc **59**, 1119–1124 (2012)
10. Xia, JJ, Wang, GW: Efficient preparation of aldoximes from arylaldehydes, ethylenediamine and Oxone® in water. Molecules **12**, 231–236 (2007)
11. Li, JT, Li, XL, Li, TS: Synthesis of oximes under ultrasound irradiation. Ultrason Sonochem **13**, 200–202 (2006)
12. Lakhinath, S, Jejiron, MB, Ashim, JT: A rapid, convenient, solventless green approach for the synthesis of oximes using grindstone chemistry. Org Med Chem Lett **1**, 12 (2011)
13. Shen, Z, Lujiang, H, Yu-Fei, S: Highly selective and efficient Lewis acid–base catalysts based on lanthanide-containing polyoxometalates for oximation of aldehydes and ketones. Eur J Inorg Chem **2013**, 1659–1663 (2013)
14. Tomilov, AP, Osadchenkol, M: Phase-transfer catalysis in synthesis of oximes. Russ J Appl Chem **75**, 511–512 (2002)
15. Zeynizadeh, B, Amjadi, E: Facile Oximation of carbonyl compounds with titanyl acetylacetonate/NH_2OH system. Asian J Chem **21**, 3611–3616 (2009)
16. Nasir Baig, RB, Rajender, SV: Magnetically retrievable catalysts for organic synthesis. Chem Commun **49**, 752–770 (2013)
17. Montazeri, H, Amani, A, Shahverdi, HR, Haratifar, E, Shahverdi, AR: Separation of the defect-free Fe_3O_4-Au core/shell fraction from magnetite-gold composite nanoparticles by an acid wash treatment. J Nanostruct Chem **3**, 25 (2013)
18. Polshettiwar, V, Luque, R, Fihri, A, Zhu, H, Bouhrara, M, Basset, JM: Magnetically recoverable nanocatalysts. Chem Rev **111**, 3036–3075 (2011)
19. Ch, MX, Cui, LH, Xu, GF: The preparation of Fe_3O_4 magnetic nanometer particle by co-precipitation. Method **320**, 271–272 (2012)

Pseudo-second-order kinetic equations for modeling adsorption systems for removal of lead ions using multi-walled carbon nanotube

Dariush Robati

Abstract

This study investigated the lead ion removal of the multi-walled carbon nanotubes (MWCNTs) and oxidized multi-walled carbon nanotubes (MWCNT-COOH). The main purpose of this work is to study the possibilities on the removal of Pb(II) ions from aqueous solutions using MWCNTs and MWCNT-COOH surfaces as adsorbents. Removal of Pb(II) ions was investigated using solutions with different concentrations in the range 10 to 100 mg/L. In this study, removal of Pb(II) ions on surfaces has been investigated by atomic absorption spectrophotometry. The microstructure of carbon nanotubes were characterized using scanning electron microscopy. Three different kinetic theories were applied to experimental data. The kinetic rates were modeled using the pseudo-first-order, four-type linear pseudo-second-order, and intraparticle diffusion. The pseudo-second-order model was found to explain the adsorption kinetics most effectively. The results indicated a significant potential of the multi-walled carbon nanotube as an adsorbent for Pb(II) ion removal.

Keywords: Oxidized multi-walled carbon nanotubes; Kinetics models; Pb(II) ions; Multi-walled carbon nanotube; Removal

Background

Exposure to heavy metal ions can cause a variety of adverse health impacts, including serious damages to the immune system, central nervous system, and reproductive system [1]. The contamination of groundwater by heavy metal ions is a major problem in many countries around the world [2]. For example, Pb(II) is a major harmful pollutant to the biosphere, and even trace amounts of it pose a detrimental risk to human health. Although lead is an essential element for the normal metabolism of many living organisms, it is believed that excess lead may be harmful to mankind. So the determination of heavy metal ions is still an intensively active research area [3-5]. Lead is one of the most harmful heavy metals. It has toxic effects on human, animals, and the plants. The main lead pollution resources are battery manufacturing, acid metal plating, ceramic and glass industries, tetraethyl lead manufacturing, and other industries. Lead has fatal effects on the kidney, the nervous system, the liver, and the brain [6]. Therefore, lead must be removed from water. Adsorption is one of the commonly used purification technologies. It is easy to apply and is an economical process. Carbon nanotubes (CNTs) are carbon allotropes consisting of sp^2-hybridized carbons, like fullerene and graphene as well as graphite. Since CNTs were discovered by Iijima in 1991, they have existed in a wide variety of forms, i.e., single-walled carbon nanotubes, double-walled carbon nanotubes, few-walled carbon nanotubes, and multi-walled carbon nanotubes (MWCNTs) [7]. Because of so many outstanding performances, CNTs exhibit a great promise for potential applications in many technological fields such as hydrogen storage, catalyst supports, chemical sensors, and nanoelectronic devices [8]. The known ability of CNTs to establish electrostatic interactions and their large surface areas can facilitate the adsorption of many kinds of pollutants from water, such as aniline, phenol, and their substitutes and sodium chloride, as well as several divalent metal ions [9-11]. Surface

Correspondence: d_robati@iiau.ac.ir
Department of Chemistry, Eslamshahr Branch, Islamic Azad University, Eslamshahr, Iran

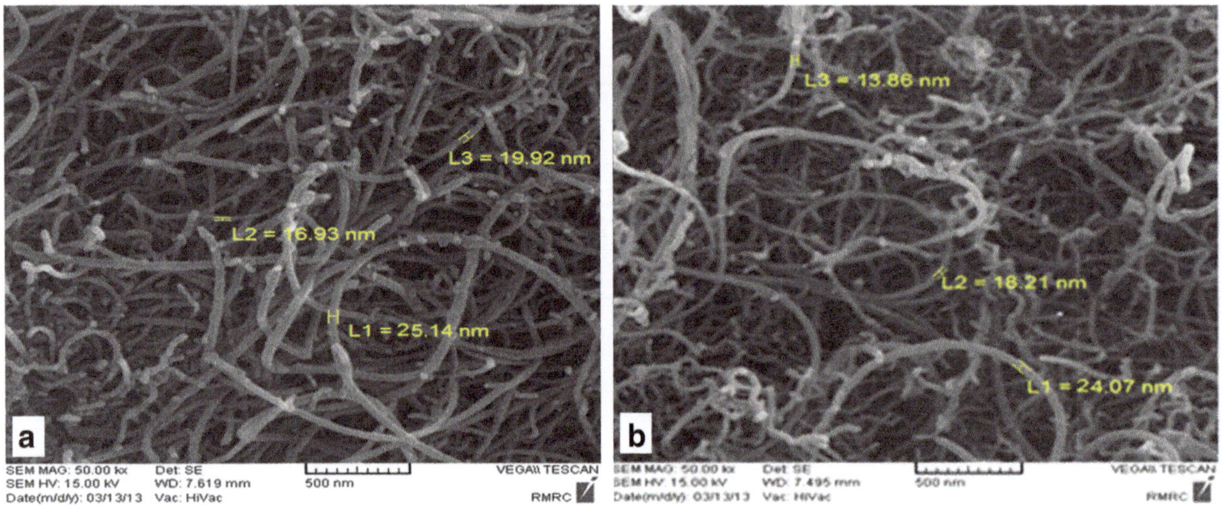

Figure 1 SEM images of multi-walled carbon nanotubes (MWCNTs) (a) and oxidized multi-walled carbon nanotubes (MWCNT-COOH) (b).

modifications of CNTs have been applied recently to enhance the dispersion property and adsorption capacities of CNTs. Oxidation of CNTs have been widely reported [12]. During oxidation, the surface characteristics are altered due to the introduction of new functional groups (e.g., COOH, OH) [12]. Many other functional groups could also be appropriate for metal ion adsorption [13,14]. The goals of this research were to evaluate the removal of behavior of Pb(II) by the MWCNTs and oxidized multi-walled carbon nanotubes (MWCNT-COOH). Three different kinetic theories were applied to experimental data. The kinetic rates were modeled using the pseudo-first-order, four-type linear pseudo-second-order, and intraparticle diffusion.

Methods

MWCNT (purity, >95%; diameter 1 to 2 nm; length, 5 to 30 nm; surface area, approximately 400 m^2/g; and

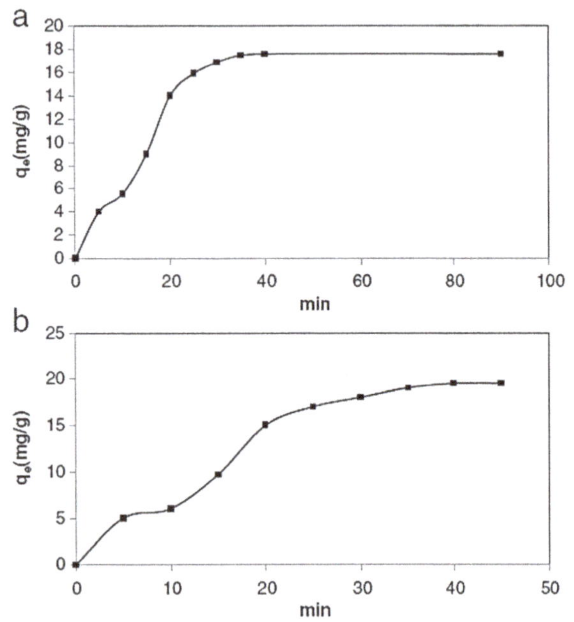

Figure 2 Effect of contact time on adsorption of Pb(II) ions by (a) MWCNT and (b) MWCNT-COOH surfaces. (a) Initial concentration, 100 mg/L; adsorbent dosage, 50 mg/L, and T = 298 K. **(b)** Initial concentration, 100 mg/L; adsorbent dosage, 50 mg/L and T = 298 K.

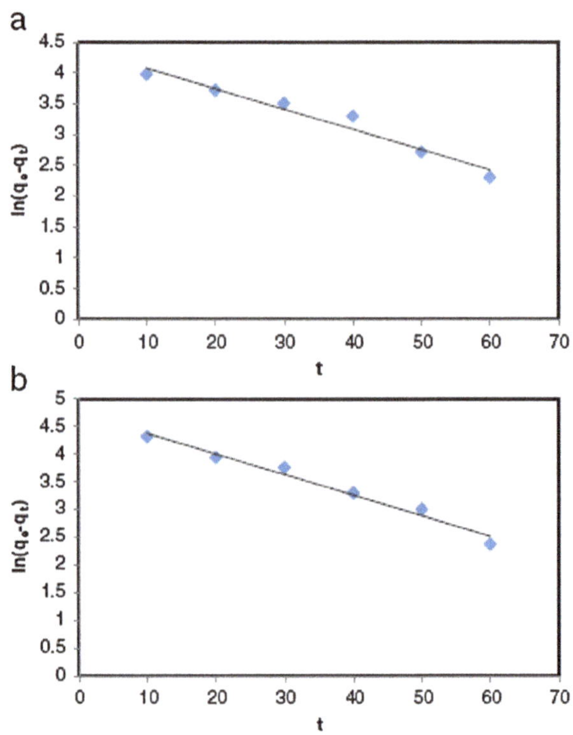

Figure 3 Pseudo-first-order kinetics for removal of lead ions by (a) MWCNT and (b) MWCNT-COOH surfaces.

Table 1 Fitted pseudo-first-order-model kinetic parameters for the removal of lead ions by MWCNTs at 25°C

C_0 (mg/L)	k_1	R^2	Equation
10	0.044	0.9869	2
50	0.046	0.9890	2
75	0.047	0.9901	2
100	0.048	0.9901	2

manufacturing method, catalytic chemical vapor deposition) and MWCNT-COOH (content of COOH: 1 to 6 wt.%; purity, >95%; average diameter 1 to 2 nm; length 5 to 30 nm, and SSA, 400 m^2/g) were purchased from NanoAmor, Nanostructured & Amorphous Materials, Inc, Houston, TX, USA. Doubly distilled water was used and all adsorbents were washed before using. Figure 1 presents SEM images of MWCNTs and MWCNT-COOH. $Pb(NO_3)_2$ (molecular weight, 331.20 g/mol) were supplied by Merck, Darmstadt, Germany (maximum purity available). All solutions were prepared with deviations of less than ±0.1% from the desired concentrations.

Adsorption studies

In this research, the 1,000 mg/L stock solution of Pb(II) was prepared by dissolving $Pb(NO_3)_2$ in deionized water and adding water to 1,000 mL. The experimental solutions were prepared by diluting the stock solution with deionized water when necessary. To evaluate the equilibrium properties, we first prepared various solutions with initial lead ion concentration from 10 to 100 mg/L and then added 0.05 g MWCNTs and MWCNT-COOH for each, as adsorbent to the solution. These samples were then mounted on a shaker (HZQ-C, Hangzhou Chincan Trading Co., Ltd, Hangzhou, China) and shaken continuously for 5 to 45 min at 298 K. Ultrasonic bath (71020-DTH-E; Model 1510 DTH, 220V; EMS Company,) was used to prevent the particles to aggregate and form a bulk. Then, the samples were centrifuged at 450 rpm, and the supernatant was filtered by 0.2-μm filtered papers for subsequent analysis of lead ion concentration. The filtrates were then immediately examined using atomic absorption spectrophotometry (Perkin-Elmer AAnalyst 700, Waltham, MA, USA) in order to measure the ion concentration. The difference between the initial and the equilibrium ion concentration determines the amount of ion adsorbed onto MWCNT-COOH and MWCNT surfaces. All the experiments were performed in triplicate, and only the mean values are reported. As shown previously, the adsorption of ion on the bottle wall is ignorable. The amount of lead adsorbed was calculated using the following equation:

Table 2 Fitted pseudo-first-order-model kinetic parameters for the removal of lead ions by MWCNT-COOH at 25°C

C_0 (mg/L)	k_1	R^2	Equation
10	0.045	0.98695	2
50	0.047	0.9869	2
75	0.049	0.9900	2
100	0.050	0.9909	2

Table 3 Pseudo-second-order kinetic model in linearized forms

Type	Linearized form	Plot	Equation
Linear 1	$\frac{t}{q_e} = \frac{1}{k_2 q_e^2} + \frac{1}{q_e} t$	t/q vs. t	3
Linear 2	$\frac{1}{q_e} = \frac{1}{q} + \left(\frac{1}{k_2 q_e^2}\right) \frac{1}{t}$	$1/q$ vs. $1/t$	4
Linear 3	$q_e = q_t - \left(\frac{1}{k q_t}\right) \frac{q}{t}$	q vs. q/t	5
Linear 4	$\frac{q_e}{t} = k q_t^2 - k q_e q_t$	q/t vs. q	6

$$q_e = \frac{(C_0 - C_e)V}{m}, \tag{1}$$

where C_0 and C_e (mg/L) are the liquid-phase concentrations of lead at initial and equilibrium concentrations, respectively. V is the volume of the solution (L) and m is the mass of adsorbent used (g). All the experiments were performed in triplicate, and the mean values have been used in calculations.

Results and discussion

Effect of contact time

In order to determine the time required for removal of lead ions by MWCNT-COOH and MWCNT surfaces, 10 mL of 100 mg/L solution was shaken with 0.05 g MWCNTs. The equilibrium studies were performed at the selected interval of time ranging from 5 to 60 min. As shown in Figure 2, after 35 min of stirring the solution, the removal efficiency did not increase. Therefore, the optimum value of stirring time was found to be 35 min. The change in the rate of lead ion removal might be due to the fact that initially all adsorbent sites were vacant and the solute concentration gradient is high. Afterwards, the lead ion uptake rate by the MWCNT-

Table 4 Fitted pseudo-second-order-model kinetic parameters for removal of lead ions by MWCNTs at 25°C

C_0 (mg/L)	k_2	R^2	Equation
75	0.0016	0.9948	3
75	0.0017	0.9970	4
75	0.0019	0.9960	5
75	0.0020	0.9960	6

Table 5 Fitted pseudo-second-order-model kinetic parameters for removal of lead ions by MWCNT-COOH at 25°C

C_0 (mg/L)	k_2	R^2	Equation
75	0.0018	0.9948	3
75	0.0019	0.9970	4
75	0.0020	0.9960	5
75	0.0021	0.9960	6

COOH and MWCNT surfaces decreased significantly due to the decrease in adsorption sites. On the basis of these results, a 35-min shaking period was selected for all further studies.

Adsorption kinetics

Kinetic modeling not only allows estimation of sorption rates but also leads to suitable rate expressions characteristic of possible reaction mechanisms. In this respect, several kinetic models including the pseudo-first-order kinetics model, pseudo-second-order kinetics models (four linear forms), and the intraparticle diffusion model were investigated.

Pseudo-first-order model

A pseudo-first-order kinetic equation is given as [15]:

$$\ln(q_e - q_t) = \ln q_e - k_1 t, \quad (2)$$

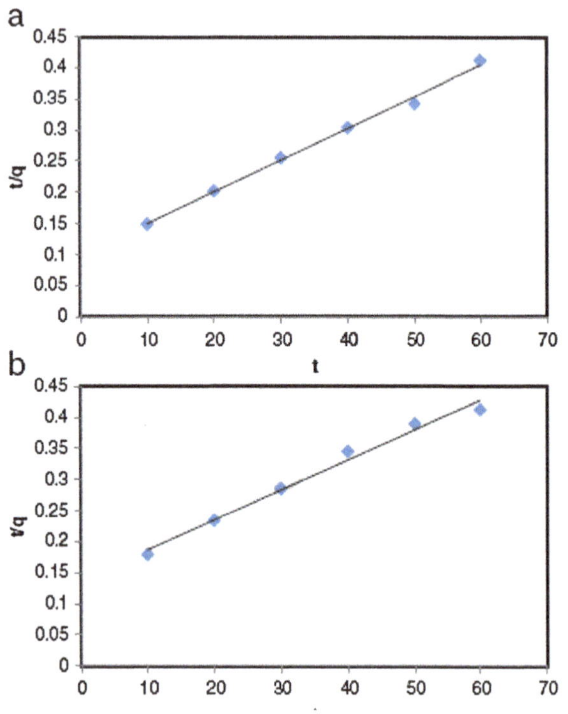

Figure 4 **Pseudo-second-order kinetic model for removal of lead ions by MWCNTs (a) and MWCNT-COOH surfaces (b).** (for Equation 2).

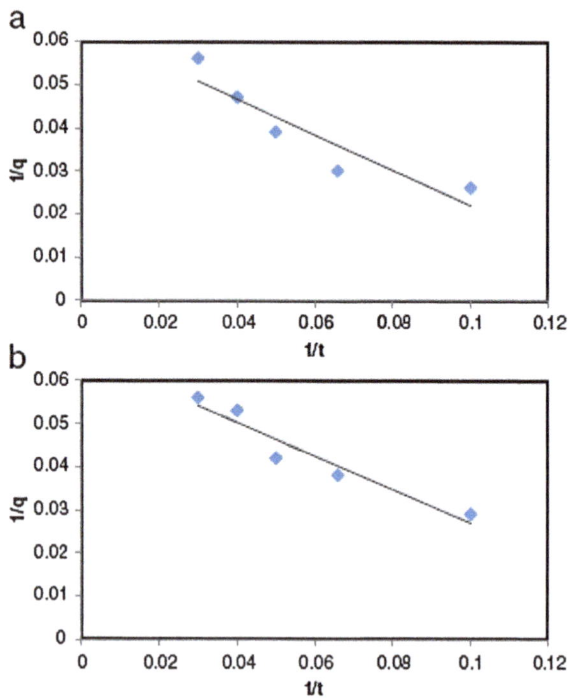

Figure 5 **Pseudo-second-order kinetic model for removal of lead ions by MWCNTs (a) and MWCNT-COOH surfaces (b).** (for Equation 3).

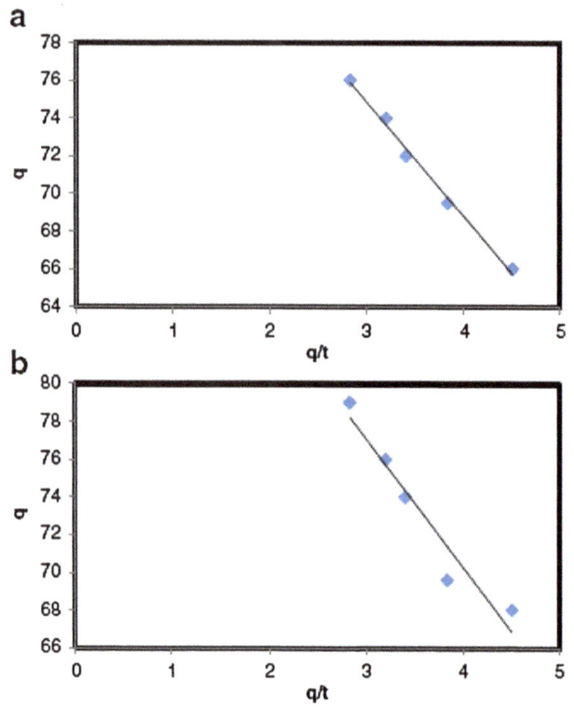

Figure 6 **Pseudo-second-order kinetic model for removal of lead ions by MWCNTs (a) and MWCNT-COOH surfaces (b).** (for Equation 4).

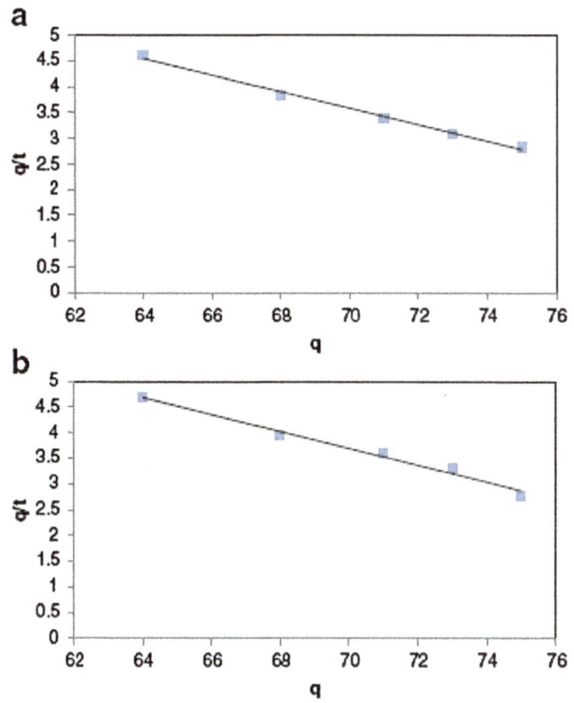

Figure 7 Pseudo-second-order kinetic model for removal of lead ions by MWCNTs (a) and MWCNT-COOH surfaces (b). (for Equation 5).

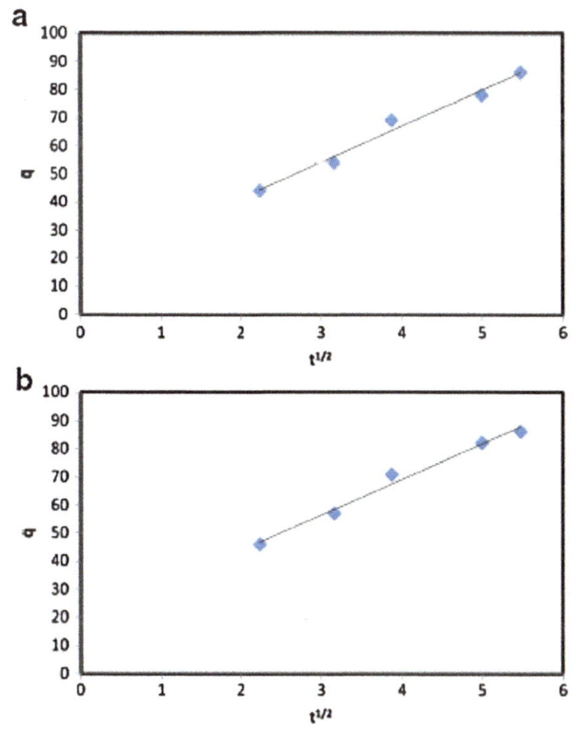

Figure 8 Intraparticle diffusion model kinetics for removal of lead by (a) MWCNT surfaces and (b) MWCNT-COOH surfaces.

Table 6 Fitted intraparticle diffusion model kinetic parameters for removal of lead ions by MWCNT surfaces at 25°C

C_0 (mg/L)	k_3	R^2	Equation
75	11.312	0.9881	7
75	11.433	0.9849	7
75	11.566	0.9899	7
75	11.877	0.9877	7

where q_t is the amount of lead ions removed at time t (mg/g), q_e is the adsorption capacity at equilibrium (mg/g), k_1 is the pseudo-first-order rate constant (1/min), and t is the contact time (min). Using Equation 2, $ln\ (q_e - q_t)$ versus t was plotted as shown in Figure 3. The pseudo-first-order rate constant (k_1) determined from the model is presented. Results are presented in Tables 1 and 2 for the removal of lead ions by MWCNTs and MWCNT-COOH, respectively, at 25°C.

Pseudo-second-order model

In this model, the rate-limiting step is the surface adsorption that involves chemisorption, where the removal from a solution is due to physicochemical interactions between the two phases [16]. The model is usually represented by its linear form as shown in Figure 2 and Table 3. Fitted pseudo-second-order-model kinetic parameters for the removal of lead ions by MWCNTs and MWCNT-COOH at 25°C are presented in Tables 4 and 5, respectively.

Where k_2 (g/mg/min) is the pseudo-second-order rate constant of adsorption. The plot is shown in Figures 4,5,6 and 7.

Intraparticle diffusion model

The intraparticle diffusion model describes adsorption processes, where the rate of adsorption depends on the speed at which adsorbate diffuses towards adsorbent (i.e., the process is diffusion-controlled), which is presented by Equation 7.

$$q_e = k_3 t^{1/2} + c, \tag{7}$$

Table 7 Fitted intraparticle diffusion model kinetic parameters for removal of lead ions by MWCNT-COOH surfaces at 25°C

C_0 (mg/L)	k_3	R^2	Equation
75	11.432	0.9890	7
75	11.489	0.9868	7
75	11.601	0.9900	7
75	11.890	0.9900	7

where, k_3 is the rate constant of the intraparticle transport (g/mg/min) and c is the intercept [17-20]. The intraparticle diffusion model as fitted with the experimental data is presented in the plot of q_t versus $t^{1/2}$ depicted in Figure 8a,b, and the values of k_i and correlation coefficients are given in Tables 6 and 7.

Comparing the results of the kinetics models reveals that the pseudo-second-order model has the best agreement with the experimental data.

Conclusions

The present investigation showed that the MWCNTs and MWCNT-COOH surfaces were effective in the removal of lead ions from the aqueous solution. The surface area of the MWCNT-COOH was relatively high with large-pore-volume MWCNT surfaces as adsorbents The adsorption of Pb(II) ion amount increased with increasing time by both surfaces. Kinetics for the removal of lead ions were obtained and fitted to different kinetics models. Kinetics data were best fitted by the pseudo-second-order model and the result indicated that this adsorbent is excellent in the removal of lead from aqueous solution at variable concentrations.

Competing interests

The author declares that he has no competing interests.

Acknowledgements

The author would like to thank the Islamic Azad, Islamshahr Branch for their financial support.

References

1. Cho, HH, Wepasnick, K, Smith, BA, Bangash, FK, Fairbrother, DH, Ball, WP, Ball, WP: Sorption of aqueous Zn[II] and Cd[II] by multiwall carbon nanotubes: the relative roles of oxygen-containing functional groups and graphenic carbon. Langmuir **26**, 967–981 (2010)
2. Hyung, H, Kim, JH: Natural organic matter (NOM) adsorption to multi walled carbon nanotubes: effect of NOM characteristics and water quality parameters. Environ. Sci. Technol. **42**, 4416–4421 (2008)
3. Xu, D, Tan, X, Chen, CL, Wang, XK: Removal of Pb(II) from aqueous solution by oxidized multiwalled carbon nanotubes. J. Hazard. Mater. **154**, 407–416 (2008)
4. Sheng, GD, Li, JX, Shao, DD, Hu, J, Chen, CL, Chen, Y, Wang, X: Adsorption of copper(II) on multiwalled carbon nanotubes in the absence and presence of humic or fulvic acids. J. Hazard. Mater. **178**, 333–340 (2010)
5. Schierz, A, Zaenker, H: Aqueous suspensions of carbon nanotubes: surface oxidation, colloidal stability and uranium sorption. Environ. Pollut. **157**, 1088–1094 (2009)
6. Giannakopoulos, E, Christoforidis, KC, Tsipis, A, Jerzykiewicz, M, Deligiannakis, Y: Influence of Pb(II) on the radical properties of humic substances and model compounds. J. Phys. Chem. A. **109**, 2223–2232 (2005)
7. Sadeghian, Z: Large-scale production of multi-walled carbon nanotubes by low cost spray pyrolysis of hexane. New. Carbon. Mater. **24**, 33–38 (2009)
8. Shao, D, Hu, J, Wang, X: Plasma induced grafting multiwalled carbon nanotube with chitosan and its application for removal of UO_2^{2+}, Cu^{2+}, and Pb^{2+} from aqueous solutions. Plasma. Process. Polym. **7**, 977–985 (2010)
9. Wang, P, Shi, QH, Liang, HJ, Steuerman, DW, Stucky, GD, Keller, AA: Enhanced environmental mobility of carbon nanotubes in the presence of humic acid and their removal from aqueous solution. Small **4**, 2166–2170 (2008)
10. Ren, XM, Chen, CL, Nagatsu, M, Wang, XK: Carbon nanotubes as adsorbents in environmental pollution management: a review. Chem. Eng. J. **170**, 395–410 (2011)
11. Petersen, EJ, Zhang, LW, Mattison, NT, O'Carroll, DM, Whelton, AJ, Uddin, N, Nguyen, T, Huang, QG, Henry, TB, Holbrook, RD, Chen, KL: Potential release pathways, environmental fate, and ecological risks of carbon nanotubes. Environ. Sci. Technol. **45**, 9837–9856 (2011)
12. Zhang, D, Pan, B, Zhang, H, Ning, P, Xing, BS: Contribution of different sulfamethoxazole species to their overall adsorption on functionalized carbon nanotubes. Environ. Sci. Technol. **44**, 3806–3811 (2010)
13. Mauter, MS, Elimelech, M: Environmental applications of carbon-based nanomaterials. Environ. Sci. Technol. **42**, 5843–5859 (2008)
14. Tian, Y, Gao, B, Wang, Y, Morales, VL, Carpena, RM, Huang, Q, Yang, L: Deposition and transport of functionalized carbon nanotubes in water saturated sand columns. J. Hazard. Mater. **213–214**, 265–272 (2012)
15. Li, YH, Wang, S, Luan, Z, Ding, J, Xu, C, Wu, D: Adsorption of Cadmium(II) from aqueous solution by surface oxidized carbon nanotubes. Carbon **41**, 1057–1062 (2003)
16. Wang, H, Zhou, A, Peng, F, Yu, H, Yang, J: Mechanism study on adsorption of acidified multi-walled carbon nanotubes to Pb(II). J. Colloid Interface Sci. **316**, 277–283 (2007)
17. Schierz, A, Zanker, H: Aqueous suspensions of carbon nanotubes: Surface oxidation, colloidal stability and uranium sorption. Envi. Pollution **157**, 1088–1094 (2009)
18. Hameed, BH, Daud, FBM: Adsorption studies of basic dye on activated carbon derived from agricultural waste: *Hevea brasiliensis* seed coat. Chem. Eng. J. **139**, 48–55 (2008)
19. Ho, YS, McKay, G: Pseudo-second order model for sorption processes. Process. Biochem. **34**, 451–465 (1999)
20. Weber, TW, Chakkravorti, RK: Pore and solid diffusion models for fixed-bed adsorbers. AIChE J. **20**, 228–232 (1974)

Effect of polyvinyl alcohol concentration on the ZnS nanoparticles and wet chemical synthesis of wurtzite ZnS nanoparticles

Abdul Kareem Thottoli* and Anu Kaliani Achuthanunni

Abstract

ZnS is one of the most important semiconducting materials, and this paper discusses the wet chemical synthesis of ZnS nanoparticles in cubic and hexagonal structures. The effect of polyvinyl alcohol concentration on the ZnS particle formation is studied, and ZnS nanocrystals smaller than the Bohr exciton radius have been prepared using the capping agent polyvinyl alcohol. ZnS nanocrystals of about 1.54 nm have been prepared, and their optical absorption shows interband transitions. It is found that the polyvinyl alcohol concentration has a tremendous effect on the particle growth, since particle sizes were varied with the concentration of polyvinyl alcohol. The important result is that the strain was seen negative due to the contraction of the particles. The important finding of this experiment is that the preparation of the hexagonal ZnS nanoparticles is possible at lower temperature when a particular concentration of trisodium citrate and polyvinyl alcohol is used simultaneously with the ZnS precursors.

Keywords: ZnS, Polyvinyl alcohol, Semiconductor

Background

When the size of a semiconductor becomes less than or comparable to the Bohr diameter of the exciton [1], various size-quantization effects such as widening of band gap and formation of discrete orbitals come into the picture as explained by the Heisenberg uncertainty principle. Among the semiconducting nanomaterials, ZnS is one of the most important semiconducting materials since it can be used in electronic or optoelectronic applications [2] such as in flat-panel displays [3], white light LEDs [4], electroluminescent devices [5], sensors [6], lasers [7], infrared windows [8], ultraviolet (UV) lasers [9], anti-reflection coating on solar cells [10], and solar cells [11]; for investigating cellular interactions [12]; for both *in vivo* and *in vitro* imaging applications [13]; and memristor applications [14]. Basically ZnS has negative conductivity [15,16] and has a small exciton Bohr radius of about 2.5 nm [17]. Particle size reduction has a tremendous effect on the properties of ZnS, such as a blueshift in the optical absorption spectrum, increased luminescence, and enhanced oscillator strength, non-linear optical effects, geometrical structure, chemical bonds, ionization potential, mechanical strength, and melting point. Furthermore, ZnS relates to the more popular ZnO in terms of atomic structure and chemical properties. Certain properties of ZnS are unique and advantageous compared to ZnO, that is, ZnS has a larger band gap than ZnO (approximately 3.4 eV [17]), and therefore, it is more suitable for visible-UV light-based devices such as sensors or photodetectors.

It is a well-known fact that bulk ZnS crystal is more stable at room temperature in zinc blende form, which transfers to a wurtzite structure at a higher temperature of about 1,020°C [18], even though nanostructures of ZnS take a wurtzite phase [19,20] at lower and higher temperatures. Further, wurtzite ZnS is much more attractive due to its optical properties than the sphalerite phase, so the low-temperature synthesis of wurtzite ZnS nanoparticles is extremely important. In this paper, through a series of designed experiments, we have synthesized wurtzite ZnS nanoparticles at low temperature by controlling the concentration of the polyvinyl alcohol and trisodium citrate. Since, polyvinyl alcohol and trisodium citrate act as surfactant and capping agent,

* Correspondence: abdulkareem.t@gmail.com
PG & Research Department of Physics, Kongunadu Arts and Science College, GN Mills PO, Coimbatore, Tamilnadu 641 029, India

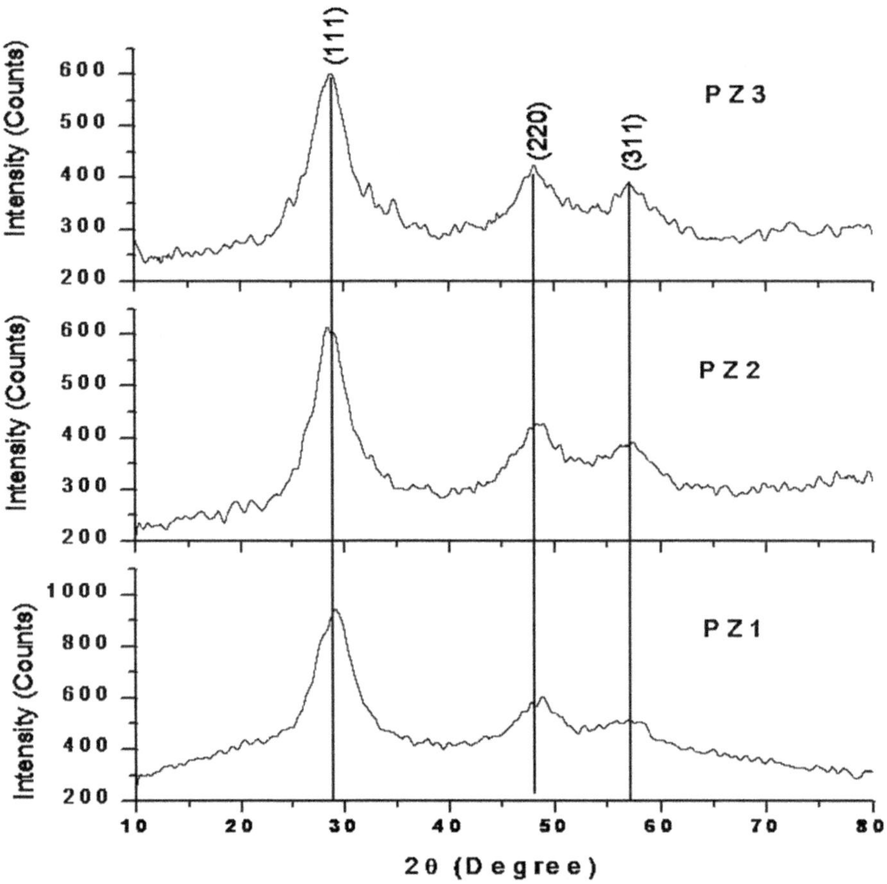

Figure 1 X-ray diffractograms of ZnS nanoparticles prepared in various concentrations of polyvinyl alcohol. PZ1, 0.03 g/ml PVA; PZ2, 0.003 g/ml PVA; PZ3, 0.0003 g/ml PVA.

they can modify the surface of ZnS particles and prevent the growth of the particles.

Results and discussion

ZnS nanocrystalline powder synthesis by varying polyvinyl alcohol concentration

The peak broadening in the X-ray diffraction (XRD) patterns of the samples shown in Figure 1 clearly indicates the formation of the ZnS nanocrystal in the samples [21], and their crystal information are tabled in Figures 2, 3, and 4. All of the data are well correlated with the American Society for Testing and Materials (ASTM) card no. 77–2100 for cubic ZnS. The increase in the diffraction angle when comparing to the ASTM data (see tables in the Figures 2, 3, and 4) is clearly a result of lattice contraction which is expected to occur because of a higher surface-to-volume ratio [22], and it is also proven from the negative strain tabulated in Table 1 from the Williamson-Hall plot [23]. Their SEM images shown in Figures 2, 3, and 4 clearly show that the sample consists of agglomerated nanocrystalline

2θ (Degree)		(hkl)	Interplanar spacing 'd' (Å)	
Experimental	ASTM 77-2100		Experimental	ASTM 77-2100
28.861	28.53	(111)	3.0910	3.1261
48.54	47.454	(220)	1.8740	1.9143
56.761	56.307	(311)	1.6205	1.6325

Figure 2 SEM micrograph of ZnS prepared in 0.03 g/ml concentration of PVA (PZ1).

2θ (Degree)		(hkl)	Interplanar spacing 'd' (Å)	
Experimental	ASTM 77-2100		Experimental	ASTM 77-2100
28.274	28.53	(111)	3.15385	3.1261
48.416	47.454	(220)	1.87855	1.9143
57.295	56.307	(311)	1.60673	1.6325

Figure 3 SEM micrograph of ZnS prepared in 0.003 g/ml concentration of PVA (PZ2).

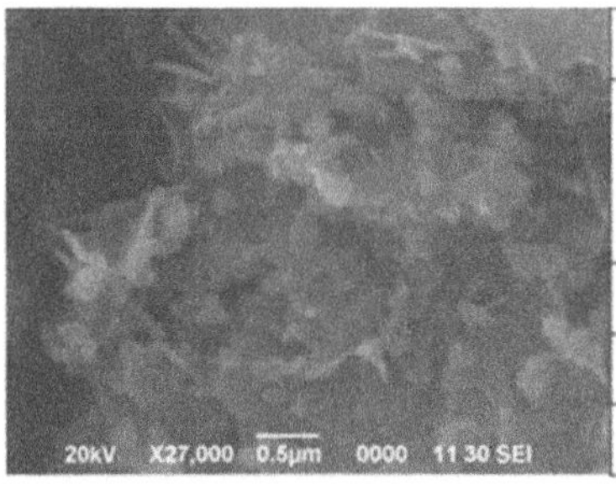

2θ (Degree)		(hkl)	Interplanar spacing 'd' (Å)	
Experimental	ASTM 77-2100		Experimental	ASTM 77-2100
28.701	28.53	(111)	3.1078	3.1261
47.953	47.454	(220)	1.8956	1.9143
56.815	56.307	(311)	1.6191	1.6325

Figure 4 SEM micrograph of ZnS prepared in 0.0003 g/ml concentration of PVA (PZ3).

Table 1 Lattice parameters and volume of crystals

Samples	Lattice parameters (ASTM value in brackets)			Average crystallite size (nm)	Average strain	Volume (10^{-28} m^3)
	a (Å)	*b* (Å)	*c* (Å)			
PZ1	5.3431 (5.4146)	-	-	1.54	−0.1490	1.5253
PZ2	5.3683 (5.4146)	-	-	2.70	−0.0061	1.5470
PZ3	5.3715 (5.4146)	-	-	2.34	−0.0351	1.5499

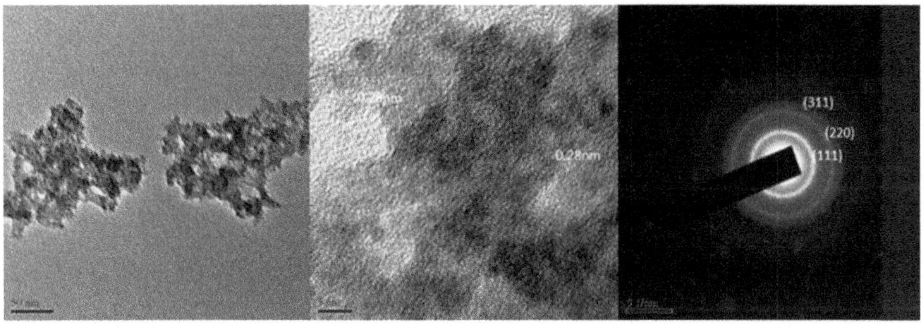

Figure 5 TEM, HRTEM, and SAED images of PZ1 sample.

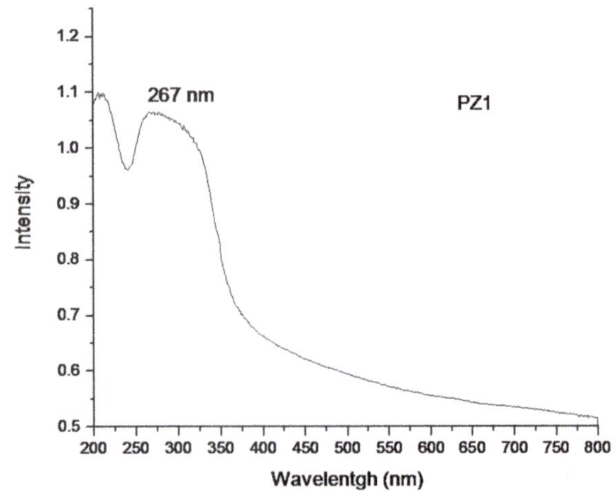

Figure 6 Absorption spectra of the ZnS prepared in PVA.

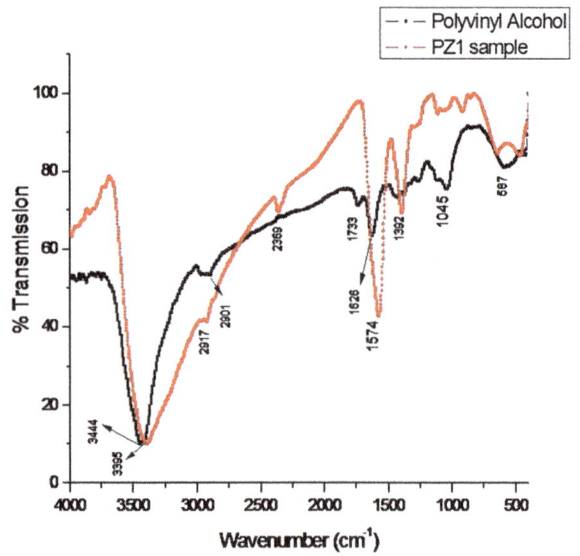

Figure 7 FTIR of the polyvinyl alcohol and PZ1 sample.

materials due to the heating of the samples for making powder. The crystallites sizes were found from the Williamson-Hall plot [23] as 1.54, 2.70, 2.34 nm for PZ1, PZ2, and PZ3 samples, respectively, (see Table 1). It is found that the sample made with 0.03 mg/ml PVA possessed smaller crystallites, so this sample was further analyzed by transmission electron microscopy (TEM) by dispersing the samples in ethanol. The TEM, high-resolution TEM (HRTEM), and selected area electron diffraction (SAED) images of the PZ1 sample are shown in Figure 5. It is very clear that the sample consists of nanosized particles, and the interplanar spacing was measured as 0.28 nm for the (111) plane. It was found that it is contracted due to the increased surface area-to-volume ratio when compared to the powder XRD data, and the polycrystalline diffraction rings were labeled to (111), (220), and (311) planes according to the ASTM card no. 77–2100 for cubic ZnS.

The absorption spectrum of the PZ1sample is shown in Figure 6 with the absorption peak. It is evident from the absorption peak that the interband transitions are happening here due to the strong quantum confinement effect since the crystallite size is smaller than the Bohr exciton radius of ZnS. The band gap of the sample is also calculated using the Brus equation [24] and shown in Table 2 with the crystallite size. It is seen that the band gap calculated from the Brus equation is larger than the band gap obtained from absorption spectra, which shows that the Brus equation cannot be expected to be quantitatively correct for very small particles [25], since the absorption in the case of small particles having a size smaller than the exciton Bohr radius is due to the interband transition. Further, the agglomeration of the nanoparticles also has an effect in their absorption spectra.

Further, the PZ1 sample was analyzed by FTIR to find the presence of PVA in the final powder sample. Figure 7 shows the infrared spectrum of pure PVA and PVA-ZnS samples between 500 and 4,000 cm^{-1}. The spectrum of PVA seems to be consistent with that previously reported in literatures [26,27]. A relatively broad and intense ν(OH) absorption stretching band is observed at 3,444 and 3,395 cm^{-1}, indicating the presence of polymeric association of the free hydroxyl groups and bonded OH stretching vibration. The slight shift of the peaks of the PVA-ZnS sample when comparing to the PVA peaks undoubtedly shows that ZnS is well incorporated in the PVA matrix. The broad O-H absorption stretching vibration is observed at around 3,444 cm^{-1} for PVA results from the superposition of multiple polymeric H bonds associated with the crystalline phase and dimeric H bonds associated with the amorphous phase [26]. The absorption bands occurring at 2,901 and 2,917 cm^{-1} resulted from anti-symmetric CH_2

Table 2 Crystallite size and band gap of ZnS prepared with PVA

Crystallite size (nm)	Band gap from the Brus equation[a] (eV)	Band gap from absorption spectra (eV)
1.54	6.45	4.64

[a]Please refer to the 'ZnS nanocrystalline powder synthesis by varying polyvinyl alcohol concentration' section.

Table 3 FTIR peak assignment of polyvinyl alcohol and ZnS prepared in polyvinyl alcohol

Polyvinyl alcohol (cm^{-1})	Band assignment	ZnS prepared in polyvinyl alcohol (cm^{-1})	Band assignment
3,444	O-H stretching	3,395	O-H stretching
2,901	C-H stretching	2,917	C-H stretching
1,733	C=O stretching	2,369	May be due to S-H interaction
1,626	C=C stretching	1,574	C=C stretching
1,428	Symmetric bending of CH_2	1,392	C-H bending
1,263	C-H bending	1,114	-C-O-H stretching
1,045	C-O stretching	915	C-C stretching
587	C-H bending	854.23	CH_2 stretching
-	-	642	OH stretching

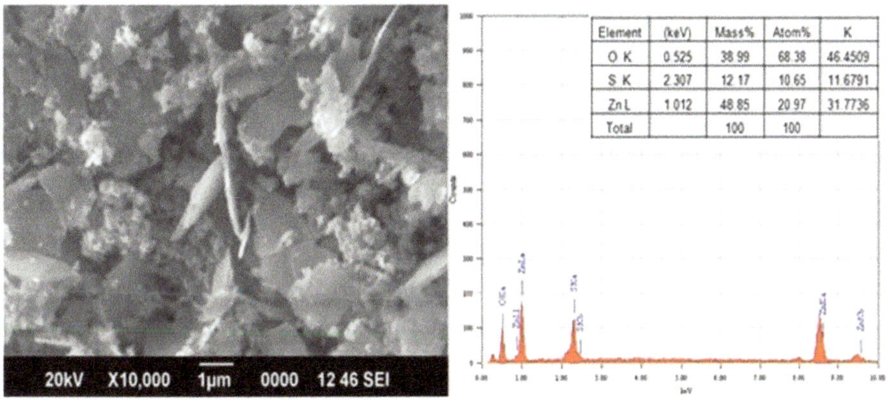

Figure 8 SEM and EDX images and data of the PC sample.

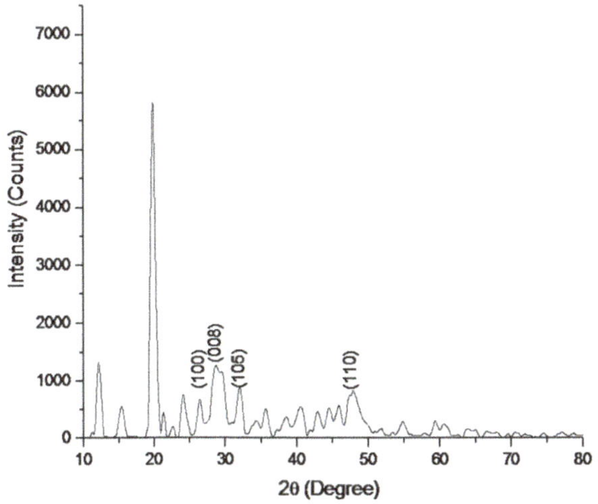

Figure 9 X-ray diffractogram of the PC sample.

stretching and C-H stretching of CH_2 groups, respectively. The bands at 1,733 and 1,626 cm^{-1} are due to the C=O stretching and C=C stretching. These carbonyl groups are due to the absorption of the residual acetate groups due to the manufacture of PVA from hydrolysis of polyvinyl acetate [26]. The symmetric bending mode of CH_2 is found at 1,428 cm^{-1}. The band at about 1,263 cm^{-1} results from the vibration of CH. The band at about 1,045 cm^{-1} is assigned to C-O stretching vibration. The IR band positions and their assignments are presented in Table 3.

Wurtzite ZnS nanoparticles

The SEM image of the prepared white powder is shown in Figure 8 with the energy-dispersive X-ray spectroscopy (EDX) data. The SEM image shows that the sample consists of nanosized particles and nanosheets. The elemental analysis shows that the sample possessed Zn, S, and oxygen. So, the sample was studied using its X-ray diffractogram (Figure 9) to confirm the compound and to find the crystal information. Also, the X-ray diffractogram of the sample

Table 4 XRD data of the PC sample

2θ (°)		(hkl)	Interplanar spacing 'd' (Å)	
Experimental	ASTM 39-1363		Experimental	ASTM 39-1363
26.399	26.914	(100)	3.3734	3.3100
28.577	28.586	(008)	3.1210	3.1201
29.788	29.257		2.9969	3.0500
32.208	32.411	(105)	2.7770	2.7601
35.597	35.597		2.5200	2.5200
46.006	46.865		1.9711	1.9370
47.7	47.727	(110)	1.9050	1.9040

was compared with the ASTM data of ZnO and ZnS due to the presence of oxygen as seen from EDX. It is found that the obtained the X-ray diffractogram correlates well with the ASTM 39–1363 for hexagonal ZnS. The crystals of ZnS were seen aligned along (100), (008), (105), and (110) planes. Further, the peaks for PVA was also seen at 19.72 [28,29] for (101) reflections due to the crystallization of PVA, and other reflection peaks from unknown planes were seen at 2θ, 11.96, 15.36, 21.16, 24.06, 34.48, 35.45, 40.53, 42.7, 44.6, 45.85, 54.81, and 59.4. Crystal data are shown in Tables 4 and 5, and it is found that the data are well correlated with the hexagonal ZnS. The average crystallite size was calculated using the Williamson-Hall plot [23] as 5.70 nm.

The sample was further analyzed by its TEM, HRTEM, and SAED images shown in Figure 10. It is seen that the sample consists of nanosized ZnS particles of about 4.17 nm. The interplanar spacing of the crystals found from the HRTEM is about 0.19 nm, and it may be along the (110) plane. The polycrystalline diffraction rings shown in the SAED images are labeled to the (110), (008), and (105) planes after comparing with the XRD data of the sample.

The absorption spectrum of the polycrystalline PC sample is shown in Figure 11, and the band gap of the sample was found at 4.07 eV from the absorption edge (305 nm) which is blueshifted when compared to the bulk value of 3.9 eV. Absorption peaks are not observed in this sample which may be due to the larger crystallite size than the Bohr exciton radius of ZnS. The band gap of the sample with particle and crystallite sizes is shown in Table 6. It is also seen that the band gap found from the Brus equation [24] is larger than the observed band gap from the absorption spectra, and the reason for the difference is already well explained before.

The FTIR spectra of the ZnS sample prepared with PVA and trisodium citrate are given in Figure 12 that shows C-H, C-O, and O-H stretching vibrations and C-H bending vibrations. All of the other observed peaks are assigned to their respective values as given in Table 7. The C=C bending vibrations are also seen in the spectra, and that may be due to the presence of esters from the manufacture of PVA.

Conclusion

ZnS nanoparticle growth using the concentration of the polymer capping agent polyvinyl alcohol was studied in this paper. ZnS nanocrystals of about 1.54 nm have been prepared, and their optical absorption shows interband transitions. Wurtzite ZnS nanocrystals have been prepared by low-temperature wet chemical method. It is found that the polyvinyl alcohol concentration has a tremendous effect in the particle growth. Particle sizes were varied with the concentration of PVA. The important factor is that the strain was seen negative due to contraction of the particles. The important finding of this experiment is that the preparation of the hexagonal ZnS nanoparticles is possible in lower temperatures when both the trisodium citrate and PVA are used simultaneously with the ZnS precursors.

Methods

ZnS nanocrystalline powder synthesis by varying polyvinyl alcohol concentration

The exact concentration of the ZnS precursor solutions and trisodium citrate for ZnS was found by fixing $ZnCl_2$ concentration and by varying Na_2S and trisodium concentration, but those experimental results are not explained

Table 5 Crystal information of prepared hexagonal ZnS

Crystallite size (nm)	Average strain	Observed lattice parameters (Å)		Lattice parameters (ASTM card no. 39–1363) (Å)		Volume of the crystal (10^{-28} m^3)	
		a	*c*	*a*	*c*	Observed	ASTM card no. 39-1363
3.2	−0.0668	3.8953	24.9687	3.8221	24.9611	3.2809	3.1578

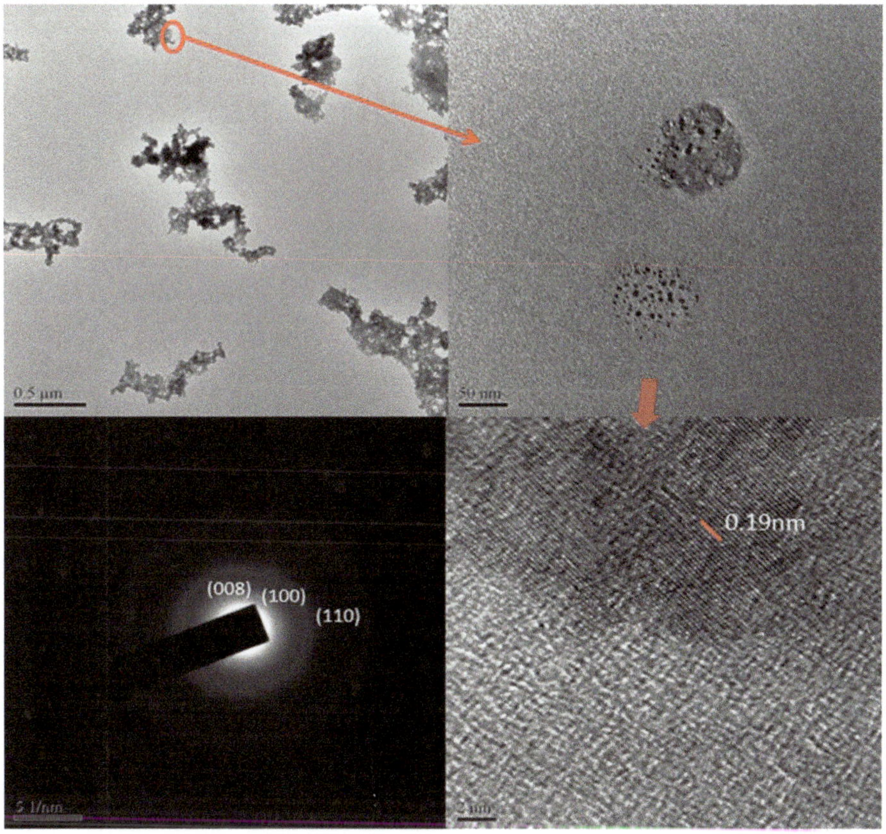

Figure 10 TEM, HRTEM, and SAED images of the PC sample.

here since they do not belong under the scope of this paper. A systematic study was performed to find the influence of polymer polyvinyl alcohol concentration on ZnS particle growth. Polyvinyl alcohol is a water-soluble polymer, and it can be used as a capping and reducing agent in nanoparticles preparation. Here, ZnS powder samples were prepared for various concentrations of polyvinyl alcohol and compared with the ZnS sample prepared without any surfactant or capping or reducing agent.

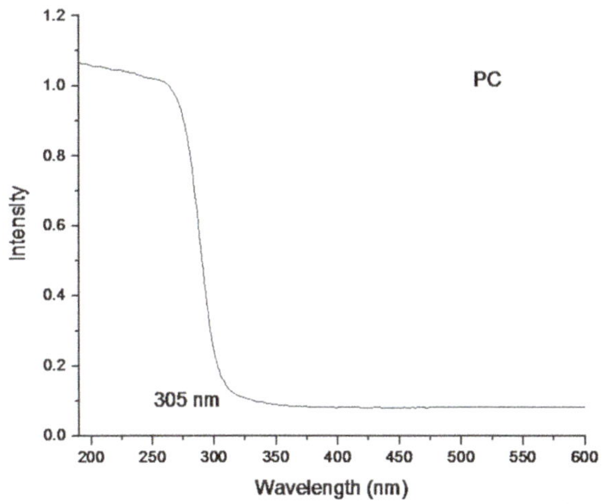

Figure 11 Absorption spectrum of the prepared wurtzite ZnS.

Polyvinyl alcohol (PVA) (80% hydrolyzed) was purchased from Sigma-Aldrich Chemicals (Bangalore, India) and used without any further purification. PVA solution of different concentrations (Table 8) was prepared at 60°C under magnetic stirring for 1 h to make it a clear solution. Then, 1 M $ZnCl_2$ solution was slowly mixed with this hot PVA solution. The stirring continued for another 1 h, and then the 0.1 M Na_2S solution was added drop by drop slowly using a burette while continuing the stirring (about 200 rpm). The formation of white powder for each drop of Na_2S was seen, and it was taken out from the heater after mixing all of the Na_2S solution. When the white precipitate settled at the bottom of the flask, the excess liquid was decanted, and the settled white powder was washed in ethanol then with deionized (DI) water four times, then

Table 6 Crystallite size, particle size, and band gap of the PC sample

Crystallite size (nm)	Particle size from TEM (nm)	Band gap from the Brus equation[a] (eV)	Band gap from absorption spectrum (eV)
3.2	4.17	4.45	4.07

[a]Please refer to the 'Wurtzite ZnS nanoparticles' section.

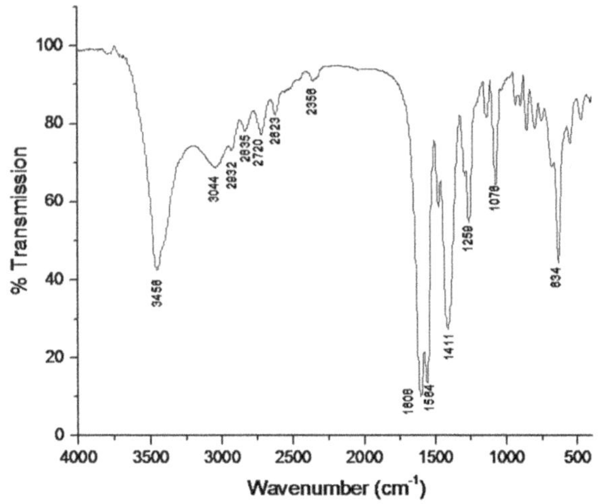

Figure 12 FTIR spectrum of ZnS prepared with PVA and trisodium citrate.

washed again with ethanol. After washing with ethanol, the precipitated white powder was taken into a heater maintained at about 60°C to dry it and to make it into a fine powder. Then, the powder samples were analyzed using a Bruker AXS D8 Advance powder X-ray diffractometer (Bruker Optik Gmbh, Ettlingen, Germany), Thermo Nicolet Avatar 370 Fourier transform infrared spectrometer (FTIR; Thermo Nicolet Corporation, Madison, WI, USA), JEOL model JSM-6390LV scanning electron microscope (JEOL Ltd., Akishima, Tokyo, Japan), JED-2300 energy-dispersive spectrometer (JEOL), and JEM-2100 transmission electron microscope (JEOL).

Table 7 FTIR peak assignment of ZnS sample prepared with PVA and trisodium citrate

Peak (cm^{-1})	Band assignment
3,456	Dimeric O-H stretch
3,044	C-H stretch
2,932	C-H asymmetrical stretching
2,835	C-H stretch
2,720	C-H stretch.
2,623	S-H stretch
2,356	May be due to S-H interaction
1,608	Conjugated C=C
1,564	C=C bending
1,480	C-H bending
1,411	C-H inplane bend
1,291	C-H in plane bend
1,259	O-H in plane bend
1,138	C-H bend
1,076	C-O symmetrical stretch
935	C-H bending
902	C-H bending
859	C-H bending
803	C-H bending
753	C-H bending
681	C-H bending
634	C-H bending

Table 8 Variation of PVA concentration on ZnS nanopowder synthesis

Name of sample	Material	Density (gm/ml)
PZ1	PVA	0.03
PZ2	PVA	0.003
PZ3	PVA	0.0003

Wet chemical synthesis of wurtzite ZnS particle

Mutual effect of the PVA and trisodium citrate concentration was studied by adding them together after finding their best concentration from the experiments explained in the above section. So, polyvinyl alcohol solution of 0.03 mg/ml concentration was made at 60°C under magnetic stirring for 1 h to make it a clear solution. Then, 1 M $ZnCl_2$ solution was slowly mixed with this hot PVA solution while continuing the stirring. Another solution of 0.1 M Na_2S and 2 M trisodium citrate were prepared and mixed together and stirred for 1 h. The Na_2S-trisodium citrate solution was added drop by drop slowly using a burette while continuing the stirring (about 200 rpm). The formation of white powder for each drop of Na_2S-trisodium citrate was seen, and it was taken out from the heater after mixing all of the Na_2S-trisodium citrate solution. When the white precipitate settled at the bottom of the flask, the excess liquid was decanted, and the settled white powder was washed in ethanol, then four times using DI water, and then washed again with ethanol. After washing with ethanol, the precipitated white powder was taken into a heater maintained at about 60°C to dry it and to make it into a fine powder. Then, the powder sample was named as PC and then analyzed using a Bruker AXS D8 Advance powder X-ray diffractometer, Thermo Nicolet Avatar 370 FTIR, JEOL model JSM-6390LV scanning electron microscope, JED-2300 energy-dispersive spectrometer, and JEM-2100 transmission electron microscope.

Competing interests

The authors declare that they have no competing interests.

Authors' contributions

AKT carried out the experiments, examined the results, and drafted the manuscript. AKAU coordinated the project and discussed the results. Both authors read and approved the final manuscript.

Authors' information
AKT has M.Sc. and M.Phil degrees and is a research scholar at the Department of Physics, Kongunadu Arts and Science College, Coimbatore. He holds a Bachelor's degree in Physics from the University of Calicut, Malappuram, Kerala, India and M. Sc. and M.Phil. in Physics from Bharathiar University, Coimbatore, Tamilnadu, India. Currently, he is doing a Ph.D. degree at the Kongunadu Arts and Science College (affiliated to Bharathiar University), Coimbatore in Physics. He specializes in the plasma electrochemical synthesis of ZnS nanoparticles and wet chemical synthesis of ZnS and Ag nanoparticles for P3HT bulk heterojunction solar cells. He is also the recipient of the Government of Tamilnadu State Collegiate Education Scholarship for Ph.D.; US National Science Foundation I2CAM International Materials Institute Travel Grant for I-CAMP Summer School 2012 held at University of Colorado, Boulder, USA.; Centre for International Co-operation in Science (promoted by the Indian National Science Academy), Chennai Travel Fellowship for I-CAMP Summer School 2012 held at University of Colorado, Boulder, USA; US National Science Foundation I2CAM International Materials Institute Travel Grant for I -CAMP Summer School 2009 held in China; and Department of Science and Technology (DST), Government of India Travel Fellowship for I-CAMP Summer School 2009 held in China. He is a life member of the Plasma Science Society of India (PSSI); full ICAM member-JNCASR, Bangalore Branch; member of the International Union of Crystallography (IUCr.); regular fellow of the Optical Society of India (F1082); and life member of the Materials Research Society of India. Currently, he is credited for ten papers in international journals, and he has visited the Kingdom of Saudi Arabia, Republic of China, and the United States of America, where he presented his research results.
AKA has M.Sc., M.Phil, and Ph.D. degrees and is an assistant professor at the Department of Physics, Kongunadu Arts and Science College, Coimbatore, India. She has more than 18 years of research experience and 11 years of teaching experience at graduation and post-graduation level. She has specialized in plasma physics and nanotechnology. She is credited with more than 20 international journal publications and has guided more than 20 M.Phil students and 5 Ph.D. students.

Acknowledgment
The authors thank STIC, Cochin University of Science and Technology, Cochin, Kerala and SAIF, North Eastern Hill University, Shillong, Meghalaya for XRD, SEM, FTIR, and TEM.

References

1. Kalele, S, Gosavi, SW, Urban, J, Kulkarni, SK: Nanoshell particles: synthesis, properties and applications. Curr. Sci. **91**, 8 (2006)
2. Chen, Z-G, Zou, J, Wang, D-W, Yin, L-C, Liu, G, Liu, Q, Sun, C-H, Yao, X, Li, F, Yuan, X-L, Sekiguchi, T, Gao Qing, L, Cheng, H-M: Field emission and cathodoluminescence of ZnS hexagonal pyramids of zinc blende structured single crystals. Adv. Funct. Mater. **19**, 484–490 (2009)
3. Sahare, S, Dhoble, SJ, Singh, P, Ramrakhiani, M: Fabrication of ZnS:Cu/PVA nanocomposite electroluminescence devices for flat panel displays. Adv. Mater. Lett. **4**(2), 169-173 (2012). http://amlett.com/uploads/157630320.pdf
4. Nizamoglu, S, Demir, HV: Excitation resolved color conversion of CdSe/ZnS core/shell quantum dot solids for hybrid white light emitting diodes. J Appl Phys **105**, 083112 (2009)
5. Muller, GO, Mach, R, Ohnishi, BSH: Efficient ZnS-like alkaline earth sulfide electroluminescence. J. Cryst. Growth **101**, 999–1003 (1990)
6. Wang, X, Xie, Z, Huang, H, Liu, Z, Chen, D, Shen, G: Gas sensors, thermistor and photodetector based on ZnS nanowires. J. Mater. Chem. **22**, 6845–6850 (2012). doi:10.1039/C2JM16523F
7. Sorokina, IT, Sorokin, E, Mirov, S, Fedorov, V, Badikov, V, Panyutin, V, Schaffers, KI: Broadly tunable compact continuous-wave Cr2+:ZnS laser. Opt. Lett. **27**(12), 1040–1042 (2002)
8. Sumitomo Electric Industries, Ltd: Environmentally-resistant ZnS lens for far-infrared cameras (new products and techniques). Sei. Tech. Rev. **71**, 113–115 (2010)
9. Onodera, C, Masaaki, Y, Tadayoshi, S, Tsunemasa, T: Threshold current density in ZnS/MgBeZnS quantum well ultraviolet lasers. Opt. Rev. **17**(3), 159–160 (2010)
10. Gangopadhyay, U, Kyunghea, K, Dhungel, SK, Mangalaraj, D, Park, JH, Yi, J: Application of CBD zinc sulfide (ZnS) film to low cost antireflection coating on large area industrial silicon solar cell. Trans. Electr. Electron Mater. **5**, 1 (2004)
11. Michael, B, Katarzyna, M, Adam, S, Jerzy, S, Anna, P-S: P3HT/ZnS: a new hybrid bulk heterojunction photo voltaic system with very high open circuit voltage. Sol. Energ. Mater. Sol. Cell **93**, 662–666 (2009)
12. Hsieh, M-F, Li, JK-J, Lin, JC-A, Huang, S-H, Sperling, RA, Parak, WJ, Chang, WH: Tracking of cellular uptake of hydrophilic CdSe/ZnS quantum dots/ hydroxyapatite composites nanoparticles in MC3T3-E1 osteoblast cells. J. Nanosci. Nanotechnol. **8**, 12 (2008)
13. Hussain, S, Won, N, Nam, J, Bang, J, Chung, H, Kim, S: One-pot fabrication of high-quality InP/ZnS (core/shell) quantum dots and their application to cellular imaging. Chemphyschem. **10**, 1466–1470 (2009)
14. Bhadra, R, Singh, VN, Mehta, BR, Datta, P: Studies on some aspects of ZnS nanocrystals for possible applications in electronics. Chalcogenide Letters **6**(5), 189–196 (2009)
15. Lan, C, Gong, J, Jiang, Y, Ding, Q: Fabrication of ZnS/SnO nanowire/ nanosheet hierarchical nanoheterostructure and its photoluminescence properties. Cryst. Eng. Comm. **14**, 8063–8067 (2012). doi:10.1039/C2CE26053K
16. Shaban, SM, Saeed, NM, Raad MS, AL-H: Fabrication and study zinc sulfide Schottky barrier detectors. Indian J. Sci. Technol. **4**, 4 (2011)
17. Coleman, VA, Jagadish, C: Basic properties and applications of ZnO. In: Jagadish, C, Pearton, S (eds.) Zinc Oxide Bulk, Thin Films and Nanostructures, Processing. Properties and Applications. Elsevier, Philadelphia (2006)
18. Qadri, SB, Skelton, EF, Hsu, D, Dinsmore, AD, Yang, J, Gray, HF, Ratna, BR: Size-induced transition-temperature reduction in nanoparticles of ZnS. Phys. Rev. B. **60**, 9191–9193 (1999). doi:10.1103/PhysRevB.60.9191
19. Yong Ding, X, Wang, D, Wang, ZL: Phase controlled synthesis of ZnS nanobelts: zinc blende vs wurtzite. Chem. Phys. Lett. **398**, 32–36 (2004)
20. Zhang, H, Qi, L: Low-temperature, template-free synthesis of wurtzite ZnS nanostructures with hierarchical architectures. Nanotechnology **17**, 3984–3988 (2006)
21. Yuan-yuan, SHE, Juan, YANG, Ke-qiang, QIU: Synthesis of ZnS nanoparticles by solid–liquid chemical reaction with ZnO and Na_2S under ultrasonic. Trans. Nonferrous Met. Soc. China **20**, s211–s215 (2010)
22. Henry, C: Size effects on structure and morphology of nanoparticles. In: Bréchignac, C, Houdy, P, Lahmani, M (eds.) Nanomaterials and Nanochemistry. Springer, Heidelberg (2006)
23. Williamson, GK, Hall, WH: X-ray line broadening from filed aluminium and wolfram. Acta Metall. **1**, 22 (1953). doi:10.1016/0001-6160(53)90006-6
24. Brus, LE: A simple model for the ionization potential, electron affinity and aqueous redox potentials of small semiconductor crystallites. J. Chem. Phys. **79**, 5566–5571 (1983)
25. Prabhu, RR, Abdul Khadar, M: Characterization of chemically synthesized CdS nanoparticles. Pramana J. Phys. **65**(5), 801–807 (2005)
26. Shehap, AM: Thermal and spectroscopic studies of polyvinyl alcohol/sodium carboxy methyl cellulose blends. Egypt. J. Solids **31**, 1 (2008)
27. Linga Raju, CH, Rao, JL, Reddy, BCV, Brahmam, KV: Thermal and IR studies on copper doped polyvinyl alcohol. Bull. Mater. Sci **30**(3), 215–218 (2007)
28. Strawhecker, KE, Manias, E: AFM of poly(vinyl alcohol) crystals next to an inorganic surface. Macromolecules **34**, 8475–8482 (2001)
29. Strawhecker, KE, Manias, E: Structure and properties of poly(vinyl alcohol)/ Na+ montmorillonite nanocomposites. Chem. Mater. **12**, 2943–2949 (2000)

Fabrication of hollow particles composed of silica containing gadolinium compound and magnetic resonance imaging using them

Yoshio Kobayashi[1*], Hikaru Morimoto[1], Tomohiko Nakagawa[2], Yohsuke Kubota[2], Kohsuke Gonda[2] and Noriaki Ohuchi[2,3]

Abstract

A preparation method for hollow particles composed of silica shell containing gadolinium compound (GdC) is proposed. GdC nanoparticles with an average size of 40.5 ± 6.2 nm were prepared with a homogeneous precipitation method at 80°C using 1.0×10^{-3} M $Gd(NO_3)_3$ and 0.5 M urea in the presence of 2.0×10^{-4} M ethylenediaminetetraacetic acid disodium salt dihydrate. Silica-coated GdC (GdC/SiO_2) nanoparticles with an average size of 100.9 ± 9.9 nm were fabricated with a sol-gel method at 35°C using 5.0×10^{-3} M tetraethylorthosilicate, 11 M H_2O, and 1.5×10^{-3} M NaOH in ethanol in the presence of 1.0×10^{-3} M GdC nanoparticles. The GdC/SiO_2 particles were aged at 80°C in water after replacement of solvent of the as-prepared GdC/SiO_2 particle colloid solution with water, which provided diffusion of GdC into the silica shell and then formation of a hollow structure. The hollow particle colloid solution revealed good MRI properties. A relaxivity value for longitudinal relaxation time-weighted imaging was as high as 3.38 $mM^{-1} \cdot s^{-1}$ that attained 80% of that for a commercial Gd complex contrast agent.

Keywords: Hollow, Core-shell, Particle, Gadolinium, Homogeneous precipitation method, Silica, Sol-gel, MRI

Background

Gadolinium compounds (GdC) can function as a contrast agent for magnetic resonance imaging (MRI) in medical diagnosis [1-6]. Typical commercial GdC-based contrast agents are solutions dissolving homogeneously gadolinium complexes at a molecular level. The GdC molecules are not strongly dragged in fluid because of their small sizes. Consequently, they cannot stay in living bodies for a long period, which provides difficulty taking steady images. Formation of particles of GdC and an increase in their apparent size are promising solutions to the problem because of their projected area which is larger than molecules or nanoparticles. Thus, their residence time will increase, which may realize steady imaging.

Growth of tumor generated in tissue produces a number of slits with a size of several hundred nanometers on the wall of the vessel near the tumor. Polymers or nanoparticles smaller than the slit size may be able to pass through the slit. Their discharge from the tumor is controlled because lymph, which is a way for the discharge, is premature in tumor. This behavior is called an enhanced permeation and retention (EPR) effect [7-9]. Accordingly, nanoparticle formation has another advantage; nanoparticles that show MRI properties may have the ability to detect the tumor by the EPR effect.

Gadolinium complexes may release free gadolinium ions through dissociation of the ions. The release of gadolinium ions may provoke adverse reactions in some patients [6,10,11]. As one of the methods for reducing the adverse reactions derived from gadolinium complexes, coating of GdC nanoparticles (core) with materials inert to living bodies (shell) can be given because contact of the GdC with living bodies may be controlled with a physical barrier of the shell materials. From this viewpoint, we have also studied on the silica coating of

* Correspondence: ykoba@mx.ibaraki.ac.jp
[1]Department of Biomolecular Functional Engineering, College of Engineering, Ibaraki University, 4-12-1 Naka-narusawa-cho, Hitachi, Ibaraki 316-8511, Japan
Full list of author information is available at the end of the article

GdC nanoparticles [12]. In a preliminary experiment, the silica-coated GdC (GdC/SiO_2) particles with a core-shell structure became hollow particles after the GdC/SiO_2 particles were washed with a centrifuge and were aged in water. In several previous studies, hollow particles are fabricated by calcining core-shell particles composed of organic compound core and inorganic compound shell to remove the core [13-15]. The hollow particles in the preliminary experiment were obtained with no calcination. Methods for producing hollow silica particles that do not use calcination process have been proposed by several researchers [16-18]. They use liquids such as aqueous basic solutions and organic solvents that dissolve core materials. The use of such liquids may damage the silica shell or give environmental load. Care of the problems is not required to be taken for the preliminary experiment since it did not use acid, base, and organic solvent to remove the core. If the components of GdC particles are left in the silica shell after the formation of hollow particles, the hollow particles are expected to function as new MRI contrast agents because many protons contained in the silica shell may interact efficiently with the GdC components compared to the GdC particles localized only at the center of the Gd/SiO_2 particles.

The present work proposes a method for preparing hollow particles composed of GdC and silica. A colloid solution of GdC nanoparticles was prepared with a homogeneous precipitation method, and then the GdC nanoparticles were silica-coated with a modified Stöber method. Elimination of GdC core was performed by replacing the solvent of the reactant with water and aging the aqueous colloid solution. The present work also studied on MRI properties of the hollow particle colloid solution.

Results and discussion

Morphology of particles

GdC nanoparticles

A photograph was taken for the colloid solution of GdC particles, as shown in Figure 1a. The colloid solution was faintly turbid but colloidally stable. Figure 1b shows a transmission electron microscopy (TEM) image of the GdC particles. Quasi-spherical particles were obtained, though they partially stuck together. Their particle size was 40.5 ± 6.2 nm. Figure 2 (pattern a) shows an X-ray diffractometry (XRD) pattern of the GdC particles. The pattern was structureless, except for broad peaks at *ca.* 30° and *ca.* 45°. Accordingly, the GdC particles were amorphous or too fine to be detected. The inductively coupled plasma (ICP) measurement showed that Gd concentration in the GdC particle colloid solution was 0.819 mM, which indicated that 81.9% of the $Gd(NO_3)_3$ used for production of GdC particles was transformed to GdC particles.

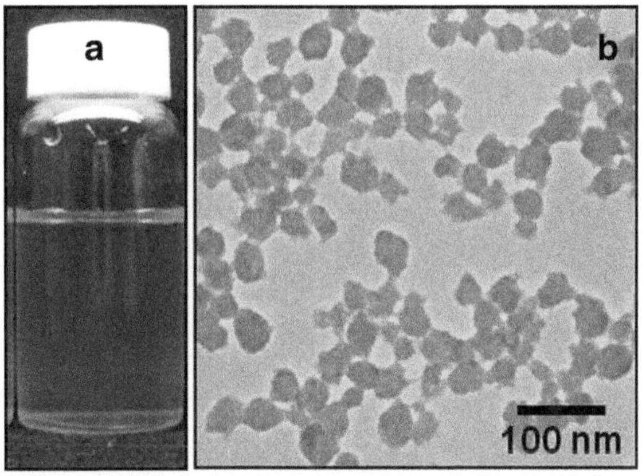

Figure 1 Photograph of as-prepared colloid solution of GdC nanoparticles (a) and their TEM image (b).

GdC/SiO_2 core-shell particles

Figure 3a shows a TEM image of the GdC/SiO_2 particles, which were as-prepared, i.e., dispersed in H_2O/ethanol solution with the H_2O concentration of 11 M. The image was taken several days after preparation. Darker and lighter parts corresponded to GdC and silica, respectively, since the electron density of GdC is significantly higher. The GdC particles were successfully coated with silica shell, though the GdC/SiO_2 particles partially stuck together. No damage was seen for the GdC/SiO_2 particles even several days after preparation. Accordingly, the GdC/SiO_2 particles were chemically stable in the as-prepared solution. The particles had a size of 100.9 ± 9.9 nm, which were composed of core particles with a size

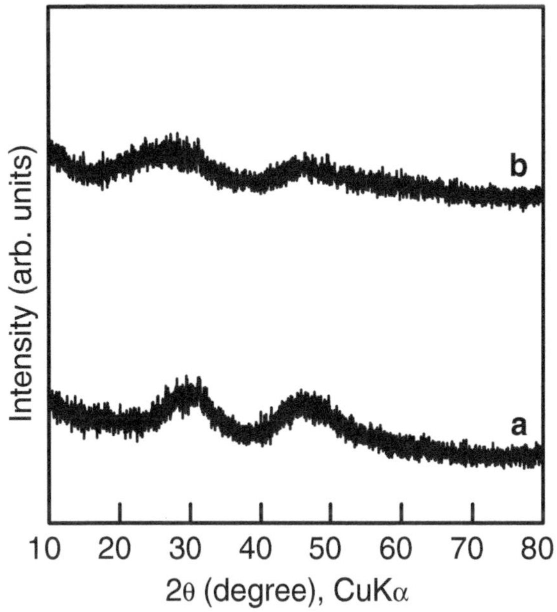

Figure 2 XRD patterns of (a) GdC particles and (b) GdC/SiO_2 particles.

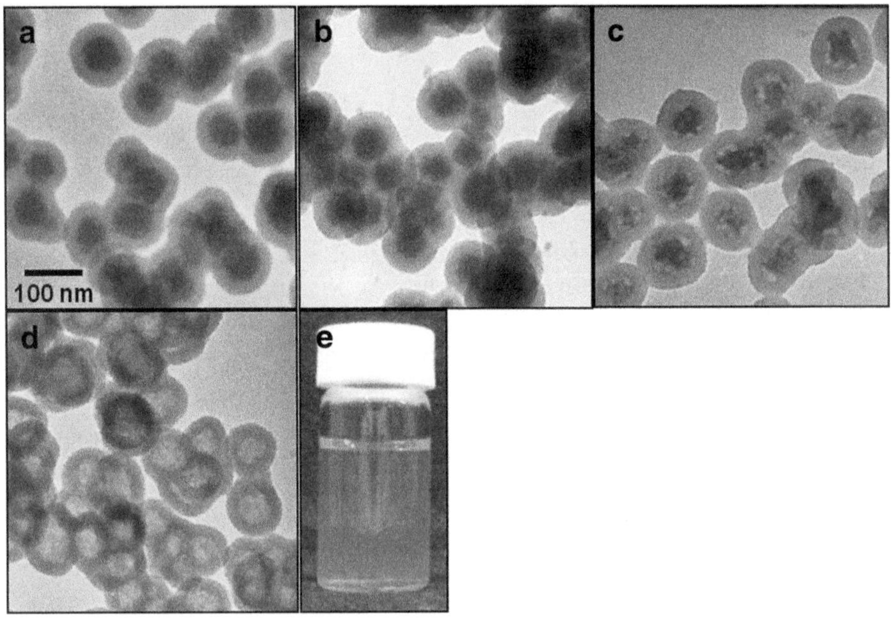

Figure 3 TEM images of various particles. (**a**) GdC/SiO_2 particles several days after preparation, (**b**) GdC/SiO_2 particles immediately after replacement of solvent to water, (**c**) GdC/SiO_2 particles aged at room temperature for 3 h after replacement of solvent to water, (**d**) GdC/SiO_2 particles aged at 80°C for 3 h after replacement of solvent to water. Image (**e**) shows a photograph of sample (**d**).

of 51.8 ± 8.2 nm and a shell with a thickness of 24.6 nm. The core size was larger than that of the GdC particles prior to the silica coating. Aggregation and particle growth were considered to take place during the silica coating. Hydrolysis of tetraethylorthosilicate (TEOS) provided an increase in ionic strength of the solution. Since an increase in the ionic strength compresses double layer on solid materials such as colloidal particles [19-21], the double-layer repulsion between the GdC particles probably became small through the hydrolysis. Thus, the GdC particles probably aggregated and grew because of the high ionic strength that would favor the particle aggregation. Figure 2 (pattern b) shows an XRD pattern of the GdC/SiO_2 particles. There was no large difference between the patterns of GdC particles and the GdC/SiO_2 particles. The GdC particles were also amorphous or too fine to be detected. Accordingly, the results of TEM observation and XRD measurement indicated that the silica coating made no chemical effect on the GdC particles. The ICP measurement revealed that an actual Gd concentration in the Gd/SiO_2 particle colloid solution was 0.698 mM, which indicated that 85.2% of Gd contained in the GdC particles was consumed for the formation of GdC/SiO_2 particles.

Hollow particles

Figure 3b,c shows TEM images of the GdC/SiO_2 particles, of which solvents were replaced with water. The GdC/SiO_2 particles immediately after the washing process (Figure 3b) appeared to be similar to those prior to the replacement (Figure 3a), which indicated that the washing process did not damage their core-shell structure mechanically. For 3 h after the washing (Figure 3c), the surface of the GdC core appeared to be partially dissolved and then became rough. The silica shell was not damaged in spite of the surface roughing. The surface roughing was promoted by heating at 80°C, as shown in Figure 3d. Then, hollow particles were produced. The silica shell was still maintained even after the heating. Figure 3e shows a photograph of the hollow particle colloid solution. The solution was colloidally stable since no sedimentation took place. According to the ICP measurement, an actual Gd concentration in the hollow particle colloid solution was 0.426 mM. This meant that 61.0% of Gd contained in the GdC/SiO_2 particles was considered to still remain in the shell of hollow particles and 39.0% of Gd was removed from the hollow particles.

Figure 4a shows a TEM image of GdC particles after the solvent of their colloid solution was replaced with water. The replacement of solvent with water did not damage the GdC particles mechanically. After the replacement, the colloid solution of GdC particles was aged at 80°C for 3 h. Their TEM image is shown in Figure 4b. The image appeared to be similar to that prior to the aging (Figure 4a). Those results indicated that both the replacement and the aging at 80°C did not affect the structure of GdC particles. Accordingly, it could be concluded that the formation of hollow structure required the GdC particles to have contact with the silica shell. The GdC was presumed to diffuse into the silica shell, which

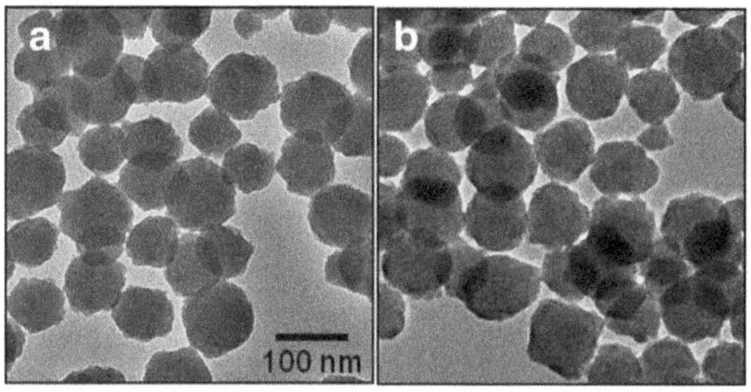

Figure 4 TEM images of GdC particles after replacement of solvent to water. (**a**) GdC particles aged at room temperature. (**b**) GdC particles aged at 80°C for 3 h.

provided the production of new compounds such as gadolinium silicate in the shell. Consequently, the hollow particles were produced. However, a precise mechanism of the hollow particle formation is still unclear.

MRI property

Longitudinal relaxation time (T_1)-weighted images of the colloid solutions of hollow particles with various actual Gd concentrations are shown in Figure 5. For T_1-weighted images, strong magnetic resonance gives positive images with light contrast. All the solutions examined were clearly imaged against a black background, and the light contrast of image increased with increasing actual Gd concentration.

Figure 6 shows a plot of longitudinal relaxation rate ($1/T_1$) of the hollow particle colloid solution as a function of the actual Gd concentration. The relaxation rate increased linearly with increasing actual Gd concentration. A value of relaxivity (r_1) is defined as a slope of relaxation rate with respect to actual Gd concentration and is commonly used as a guideline on the performance of positive contrast agents. The r_1 value, which was calculated by linear fitting, was 3.38 $mM^{-1} \cdot s^{-1}$. This value attained 80% of that of Magnevist since an r_1 value of Magnevist, which was measured in our previous work [22], was 4.23 $mM^{-1} \cdot s^{-1}$. Our previous work also gave that a colloid solution of GdC/SiO_2 particles revealed an r_1 value of 3.11 $mM^{-1} \cdot s^{-1}$ [22], which indicated that the r_1 value of the hollow particle colloid solution was larger than that of the GdC/SiO_2 particle colloid solution. Active exchange of the protons with the contrast agent shortens T_1, i.e., increases r_1 [23,24]. Since the GdC present in the shell lay close to the water protons compared to the GdC inside the GdC/SiO_2 particles, the interaction between water protons and gadolinium ions was more significant than that for the GdC/SiO_2 particles. Accordingly, the formation of hollow particles with shell-containing GdC was found to improve imaging ability based on longitudinal relaxation.

Conclusions

A method for producing hollow particles with silica shell containing GdC was proposed. The GdC nanoparticles with the average size of 40.5 nm were prepared by means of the homogeneous precipitation method using aqueous solution of 1.0×10^{-3} M $Gd(NO_3)_3$, 0.5 M urea, and 2.0×10^{-4} M ethylenediaminetetraacetic acid disodium salt dihydrate (EDTA). The silica coating of the GdC particles was achieved with the sol-gel method using 5.0×10^{-3} M TEOS, 11 M H_2O, and 1.5×10^{-3} M NaOH in the presence of the GdC particles at the Gd concentration of 1.0×10^{-3} M, which produced the GdC/SiO_2 particles with the average size of 100.9 nm. The replacement of the solvent of GdC/SiO_2 particle colloid solution with water and the aging of GdC/SiO_2 particles in water at 80°C transformed the GdC/SiO_2 particles to hollow particles due to the diffusion of GdC into the silica shell. The hollow particle colloid solution showed high-contrast T_1-weighted magnetic resonance images. Its relaxivity value, which was 3.38 $mM^{-1} \cdot s^{-1}$, was recorded at the value as high as 80% of that for Magnevist, the commercial Gd complex contrast agent. These results obtained in the present work indicated that the hollow particle colloid solution had an ability of MRI contrast agent.

Methods

Chemicals

Gadolinium nitrate hexahydrate ($Gd(NO_3)_3 \cdot 6H_2O$, 99.5%) and urea (99.0%) were used as a raw chemical for GdC nanoparticles and a precipitation inducer in the preparation of GdC nanoparticles, respectively. The stabilizer used in the preparation of GdC nanoparticles was EDTA (99.0% to 101.0%). A raw chemical for silica shell and a

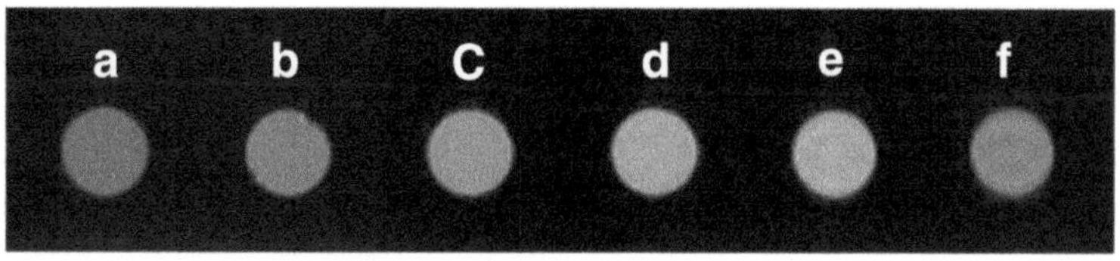

Figure 5 T_1-weighted images. Of (**a**) water and hollow particle colloid solutions with actual Gd concentrations of (**b**) 0.01, (**c**) 0.03, (**d**) 0.1, (**e**) 0.2 and (**f**) 0.4 mM.

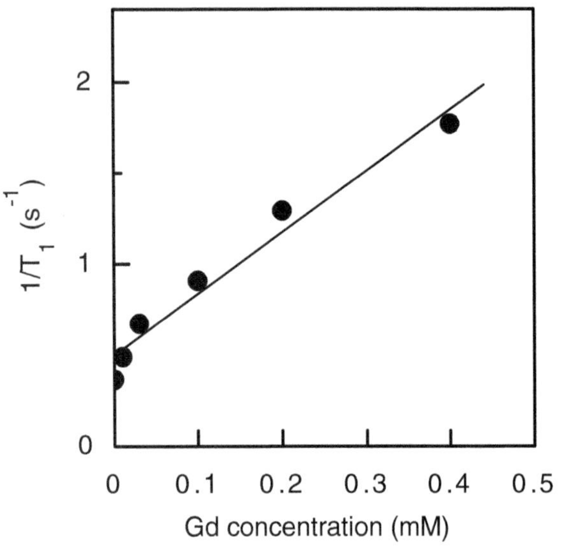

Figure 6 **Relaxation rates of hollow particle colloid solutions as a function of actual Gd concentration.**

solvent for silica coating were TEOS (95%) and ethanol (99.5%), respectively. A sol-gel reaction of TEOS in the silica coating was catalyzed with sodium hydroxide (NaOH) solution (5 M). The commercial MRI contrast agent used for comparison to the particle colloid solutions in imaging examinations was Magnevist (0.5 M Gd, Bayer Co., Ltd., Leverkusen, Germany). All the chemicals except for the commercial contrast agents were purchased from Kanto Chemical Co., Inc. (Tokyo, Japan) and used as received. The water that was ion-exchanged and distilled with Shimadzu SWAC-500 (Kyoto, Japan) was used in all the preparations.

Preparation of materials

GdC particles

A colloid solution of GdC nanoparticles was prepared in water containing EDTA by a homogeneous precipitation method, according to our previous work [22]. $Gd(NO_3)_3$ aqueous solution was added to aqueous solution dissolving urea and EDTA in a hermetically sealed flask reactor. The mixture was stirred at 80°C for 3 h. Initial concentrations of $Gd(NO_3)_3$, urea, and EDTA were 1.0×10^{-3}, 0.5, and 2.0×10^{-4} M, respectively. The GdC particles in the colloidal solution were washed by repeating centrifugation, removal of supernatant, addition of water, and sonication three times. The solvent was replaced with H_2O/ethanol solution by adding the H_2O/ethanol solution at the last washing process.

GdC/SiO$_2$ core-shell particles

GdC/SiO_2 core-shell particles were fabricated by the silica coating of the GdC particles, which was performed in the H_2O/ethanol solution by a sol-gel method, according to our previous work [22]. TEOS and 0.1 M NaOH aqueous solution were successively added to the GdC particle colloid solution at 35°C. The reaction time was 24 h. Initial concentrations of Gd, TEOS, H_2O, and NaOH were 1.0×10^{-3}, 5.0×10^{-3}, 11, and 1.5×10^{-3} M, respectively.

Hollow particles

Hollow particles were fabricated from the Gd/SiO_2 core-shell particles. The colloidal suspensions of GdC/SiO_2 particles were washed by repeating centrifugation, removal of supernatant, addition of water, and sonication three times, which resulted in the replacement of solvent with water. Then, the washed colloid suspensions were aged in water.

Characterization

The samples were characterized by TEM and XRD. TEM was performed with a JEOL JEM-2000FX II microscope (Akishima, Japan) operating at 200 kV. Samples for TEM were prepared by dropping and evaporating the nanoparticle suspensions on a collodion-coated copper grid. Dozens of particle diameters in TEM images were measured to determine volume-averaged particle size, d_v, and standard deviation of particle size distribution, σ, defined by the following equations:

$$d_v = \left(\sum_i n_i d_i^3 / \sum_i n_i\right)^{1/3}, \quad (1)$$

$$\sigma = \left(\sum_i (d_i - d_v)^2 / \sum_i n_i\right)^{1/2}, \quad (2)$$

where n_i is the number of particles with a size of d_i. XRD measurements were carried out with a Rigaku RAD-B (Tokyo, Japan) operating at 7.5 kW Cu Kα radiation using a wide-angle goniometer. Samples for XRD were obtained with centrifugation of the solution containing the particles, removal of the supernatant, and drying of the residue at room temperature for 24 h in vacuum.

Actual Gd concentrations in particle colloid solutions were measured by ICP emission spectroscopy. ICP measurement was performed with a Shimadzu ICPS-7510 atom emission spectrometer (Kyoto, Japan). The emission was detected at a wavelength of 342.247 nm. Samples for ICP were prepared by dissolving the particles with aqua regia completely and then diluting the obtained solution with water to adjust its concentrations to concentrations suitable for the measurements.

For the estimation of MR signal intensity, T_1-weighted images of colloid solutions with different Gd concentrations were taken with a Bruker AVANCE III 400WB magnetic resonance imaging system (Madison, WI, USA) at a static magnetic field of 9.4 T. Times of echo and repetition were 8.5 and 1,500 ms, respectively. The colloid solutions were prepared by diluting the as-prepared colloid solutions with water.

Competing interests

The authors declare that they have no competing interests.

Authors' contributions

YK (Kobayashi) drafted the manuscript. KG and NO modified it. HM, TN, and YK (Kubota) participated in some practical work. All authors read and approved the final manuscript.

Acknowledgements

We express our thanks to Prof. T. Noguchi in the College of Science of Ibaraki University, Japan, for his help in TEM observation.

Author details

[1]Department of Biomolecular Functional Engineering, College of Engineering, Ibaraki University, 4-12-1 Naka-narusawa-cho, Hitachi, Ibaraki 316-8511, Japan. [2]Department of Nano-Medical Science, Graduate School of Medicine, Tohoku University, Seiryo-machi, Aoba-ku, Sendai 980-8575, Japan. [3]Department of Surgical Oncology, Graduate School of Medicine, Tohoku University, Seiryo-machi, Aoba-ku, Sendai 980-8574, Japan.

References

1. Yim, H, Yang, SG, Jeon, YS, Park, IS, Kim, M, Lee, DH, Bae, YH, Na, K: The performance of gadolinium diethylene triamine pentaacetate-pullulan hepatocyte-specific T1 contrast agent for MRI. Biomater. **32**, 5187–5194 (2011)
2. Wang, CL, Chea, YW, Boll, DT, Samei, E, Neville, AM, Dale, BM, Merkle, EM: Effect of gadolinium chelate contrast agents on diffusion weighted MR imaging of the liver, spleen, pancreas and kidney at 3 T. Europ. J. Radiol. **80**, e1–e7 (2011)
3. Chandrasekharan, P, Yong, CX, Poh, Z, He, T, He, Z, Liu, S, Robins, EG, Chuang, KH, Yang, CT: Gadolinium chelate with DO3A conjugated 2-(diphenylphosphoryl)-ethyldiphenylphosphonium cation as potential tumor-selective MRI contrast agent. Biomater. **33**, 9225–9231 (2012)
4. Shen, C, New, EJ: Promising strategies for Gd-based responsive magnetic resonance imaging contrast agents. Curr. Opin. Chem. Biol. **17**, 1–9 (2012)
5. Akai, H, Matsuda, I, Kiryu, S, Tajima, T, Takao, H, Watanabe, Y, Imamura, H, Kokudo, N, Akahane, M, Ohtomo, K: Fate of hypointense lesions on Gd-EOB-DTPA-enhanced magnetic resonance imaging. Europ. J. Radiol. **81**, 2973–2977 (2012)
6. Telgmann, L, Sperling, M, Karst, U: Determination of gadolinium-based MRI contrast agents in biological and environmental samples: a review: Anal. Chim. Acta. **764**, 1–16 (2013)
7. Shalviri, A, Foltz, WD, Cai, P, Rauth, AM, Wu, XY: Multifunctional terpolymeric MRI contrast agent with superior signal enhancement in blood and tumor. J. Control. Release **167**, 11–20 (2013)
8. Yamashita, F, Hashida, M: Pharmacokinetic considerations for targeted drug delivery. Adv. Drug Deliver. Rev. **65**, 139–147 (2013)
9. Doane, T, Burda, C: Nanoparticle mediated non-covalent drug delivery. Adv. Drug Deliver. Rev. (2012). doi:10.1016/j.addr.2012.05.012
10. Marshall, G, Kasap, C: Adverse events caused by MRI contrast agents: implications for radiographers who inject. Radiography **18**, 132–136 (2012)
11. Mi, P, Cabral, H, Kokuryo, D, Rafi, M, Terada, Y, Aoki, I, Saga, T, Takehiko, I, Nishiyama, N, Kataoka, K: Gd-DTPA-loaded polymer-metal complex micelles with high relaxivity for MR cancer imaging. Biomater. **34**, 492–500 (2013)
12. Morimoto, H, Minato, M, Nakagawa, T, Sato, M, Kobayashi, Y, Gonda, K, Takeda, M, Ohuchi, N, Suzuki, N: X-ray imaging of newly-developed gadolinium compound/silica core-shell particles. J Sol-Gel. Sci. Technol. **59**, 650–657 (2011)
13. Pi, M, Yang, T, Yuan, J, Fujii, S, Kakigi, Y, Nakamura, Y, Cheng, S: Biomimetic synthesis of raspberry-like hybrid polymer–silica core–shell nanoparticles by templating colloidal particles with hairy polyamine shell. Colloids Surf. B **78**, 193–199 (2010)
14. Liu, D, Sasidharan, M, Nakashima, K: Micelles of poly(styrene-b-2-vinylpyridine-b-ethylene oxide) with blended polystyrene core and their application to the synthesis of hollow silica nanospheres. J. Colloid Interface Sci. **358**, 354–359 (2011)
15. Takai, C, Watanabe, H, Asai, T, Fuji, M: Determine apparent shell density for evaluation of hollow silica nanoparticle. Colloids Surf. A **404**, 101–105 (2012)
16. Fuji, M, Shin, T, Watanabe, H, Takei, T: Shape-controlled hollow silica nanoparticles synthesized by an inorganic particle template method. Adv. Powder Technol. **23**, 562–565 (2012)
17. Yang, H, Min, Y, Kim, YJ, Jeong, U: Preparation of $Cu_2O@SiO_2$ particles and their evolution to hollow SiO_2 particles. Colloids Surf. A **420**, 30–36 (2013)
18. Kim, J, Lee, J, Chang, H, Choi, JW, Jang, HD: Synthesis of hollow silica particles with tunable size, shell thickness, and morphology. J. Cryst. Growth (2012). doi:10.1016/j.jcrysgro.2012.09.047
19. Singh, G, Song, L: Experimental correlations of pH and ionic strength effects on the colloidal fouling potential of silica nanoparticles in crossflow ultrafiltration. J. Memb. Sci. **303**, 112–118 (2007)
20. Yilmaz, H, Sato, K, Watari, K: AFM interaction study of α-alumina particle and c-sapphire surfaces at high-ionic-strength electrolyte solutions. J. Colloid Inter. Sci. **307**, 116–123 (2007)
21. Li, SZ, Xu, RK: Electrical double layers' interaction between oppositely charged particles as related to surface charge density and ionic strength. Colloids Surf. A **326**, 157–161 (2008)
22. Kobayashi, Y, Morimoto, H, Nakagawa, T, Gonda, K, Ohuchi, N: Preparation of silica-coated gadolinium compound particle colloid solution and its application in imaging. Adv Nano Res under review
23. Bagher-Ebadian, H, Paudyal, R, Nagaraja, TN, Croxen, RL, Fenstermacher, JD, Ewing, JR: MRI estimation of gadolinium and albumin effects on water proton. NeuroImage **54**, S176–S179 (2011)
24. Liu, Y, Chen, Z, Liu, C, Yu, D, Lu, Z, Zhang, N: Gadolinium-loaded polymeric nanoparticles modified with anti-VEGF as multifunctional MRI contrast agents for the diagnosis of liver cancer. Biomater. **32**, 5167–5176 (2011)

Fruit-mediated synthesis of silver nanoparticles using *Vitis vinifera* and evaluation of their antimicrobial efficacy

Gnanadhas Gnanajobitha[1], Kanniah Paulkumar[1], Mahendran Vanaja[1], Shanmugam Rajeshkumar[1], Chelladurai Malarkodi[1], Gurusamy Annadurai[1*] and Cellapandian Kannan[2]

Abstract

Nanobiotechnology is a rapidly growing scientific field of producing and constructing devices utilizing nanosized particles of about nanometer scale level (1 to 100 nm). Nowadays, nanoparticles have potential effects in life sciences and human health care applications. Among the nanoparticles, silver nanoparticles are playing a major role in the field of biomedical nanotechnology and nanomedicine. Silver is a naturally occurring precious metal, most often as a mineral ore in association with other elements. Silver nanoparticle has a natural antimicrobial effect against many pathogens such as bacteria, fungus, viruses, and yeast. In this present study, the silver nanoparticle is produced using *Vitis vinifera* fruit extract, and the purified nanoparticles will be used as an antibacterial agent against *Bacillus subtilis* and *Klebsiella planticola*. This green chemistry for the biosynthesis of silver nanoparticles has several advantages such as cost-effectiveness and compatibility for biomedical and pharmaceutical applications as well as for large-scale commercial production. Apart from this, it is an eco-friendly process, and compared with microorganisms, plant extract biorecovery unit has an added benefit of ease handling.

Keywords: *Vitis vinifera*; Biosynthesis; Silver nanoparticles; SEM; XRD

Background

Nanotechnology is a developing interdisciplinary field of research interspersing material science, bionanoscience, and technology. Remarkable advances are made in the field of biotechnology and nanotechnology to harness the benefit of life sciences, health care and industrial biotechnology [1-3]. A reliable and eco-friendly process for synthesis of metallic nanoparticles is an important step in the field of nanotechnology. In recent years, noble metal nanoparticles have been the subject of focused research due to their unique optical, electronic, mechanical, magnetic, and chemical properties that are significantly different from those of bulk materials [4]. There is particular interest in nanoparticulate Ag, due to its ability to act as both an electron sink as well as a redox catalyst [5]. Numerous microorganisms and plant extracts have been applied to synthesize inorganic nanostructures either intracellularly or extracellularly. It can be realized by electrostatic interaction between Ag^+ and negatively charged carboxylate groups on the cell surface. In a microorganism the reduction of metal ions occurs on its cell surface by enzymes present in the cell wall [6].The development of green processes for the synthesis of nanoparticles is evolving into an important branch of nanotechnology [7,8]. Several plant biomass or plant extracts have been successfully used for extracellular biosynthesis of silver nanoparticles. The metal ions reduction occurs very rapidly, and the reduction of Ag ions will be completed within hours. Rapid synthesis and excellent yield of silver nanoparticles through this plant-mediated [9-12] biosynthesis have a time-related (2 to 4 h) advantage in comparison with microbial synthesis (24 to 120 h) [13,14]. Therefore green chemistry approaches in nanotechnology for nanoparticle synthesis is used in medical applications, targeted drug delivery, imaging, and dye and heavy metal adsorption. The application of nanoparticles as delivery vehicles for

* Correspondence: annanateam@gmail.com
[1]Environmental Nanotechnology Division, Sri Paramakalyani Centre for Environmental, Sciences, Manonmaniam Sundaranar University, Alwarkurichi 627412, Tamil Nadu, India
Full list of author information is available at the end of the article

bactericidal agents represents a new paradigm in the design of antibacterial therapeutics.

In this present investigation, we developed an inexpensive, versatile, and very reproducible method for large-scale synthesis of silver nanoparticles by reduction process using fruit extract of *Vitis vinifera*. This fruit extract can act both as reducing and stabilizing agents. *V. vinifera* is an important fruit which has phytochemicals such as resveratrol (a polyphenol), and it may have potential use in biomedical applications due to its high antibacterial activity.

Results and discussion

Visual identification

V. vinifera fruit-mediated synthesis of silver nanoparticles is a slow process, when compared with other natural plant materials. Here, the silver nanoparticles synthesis process was started in 1 h and completed in 4 h of incubation (Figure 1). This is due to the increasing concentration of silver nanoparticles as well as the particles' growth in size [15]. The intensity of color is increased when the time of incubation was increased. The color exhibited by metallic nanoparticles is due to the coherent excitation of all "free" electrons within the conduction band, leading to an in-phase oscillation, which is known as surface plasmon resonance [9]. In *V. vinifera* fruit synthesized silver nanoparticles, there are no significant changes beyond 4 h, indicating the completion of the reduction reaction. This result correlated with the earlier reports of *Punica granatum* [16], and *Dioscorea oppositifolia* [17]. At the end of reduction process, there is a dark brown color of silver nanoparticles that settled at the bottom of the conical flask.

UV-visible spectroscopic analysis

In fruit-mediated synthesis, after mixing *V. vinifera* fruit extract in aqueous solution of the silver ion complex, the solution slightly started to change in color from yellow to brown in 1 h of incubation at 37°C. Colloidal solutions of Ag nanoparticles show a very intense color while the time interval was increased from 1 to 4 h. This result correlated with the previous report of Chauhan et al. [16]. It is an indication of silver nanoparticles formation as the color change observed is due to the excitation of surface plasmon vibrations in silver nanoparticles [18]. The broad surface plasmon resonance band observed around 450 to 470 nm (Figure 2) indicates that the particles are large and polydispersed [18,19].

SEM analysis

After the reduction process, the silver nanoparticles are separated by centrifugation of the solution at 15000 rpm for 15 min. Silver nanoparticles were purified by repeated centrifugation for three to four times, and the pellet was dried in hot air oven. Further, the air-dried powder was used to identify the shape and morphology of the nanoparticles. SEM characterizations of the synthesized Ag nanoparticles are shown in Figure 3. It shows relatively spherical-shaped nanoparticles formed with diameters that range from 30 to 40 nm. The nanoparticles were examined under various magnifications of ×16,000, ×25,000, and ×45,000. This result correlated with a previous report obtained using papaya fruit extract which was at a range of 25 to 50 nm [20].

EDX analysis

The fruit extract of *V. vinifera*-synthesized silver nanoparticles also produces a strong signal at 3 keV which reveals the presence of silver nanoparticles. The presence of the strong signal from silver (59.15%) atoms in the nanoparticles and weaker signals from oxygen

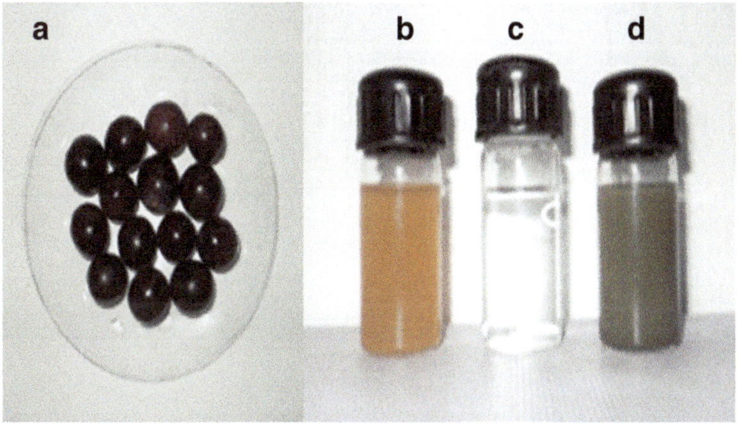

Figure 1 *Vitis vinifera* **fruit (a), fruit extract without (b) and without (d)** $AgNO_3$**, and** $AgNO_3$ **solution (c).**

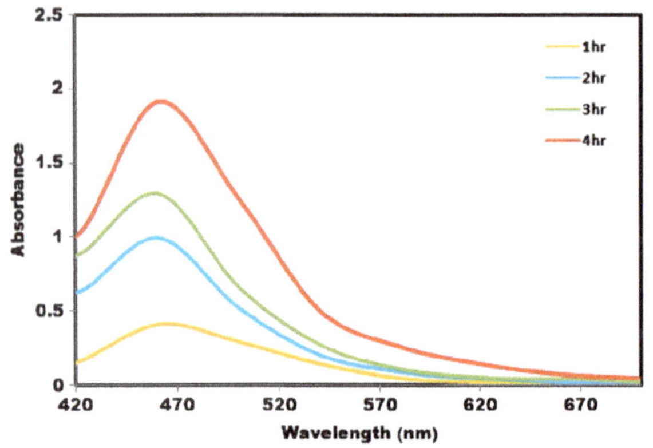

Figure 2 UV-visible absorption spectra of biosynthesized silver nanoparticles from *V. vinifera* depicting a peak at 460 nm.

(27.30%), chlorine (12.49%), and calcium (1.09%) atoms is confirmed (Figure 4). The weaker signals are recorded possibly due to elements (Cl, Ca, and O) from organic moieties like enzymes or proteins present within the extract of the pomegranate fruit [21].

X-ray diffraction analysis

The X-ray diffraction (XRD) pattern clearly shows that synthesized Ag nanoparticles formed are crystalline in nature. The fruit extract-synthesized (*Vitis vinifera*) Ag nanoparticles have shown five intense peaks (Figure 5) in the whole spectrum of 2θ values ranging from 30 to 80. The typical XRD pattern revealed that the sample contains a spherical structure of silver nanoparticles. A number of Bragg reflections corresponding to (0 0 4), (1 1 1), (2 0 0), (2 2 0) and (3 1 1) sets of lattice planes are observed which may be indexed, based on the FCC structure of silver. Peaks are also observed, suggesting that crystallization in the bio-organic phase occurs on the surface of the nanoparticles [22].

Fourier transform infrared spectroscopy

The fruit-synthesized (*V. vinifera*) silver nanoparticles obtained obvious absorption peaks at 3,314, 3,199, 1,671, 1,402, 1,399 and 1,122 cm^{-1}(Figure 6) as well as some intensity peaks that decreased at 2,967, 2,879 and 2,362 cm^{-1}. Figure 6 shows that bands at 3,199 to 3,314 cm-1 correspond to N-H, and O-H stretching vibrations of 1°, and 2° amines, amides, alcohol and H–bonded to phenols. The peak at 1,399 to 1,671 cm^{-1} indicates C-H stretching vibrations of alkene. The peak at 1,122 cm^{-1} represents the C-H in-plane bending of alkenes, alcohols, carboxylic acids, esters, and ethers. The weak band at 2,362 to 2,967 cm^{-1} corresponds to C-H stretching vibrations of alkanes.

Antibacterial activity

The inhibition zone was observed against *Bacillus subtilis* and *Klebsiella planticola*. In the quantitative assay, 10, 20, 30, 40, and 50 μl of Ag nanoparticles were added in five respective wells. The zone of inhibition was measured as 6, 7, 8, 9 and 10 mm respectively in diameter. The zone of inhibition measured is summarized in Table 1. From the table, it is evident that the synthesized nanoparticles are good candidates for their usage in antibacterial drugs. Similarly, the silver nanoparticles synthesized from the extracts of the plant

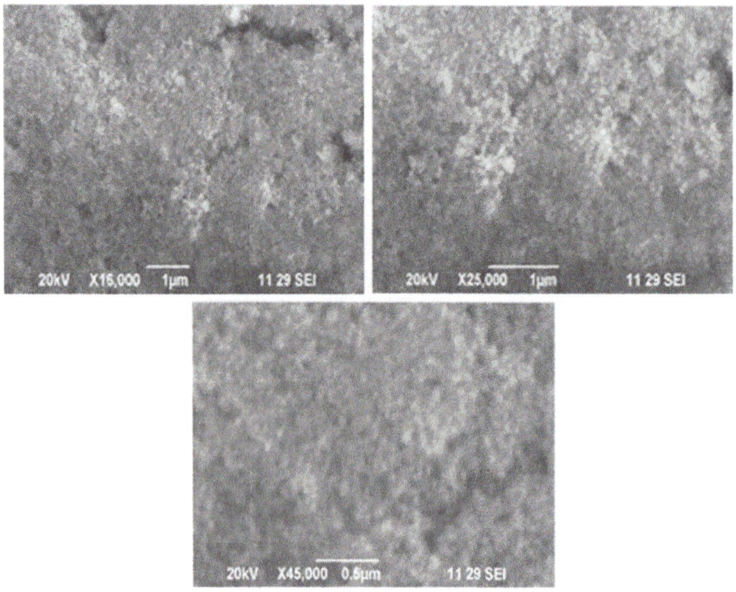

Figure 3 SEM images of silver nanoparticles synthesized from *V.s vinifera* fruit extracts at various magnifications.

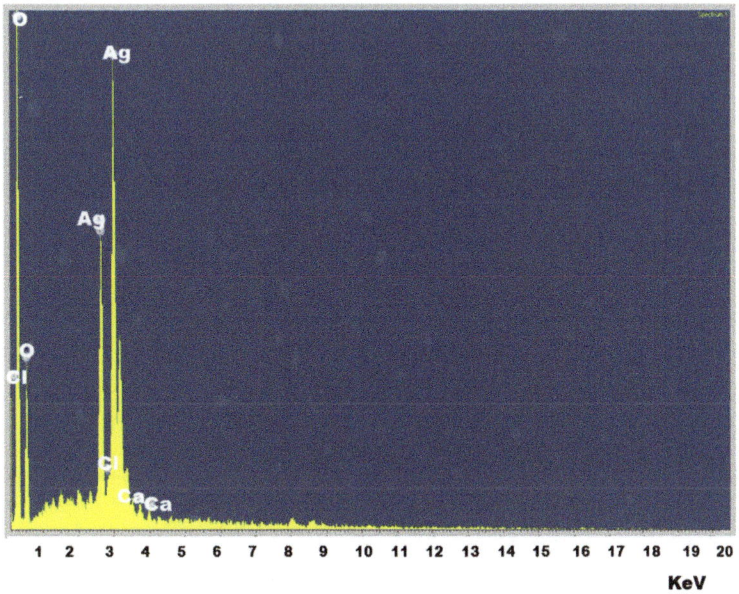

Figure 4 EDX spectra of prepared silver nanoparticles by *V. vinifera*.

barks of *Cinnamomum camphora* plants, are toxic to multi-drug-resistant bacteria, which shows their great potential effect in biomedical applications. Similar observations were found with the *Allium cepa* [23] and *Argimone mexicana* [24].

Conclusions

In the present study, we found that *V. vinifera* is a good source for the synthesis of silver nanoparticles. It was confirmed by the brown color formation of the extract. Silver nanoparticles were synthesized using the present method having 30 to 40-nm average mean sizes with spherical shape. The reduction of silver ions and the stabilization of silver nanoparticles were thought to occur through the participation of fruit proteins and metabolites. Most importantly, the reaction was simple and convenient to handle; it is believed that the *in vitro* phytosynthesis of silver nanoparticles has more advantages over other biological syntheses.

Methods

Chemicals

$AgNO_3$ was purchased from Himedia, Mumbai, India. The bacterial cultures of *B. subtilis* and *K. planticola* were obtained from Microbial Type Culture Collection, Chandigarh, India.

Preparation of fruit extract

Ripened fruit of *V. vinifera* was used for the preparation of the extract. This ripened fruit weighing 25 g was thoroughly washed in distilled water and dried, cut into

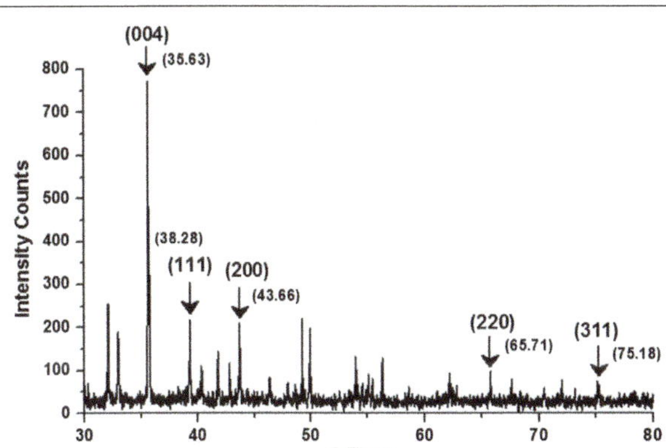

Figure 5 XRD patterns of silver nanoparticles synthesized using *V. vinifera* fruit broth.

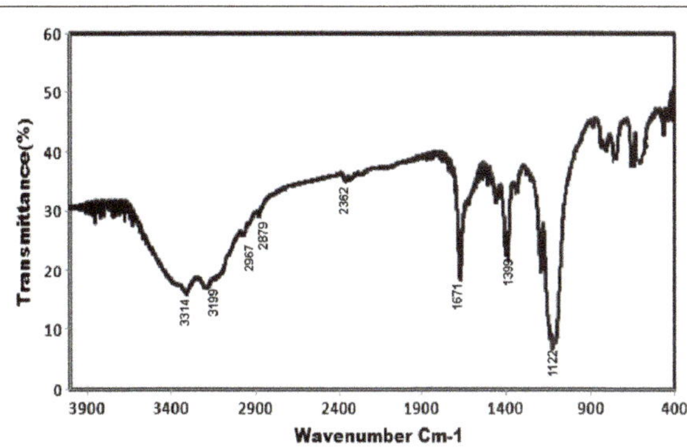

Figure 6 FT-IR spectrum of silver nanoparticles synthesized by reacting $AgNO_3$ with *V. vinifera* fruit extract.

Table 1 Antimicrobial activity of Ag nanoparticles synthesized using the fruit extract of *V. vinifera* (mm)

Concentration of silver nanoparticles (μl)	Bacterial strain (zone of inhibition)	
	B. subtilis	*K. planticola*
10	8	5
20	8	6
30	8.5	8
40	11	9
50	13	11

small pieces, crushed into 100 ml of sterile water, and filtered using Whatman filter paper no. 1; the filtrate was collected and underwent a second filtration using the same type of filter paper, but the pore size was smaller. The filtrate was collected, and this filtrate was used as the sample.

Formation of silver nanoparticle

One millimolar of 95-ml silver nitrate (0.016 g) solution was prepared and kept in a 250-ml Erlenmeyer flask. *V. vinifera* fruit extract (5 ml) was added to the silver nitrate solution. Ninety-five percent of the bioreduction of $AgNO_3$ ions occurred within 4 h. The yellow colored solution which slowly turned brown indicated the formation of silver nanoparticles.

Antibacterial activity

Agar disk diffusion method was used to assay the bactericidal effect of green synthesized silver nanoparticles on Muller-Hinton agar plates against the strains of *B. subtilis* and *K. planticola*. A single colony of test strain was grown overnight in LB broth medium on a rotary shaker (200 rpm) at 35°C. After 24 h of incubation, a loop full of bacterial culture was placed on the Muller-Hinton broth medium. The silver nanoparticles are taken at different concentrations of 10 to 50 μl and are impregnated with a sterile disk placed over the surface of the medium to assess the bactericidal effect of the silver nanoparticles. After incubation at 35°C for 24 to 48 h, the zones of inhibition were measured.

Characterization study of synthesized silver nanoparticles

After completion of the nanoparticle synthesis process, the particle-containing solution was centrifuged to separate the nanoparticles, and this process was repeated two or three times to remove unwanted garbage. It was then dried in a hot air oven at 70°C. The synthesized nanoparticles indicated by UV-visible spectroscopy, were carried out in a Perkin-Elmer spectrophotometer (Branford, CT, USA) operating in a wavelength from 300 to 900 nm. The morphology of the nanoparticles was identified using a scanning electron microscope by Philips (Amsterdam, The Netherlands). The presence of elemental silver in the solution mixture was identified using an energy dispersive X-ray spectrophotometer, and it was operated at an accelerating voltage of 20 kV. The crystallinity of the nanoparticles was evident from the XRD measurements. FT-IR data indicate a bonding of Ag nanoparticles with the functional group of fruit extract through bridging linkage. The antimicrobial activities were done using bacteria like *B. subtilis, K. planticola* for biomedical application.

Competing interests

The authors declare that they have no competing interests.

Authors' contributions

GG carried out the nanoparticles synthesis and characterization. KP carried out the manuscript preparation, MV, SR and CM carried out the antibacterial activity. GA and CK supervised the research work and corrected the manuscript. All authors read and approved the final manuscript.

Authors' information

GG completed her M.Sc. in Biotechnology in Periyar University and obtained her Ph.D. degree in Nanotechnology under the guidance of GA. Her research interests include green-meditated synthesis of nanoparticles and its agricultural applications for controlling plant diseases. KP received his B.Sc. (2005) and M.Sc. in Biotechnology (2007) from Madurai Kamaraj University and Periyar University, India, respectively. He is currently working to obtain his Ph.D. degree at Manonmaniam Sundaranar University in India. He is interested in the green and immobilized microbe-mediated synthesis process of nanoparticles and nanocomposites and their biomedical and textile industry applications. MV obtained her M.Sc. in Environmental Biotechnology in Sri Paramakalyani Centre for Environmental Sciences, Manonmaniam Sundaranar University, Alwarkurichi, Tirunelveli. She obtained her Ph.D. degree under the guidance of GA in the field of nanotechnology. Her research interests include eco-friendly synthesis of nanoparticles and its application in the environment. SR completed his M.Sc. in Biotechnology in Periyar University and obtained his Ph.D. degree in Nanotechnology under the guidance of GA. He published 11 research articles in nanoparticle synthesis. He is interested in the metallic nanoparticles synthesis using algae and algae-derived compounds and their potential applications in the biomedical field. CM completed her B.Sc. in Zoology and M.Sc. in Environmental Biotechnology in SPKCES, Manonmaniam Sundaranar University, Alwarkurichi, Tirunelveli. She achieved her Ph.D. degree under the guidance of GA in the field of nanotechnology. Her research interests include synthesis of semiconductor nanomaterials and metal nanoparticles and their biomedical applications. CK had done his master's degree and doctoral degree in the discipline of Chemistry. His research interests are in the field of green chemistry, environmental chemistry, and synthesis of nanoparticles. Currently, he works as an associate professor in MSU, Tirunelveli. GA received his M.Sc. in Applied Chemistry (1992) and Ph.D. in Environmental Biotechnology (1997) from Anna University, India. He had 10 years (1999 to 2008) of post-doctoral experiences from the National Taiwan University in Taiwan, National Institute of Advanced Industrial Science and Technology in Japan, and National Central University in Taiwan. He has received many research awards from the Indian and other country governments. He is an associate editor in five international journals. At present, he is an associate professor of Environmental Biotechnology and the leader of the Environmental Nanobiotechnology Division at SPKCES, MS University, India. His main research interest is on biosynthesis of nanoparticles and nanomaterials, nanobiocatalyst, and environmental chemistry.

Acknowledgements

The authors gratefully acknowledge the DST-FIST-sponsored program of the Department of Science and Technology, New Delhi, India for funding the research development (DST/FST/ESI-101/2010) to carry out of this work.

Author details

[1]Environmental Nanotechnology Division, Sri Paramakalyani Centre for Environmental, Sciences, Manonmaniam Sundaranar University, Alwarkurichi 627412, Tamil Nadu, India. [2]Department of Chemistry, Manonmaniam Sundaranar University, Tirunelveli 627012, Tamilnadu, India.

References

1. Gardea-Torresdey, JL, Parsons, JG, Gomez, E, Videa-Peralta, J, Troiani, HE, Santiago, P: Formation and growth of Au nanoparticles inside live alfalfa plants. Nano Lett. **2**, 397–401 (2002)
2. Stephen, JR, Macnaughton, SJ: Developments in terrestrial bacterial remediation of metal. Curr Opin Biotchnol. **10**, 230–233 (1999)
3. Lee, HJ, Yeo, SY, Jeong, SH: Antibacterial effect of nanosized silver colloidal solution on textile fabrics. J. Mater. Sci. **38**, 2199–2204 (2003)
4. Mazur, M: Electrochemically prepared silver nanoflakes and nanowires. Electrochem. Commun. **6**, 400–403 (2004)
5. Askari, S, Halladj, R, Nasernejad, B: Characterization and preparation of sonochemically synthesized silver–silica nanocomposites. Mater. Sci. Pol. **2**, 27 (2009)
6. Mukherjee, P, Ahmad, A, Mandal, D, Senapati, S, Sainkar, RS, Khan, IM, Parishcha, R, Ajaykumar, VP, Alam, M, Kumar, R, Sastry, M: Fungus-mediated synthesis of silver nanoparticles and their immobilization in the mycelia matrix. a novel biological approach to nanoparticle synthesis. Nano Lett. **1**, 515–519 (2001)
7. Raveendran, P, Fu, J, Wallen, SL: A simple and green method for the synthesis of Au, Ag, and Au-Ag alloy nanoparticles. Green Chem **8**, 34–38 (2006)
8. Armendariz, V, Gardea-Torresdey, JL, Jose Yacaman, M, Gonzalez, J, Herrera, I, Parsons, JG: Gold nanoparticle formation by oat and wheat biomasses. In: Proceedings of Conference on Application of Waste Remediation Technologies to Agricultural Contamination of Water Resources. Marriott – Country Club Plaza, Kansas City (2002). 30 July 1 August
9. Ankanna, S, Prasad, TK, Elumalai, EK, Savithramma, N: Production Of biogenic silver nanoparticles using *Boswellia ovalifoliolata* stem bark. Digest Journal of Nanomaterials and Biostructures **5**, 369–372 (2010)
10. Basavaraja, S, Balaji, SD, Lagashetty, A, Rajasab, AH, Venkataraman, A: Extracellular biosynthesis of silver nanoparticles using the fungus *Fusarium semitectum*. Mater. Res. Bull. **43**, 1164–1170 (2008)
11. Balaji, DS, Basavaraja, S, Raghunandan, D, Mahesh, D, Venkataraman, A: Biosynthesis and stabilization of Au and Au-Ag alloy nanoparticles by fungus *Fusarium semitectum*. Sci. Technol. Adv. Mater. **9**, 035012 (2008)
12. Balaji, DS, Basavaraja, S, Raghunandan, D, Mahesh, B, Prabhakar, BK, Venkataraman, A: Extracellular biosynthesis of functionalized silver nanoparticles by strains of *Cladosporium cladosporioides* fungus. Coll Surf B: Biointerfaces **68**, 88 (2009)
13. Gardea-Torresdey, JL, Gomez, E, Peralta- Videa, J, Parsons, JG, Troiani, HE, Jose-, Y: Synthesis of gold nanotriangles and silver nanoparticles using *Aloe vera* plant extract. Langmuir **13**, 1357 (2003)
14. Shankar, SS, Rai, A, Ahmad, A, Sastry, M: Rapid synthesis of Au, Ag, and bimetallic Au core–Ag shell nanoparticles using neem (*Azadirachta indica*) leaf broth. J. Colloid Interface Sci. **275**, 496–502 (2004)
15. Fayaz, AM, Balaji, K, Kalaichelvan, PT, Venkatesan, R: Fungal based synthesis of silver nanoparticles-an effect of temperature on the size of particles. Colloids Surf B **74**, 123–6 (2009)
16. Chauhan, S, Upadhyay, MK, Rishi, N, Rishi, S: Phytofabrication of silver nanoparticles using pomegranate fruit seeds. International Journal of Nanomaterials and Biostructures **1**, 17–21 (2011)
17. Maheswari, RU, Lakshmi Prabha, AL, Nandagopalan, V, Anburaja, V, Barath, J: Synthesis of silver nanoparticles by using fruit extract of *Dioscorea oppositifolia* L. Journal of Research in Nanobiotechnology **1**, 009–013 (2012)
18. Kamat, PV, Flumiani, M, Hartland, GV: Picosecond dynamics of silver nanoclusters photoejection of electrons and fragmentation. J. Phys. Chem. B **102**, 3123–3128 (1998)
19. Rani, PU, Rajasekharreddy, P: Green synthesis of silver -protein (core-shell) nanoparticle using *Piper betle* L leaf extract and its ecotoxicological studies on *Daphnia magna*. Colloids and surf A **389**, 188–194 (2011)
20. Jain, D, Daima, HK, Kachhwaha, S, Kothari, SL: Synthesis of plant-mediated silver nanoparticles using papaya fruit extract and evaluation of their anti microbial activities. Digest Journal of Nanomaterials and Biostructures **4**, 557–563 (2009)
21. Xu, H, Kall, M: Modeling the optical response of nanoparticle-based surface plasmon resonance sensors. Sensors and Actuators B: Chemical **4**, 244–249 (2002)
22. Sathyavathi, R, Krishna, MB, Rao, SV, Saritha, R, Rao, N: Biosynthesis of silver nanoparticles using *Coriandrum sativum* leaf extract and their application in nonlinear optics. Adv Sci Letters **3**, 1 (2010)
23. Saxena, A, Tripathi, RM, Singh, RP: Biological synthesis of silver nanoparticles by using onion (*Allium cepa*) extract and their antibacterial activity. Digest. J Nanomater Biostruct **5**, 427–432 (2010)
24. Khandelwal, N, Singh, A, Jain, D, Upadhyay, MK, Verma, HN: Green synthesis of silver nanoparticles using *Argimone mexicana* leaf extract and evaluation of their antimicrobial activities. Digest J Nanomater Biostruct **5**, 483–489 (2010)

A novel method for the production of highly adherent Au layers on glass substrates used in surface plasmon resonance analysis: substitution of Cr or Ti intermediate layers with Ag layer followed by an optimal annealing treatment

Mohammad Ghorbanpour[1] and Cavus Falamaki[2*]

Abstract

The adherence of sputtered Au layers on glass substrates used for surface plasmon resonance (SPR) analysis is very weak. This is usually compensated using a thin Cr or Ti intermediate layer. In this study, a 10-nm sputtered Ag layer (instead of usual Cr or Ti layer) was used as an adhesive intermediate layer between the 40-nm Au layer and the glass substrate. The effect of successive heat treatment at different temperatures on the Au layer adherence and SPR resonance characteristics has been studied. Pull off tests and stability against treatment in boiling water tests have been used for the assessment of Au layer adherence. Based on SPR analysis and pull off and boiling water tests, it is shown that an optimum annealing temperature of 250°C exists for which the Au/Ag/glass substrate composite layer possesses a reasonable adherence along SPR sensitivity.

Keywords: Surface plasmon resonance; Gold; Silver; Annealing

Background

The Au nanofilms deposited on borosilicate glass substrates have gained ever increasing application of modern technologies such as surface plasmon resonance (SPR) biosensors because of their high conductivity and chemical inertness. Fifty nanometers is a conventional thickness for obtaining a proper SPR response from Au/glass composite layers [1,2]. Such nanofilms are usually produced by physical evaporation [3] and chemical vapor deposition of the metal [4], among the other methods.

It is well known that such coating processes suffer from a severe drawback: weak adherence of the Au layer to the substrate. To overcome the aforementioned drawback, two main techniques that have been reported in the literature are used sequentially: (a) implementation of Cr or Ti thin intermediate layers, usually by radio frequency (RF) sputtering and then (b) thermal annealing of the composite Cr(Ti)/Au layer.

To enhance the adherence of the Au layer to the gold substrate, usually a 1- to 5-nm Cr [5-8] or Ti [9-12] layer is used as an intermediate layer. Sexton et al. [9] showed that Ti is preferred over chromium as it less strongly absorbs light and less prone to variations in transmission.

Thermal treatment has been superimposed to the fabrication procedure above to enhance the ultimate layer properties. It is well known that thermal treatment of the gold nanolayer/glass composites decreases the SPR response, especially at high annealing temperatures [2]; this thermal treatment results in the coalescence of the gold particles, increasing the mechanical strength of the gold layer. SPR depends strongly on the nanoparticle layer structure and morphology [2,13,14]. Annealing has a strong influence on the film properties [15], and this may also be used to alter the film properties in a controlled way [16]. Markert et al. [16] showed how the substrate temperature during sputter deposition; a subsequent isochronous annealing process determined the

* Correspondence: c.falamaki@aut.ac.ir
[2]Chemical Engineering Department, Amirkabir University of Technology, P.O. Box 15875-4413 Tehran, Iran
Full list of author information is available at the end of the article

surface topography and the electrical resistivity of thin polycrystalline gold films deposited on float glass. Snopok et al. [8] optimized the thermal treatment procedure for Au films (40 to 60 nm) sputtered on glass substrates containing a 1-nm chromium under layer [8]. They did not perform any stability test on the annealed layers. They realized that the composite layer should not be treated at temperatures exceeding 120°C in order to preserve a reasonable SPR response signal.

3-Mercaptopropyltrimethoxysilane has also been used for improving the adherence of gold to glass substrates [7] due to the chemical affinity of the -SH group with Au. In the latter work, the Au/glass substrate composite layer is further thermally treated at 300°C for 5 to 10 min to reduce the surface roughness and increase the mechanical stability of the Au film.

Zynio et al. [17] studied the effect of using Au/Ag bimetallic layers on the sensitivity of affinity sensors based on surface plasmon resonance. They did not use any Cr or Ti intermediate layer. They aimed at obtaining a higher signal-to-noise ratio by introducing Ag as a primary layer, a high shift of the resonance angle using the second Au layer for a total composite layer thickness of 50 nm. Their work lacks a proper study of the adherence of the bimetallic layer to the substrate under study. We show in the present work that such bimetallic layers suffer from a low adherence to the glass substrate. This is while such composite layers are usually further treated with different chemical agents at different temperatures and may be eroded during the washing in rinsing steps. In other words, it is imperative to improve the mechanical adherence of such bimetallic layers in order to make them industrially attractable.

The aim of the present work is to introduce a new strategy for increasing the adherence and mechanical stability of Au thin films (50-nm thickness) on glass substrates while keeping a reasonable SPR signal response. We show that it is possible to enhance the adherence of the Au layer by implementing an intermediate 10-nm Ag layer, employing an optimal further heat treatment step. As far as the SPR response characteristics are concerned, application of Ag intermediate films is supposed to be superior to Cr and Ti. This is due to the higher value of $|\varepsilon_r/\varepsilon_i|$ (the ratio of the real to imaginary part of the dielectric constants) of Ag with respect to other metals [3].

We have investigated the effect of thermal annealing on the SPR response characteristics and adherence of the Au/Ag composite layer on the glass substrate. To the knowledge of the authors of this work, the adherence of the treated gold layer to its substrate has not been considered in a thorough fashion so far. Szunertis et al. [4] studied the effect of treating the annealed layers in basic and acidic solutions at room temperature on the change in the SPR behavior before and after treatment. This is a rather simple test and does not give a proper insight about the adherence of the layer. To obtain more reliable information on the adherence of the Au layer on the glass substrate, two kinds of layer stability tests have been devised: (a) resistance to boiling water for a predetermined time and (b) pull off tests. The philosophy behind the first test is that in many applications, the gold-coated chips are subjected to thermal treatments therefore creating thermal shocks in aqueous solutions. It is then imperative that the layer withstands thermal shocks as the linear expansion coefficient of the Au layer is substantially different from that of the glass substrate. After treatment with boiling water, the layer should not totally or partially detach from the glass substrate and should keep its SPR response. It is a qualitative test but of relevant importance. The results of the pull off test involve the adherence characteristic of the nanolayer on the glass substrate. Pull off tests are widely used for assessing the interfacial delamination properties of thin layers and coatings of optical devices [18], and the theory behind has been thoroughly elaborated. In our case, they provide a meaningful interpretation of the adherence of the treated nanolayer for withstanding conventional mechanical manipulations.

The SPR behavior of Au/glass substrate composite layers (as-synthesized and annealed at different temperatures), and their stability is presented in the first part of the present work. This is essential for a rational discussion of the second part of this work. In the second part, the same characteristics of the Au/Ag/glass substrate composite layers are presented and compared with the results of the first part.

Results and discussion

Figure 1a,b,c shows the AFM topologies of the as-deposited and annealed Au layer films at 250°C and 400°C for 60 min. The scan area of the images is $2{,}500\times2{,}500$ nm^2. Table 1 shows the corresponding roughness values. It is observed that the average surface roughness of the gold film gradually increases with increasing the annealing temperature. This is mainly attributed to some sort of sintering of the original gold particles and the creation of larger particles. An eightfold increase of the average surface roughness is observed in the sample treated at 400°C with respect to the as-deposited sample. However, for an annealing temperature of 100°C, the surface roughness remains practically intact. A clear change in the surface roughness and surface topology is observed for annealing temperatures higher than 200°C. The trend described above is well known in the art. Grain mobility and coalescence is considered the predominant mechanism responsible for the morphological change induced by annealing [19].

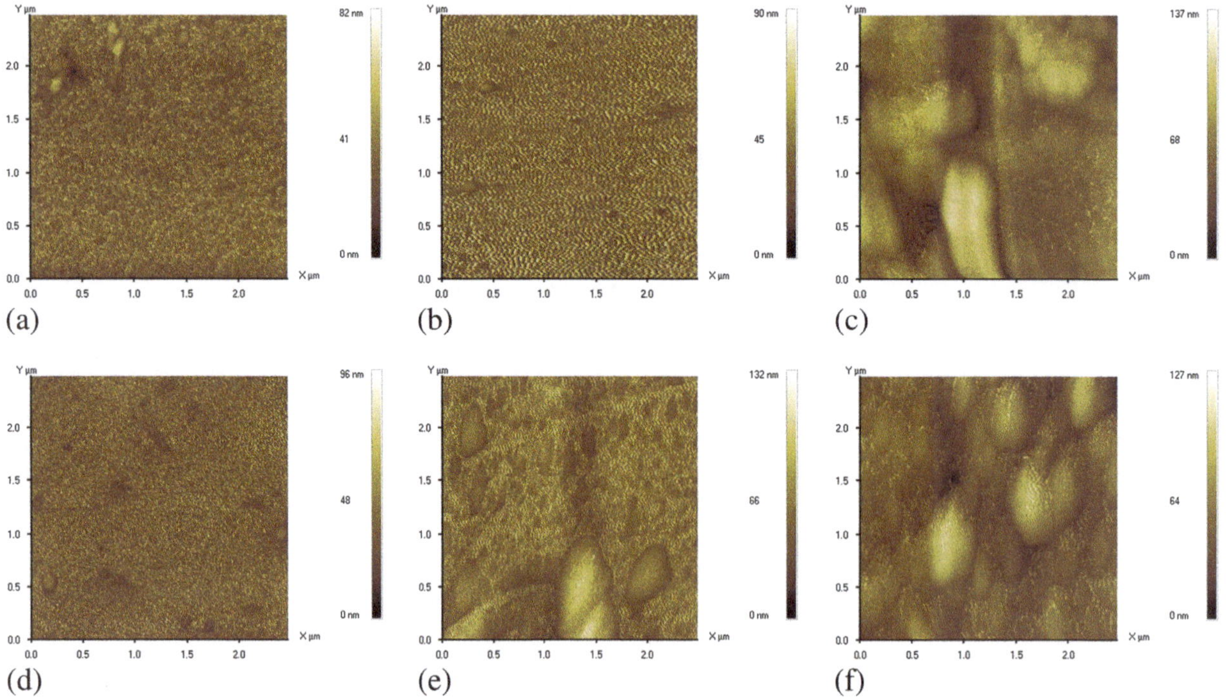

Figure 1 AFM images of (a,b,c) Au/glass and (d,e,f) Au/Ag/glass composite. **(a)** as-prepared, **(b)** annealed at 250°C, and **(c)** annealed at 400°C; **(d)** as-prepared **(e)** annealed at 250°C, and **(f)** annealed at 400°C.

Apart from the mere sintering of the Au particles, the authors of this work present an additional mechanism for explaining the improvement of the adherence of Au to the glass substrate. This is schematically shown in Figure 2. Referring to this figure, the effect of heat treatment is not merely the coalescence of the gold particles. Another phenomenon also takes place: the gold particles undergo a better wetting of the hills of the glass substrate (Ra ≈ 0.34 nm), resulting in a better 'entanglement' of gold particles with the glass protuberances. The latter phenomenon results in a better adherence of the gold thick film to the glass substrate.

The effect of annealing temperature on the SPR response and transmission mode absorbance of the Au coated samples is shown in Figure 3. It is observed that

Table 1 Average roughness (Ra) of metal surface for the Au/glass and Au/Ag/glass composite layers in their original and annealed states

State	Au/glass (nm)	Au/Ag/glass (nm)
Original	9.65	9.76
Annealed at 100°C	9.86	9.94
Annealed at 150°C	10.11	10.90
Annealed at 200°C	11.19	13.03
Annealed at 250°C	12.99	13.39
Annealed at 300°C	14.24	15.02
Annealed at 400°C	16.33	15.58

the resonance angle and percent absorbance remain approximately constant upon heat treatment at 100°C and 150°C. Actually, the change in the depth of the SPR curve is 0.0% and 1.8% for annealing temperatures of 100°C and 150°C, respectively (Figure 3b). Further increase of the annealing temperature reduces sharply the SPR response. This is in accordance with the trend of topology change with increasing annealing temperature. The 'sintering' of the gold particles is pronounced at temperatures equal or higher than 200°C. The absorbance undergoes significant reduction at an annealing temperature of 300°C and assumes its smallest value at 400°C throughout the wavelength span under consideration. Such a behavior has been reported before [2] and is attributed to the formation of hills and valleys due to the coalescence of gold particles. The hills provide 'low absorbance' regions for the passing light beam thus resulting in lower absorption values.

At this stage, it is noteworthy to consider the optical characteristics of the samples coated with 10 nm Ag and 40 nm Au. The SPR response curves (Figure 3) show the same trend as the Au film as a function of annealing temperature. The change in the SPR response is 0.0 and 4.8% at 100°C and 150°C, respectively. The attenuation of the SPR response is somewhat stronger at 150°C for the composite layer compared to the Au/glass composite layer (1.8%). The SPR signal undergoes a 28% reduction at 200°C, being still sharp. The SPR response vanishes at

Figure 2 Scheme showing proposed mechanism for explaining improved adherence of Au to substrate due to annealing.

400°C. The optical behavior as a function of temperature has a peculiarity not observed for the Au/glass composite layer. The SPR angle of Au/Ag/glass composite changes significantly with increasing the annealing temperature. Figure 3b shows the latter phenomenon. Interestingly, the rate of change of the SPR angle has a maximum at 200°C.

Figure 1d,e,f shows the surface morphology of the as-deposited, annealed at 250°C, and annealed at 400°C Au/Ag composite layers, respectively. When compared with Figure 1a,b,c, it is observed that the introduction of the Ag as an intermediate layer results in a completely different sintering behavior of the gold particles. Diffusion of Ag atoms into the Au film prevents the formation of large Au particles and results in the formation of a well-patterned surface lattice-like collection of spindly entities. It is not clear at this stage if the elimination of the SPR signal is due to a different surface

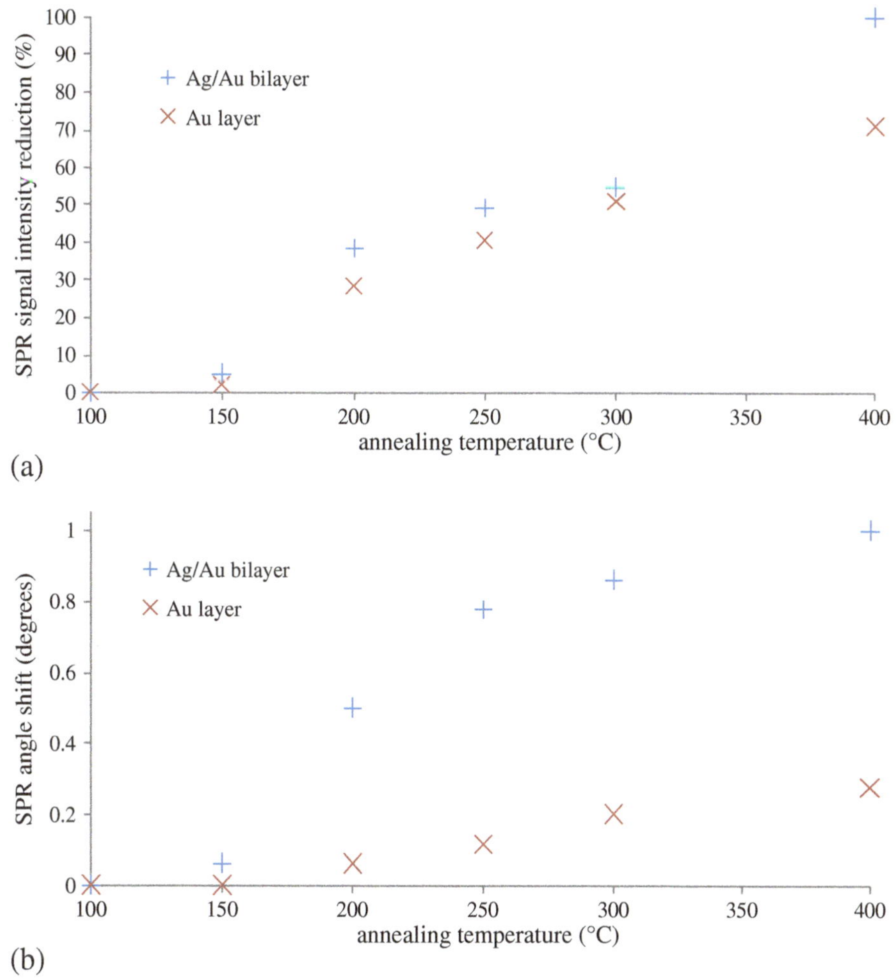

Figure 3 SPR angle (a) and depth (b) of Au/glass and Au/Ag/glass composite layers. Composite layers as a function of annealing temperature.

topology or because of different bulk properties of the composite layer.

The layer stability characteristics of the Au/glass composite layers as a function of annealing temperature is summarized in Figure 4. It is observed that the original Au/glass composite layer does not pass the boiling water stability test. Heat treatment in the temperature range of 100°C to 400°C results in Au/glass composite layers that pass this test while the original Ag/Au composite film passed the latter test. In other words, the mere introduction of Ag as an intermediate layer enhances significantly the adherence of Au to the glass substrate. This mechanical characteristic is maintained after thermal annealing up to 400°C.

Referring to Figure 5, it is observed that the SPR response of the Au/Ag/ glass composite layers follow a different trend compared to the Au/glass layers. For the Au/glass composite layers, only the sample treated at 400°C passes this test and shows no SPR response (Figure 5a). According to Figure 5, heat treatment of the Au/Ag/glass composite layers in the temperature range 100°C to 200°C enhances the adherence of the layer while keeping a reasonable SPR output signal characteristic. It is observed that an annealing temperature of 250°C results in intact composite layers after performing the pull off test. This is while the SPR signal is still of a reasonable intensity ca. 40% reduction with respect to no annealing (Figure 5b). Composite layers heat-treated at temperatures higher than 250°C posses a reasonable adherence but show a weak SPR characteristic.

Conclusion

Summed up, it may be stated that the mere introduction of a 10-nm intermediate layer enhances significantly the Au layer adherence to the glass substrate while the Au/Ag/substrate composite shows acceptable SPR characteristics. However, this is not enough as the composite layer chip should withstand sever eventual physical successive treatments during fictionalization processes prior to SPR analysis. The last results showed that such a resistance may be achieved only by an optimal treatment method of the as-synthesized Au/Ag/glass composite layer.

State	Au/Ag/glass		Au/glass	
	Boiling water test	Pull off test	Boiling water test	Pull off test
Original	Passed		Not passed	
Annealed at 100°C	Passed		Passed	
Annealed at 150°C	Passed		Passed	
Annealed at 200°C	Passed		Passed	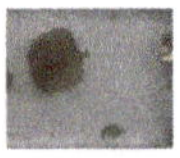
Annealed at 250°C	Passed		Passed	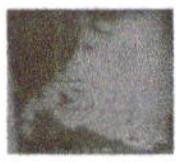
Annealed at 300°C	Passed		Passed	
Annealed at 400°C	Passed		Passed	

Figure 4 Stability tests of metal surface for Au/glass and Au/Ag/glass composite layers in original and annealed states.

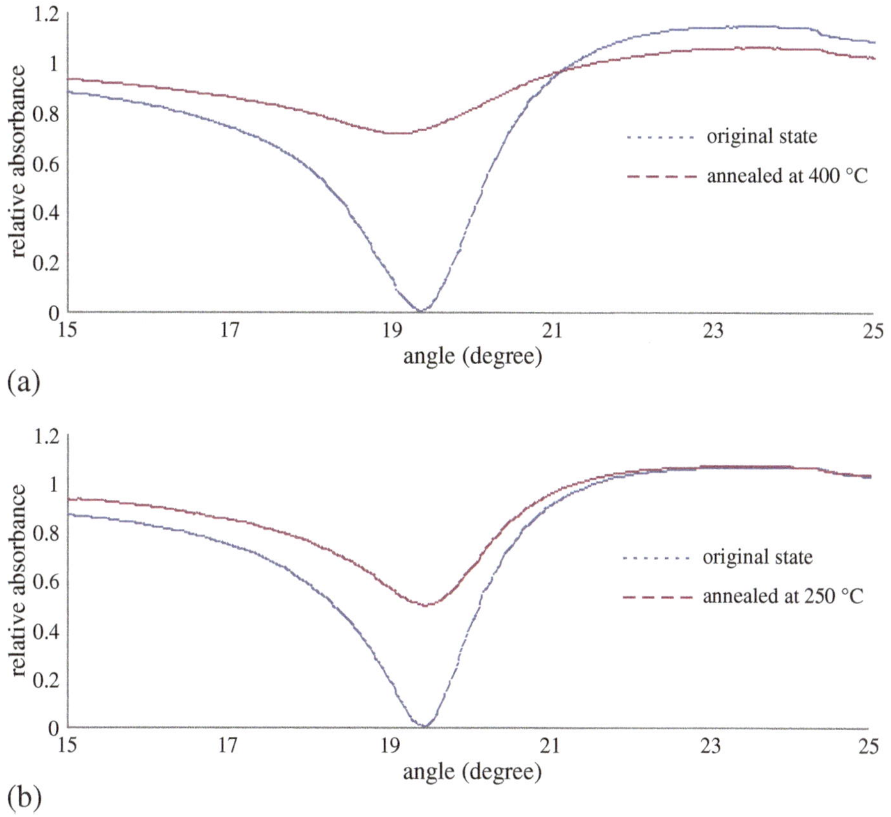

Figure 5 SPR response of Au/glass composite layers annealed at 400°C (a) and Au/Ag/glass composite layers annealed at 250°C (b).

Methods

H_2O_2 (30 wt.%), H_2SO_4 (94 wt.%) were purchased from Merck Company (Whitehouse Station, NJ, USA). Triply distilled water was used. The substrate surfaces were initially and thoroughly treated in piranha solution. This pretreatment results in an approximately complete cleaning of the surface. In addition, it results in the change of the chemical structure of the exposed glass surface, eventually increasing the concentration of silanol surface groups. Glass substrates (microscope slides, soda-lime glass) were cleaned with piranha solution (3:1 vol. ratio, H_2SO_4/H_2O_2) for 30 min at 80°C. After cooling, the samples were rinsed with water and dried under nitrogen gas flow.

Au layers with a thickness of about 50 nm were sputter-deposited on cleaned glass slides (7×7×1 mm^3 pieces) using a BAL-TEC SCD 005 DC sputtering system. The surface roughness (Ra) of the initial slides was 0.34 nm. No intermediate Cr or Ti layer was used. Sputtered metal thickness was determined using a Stylus Profilometer DEKTAK3 (Veeco, Plainview, NY, USA) instrument. In the case of Au/Ag films, the glass substrates were coated with a 10-nm Ag film in the first step and with a 40-nm Au film in the second step.

The samples were heated to the desired temperature (100°C, 150°C, 200°C, 250°C, 300°C, 400°C) using an electrical furnace with a heating rate of 5°C/min in ambient air and treated for 1 h.

The analysis of the surface topology of the original and treated gold films was performed by a Nanoeducator AFM apparatus using a non-contact mode. Transmission-mode absorption spectra were obtained using an Ocean Optics UV-visible spectrometer (Dunedin, FL, USA). An Iranian SPR instrument (Nano SPR) was used to investigate the SPR characteristics of the original and treated samples [20]. The resolution of this version of the SPR determination apparatus was 0.01°. The goniometer of the nano SPR system worked with a θ to 2θ geometry. A 2-mW He-Ne gas source was used for producing the laser beam (632.8 nm). Glycerin was used as an interface layer between the borosilicate glass prism and the glass slides.

The layer stability was evaluated by placing the sample in a 200-cm^3 beaker containing 50 cm^3 boiling water for 15 min. Partial or full detachment of the gold layer was considered as a failed experiment. If the sample remained constant and produced the original SPR response with less than 0.02° change in the SPR angle, the experiment was considered successful. Each stability test was repeated twice.

The pull off test was performed as follows: an adhesive tape was put horizontally on a holder, then the composite layer ($7 \times 7 \times 1$ mm^3) was put on it so that the metal came into contact with the tape. Afterwards, a 100-g standard weight was put on the glass side of the slide for 30 s. Afterwards, the weight was raised and the slide was detached manually from the tape. Each pull off test was repeated twice.

Competing interests
The authors declare that they have no competing interests.

Authors' contributions
Both authors read and approved the final manuscript.

Author details
[1]Chemical Engineering Department, University of Mohaghegh Ardabili, Ardabil, Iran. [2]Chemical Engineering Department, Amirkabir University of Technology, P.O. Box 15875-4413 Tehran, Iran.

References

1. Huanga, Y, Qiua, H, Wanga, F, Pana, L, Tiana, Y, Wu, P: Effect of annealing on the characteristics of Au/Cr bilayer films grown on glass. Vacuum. **71**, 523–528 (2003)
2. Serrano, A, Rodríguez de la Fuente, O, García, MA: Extended and localized surface plasmons in annealed Au films on glass substrates. J Appl Phys **108**, 074303–074310 (2010)
3. Mitsushio, M, Miyashita, K, Higo, M: Sensor properties and surface characterization of the metal-deposited SPR optical fiber sensors with Au, Ag, Cu, and Al Sens. Actuators A **125**, 296–303 (2006)
4. Szunertis, S, Coffinier, Y, Janel, S, Boukherroub, R: Stability of the gold/silica thin film interface, electrochemical and surface plasmon resonance studies. Langmuir. **22**, 10716–10722 (2006)
5. Parisi, A, Cino, AC, Busacca, AC, Cherchi, M, Riva-Sanseverino, S: Integrated optic surface plasmon resonance measurements in a borosilicate glass substrate. Sensors. **8**, 7113–7124 (2008)
6. Jang, LS, Liu, HJ: Fabrication of protein chips based on 3-aminopropyltriethoxysilane as a monolayer. Microdevices. **11**, 331–338 (2009)
7. Hirlekar Schmid, A, Stanca, SE, Thakur, MS, Ravindranathan Thampi, K, Raman Suri, C: Site-directed antibody immobilization on gold substrate for surface plasmon resonance sensors Sens. Actuators **113**, 297–303 (2006)
8. Snopok, BA, Kostyukevich, KV, Lysenko, SI, Lytvyn, PM, Lytvyn, OS, Mamykin, SV, Zynyo, SA, Shepeliavyij, PE, Kostyukevich, SA, Shirshov Yu, M, Venger, EF: Optical biosensors based on the surface plasmon resonance phenomenon: optimization of the metal layer parameters. Semicond Phys Quantum Electron Optoelectron **4**, 56–69 (2001)
9. Sexton, BA, Feltis, BN, Davis, TJ: Characterisation of gold surface plasmon resonance sensor substrates Sens. Actuators A **141**, 471–475 (2008)
10. Hoogvliet, JC, van Bennekom, WP: Gold thin-film electrodes: an EQCM study of the influence of chromium and titanium adhesion layers on the response. Electrochimica Acta. **47**, 599–611 (2001)
11. Szunerits, S, Boukherroub, R: Electrochemical investigation of gold/silica thin film interfaces for electrochemical surface plasmon resonance studies. Electrochemistry Communications. **8**, 439–444 (2006)
12. Kim, JM, Kim, SH, Ohashi, T, Muramatsu, H, Chang, SM, Kim, WS: Construction of simultaneous SPR and QCM sensing platform bioprocess. Biosyst Eng **33**, 39–45 (2010)
13. Yu, K, Lance, KK, Nobuyuki, S, Tetsu, T: Morphologies and surface plasmon resonance properties of monodisperse bumpy gold nanoparticles. Langmuir. **24**, 5849–5854 (2008)
14. Armelao, L, Barreca, D, Bottaro, G, Bruno, G, Gasparotto, A, Losurdo, M, Tondello, E: RF-sputtering of gold on silica surfaces: Evolution from clusters to continuous films. Mat Sci Eng **2005**(25), 599–603 (2005)
15. Švor˘cík, V, Siegel, J, Šutta, P, Mistrík, J, Janí˘cek, P, Worsch, P, Kolská, Z: Annealing of gold nanostructures sputtered on glass substrate. Appl Phys A **102**, 605–610 (2011)
16. Markert, C, Lutzenkirchen-Hecht, D, Fraham, R: Structural and electrical properties of thin d.c.magnetron-sputtered gold films deposited on float glass Structural. Surf Interface Anal **38**, 715–718 (2006)
17. Zynio, SA, Samoylov, AV, Surovtseva, ER, Mirsky, VM, Shirshov, YM: Bimetallic layers increase sensitivity of affinity sensors based on surface plasmon resonance. Sensors. **2**, 62–70 (2002)
18. Zuo, S, Wan, KT, Dillard, DA: A theoretical and numerical study of thin film delamination using the pull-off test. Int. J Sol Struc **4**, 717–730 (2004)
19. Lee, KC, Lin, SJ, Lin, CH, Tsai, CS, Lu, YJ: Size effect of Ag nanoparticles on surface plasmon resonance. Surf. Coat Techn **202**, 5339–5342 (2008)
20. Sahebghadam Lotfi, A, Falamaki, C, Pourghafoor, B, Sadeghi, A, Faraji, F: Iranian Patent 68727. (2011)

Synthesis, characterization, and investigation of optical and magnetic properties of cobalt oxide (Co_3O_4) nanoparticles

Saeed Farhadi*, Jalil Safabakhsh and Parisa Zaringhadam

Abstract

Spinel-type cobalt oxide (Co_3O_4) nanoparticles have been easily prepared through a simple thermal decomposition route at low temperature (175°C) using carbonatotetra(ammine)cobalt(III) nitrate complex, $[Co(NH_3)_4CO_3]NO_3{\cdot}H_2O$, as a new precursor. The structure and morphology of as-prepared Co_3O_4 nanoparticles were characterized by Fourier transform infrared (FT-IR) spectroscopy, X-ray diffraction (XRD), transmission electron microscopy (TEM), energy-dispersive X-ray spectroscopy (EDS), UV–vis spectroscopy, Brunauer-Emmett-Teller specific surface area measurement and magnetic measurements, and thermogravimetry/differential thermal analysis. The FT-IR, XRD, and EDS results indicated that the product was highly pure well-crystallized cubic phase of Co_3O_4. The TEM images showed that the product powder consisted of dispersive quasi-spherical particles with a narrow size distribution ranged from 6 to 16 nm and an average size around 11 nm. The magnetic measurements confirmed that the Co_3O_4 nanoparticles show a little ferromagnetic behavior which could be attributed to the uncompensated surface spins and/or finite size effects. The ferromagnetic order of the Co_3O_4 nanoparticles is raised with increasing the decomposition temperature. Using the present method, Co_3O_4 nanoparticles can be produced without the need of expensive organic solvents and complicated equipments.

Keywords: Co_3O_4 nanoparticles; Soft chemical methods; Co(III) amine complex; Thermal decomposition; Optical properties; Magnetic properties

Background

Nanometer-scale materials with the size of 1 to 100 nm have attracted considerable interest in recent years due to the departure of properties from bulk phases arising from quantum size effects [1]. Spinel-type cobalt oxide (Co_3O_4) is a technologically important material with applications in lithium ion batteries, heterogeneous catalysts, gas sensing, ceramic pigments, and electrochemical devices [2-11]. This transition metal oxide, when falling in the nanosized regime, is expected to lead to even more attractive applications in the conjunction of their traditional arena and nanotechnology.

In recent years, many efforts have been devoted to the synthesis of Co_3O_4 nanostructures with different morphologies such as nanoparticles, hollow spheres, nanorods, nanoplates, nanowires, nanotubes, and nanocubes, and nanoporous structures have been prepared [12-22]. Among them, Co_3O_4 nanoparticles have been prepared by various physical and chemical techniques such as combustion method, microwave irradiation, hydrothermal/solvothermal method, sol–gel process, chemical spray pyrolysis, sonochemical method, polyol method, and so on [23-37]. Most of these methods need some special instruments, harsh conditions, and relatively high processing temperature higher than 350°C. In addition, these methods are either time-consuming or require expensive instruments.

Among various soft chemical methods for preparing nanoscale materials, the thermal decomposition method is widely used due to the process simplicity [38-40]. This technique offers several unique advantages over other methods including easy work-up, low temperature processing, short reaction time, and production of inorganic nanomaterials with narrow size distribution. In recent years, several precursors have been used to synthesize Co_3O_4 nanoparticles via the thermal decomposition

* Correspondence: sfarhadi48@yahoo.com
Department of Chemistry, Lorestan University, Khoramabad 68135-465, Iran

technique [41-45]. However, the most important issue in this technique is to design a precursor which would allow the synthesis of nanomaterials at a low temperature.

In this work, we wish to describe the thermal decomposition of the carbonatotetra(ammine)cobalt(III) nitrate complex, $[Co(NH_3)_4CO_3]NO_3{\cdot}H_2O$, which resulted in the synthesis of Co_3O_4 nanoparticles at rather low temperature (175°C). The product was identified by various instrumental techniques such as X-ray diffraction (XRD), Fourier transform infrared (FT-IR), transmission electron microscopy (TEM), energy-dispersive X-ray spectroscopy (EDS), thermogravimetry/differential thermal analysis (TG/DTA), UV–vis spectroscopy, Brunauer-Emmett-Teller (BET) surface area measurement, and magnetic measurement. This approach provides a one-step, simple, general, and inexpensive method for the preparation of the Co_3O_4 nanoparticles.

Results and discussion

The FT-IR spectra of the $[Co(NH_3)_4CO_3]NO_3{\cdot}H_2O$ complex and its decomposition products at different temperatures are shown in Figure 1. For the complex (Figure 1, curve a), the characteristic stretching bands of NH_3, CO_3, and NO_3 groups are observed at approximately 3,250 to 3,500, 1,600, and 1,350 cm^{-1}, respectively [46]. As shown in Figure 1, curve b, the intensity of these bands decreases when the complex is heated at 150°C. At this temperature, there are two small absorption bands (black square) at about 663.47 and 572.82 cm^{-1}, providing clear evidence for the presence of the crystalline Co_3O_4 [12]. This observation confirms that the formation of Co_3O_4 nanocrystals begins at approximately 150°C. As can be clearly seen in Figure 1, curve c, with increasing the decomposition temperature to 175°C, only two strong bands assigned to the Co-O stretching of the cubic Co_3O_4 structure are observed [47]. As shown in Figure 1, curves d,e,f, FT-IR spectra of the samples that were decomposed in the 200°C to 300°C range show only the bands related to the Co_3O_4 without obvious changes. It is noted that the bands at approximately 3,550 and 1,650 cm^{-1} in the FT-IR spectrum of some samples should be assigned to the stretching and bending vibrations of the water molecules

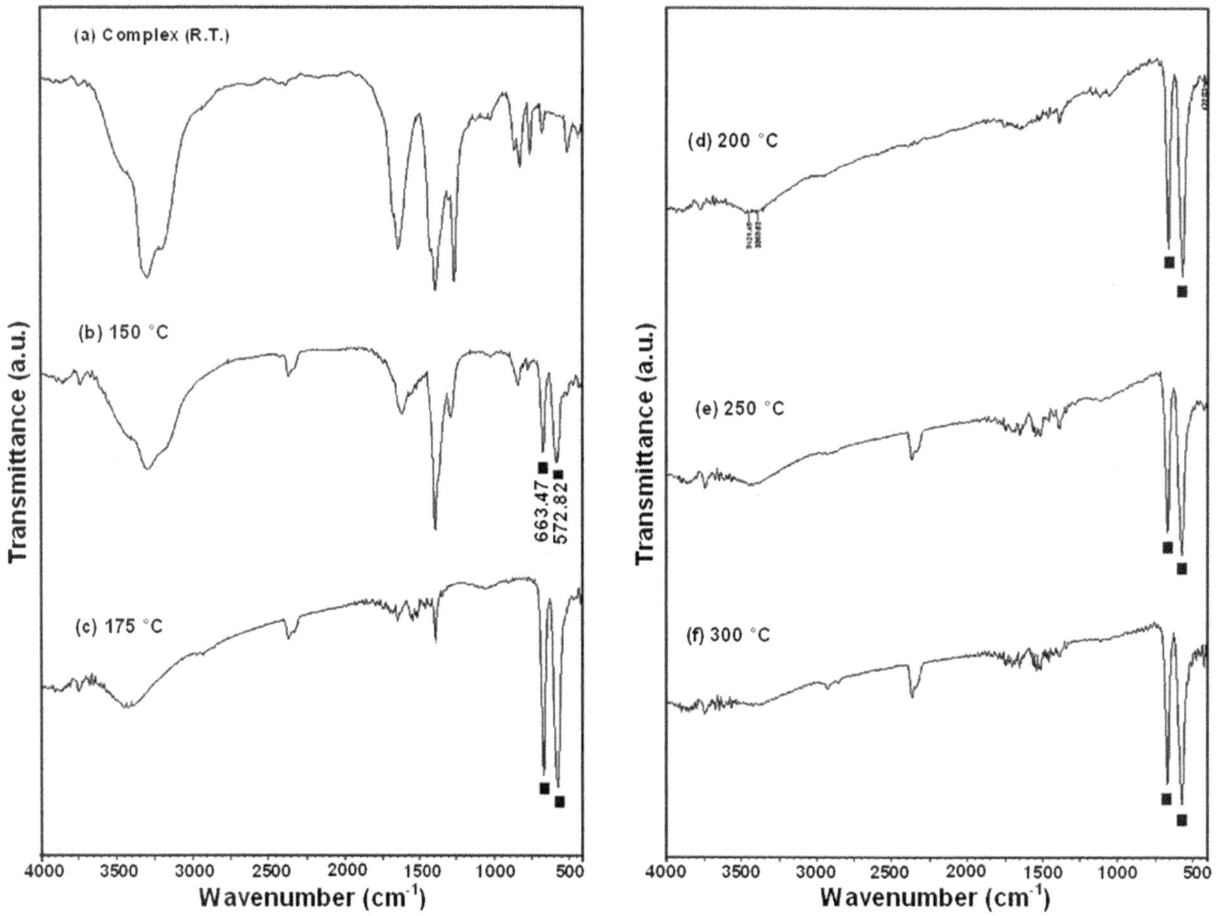

Figure 1 FT-IR spectra of the $[Co(NH_3)_4CO_3]NO_3{\cdot}H_2O$ complex and its decomposition products at given temperatures. Black squares show the bands related to the Co_3O_4 phase.

absorbed by the samples or KBr. Also, there is a tiny band at approximately 2,360 cm^{-1} on the spectrum of some samples due to the presence of atmospheric CO_2.

The XRD patterns of the $[Co(NH_3)_4CO_3]NO_3{\cdot}H_2O$ complex and its decomposition products at various temperatures are shown in Figure 2. Figure 2a shows the XRD pattern of $Co(NH_3)_4CO_3]NO_3{\cdot}H_2O$ precursor. All diffraction peaks appeared in this pattern match very well with those reported in literature for the pure $[Co(NH_3)_4CO_3]NO_3{\cdot}H_2O$ (JCPDS 50–1674). As can be seen in Figure 2b, all diffraction peaks related to $[Co(NH_3)_4CO_3]NO_3{\cdot}H_2O$ disappeared at 150°C and new broad and weak peaks with 2θ values at 19.50°, 31.37°, 37.02°, 39.10°, 44.97°, 55.84°, 59.58°, 65.46°, and 77.62° appeared. These diffraction peaks can be indexed to the crystalline cubic phase Co_3O_4 with lattice constants of a = 8.076 Å and a space group of Fd3 m, which are in agreement with the reported values (JCPDS card no. 76–1802). As shown in Figure 2c, the intensity of the characteristic peaks of the Co_3O_4 phase increases markedly as the temperature increases to 175°C, confirming the complete formation of the Co_3O_4 phase in good agreement with the FT-IR result. No characteristic peaks of other impurity phases have been detected, indicating that the final product is of high purity. The considerable broadening of the diffraction peaks demonstrates the nanometric character of the particles. The average size of particles prepared was estimated to be about 11.5 nm by the Debye-Scherrer equation [48]: $D_{XRD} = 0.9\lambda/(\beta\cos\theta)$ where D_{XRD} is the average crystalline size, λ is the wavelength of CuKα, β is the full width at half maximum of the diffraction peak, and θ is the Bragg's angle. As we can see in Figure 2d, e,f, no new phase is observed when the decomposition temperature increases to 200°C, 250°C, and then 300°C, but the width of the Co_3O_4 peaks decreases because of crystallite growth. Further, the increase in the sharpness of peaks with increasing temperature indicates the well crystallization of the product.

The size and shape of the Co_3O_4 particles prepared by the thermal decomposition of the $[Co(NH_3)_4CO_3]$

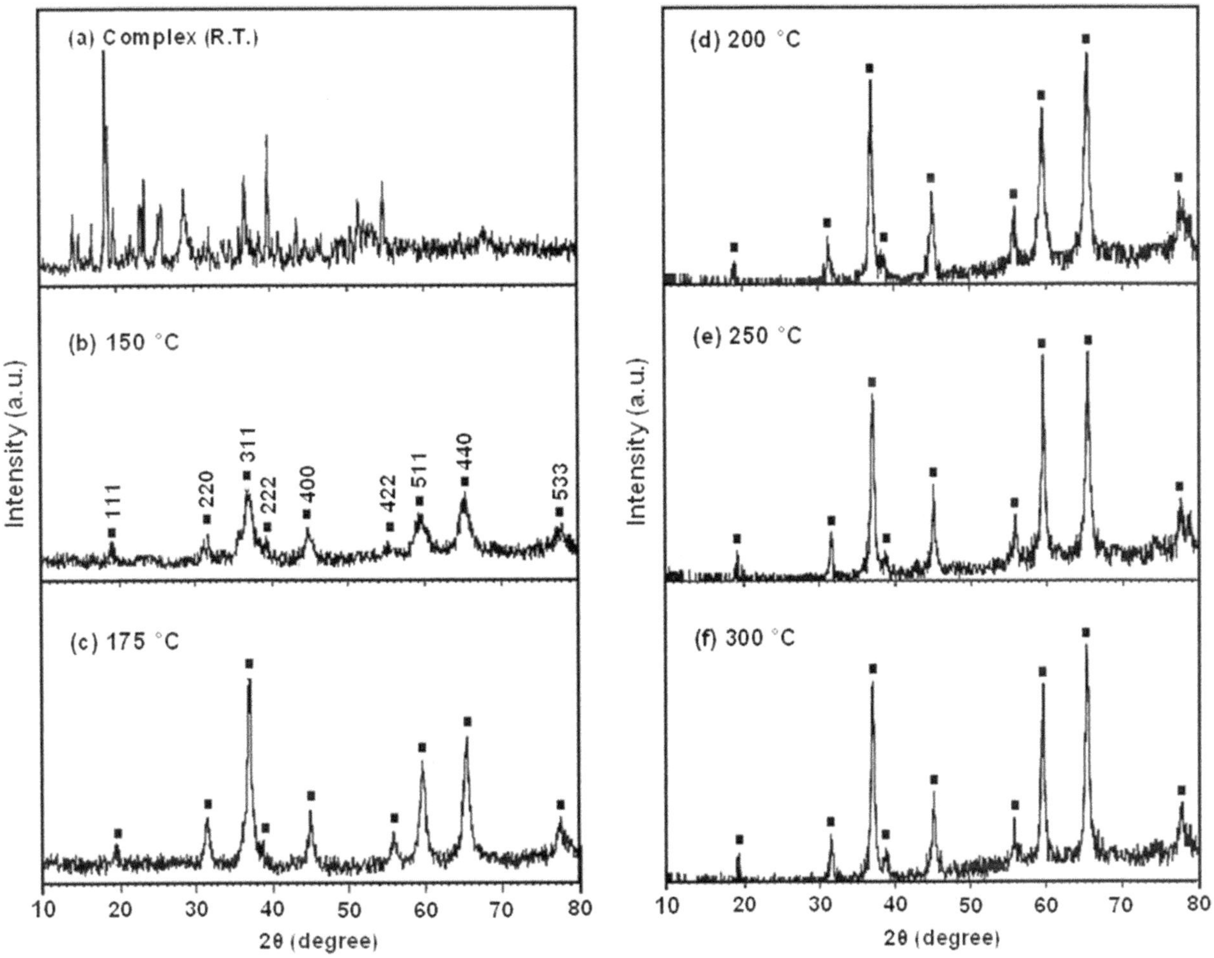

Figure 2 XRD patterns of the $[Co(NH_3)_4CO_3]NO_3{\cdot}H_2O$ complex and its decomposition products at given temperatures. Black squares show the diffraction peaks of the Co_3O_4 phase.

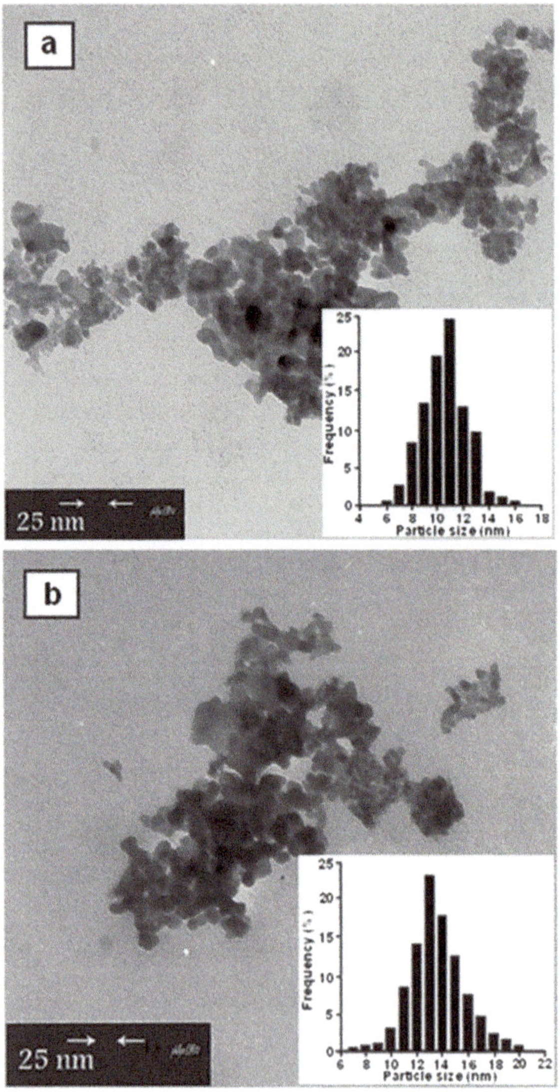

Figure 3 TEM images of the Co_3O_4 nanoparticles prepared at (a) 175°C and (b) 200°C. Insets show particle size distribution histograms determined from TEM images.

$NO_3{\cdot}H_2O$ complex at 175°C and also at 200°C were investigated by TEM, as shown in Figure 3a,b. In both cases, the TEM analysis revealed that the samples were formed from extremely fine particles with the sizes of less than 20 nm. It is evident that the particles have uniform size, a homogeneous sphere-like morphology and a narrow size distribution. From the TEM images, it could be concluded that this preparation method is appropriate to obtain the Co_3O_4 nanoparticles with very small size. The size distribution of the Co_3O_4 nanoparticles has also been investigated from the particles visualized under TEM analysis. The particle size histograms were determined by counting more than 100 particles in randomly selected regions on the TEM copper grid. The histograms based on TEM analysis are shown in the

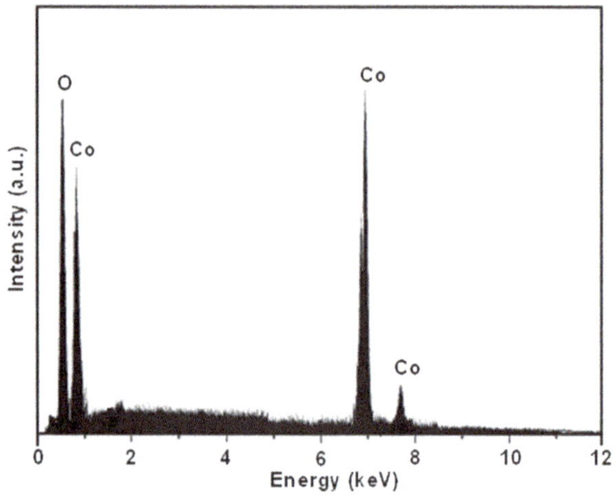

Figure 4 EDS spectrum of the Co_3O_4 nanoparticles prepared at 175°C.

insets of Figure 3. The inset of Figure 3a shows the particle size distribution of the Co_3O_4 nanoparticles prepared at 175°C. It can be seen that the particles possess a narrow size distribution in the range of 6 to 16 nm, and the mean particle diameter is approximately 11 nm. Actually, the mean particle size determined by TEM is very close to the average particle size calculated by the Debye-Scherrer formula from the XRD pattern. The particle size distribution was also presented for the Co_3O_4 nanoparticles prepared at 200°C in the inset of Figure 3b. The size of particles is in the range of 7 to 20 nm with an average particle size of around 13 nm. This value is slightly greater than that of the Co_3O_4 nanoparticles prepared at 175°C probably due to the crystallite growth at the higher temperature.

The EDS analysis was employed to determine the composition of the product prepared at 175°C. As shown

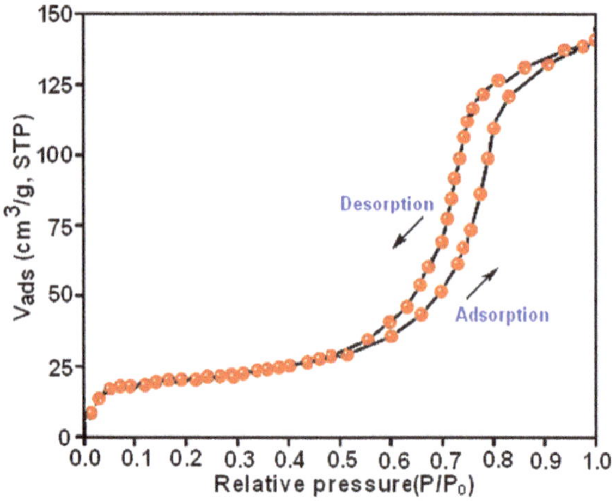

Figure 5 BET nitrogen adsorption-desorption isotherm curve for the Co_3O_4 nanoparticles prepared at 175°C.

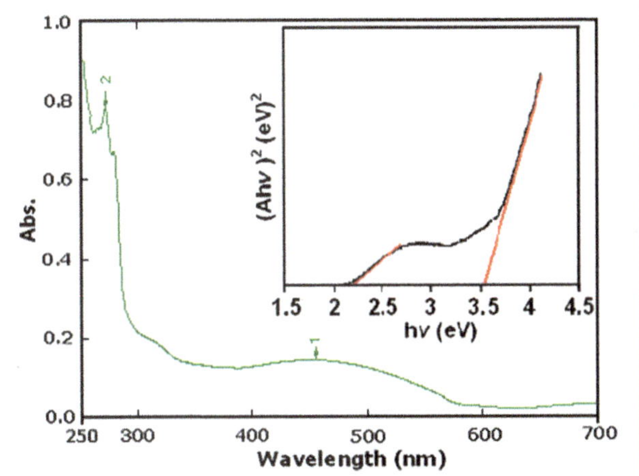

Figure 6 UV–vis spectrum and $(Ahv)^2$ – *hv* curve (inset) of the Co_3O_4 nanoparticles prepared at 175°C.

in Figure 4, only cobalt and oxygen elements existed in the product with a Co/O atomic ratio of about 3:3.97 which is consistent with the theoretical value of Co_3O_4. No other elements can be detected, indicating the high purity of the Co_3O_4 nanoparticles.

BET surface area measurements were also made on the Co_3O_4 nanoparticles prepared at 175°C. Figure 5 shows N_2 adsorption-desorption which is close to type IV of the IUPAC classification with an evident hysteresis loop in the 0.5 to 1.0 range, suggesting that the sample under study is basically mesoporous. The specific surface area of the sample calculated by the BET method is 73.70 m^2 g^{-1}. The relatively high specific surface area of the product is related to the nanometric size of its particles. Assuming that the Co_3O_4 nanoparticles are almost spherical, as confirmed by TEM, the surface area can be used to estimate the particle size according to the equation $D_{BET} = 6000/(\rho \times S_{BET})$, where D_{BET} is the diameter of a spherical particle (in nm), ρ is the theoretical density of Co_3O_4 (6.08 g cm^{-3}), and S_{BET} is the specific surface area of the Co_3O_4 powder in meter squared per gram. The particle size calculated from the surface area data is approximately 13.4 nm, which is in good agreement with the XRD and TEM results.

Optical absorption properties of the Co_3O_4 nanoparticles prepared at 175°C were investigated at room temperature by UV–vis spectroscopy. Figure 6 shows the absorbance spectrum of the Co_3O_4 sample with two absorption bands

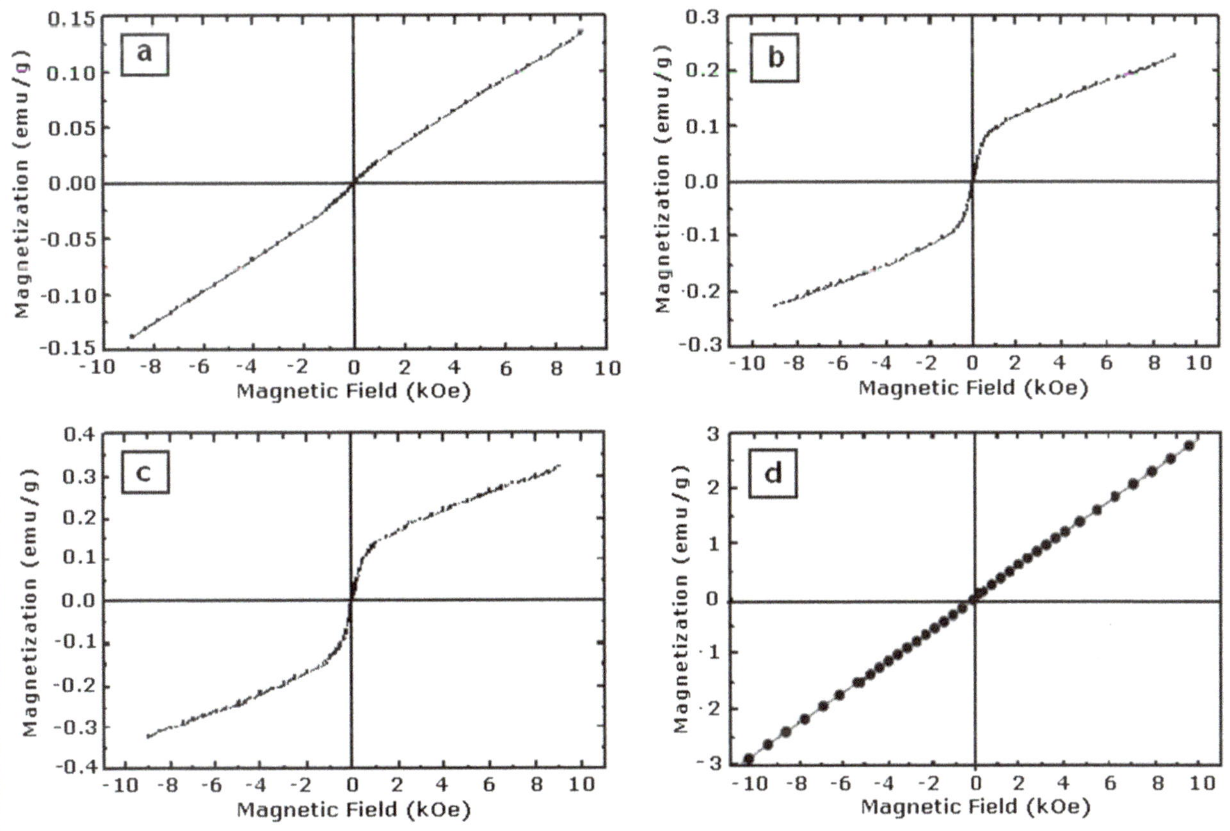

Figure 7 Room temperature magnetization curves of the Co_3O_4 nanoparticles prepared at: (a) 175°C (b) 200°C and (c) 250°C. **(d)** Magnetization curve of a bulk Co_3O_4 sample.

in 250 to 350 and 400 to 580 nm wavelength ranges. The first band can be assigned to the $O^{2-} \rightarrow Co^{2+}$ charge transfer process while the second one to the $O^{2-} \rightarrow Co^{3+}$ charge transfer [49]. Co_3O_4 is a p-type semiconductor, and the absorption band gap (E_g) can be determined by the following equation [24]: $(Ahv)^2 = K\ (hv - E_g)$, where hv is the photon energy (eV), A is the absorption coefficient, K is a constant, and E_g is the band gap. The band gap can be estimated by extrapolating the linear region in the plot of $(Ahv)^2$ versus photon energy as shown in the inset of Figure 6. Two absorption peaks give two E_g values for the product, 3.55 and 2.2 eV. As it has been reported in the literatures [50], the E_g values of Co_3O_4 nanoparticles prepared in this study are greater than those of the bulk Co_3O_4 (E_g = 3.17 and 1.77 eV, respectively). The increase in the band gaps can be related to the quantum confinement effects and/or small size effects of the Co_3O_4 nanoparticles [33,50].

The magnetic measurements for the Co_3O_4 samples prepared at different temperatures were carried out at room temperature. As shown in Figure 7a, the magnetization curve for the Co_3O_4 nanoparticles prepared at 175°C exhibits a weak ferromagnetic behavior with a saturation magnetization of 0.137 emu g^{-1} at the maximum field of 9 kOe applied while the magnetization curves for the Co_3O_4 samples prepared at 200°C and 250°C in the same figure (Figure 7b,c) display higher ferromagnetic properties with saturation magnetization values of 0.225 and 0.325 emu g^{-1} at the applied field of 9 kOe, respectively. To confirm that the ferromagnetic behavior originates from the nanoparticles, this measurement was also conducted on a bulk sample. As shown in Figure 7d, the curve exhibits an antiferromagnetic behavior. The ferromagnetic behavior of the nanoparticles can be explained as follows: bulk Co_3O_4 has a normal spinel structure with antiferromagnetic exchange between ions which occupy the tetrahedral and octahedral sites [51]. It has zero net magnetization owing to the complete compensation of sublattice magnetizations. Hence, the change from an antiferromagnetic state for bulk Co_3O_4 to a weakly ferromagnetic state for the Co_3O_4 nanoparticles can be ascribed to the uncompensated surface spins and/or finite size effects [11,41,52,53]. It is well known that the magnetic properties of nanomaterials are strongly dependent on the shape and sizes of their particles, crystallinity, magnetization direction, and so on.

In order to obtain further insight into the nature of the reactions involved in the decomposition pathway of the complex, its thermal behavior was investigated by TG/DTA. Figure 8 shows TG/DTA curves recorded for $[Co(NH_3)_4CO_3]NO_3{\cdot}H_2O$ in the 25°C to 600°C temperature range. The TG curve shows that the decomposition of complex proceeds in three stages. The first stage occurred at about 95°C to 110°C and shows 6.68% weight loss which is consistent with the theoretical value of 6.75% caused by the loss of 1 mol of H_2O per mole of the complex. The second stage occurs at approximately 135°C to 150°C and shows a 16.75% weight loss, which is consistent with the theoretical value of 16.50% caused by the loss of 1 mol of CO_2 per mole of the complex. In the third stage, an extensive weight loss is observed in the range of 160°C to 175°C, which related to the decomposition of the residue complex. Above 175°C, the weight remained constant, confirming the complete decomposition of the complex. The weight loss of all steps that is about 70% is consistent with the theoretical value (70.10%) calculated for the formation of Co_3O_4 from the complex. The DTA curve for the $[Co(NH_3)_4CO_3]NO_3{\cdot}H_2O$ complex as shown in the inset of Figure 8 gave three characteristic peaks in consistent with TG data. The small endothermic peaks at about 100°C can be explained by freeing one H_2O molecule. The decomposition of the dehydrated complex was confirmed by one endothermic peak at about 150°C which is immediately followed by a sharp exothermic peak at about 175°C. The endothermic peak at 150°C can be related to the decomposition of unstable bidentate

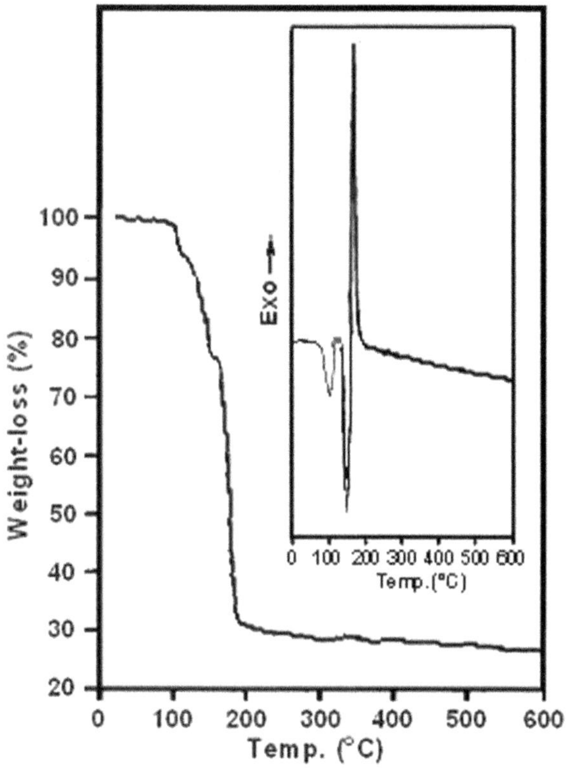

Figure 8 TG and DTA (inset) curves of the $[Co(NH_3)_5CO_3]NO_3{\cdot}H_2O$ complex in air.

carbonato ligand, and the sharp exothermic peak can be explained by the explosive decomposition of the complex via a redox process taking place between the NH_3 ligands as the reductants and NO_3^- and/or O_2 as oxidants. According to the TG/DTG data, the decomposition reactions of the complex can be written as follows:

- Step 1: $[Co(NH_3)_4CO_3]NO_3{\cdot}H_2O(s) \rightarrow [Co(NH_3)_4CO_3]NO_3\ (s) + H_2O\ (g)$
- Step 2: $[Co(NH_3)_4CO_3]NO_3\ (s) \rightarrow [Co(NH_3)_4O]NO_3\ (s) + CO_2\ (g)$
- Step 3: $[Co(NH_3)_4O]NO_3\ (s) + 13/3O_2 \rightarrow 1/3Co_3O_4\ (s) + 6H_2O\ (g) + NO\ (g)\ (or\ NO_2) + 2\ N_2\ (g)$

Conclusions

In summary, pure and nanosized Co_3O_4 particles with an average particle size of 11 nm were successfully synthesized by the thermal decomposition of the $[Co(NH_3)_4CO_3]NO_3{\cdot}H_2O$ complex as a new precursor at 175°C. Co_3O_4 nanoparticles are probably formed via the elimination of H_2O and CO_2 molecules and then explosive redox reaction between the NH_3 ligands as the reducing agent and the NO_3^- ions as the oxidizing agent. By this method, uniform and sphere-like Co_3O_4 nanoparticles with a narrow size distribution and a weak ferromagnetic behavior can be obtained. The optical absorption band gaps of the Co_3O_4 nanoparticles were estimated to be approximately 2.20 and 3.55 eV, which are blue shifted in comparison with previously reported values. This method is simple, low cost, safe, and suitable for the industrial production of high-purity Co_3O_4 nanoparticles for various applications.

Methods

Materials

All materials were of analytical grade and obtained from Merck Company, Merck KGaA, Darmstadt, Germany. Solvents used throughout the reactions were of high purity and used without further purification.

Synthesis of Co_3O_4 nanoparticles

The precursor complex, $[Co(NH_3)_4CO_3]NO_3{\cdot}H_2O$, was synthesized according to the literature method [54]. To prepare Co_3O_4 nanoparticles, an appropriate amount of the precursor complex (1 to 2 g) was added to a porcelain crucible and then was placed in an electric furnace. The sample was heated at the rate of 10°C min^{-1} from room temperature to 150°C in an air atmosphere and then was maintained at this temperature for 1 h. Similar experiments were performed for the samples decomposed in the range of 175°C to 300°C. The decomposition products of the complex at various temperatures were cooled to room temperature and collected for the characterization.

Characterization

The XRD patterns were recorded on a Rigaku D-max C III, X-ray diffractometer (Rigaku Corporation, Shibuya-ku, Japan) using Ni-filtered CuKα radiation (λ = 1.5406 Å) to determine the phases present in the decomposed samples. Infrared spectra were recorded on a Shimadzu system FT-IR 8400S spectrophotometer (Shimadzu Corporation, Kyoto, Japan) using KBr pellets. The optical absorption spectrum was recorded on a Shimadzu 1650PC UV–vis spectrophotometer in the 250 to 700 nm wavelength range at room temperature. The sample for UV–vis studies was well dispersed in distilled water to form a homogeneous suspension by sonication for 25 min. The particle size was determined by a transmission electron microscope (Philips CM10, Philips, Amsterdam, The Netherlands) equipped with a link EDS analyzer. The powders were ultrasonicated in ethanol, and a drop of the suspension was dried on a carbon-coated copper microgrid for the TEM measurements. The specific surface area of the product was measured by the BET method using an N_2 adsorption-desorption isotherm carried out at −196°C on a surface area analyzer (Micromeritics ASAP 2010, Micromeritics, Norcross, GA, USA). Before each measurement, the sample was degassed at 150°C for 2 h. Magnetic measurements were carried out at room temperature using a vibrating sample magnetometer (Meghnatis Daghigh Kavir Co. Kashan, Iran). The thermal behavior of the precursor complex was studied using a Netzsch STA 409 PC/PG thermal analyzer (Netzsch, Burlington, MA, USA) at a heating rate of 5°C min^{-1} in air.

Abbreviations

BET: Brunauer-Emmett-Teller; EDS: energy-dispersive X-ray spectroscopy; FT-IR: Fourier transformed infrared spectroscopy; TEM: transmission electron microscopy; TG/DTA: thermogravimetry/differential thermal analysis; XRD: X-ray diffraction.

Competing interests

The authors declare that they have no competing interests.

Authors' contributions

SF proposed the idea of the study, revised the manuscript critically, and gave final approval for submission. JS was involved in the synthesis and the physicochemical characterization of the Co_3O_4 nanoparticles and early drafted the manuscript. PZ was involved in the analysis and interpretation of spectral and thermal analysis data. All authors read and approved the final manuscript.

Authors' information

SF received his BSc degree in Chemistry from Shahid Chamran University in 1991, his MSc degree in Inorganic Chemistry from Tehran University in 1994, and his Ph. D in Inorganic Chemistry from Isfahan University, Iran, in 2000. He is now a professor in inorganic chemistry and the head of the Chemistry Department at Lorestan University, Iran. His research area interests have concentrated on synthesis and characterization of metal and metal oxide nanostructures and their catalytic applications. JS got his BSc in Chemistry in

2009 from the Faculty of Science, Kashan University, Iran. He obtained his MSc in inorganic chemistry in 2012 with the thesis entitled 'Low-temperature synthesis of Co_3O_4 nanoparticles from thermal decomposition of the $[Co(NH_3)_4(L)](NO_3)_n$ complexes ($L = CO_3$, NO_3, H_2O; n = 1–3) and their characterization' from the Faculty of Science, Lorestan University, Iran. PZ obtained her BSc and MSc degrees in Chemistry from the Faculty of Science, Lorestan University, Iran, in 2003 and 2006, respectively. During her MSc course, she has been involved in synthesizing mullite and magnesium aluminate spinel nanopowders by sol–gel process.

Acknowledgment
We wish to acknowledge the financial support from Lorestan University Research Council and Iran Nanotechnology Initiative Council (INIC).

References

1. Klabunde, KJ, Richards, RM: Nanoscale Materials in Chemistry, 2nd edn. Wiley, New York (2012)
2. Mate, VR, Shirai, M, Rode, CV: Heterogeneous Co_3O_4 catalyst for selective oxidation of aqueous veratryl alcohol using molecular oxygen. Catal Commun **33**, 66–69 (2013)
3. Warang, T, Patel, N, Santini, A, Bazzanella, N, Kale, A, Miotello, A: Pulsed laser deposition of Co_3O_4 nanoparticles assembled coating: role of substrate temperature to tailor disordered to crystalline phase and related photocatalytic activity in degradation of methylene blue. Appl. Catal. A: Gen. **423–424**, 21–27 (2012)
4. Casas-Cabanas, M, Binotto, G, Larcher, D, Lecup, A, Giordani, V, Tarascon, JM: Defect chemistry and catalytic activity of nanosized Co_3O_4. Chem Mater **21**, 1939–1947 (2009)
5. Askarinejad, A, Bagherzadeh, M, Morsali, A: Catalytic performance of Mn_3O_4 and Co_3O_4 nanocrystals prepared by sonochemical method in epoxidation of styrene and cyclooctene. Appl. Surface Sci. **256**, 6678–6682 (2010)
6. Lou, XW, Deng, D, Lee, JY, Feng, J, Archer, LA: Self-supported formation of needlelike Co_3O_4 nanotubes and their application as lithium-ion battery electrodes. Adv Mater **20**, 258–262 (2008)
7. Chou, S-L, Wang, J-Z, Liu, H-K, Dou, S-X: Electrochemical deposition of porous Co_3O_4 nanostructured thin film for lithium-ion battery. J Power Sources **182**, 359–364 (2008)
8. Li, YG, Tan, B, Wu, YY: Mesoporous Co_3O_4 nanowire arrays for lithium ion batteries with high capacity and rate capacity. Nano Lett **8**, 265–270 (2008)
9. Li, W-Y, Xu, L-N, Chen, J: Co_3O_4 nanomaterials in lithium-ion batteries and gas sensors. Adv Funct Mater **15**, 851–857 (2005)
10. Sugimoto, T, Matijevic, E: Colloidal cobalt hydrous oxides, preparation and properties of monodispersed. Co_3O_4. J Inorg Nucl Chem **41**, 165–172 (1979)
11. Makhlouf, SA: Magnetic properties of Co_3O_4 nanoparticles. J Magn Magn Mater **246**, 184–190 (2002)
12. Sun, L, Li, H, Ren, L, Hu, C: Synthesis of Co_3O_4 nanostructures using a solvothermal approach. Solid State Sci **11**, 108–112 (2009)
13. Chen, Y, Zhang, Y, Fu, S: Synthesis and characterization of Co_3O_4 hollow spheres. Mater Lett **61**, 701–705 (2007)
14. Lai, T, Lai, Y, Lee, C, Shu, Y, Wang, C: Microwave-assisted rapid fabrication of Co_3O_4 nanorods and application to the degradation of phenol. Catal Today **131**, 105–110 (2008)
15. Wang, WW, Zhu, YJ: Microwave-assisted synthesis of cobalt oxalate nanorods and their thermal conversion to Co_3O_4 rods. Mater Res Bull **40**, 1929–1935 (2005)
16. Li, L, Chu, Y, Liu, Y, Song, JL, Wang, D, Du, XW: A facile hydrothermal route to synthesize novel Co_3O_4 nanoplates. Mater Lett **62**, 1507–1510 (2008)
17. Du, J, Chai, L, Wang, G, Li, K, Qian, Y: Controlled synthesis of one-dimensional single-crystal Co_3O_4 nanowires. Aust J Chem **61**, 153–158 (2008)
18. Wang, RM, Liu, CM, Zhang, HZ, Chen, CP, Guo, L, Xu, HB, Yang, SH: Porous nanotubes of Co_3O_4: synthesis, characterization and magnetic properties. Appl Phys Lett **85**, 2080–2082 (2004)
19. Li, Y, Zhao, J, Dan, Y, Ma, D, Zhao, Y, Hou, S, Lin, H, Wang, Z: Low temperature aqueous synthesis of highly dispersed Co_3O_4 nanocubes and their electrocatalytic activity studies. Chem Eng J **166**, 428–434 (2011)
20. Sun, H, Ahmad, M, Zhu, J: Morphology-controlled synthesis of Co_3O_4 porous nanostructures for the application as lithium-ion battery electrode. Electrochim Acta **89**, 199–205 (2013)
21. Ren, M, Yuan, S, Su, L, Zhou, Z: Chrysanthemum-like Co_3O_4 architectures: hydrothermal synthesis and lithium storage performances. Solid State Sci **14**, 451–455 (2012)
22. Yang, LX, Zhu, YJ, Li, L, Zhang, L, Tong, H, Wang, WW: A facile hydrothermal route to flower-like cobalt hydroxide and oxide. Eur J Inorg Chem **23**, 4787–4792 (2006)
23. Jiu, J, Ge, Y, Li, X, Nie, L: Preparation of Co_3O_4 nanoparticles by a polymer combustion route. Mater Lett **54**, 260–263 (2002)
24. Gu, F, Li, C, Hu, Y, Zhang, L: Synthesis and optical characterization of Co_3O_4 nanocrystals via a facile combustion method. J. Cryst. Growth **304**, 369–373 (2007)
25. Gardey-Merino, MC, Palermo, M, Belda, R, Fernández De Rapp, ME, Lascalea, GE, Vázquez, PG: Combustion synthesis of Co_3O_4 nanoparticles: fuel ratio effect on the physical properties of the resulting powders. Proced. Mater. Sci **1**, 588–593 (2012)
26. Ai, L-H, Jiang, J: Rapid synthesis of nanocrystalline Co_3O_4 by a microwave-assisted combustion method. Powder Tech. **195**, 11–14 (2009)
27. Li, L, Ren, J: Rapid preparation of spinel Co_3O_4 nanocrystals in aqueous phase by microwave irradiation. Mater Res Bull **41**, 2286–2290 (2006)
28. Bhatt, AS, Bhat, DK, Tai, C-W, Santosh, MS: Microwave-assisted synthesis and magnetic studies of cobalt oxide nanoparticles. Mater Chem Phys **125**, 347–350 (2011)
29. Ma, J, Zhang, S, Liu, W, Zhao, Y: Facile preparation of Co_3O_4 nanocrystals via a solvothermal process directly from common Co_2O_3 powder. J Alloys Compd **490**, 647–651 (2010)
30. Lester, E, Aksomaityte, G, Li, J, Gomez, S, Gonzalez-Gonzalez, J, Poliakoff, M: Controlled continuous hydrothermal synthesis of cobalt oxide (Co_3O_4) nanoparticles Prog. Cryst. Growth Charact. Mater. **58**, 3–13 (2012)
31. Baydi, ME, Poillerat, G, Rehspringer, JL, Gautier, JL, Koenig, JF, Chartier, P: A sol–gel route for the preparation of Co_3O_4 catalyst for oxygen electrocatalysis in alkaline medium. J. Solid State Chem. **109**, 281–288 (1994)
32. Kim, DY, Ju, SH, Koo, HY, Hong, SK, Kangf, YC: Synthesis of nanosized Co_3O_4 particles by spray pyrolysis. J Alloys Compd **417**, 254–258 (2006)
33. Kumar, RV, Diamant, Y, Gedanken, A: Sonochemical synthesis and characterization of nanometer-size transition metal oxides from metal acetates. Chem Mater **12**, 2301–2305 (2000)
34. Wang, X, Chen, XY, Gao, LS, Zheng, HG, Zhang, Z, Qian, YT: One-dimensional arrays of Co_3O_4 nanoparticles: synthesis, characterization, and optical and electrochemical properties. J Phys Chem B **108**, 16401–16404 (2004)
35. Fan, S, Liu, X, Li, Y, Yan, E, Wang, C, Liu, J, Zhang, Y: Non-aqueous synthesis of crystalline Co_3O_4 nanoparticles for lithium-ionbatteries. Mater Lett **91**, 291–293 (2013)
36. Jiang, J, Li, L: Synthesis of sphere-like Co_3O_4 nanocrystals via a simple polyol route. Mater Lett **61**, 4894–4896 (2007)
37. Zou, D, Xu, C, Luo, H, Wang, L, Ying, T: Synthesis of Co_3O_4 nanoparticles via an ionic liquid-assisted methodology at room temperature. Mater Lett **62**, 1976–1978 (2008)
38. Traversa, E, Sakamoto, M, Sadaoka, Y: A chemical route for the preparation of nanosized rare earth Perovskite-type oxides for electroceramic applications. Part Sci Technol **16**, 185–214 (1998)
39. Farhadi, S, Rashidi, N: Preparation and characterization of pure single-phase $BiFeO_3$ nanoparticles through thermal decomposition of the heteronuclear $Bi[Fe(CN)_6] \cdot 5H_2O$ complex. Polyhedron **29**, 2959–2965 (2010)
40. Farhadi, S, Roostaei-Zaniyani, Z: Simple and low-temperature synthesis of NiO nanoparticles through solid-state thermal decomposition of the hexa(ammine) Ni(II) nitrate, $[Ni(NH_3)_6](NO_3)_2$, complex. Polyhedron **30**, 1244–1249 (2011)
41. Mohandes, F, Davar, F, Salavati-Niasari, M: Preparation of Co_3O_4 nanoparticles by nonhydrolytic thermolysis of $[Co(Pht)(H_2O)]_n$ polymers. J Magn Magn Mater **322**, 872–877 (2010)
42. Ren, L, Wang, P, Han, Y, Hu, C, Wei, B: Synthesis of $CoC_2O_4 \cdot 2H_2O$ nanorods and their thermal decomposition to Co_3O_4 nanoparticles. Mater. Phys. Lett. **476**, 78–83 (2009)
43. Thangavelu, K, Parameswari, K, Kuppusamy, K, Haldorai, Y: A simple and facile method to synthesize Co_3O_4 nanoparticles from metal benzoate dihydrazinate complex as a precursor. Mater Lett **65**, 1482–1484 (2011)
44. Salavati-Niasari, M, Khansari, A, Davar, F: Synthesis and characterization of cobalt oxide nanoparticles by thermal treatment process. Inorg Chim Acta **362**, 4937–4942 (2009)

45. Farhadi, S, Pourzare, K: Simple and low-temperature preparation of Co_3O_4 sphere-like nanoparticles via solid-state thermolysis of the $[Co(NH_3)_6](NO_3)_3$ complex. Mater Res Bull **47**, 1550–1556 (2012)
46. Nakamoto, K: Infrared and Raman Spectra of Inorganic and Coordination Compounds. Part B: Applications in Coordination, Organometallic, and Bioinorganic Chemistry, 6th edition, Wiley, New York (2009)
47. Pejova, B, Isahi, A, Najdoski, M: Fabrication and characterization of nanocrystalline cobalt oxide thin films. Mater Res Bull **36**, 161–170 (2001)
48. Klug, HP, Alexander, LE: X-ray Diffraction Procedures, 2nd edn. Wiley, New York (1964)
49. He, T, Chen, DR, Jiao, XL, Wang, YL, Duan, YZ: Solubility-controlled synthesis of high-quality Co_3O_4 nanocrystals. Chem Mater **17**, 4023–4030 (2005)
50. Gulino, A, Dapporto, P, Rossi, P, Fragala, I: A novel self-liquid MOCVD precursor for Co_3O_4 thin films. Chem Mater **15**, 3748–3752 (2003)
51. Ichiyanagi, Y, Kimishima, Y, Yamada, S: Magnetic study on Co_3O_4 nanoparticles. J Magn Magn Mater **272–276**, e1245–e1246 (2004)
52. Kodama, RH, Makhlouf, SA, Berkowitz, AE: Growth mechanism and magnon excitation in NiO nanowalls. Phys Rev Lett **79**, 1393–1396 (1997)
53. Ozkaya, T, Baykal, A, Toprak, MS, Koseoglu, Y, Durmus, Z: Reflux synthesis of Co_3O_4 nanoparticles and its magnetic characterization. J Magn Magn Mater **321**, 2145–2149 (2009)
54. Schlessinger, G: Synthesis of carbonatotetra(ammine)cobalt(III) nitrate. Inorg Synth **6**, 173–175 (1960)

Dipeptide derived from benzylcystine forms unbranched nanotubes in aqueous solution

Biswadip Banerji[1*], Sumit Kumar Pramanik[1], Uttam Pal[2] and Nakul Chandra Maiti[2]

Abstract

The essence of modern nanotechnology is manifested in the formation of well-ordered nanostructures by a process of self-association. Peptides are among the most useful building blocks for organic bionanostructures such as nanotubes, nanospheres, nanotapes, nanofibrils, and other different ordered structures at the nanoscale. Peptides are biocompatible, chemically diverse, and much more stable and can be readily synthesized on a large scale. Also, they have diverse application in biosensors, tissue engineering, drug delivery, etc. Here, we report a short cystine-based dipeptide, which spontaneously self-associates to form straight, unbranched nanotubes. Such self-assembled nanobiomaterials provide a novel possibility of designing new functional biomaterials with potential applications in nanobiotechnology. The formation of nanotubes in solution state has been demonstrated by atomic force microscopy and scanning electron microscopy. Infrared absorption and circular dichroism demonstrated the intermolecular β-sheet-like backbone hydrogen bonding in juxtaposing and stacking of aromatic side chains.

Keywords: Nanotube, Dipeptide, π-π stacking, Cystine, Self-assembly

Background

There has been rapid advancement in the development of self-assembled nanobiomaterials, such as nanotubes, nanocrystals, and nanowires, which have potential application in electronics, biosensors, catalysis, drug delivery, and tissue engineering [1-4]. The physical and chemical properties of these nanomaterials are tunable by controlling their shapes and sizes [5]. There are various designing rules of the synthesis of these biomaterials, where secondary structure of the self-assembled fiber, the thickness of fiber, and hydrogel porosity, and different mechanical properties can all be varied predictably simply by changing the amino acid sequence, its concentration, the surrounding media, and its processing route [6-8]. Over the years, various natural self-assembling systems have served as inspiration for the design of novel building blocks on the nanoscale [9-11]. It has been already reported that cyclic peptides, amphiphilic peptides, and amyloid-inspired peptides can form ordered nanostructures with different morphologies including nanowires, nanotubes, nanovesicles, nanofibrils, and nanosheets [12-14]. The formation of these nanostructures is facilitated by the combination of different non-covalent forces that include electrostatic interactions, hydrogen bonds, hydrophobic interactions, and aromatic stacking interactions [9,13,15]. Unlike inorganic and carbon-based nanomaterials, the peptides undergo self-assembly process at ambient temperature and atmospheric pressure [16]. Besides, the peptides can be produced in large scale by simpler experimental methods. The self-assembled peptide nanostructures can further organize to form nanoscale devices [17]. Therefore, the application of self-assembling synthetic peptides as the building blocks of nanoscale devices is practical, robust, and affordable [18-20].

In this report, we present a novel peptide that, in solution, readily undergoes the formation of a linear nanotube with β-sheet-like backbone hydrogen-bonding pattern. The dipeptide constitutes two-side-chain benzyl-protected cystine moieties. This is possibly the first report for the formation of a cystine-based nanotube. First, this dipeptide has been synthesized, purified, and characterized. The β-sheet-like backbone arrangement of the peptide in the solid state has been confirmed by Fourier transform infrared (FT-IR) spectroscopy. The computational model calculation and density functional theory (DFT) study also support these data.

* Correspondence: biswadip.banerji@gmail.com
[1]Department of Chemistry, CSIR-Indian Institute of Chemical Biology, 4, Raja S.C. Mullick Road, Kolkata 700032, India
Full list of author information is available at the end of the article

Results and discussions

Synthesis of the dipeptide

The dipeptide was synthesized by conventional solution-phase methodology (Scheme 1). The N-terminus was protected by the Bocgroup, and the C-terminus was protected as methyl ester. Couplings were mediated using 1-ethyl-3-(3-dimethyllaminopropyl) carbodiimide hydrochloride/hydroxybenzotriazole (EDC·HCl/HOBt). The deprotection of methyl ester leading to the acid was performed by saponification (aq. LiOH) method, and the Boc group was removed using 4 M HCl in 1,4-dioxane. The final tetrapeptide was fully characterized by ^{1}H-NMR, HRMS, and infrared (IR) spectroscopy.

FT-IR study

FT-IR spectroscopy studies were performed to determine the secondary structures of the peptide (Figure 1). In the case of β-sheet structure, the FT-IR spectra show two amide I transitions, one is a strong transition between 1,610 and 1,640 cm^{-1} and the other one is a weak transition between 1,680 and 1,700 cm^{-1}. Here, the FT-IR spectra of the peptide show characterized CO stretching bands at around 1,631 and 1,680 cm^{-1} and the NH stretching band at around 3,304 cm^{-1}, typical for intermolecularly hydrogen-bonded β-sheet structures in solid. Furthermore, the NH bending frequencies of this dipeptide appear at 1,512 cm^{-1}, suggesting the formation of a β-sheet structure, whereas absorption in the range 1,610 to 1630 cm^{-1}, which is a characteristic of aggregated amyloid-like peptides, was not observed. IR frequencies were calculated using DFT level of theory. NH and COstretching frequencies were found to be 1,635 and 1,681 cm^{-1}. It nicely corroborates with the experimentally obtained data.

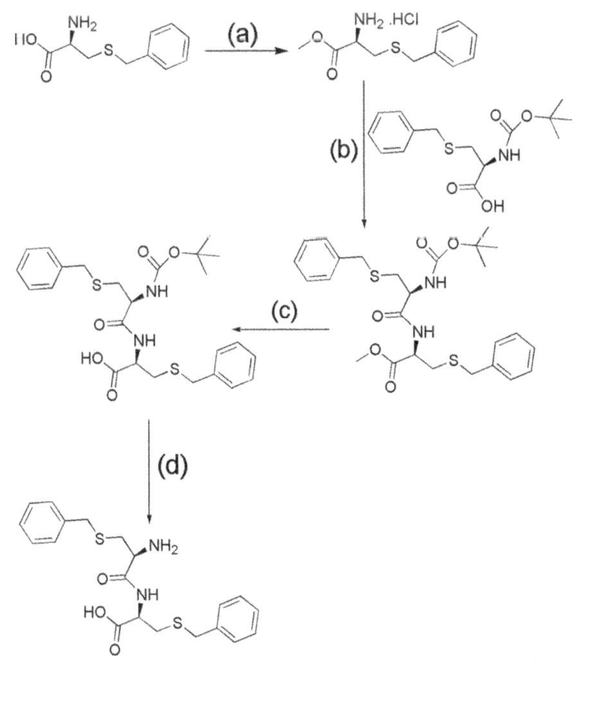

Scheme 1 Reagent and conditions. (a) $SOCl_2$, MeOH, 0°C to reaction time (rt), 4 h; **(b)** EDC·HCl (1.5 equivalency (equiv.)), HOBt (1.2 equiv.), TEA (3 equiv.), 0°C to rt, 7.0 h; **(c)** lithium hydroxide (1.5 equiv.), MeOH/water (5:1), rt, 1.5 h; **(d)** 4 M HCl in 1,4-dioxane, 0°C to rt, 2.0 h.

SEM and AFM imaging

The scanning electron microscopy (SEM) images of the dipeptide (Figure 2A) show that it forms nanofibrils with a diameter of 80 nm and length of 30 μm. They are almost uniform in nature. The atomic force microscopy (AFM) image of the dipeptide also confirmed the formation of nanotubes with a diameter nearly 80 nm and length of 30 μm.

Model of self-assembly process

The dipeptide showed intermolecular β-sheet-like backbone hydrogen bonding in juxtaposing and stacking of aromatic benzyl side chains (Figure 3). They have a hydrophobic head with an array of overlapping benzyl groups similar to what is found in silk fibroin or spider silk assemblies [21]. The benzyl groups form packed hydrophobic interactions in water, which can be disrupted mechanically during sonication. However, these hydrophobic cohesive ends can find each other quickly in water since the exposure of hydrophobic alanine arrays to water is energetically unfavorable. Since the hydrophobic alanines' interaction is non-specific, they can slide diffuse along the nanostructure, like trains sliding along the rails [21]. Such fragments could form various assemblies readily through hydrophobic and intermolecular hydrogen-

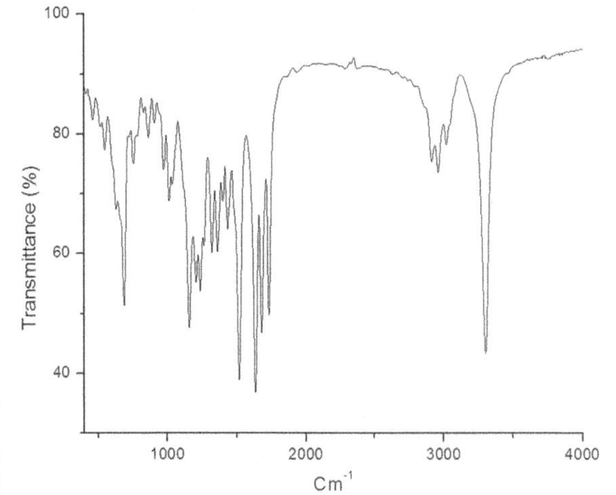

Figure 1 FT-IR spectra of the benzyl-protected cystine dipeptide.

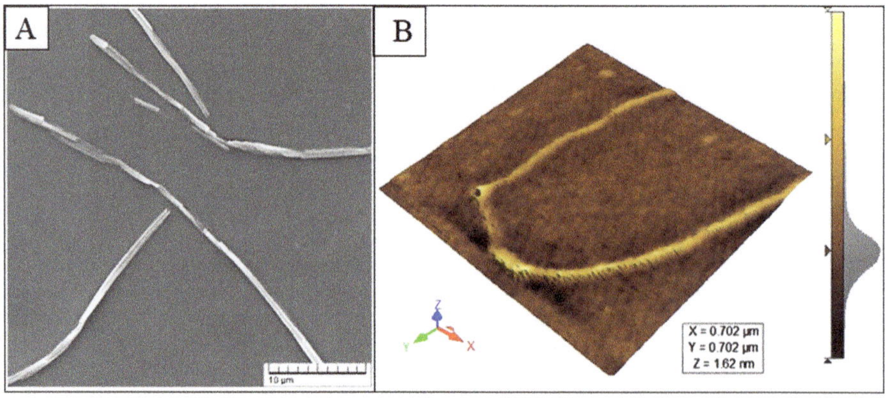

Figure 2 SEM image (A) and AFM image (B) of the dipeptide.

bonding interactions. The model of self-assembly process is shown in Figure 4.

Here, we have shown a model for plausible intermolecular interactions where the benzene ring of the benzyl groups shows π-π stacking interaction with the benzene ring of the benzyl group of another dipeptide. The hydrogen bonding occurs between the N-H of the amide group and the oxygen atom of the backbone carbonyl. The importance of hydrogen bonds and π-π stacking interaction in the formation of nanotubes has been recognized and emphasized in the literature. It was pointed out that the formation of the hydrogen bonds is the necessary controlling factor in the formation of cylindrical shape. In this present system, the assembled phenyl rings could generate the possibilities of π-π stacking interaction between the nearby dipeptide, which leads to the circular ring of the self-assemblies, though no β-sheet structure formed in solution phase. These two forces help form the nanotubes.

CD spectroscopy

The solution-state conformation of the peptide was also defined by circular dichroism (CD) spectroscopy. Circular dichroism revealed an extended conformation. The CD spectra of the dipeptide in water are shown in Figure 5. The peptide in dilute conditions showed a negative CD band at 201 nm. However, with an increase in peptide concentration, the CD absorption band red-shifted (approximately 3 nm) to longer wavelengths, which indicates that at a higher concentration of the peptide due to the formation of oligomer, the amide geometry may be changed.

Ultraviolet spectroscopy

Two distinct absorption peaks in Figure 6 were observed due to the benzyl groups present in the dipeptide molecule: one at 260 nm and the other at 267 nm. Scattering was also observed which indicates self-aggregation and higher oligomer formation. With the increasing concentration, scattering was also found to be increased.

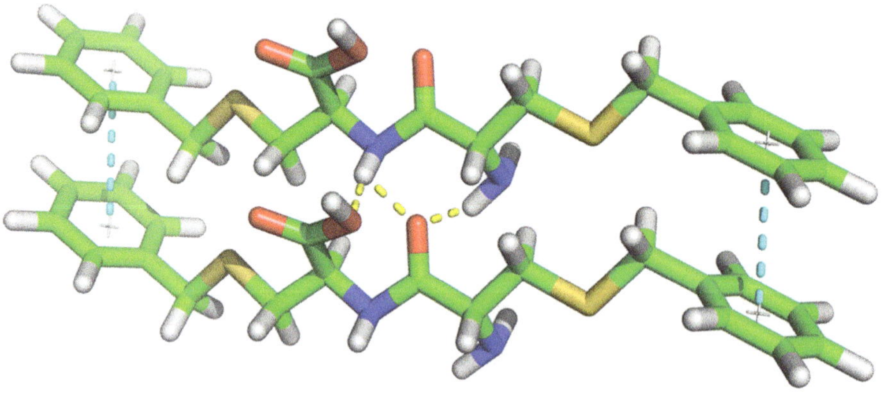

Figure 3 Plausible intermolecular interaction pattern. Illustrating the role of intermolecular β-sheet-like backbone hydrogen bonding in juxtaposing and stacking of phenyl rings in the benzyl group side chains. Pi stacking is shown with cyan dotted lines, and the hydrogen bonding is shown in yellow dotted lines.

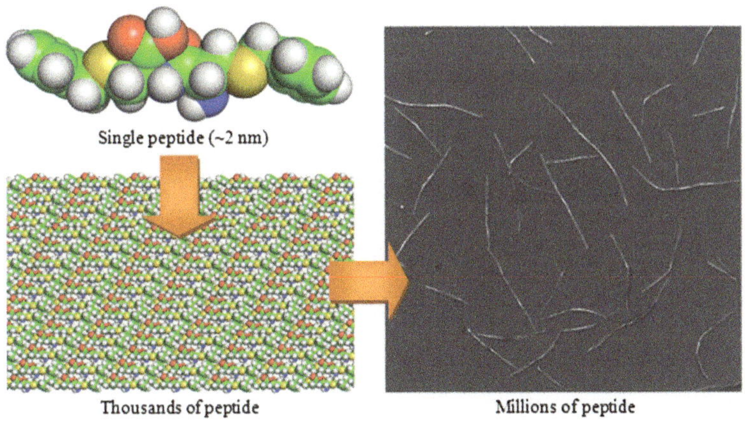

Figure 4 Schematic presentation of nanotube formation using cystine dipeptide as template. The peptide is about 2 nm long; tens and hundreds of thousands of individual peptides self-assemble into nanotubes.

Conclusion

This study clearly demonstrates that short water-soluble benzylcystine dipeptide self-assembles to form a supramolecular extended structure with β-sheet-like intermolecular hydrogen-bonding pattern in solid state, and this dipeptide also forms an ordered nanostructure from an aqueous solution under the proper conditions. The molecular arrangement of the self-association of this fibril-forming peptide is explained using molecular modelling studies. This work may help guide the design of biomaterial scaffoldings in thefuture by inscribing biological signals in the self-assemblies. Moreover, biocompatibility of peptide nanotubes makes them valuable for applications as biosensors and in tissue engineering. The field of nanoscale device development based on peptide nanotubes is still very nascent. Our efforts had continually been in developing such exotic nanomaterials for further advancement of this field.

Methods

Synthesis of the dipeptide

The starting materials, Boc-S-benzyl-L-cysteine, were purchased from Across Organics, and S-benzyl-L-cysteine was purchased from Alfa Aesar (Ward Hill, MA, USA) and were used without further purification. Solvents were freshly distilled by standard procedures prior to use. The dipeptide was purified by column chromatography. All ^{1}H and ^{13}C-NMR spectra were recorded on a Bruker 600-MHz spectrometer (Bruker AXS, Inc., Madison, MI, USA). For ^{1}H NMR, tetramethylsilane served as internal standard ($\delta = 0$), and data are reported as follows: chemical shift, integration, multiplicity (s = singlet, d = doublet, t = triplet, q = quartet, m = multiplet), and coupling constant(s) in hertz. High-resolution mass spectra were obtained on a Jeol MS station 700 (JEOL Ltd., Akishima-shi, Japan).

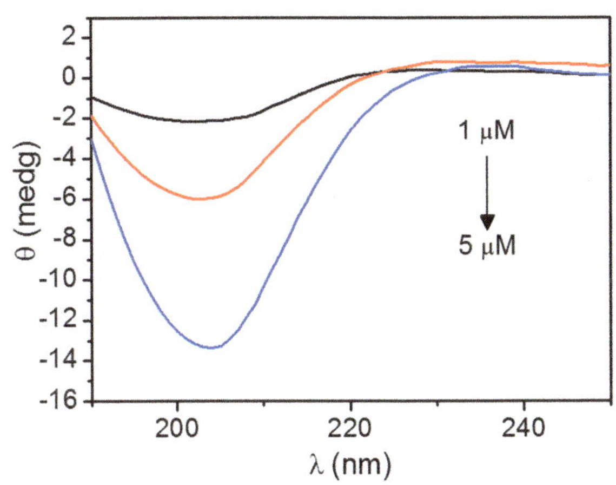

Figure 5 Circular dichroism spectra of the dipeptide at different dilutions. Approximately 3-nm shift in the CD band was observed.

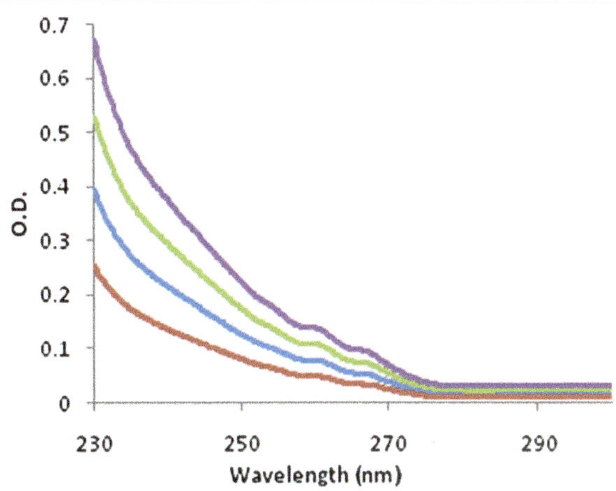

Figure 6 Absorption spectra of dipeptide at four different concentrations. Band for the benzyl groups appeared at 260 and 267 nm. Scattering was observed due to aggregation of the peptide. OD, optical density.

FT-IR study

The FT-IR spectra of the samples were recorded on a JASCO FT-IR 4200 spectrometer (Jasco, Easton, MD, USA) by KBr disc technique. The peptide was mixed with KBr in a clean glass pestle and compressed to obtain a pellet. The spectra were recorded from 400 to 4,000 cm^{-1}. Background spectra were obtained with KBr pellet for each sample. JASCO software was used for data processing.

SEM imaging

The dipeptide sample slides were dried completely under a critical point dryer (Quorum Technologies Inc., Guelph, Canada), and after drying, the samples were kept on the sample holder using carbon tape. Gold coating was done with the current of 10 mA at 10–6 to 10–8 mbar/Pa (model no. SC7620, Quorum Technologies Inc.). SEM imaging was performed using TESCAN Vega II LSU (TESCAN Digital Microscopy Imaging, TESCAN, Brno, Czech Republic). Imaging and measurements were done using Vega TC software (TC Software, Hampton, VA, USA).

AFM imaging

Twenty microliters of the dipeptide sample (5 mg dissolved in 1 mL water) was deposited onto freshly cleaved muscovite Ruby mica sheet (ASTM V1 Grade Ruby Mica, Micafab India Pvt. Ltd., Chennai, India) for 5 to 10 min. Mica sheets are basically negatively charged, so the dipeptide molecule binds strongly on the mica surface. After 10 min, the sample was dried using a vacuum dryer. AAC-mode atomic force microscopy was performed using a Pico Plus 5500 AFM (Agilent Technologies, Inc., Santa Clara, CA, USA) with a piezo scanner maximum range of 9 μm. Microfabricated silicon cantilevers of 225 μm in length with a nominal spring force constant of 21 to 98 N/m were used from nanosensors. Cantilever oscillation frequency was tuned into resonance frequency. The cantilever resonance frequency was 150 to 300 kHz. The images (512 × 512 pixels) were captured with a scan size between 0.5 and 5 μm at the scan speed rate of 0.5 rpm. The images were processed by flattening using Pico view software (Molecular Imaging Inc., Ann Arbor, MI, USA). The image presented in this paper was derived from the original data. Length, height, and width were measured manually using Pico view software.

CD spectroscopy

CD spectra (190 to 400 nm) were acquired at 200-nm/min scan speed, with 1-nm bandwidth on a Jasco J-810 spectrometer using a 1-mm-path-length quartz cell. Peptide stock solutions (2 mg/mL) were prepared in water and appropriately diluted during CD measurements. Five spectra were averaged to improve the signal-to-noise ratio and smoothed using the noise-reducing option. The results were represented as machine unit (θ, in units of mdeg).

UV spectroscopy

Absorption spectra were acquired using a Shimadzu UV-2401PC spectrophotometer (Shimadzu Corporation, Kyoto, Japan) with 10-mm-path-length quartz cuvette. A very dilute concentration of the peptide was used (5 to 20 μM).

DFT calculations

Structure of the dipeptides was drawn on Schrodinger Maestro molecular modeling environment (http://www.schrodinger.com/). Geometry was optimized using simulated annealing algorithm with the help of Desmond molecular dynamics simulations software tool. Simulation was run for 5.2 ns in SPC water environment in an orthorhombic periodic boundary condition. The model system was relaxed before simulation. Number of particle (N), and volume of system (V) in the ensemble were constant and the system had a well defined temperature (T). The molecule was cooked at a high temperature and then slowly cooled down to 300 K. The process was repeated several times [22]. The lowest energy conformation at 300 K was chosen for further analysis using DFT implemented in Gaussian 09 software (Gaussian Inc., Wallingford, CT, USA). In Gaussian 09, geometry optimizations and frequency calculations were performed *in vacuo* with B3LYP density functional using 6-311G + (2d,p) basis set [23,24].

Competing interests

The authors declare that they have no competing interests.

Authors' contributions

BB, SKP, and UP conceived and designed the experiments. SKP and UP performed the experiments. All the authors analyzed the data. SKP and UP drafted the manuscript. All authors read and approved the final manuscript.

Authors' information

Biswadip Banerji had his early education in the Ramakrishna Mission School & Colleges, Rahara, West-Bengal and post-graduation studies in Pure Chemistry from Calcutta University. He joined Prof. Javed Iqbal's group at Indian Institute of Technology-Kanpur (IIT-K) to do PhD studies in the field of Bio-Organic Chemistry. Soon after his graduation, he worked in Regional Research Laboratory-Trivandrum (RRL-T) as a research associate for 1 year and thereafter joined Prof. Christopher J. Schofield's group as a postdoctoral researcher at the OCMS, Oxford University. After spending more than 3 years at Oxford, he joined Prof. K. C. Nicolaou's group at ICES-Singapore in 2006, where he was a research fellow working in the area of total synthesis of complex natural products. Soon after the completion of his postdoctoral studies, in May 2008, he returned to India and joined 'Chembiotek International' as a team-leader. He has taken up his current senior scientist position at IICB, one of the renowned research institutes under the umbrella of CSIR, from January 2009. His current research interests are broadly on Chemical biology, mainly on the structure-based drug design/enzyme inhibition focusing natural product-based privileged scaffolds. He also has great interest in the cutting edge research on nano- and biotechnology, particularly on the synthesis of 'smart molecules' and their application in therapeutics and pharmaceuticals.

Sumit Kumar Pramanik obtained his B.Sc. in Chemistry from Vidyasagar University, India. He earned his M.Sc. in Applied Chemistry from Bengal Engineering and Science University, Shibpur, India. He is a Ph.D. student from the Chemistry Division, Indian Institute of Chemical Biology, Kolkata, India. His research area interests include nanobiomaterials and peptide-based drug design and biophysical chemistry.

Uttam Pal earned his B.Sc. in Physiology from the Presidency College, Kolkata, India. He is a M.Sc. degree holder of Biophysics and Molecular Biology from University of Calcutta, Kolkata, India. He is a Ph.D. student from the Structural Biology and Bioinformatics Division, Indian Institute of Chemical Biology, Kolkata, India. His research area covers structural biology and bioinformatics.

Nakul Chandra Maiti achieved M.Sc. in Chemistry, University of Calcutta, Kolkata, India; Ph.D. From Tata Institute of Fundamental Research, Mumbai, India; Postdoctoral JSPS visiting scientist, Institute for Molecular Science, Japan; Postdoctoral Senior Research Associate, Biochemistry, Case, Cleveland, Ohio, USA; Postdoctoral Research Associate/lecturer, California State University, Los Angeles, USA. He is a senior scientist from the Indian Institute of Chemical Biology, Kolkata, India. His research area interests cover structure-based amyloid research, structural aspects and *in vitro* behavior of natively unfolded proteins and peptides those are linked to human diseases, applications of NMR, fluorescence and Raman spectroscopy to biological systems, computational biochemistry, and bioinformatics.

Acknowledgements

SP gave thanks to CSIR-India, UP thanks INSPIRE Fellowship Programme, DST, India, for financial support. We would also like to thank CSIR-IICB, India, for providing financial assistance.

Author details

[1]Department of Chemistry, CSIR-Indian Institute of Chemical Biology, 4, Raja S.C. Mullick Road, Kolkata 700032, India. [2]Department of Structural Biology and Bioinformatics, CSIR-Indian Institute of Chemical Biology, 4, Raja S.C. Mullick Road, Kolkata 700032, India.

References

1. Cui, Y, Wei, Q, Park, H, Lieber, CM: Nanowire nanosensors for highly sensitive and selective detection of biological and chemical species. Science **293**(5533), 1289–1292 (2001)
2. Diehl, MR, Yaliraki, SN, Beckman, RA, Barahona, M, Heath, JR: Self-assembled, deterministic carbon nanotube wiring networks. Angew Chem Int Ed Engl **41**(2), 353–356 (2002)
3. Collins, PG, Arnold, MS, Avouris, P: Engineering carbon nanotubes and nanotube circuits using electrical breakdown. Science **292**(5517), 706–709 (2001)
4. Gao, X, Matsui, H: Peptide-based nanotubes and their applications in bionanotechnology. Adv. Mater. **17**(17), 2037–2050 (2005)
5. Wang, Y, Dai, Q, Zou, B, Yu, WW, Liu, B, Zou, G: Facile assembly of size- and shape-tunable IV–VI nanocrystals into superlattices. Langmuir **26**(24), 19129–19135 (2010)
6. Whitesides, GM: Nanoscience, nanotechnology, and chemistry. Small **1**(2), 172–179 (2005)
7. Martin, CR, Kohli, P: The emerging field of nanotube biotechnology. Nat Rev Drug Discov **2**, 29–37 (2003)
8. Bong, DT, Clark, TD, Granja, JR, Ghadiri, MR: Self-assembling organic nanotubes. Angew. Chem. Int. Ed. **40**(6), 988–1011 (2001)
9. Gazit, E: Self-assembled peptide nanostructures: the design of molecular building blocks and their technological utilization. Chem. Soc. Rev. **36**(8), 1263–1269 (2007)
10. Lv, C, Xue, Q, Shan, M, Jing, N, Ling, C, Zhou, X, Jiao, Z, Xing, W, Yan, Z: Self-assembly of double helical nanostructures inside carbon nanotubes. Nanoscale (2013). doi:10.1039/C2NR33157H
11. Colombo, G, Soto, P, Gazit, E: Peptide self-assembly at the nanoscale: a challenging target for computational and experimental biotechnology. Trends Biotechnol. **25**(5), 211–218 (2007)
12. Brea, RJ, Granja, JR: Self-assembly of cyclic peptides in hydrogen-bonded nanotubes. In: Schwarz, JA, Contescu, CI, Putyera, K (eds.) Dekker Encyclopedia of Nanoscience and Nanotechnology, 1st edn, pp. 3439–3457. Marcel Dekker, New York (2004)
13. Ghadiri, MR, Granja, JR, Milligan, RA, McRee, DE, Khazanovich, N: Self-assembling organic nanotubes based on a cyclic peptide architecture. Nature **366**(6453), 324–327 (1993)
14. Khazanovich, N, Granja, JR, Milligan, RA, McRee, DE, Ghadiri, MR: Nanoscale tubular ensembles with specified internal diameters. Design of a self-assembled nanotube with a 13 Å pore. J. Am. Chem. Soc. **116**(13), 6011–6012 (1994)
15. Ghadiri, MR, Granja, JR, Buehler, LK: Artificial transmembrane ion channels from self-assembling peptide nanotubes. Nature **369**(6478), 301–304 (1994)
16. Panciera, M, Amorín, M, Castedo, L, Granja, JR: Design of stable β-sheet -based cyclic peptide assemblies assisted by metal coordination: selective homo- and heterodimer formation. Chemistry (2013). doi:10.1002/chem.201203213
17. Amin, R, Kim, S, Park, SH, LaBean, T: Artificially designed DNA nanostructures. NANO: Brief Reports and Reviews **4**(3), 119–139 (2009)
18. Maity, S, Jana, P, Maity, SK, Haldar, D: Fabrication of hollow self-assembled peptide microvesicles and transition from sphere-to-rod structure. Langmuir **27**(7), 3835–3841 (2011)
19. Tao, K, Wang, J, Zhou, P, Wang, C, Xu, H, Zhao, X, Lu, JR: Self-assembly of short Aβ(16–22) peptides: effect of terminal capping and the role of electrostatic interaction. Langmuir **27**(6), 2723–2730 (2011)
20. Tian, F, Cziferszky, M, Jiao, D, Wahlströ m, K, Geng, J, Scherman, OA: Peptide separation through a CB[8]-mediated supramolecular trap-and-release process. Langmuir **27**(4), 1387–1390 (2010)
21. Hauser, CAE, Zhang, S: Designer self-assembling peptide nanofiber biological materials. Chem. Soc. Rev. **39**(8), 2780–2790 (2010)
22. Banerji, B, Pramanik, SK, Pal, U, Maiti, NC: Conformation and cytotoxicity of a tetrapeptide constellated with alternative d- and l-proline. RSC Advances **2**(17), 6744–6747 (2012)
23. Banerji, B, Pramanik, SK, Mandal, S, Maiti, NC, Chaudhuri, K: Synthesis, characterization and cytotoxicity study of magnetic (Fe_3O_4) nanoparticles and their drug conjugate. RSC Advances **2**(6), 2493–2497 (2012)
24. Rudra, DS, Pal, U, Maiti, NC, Reiter, RJ, Swarnakar, S: Melatonin inhibits matrix metalloproteinase-9 activity by binding to its active site. J Pineal Res (2012). doi:10.1111/jpi.12034

Simple preparation of ferromagnetic Co_3O_4 nanoparticles by thermal dissociation of the $[Co^{II}(NH_3)_6](NO_3)_2$ complex at low temperature

Saeid Farhadi*, Kolsoum Pourzare and Shokooh Sadeghinejad

Abstract

Co_3O_4 nanoparticles were prepared using $[Co(NH_3)_6](NO_3)_2$ as a starting material via a solid-state thermal decomposition route at low temperature (150°C). The product was characterized by thermal analysis (thermogravimetric/derivative thermogravimetric/differential thermal analysis), X-ray diffraction, Fourier transform infrared spectroscopy, Raman spectroscopy, Brunauer-Emmett-Teller specific surface area measurement, UV-visible spectroscopy, transmission electron microscopy (TEM), energy-dispersive X-ray spectroscopy, and magnetic measurements. The results confirmed that the nanoparticles are highly pure Co_3O_4 with weak ferromagnetic properties. TEM images showed that the Co_3O_4 nanoparticles have an average diameter size of around 13 nm. The optical spectrum indicated two direct bandgaps at 2.3 and 3.2 eV which are blueshifted relative to reported values for the bulk sample. By this fast and simple method, Co_3O_4 nanoparticles can be produced without expensive and toxic solvents or complicated equipment.

Keywords: Thermal dissociation, Co(II) complex, Nanoparticles, Cobalt oxide, Magnetic properties, Optical properties

Background

Transition metal oxide nanoparticles represent a broad class of materials that have been investigated extensively due to their interesting catalytic, electronic, and magnetic properties relative to those of the bulk counterparts, and the wide scope of their potential applications [1-4]. Among metal oxides, special attention has been focused on the synthesis and properties of spinel-type Co_3O_4 which is important as a heterogeneous catalyst, solid-state sensor, anode material in Li-ion rechargeable batteries, pigment, electrochromic sensor, and magnetic material and in solar energy storage [5-12]. For these diverse applications, it is of great importance to prepare Co_3O_4 with well-controlled dimensionality, sizes, and crystal structure. Up to now, several methods have been reported in order to synthesize Co_3O_4 nanoparticles, including sol–gel method [13], hydrothermal method [14,15], combustion method [16], microemulsion method [17], chemical spray pyrolysis [18], chemical vapor deposition [19,20], thermal decomposition of cobalt precursors [21-26], sonochemical route [27,28], co-precipitation [29], microwave irradiation [30], and mechanochemical processing [31]. However, most of these methods have one or more drawbacks, such as prolonged reaction times, the use of toxic and expensive solvents/reagents, complicated synthetic steps, the use of expensive equipment, and high synthetic temperatures. Therefore, the development of simple, inexpensive, and nontoxic methods for the preparation Co_3O_4 nanoparticles at relatively low temperature is still demanded.

Among the numerous methods developed for preparing metal oxide nanomaterials, the molecular precursor route has been regarded as one of the most convenient and practical techniques because it not only enables to avoid special instruments, complicated processes, and severe preparation conditions but also provides good control over purity, homogeneity, composition, phase, and microstructure of the resultant products [32-39]. By choosing a proper molecular precursor, coupled with a rational calcining procedure or other decomposition processes [40-43], nanocrystalline products could be obtained usually under the conditions significantly milder than those employed in the conventional solid-state synthesis.

Herein, we report on the preparation of Co_3O_4 nanoparticles by direct solid-state thermal dissociation of the

* Correspondence: sfarhadi48@yahoo.com
Department of Chemistry, Lorestan University, Khoramabad 68135-465, Iran

labile $[Co(NH_3)_6](NO_3)_2$ complex as a new precursor, as well as the characterization of the obtained product by X-ray diffraction (XRD), Fourier transform infrared spectroscopy (FT-IR), Raman spectroscopy, UV-visible (UV–vis) spectroscopy, Brunauer-Emmett-Teller (BET) specific surface area measurement, energy-dispersive X-ray spectroscopy (EDX), transmission electron microscopy (TEM), thermal analysis (thermogravimetric(TG)/derivative thermogravimetric (DTG)/differential thermal analysis (DTA)), and magnetic measurements.

Results and discussion

First, the thermal behavior of the $[Co(NH_3)_6](NO_3)_2$ precursor was studied by thermal analysis. Figure 1 shows TG, DTG, and DTA curves recorded for $[Co(NH_3)_6]$ $(NO_3)_2$ at a constant heating rate of 10°C min^{-1}in the temperature range of 25°C to 600°C. The TG and DTG curves show that the decomposition of the complex proceeds in two main stages. The first stage that occurred at about 100°C shows 12.20% weight loss which is consistent with the theoretical value of 11.95% caused by the loss of two molecules of NH_3 per one molecule of the complex. In the second stage, an extensive weight loss (60%) is observed at about 150°C, which related to the decomposition of the residue $[Co(NH_3)_4](NO_3)_2$ complex. Above 150°C, the weight remained constant, confirming the complete decomposition of the complex. The weight loss of all steps is about 72% which is consistent with the theoretical value (71.85%) calculated for the formation of Co_3O_4 from the complex. The DTA curve in the inset of Figure 1 shows two characteristic peaks. The small endothermic peak at about 100°C can be explained by freeing two NH_3 molecules in consistent with TG and DTG data. The residue gives a sharp exothermic peak at about 150°C. This exothermic peak related to the explosive decomposition of the complex via an intramolecular redox process occurring between the reductants (NH_3 ligands) and the oxidants (NO_3^-). The explosive decomposition of the complex resulted in the solid Co_3O_4 and gaseous products, i.e., NH_3, N_2, NO, N_2O, and H_2O. Although not all of the reaction products were determined in the analyses, the decomposition of the complex and formation of the Co_3O_4 nanoparticles can be expressed as follows:

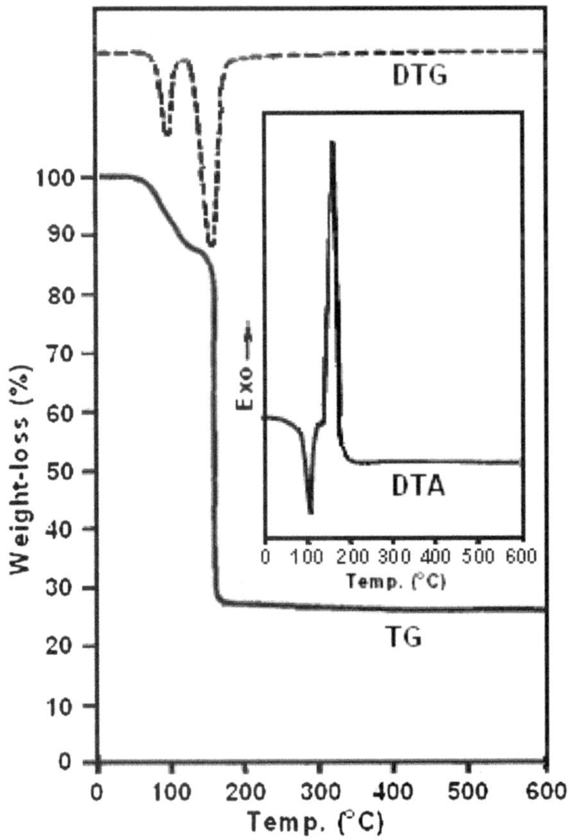

Figure 1 TG, DTG, and DTA curves of the $[Co(NH_3)_6](NO_3)_2$ complex.

$$9[Co(NH_3)_6](NO_3)_2(s) \rightarrow 3Co_3O_4(s) + 30NH_3(g) + 4NO(g) + 2N_2O(g) + 17N_2(g) + 36H_2O(g)$$

Figure 2 shows FT-IR spectra of the $[Co(NH_3)_6](NO_3)_2$ complex and its decomposition products at different temperatures. In Figure 2(spectrum a), the characteristic stretching bands of NH_3 and NO_3 groups are observed at about 3,500 to 3,000, 1,600, 1,350, and 800 cm^{-1}[44]. As shown in Figure 2(spectrum b), all the bands associated with the complex clearly disappeared when the complex was decomposed at 150°C. At this temperature, only two characteristic strong bands of the spinel-type Co_3O_4 structure at about 663.47 and 570.89 cm^{-1} are observed [45], confirming that the complex was decomposed completely at 150°C to the Co_3O_4 phase as indicated by the TG/DTA results. In Figure 2(spectrum c), the FT-IR spectrum of the sample decomposed at 175°C shows the strong bands related to Co_3O_4 without obvious change.

Figure 3 presents the XRD patterns of the decomposition products of the $[Co(NH_3)_6](NO_3)_2$ complex at 150°C and 175°C. Well-defined diffraction peaks at about 19.52°, 31.50°, 37.05°, 38.77°, 44.96°, 55.80°, 59.53°, 65.30°, 74.55°, 77.50°, and 78.60° are observed, corresponding to the (111), (220), (311), (222), (400), (422), (511), (440), (620), (533), and (622) planes of Co_3O_4 crystals, respectively (JCPDS card no. 76–1802). This result confirms that the complex was decomposed completely into the Co_3O_4 phase at 150°C, in good agreement with the thermal analysis and FT-IR results. No impurity peaks were found in the XRD patterns, confirming that the product is well-crystallized Co_3O_4 with high purity. As can be seen, the diffraction peaks are broadened because of the small size effect of the nanoparticles. The average crystallite size (d) was calculated to be approximately 13 nm by the Scherrer

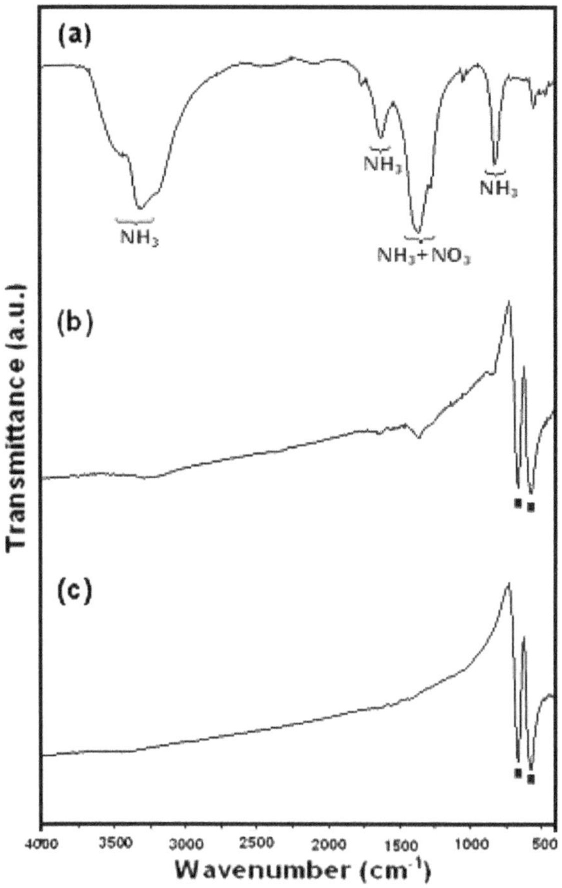

Figure 2 FT-IR spectra of (a) the $[Co(NH_3)_6](NO_3)_2$ complex and its decomposed products at (b) 150°C and (c) 175°C. The black-filled square indicates the Co_3O_4 phase.

relation [46]: $d = 0.9\lambda/(B\cos\theta)$, where λ is the wavelength of Cu Kα radiation, B is the corrected full width at half maximum of the diffraction peak, and θ is the Bragg angle. As shown in Figure 3 (pattern b), no new phase was observed when the decomposition temperature increased to 175°C, but the width of the Co_3O_4 peaks decreased because of crystallite growth.

Figure 4 displays the Raman spectrum of the Co_3O_4 nanoparticles. As shown in Figure 4, the Raman spectrum of the Co_3O_4 nanoparticles in the range of 300 to 800 nm shows four obvious peaks located at around 468, 510, 605, and 670 cm^{-1}, corresponding to the four Raman-active modes (A_{1g}, E_g, and $2F_{2g}$) of Co_3O_4. The Raman shifts are consistent with those of pure crystalline Co_3O_4[47,48], indicating that the Co_3O_4 nanoparticles have a similar crystal structure to bulk Co_3O_4. However, compared with those for bulk Co_3O_4, the peak positions of the four active modes shift to low wavenumbers by about 10 to 20 cm^{-1} [49]. This phenomenon is attributed to the optical phonon confinement effect in nanostructures, which can cause uncertainty in the phonon wave vectors and thus a

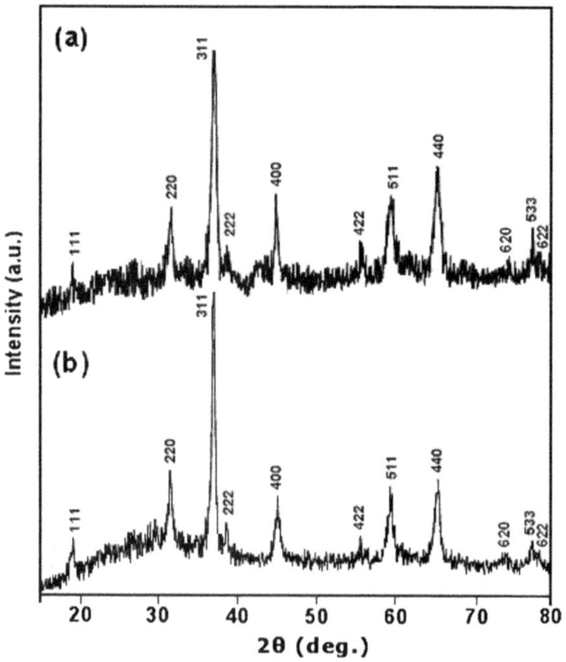

Figure 3 XRD patterns of the $[Co(NH_3)_6](NO_3)_2$ complex decomposed at (a) 150°C and (b) 175°C.

downshift in the Raman peaks [50]. This result further confirms the formation of the Co_3O_4 nanoparticles.

Figure 5 shows TEM images of the Co_3O_4 powder prepared by thermal decomposition of the $[Co(NH_3)_6](NO_3)_2$ complex at 150°C. The TEM sample was prepared by dispersing the powder in ethanol by ultrasonic vibration. It can be seen that the product was formed from extremely fine spherical particles which were loosely aggregated. The uniform Co_3O_4 particles have sphere-like shapes with weak agglomeration. As can be seen in the inset of Figure 5, the particle sizes possess a narrow distribution in a range of 6 to 20 nm, and the mean particle diameter is about 13 nm. In fact, the mean particle size determined by

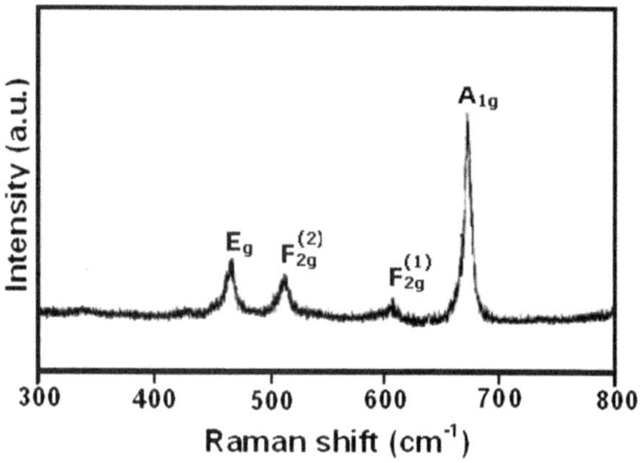

Figure 4 Raman spectrum of the Co_3O_4 nanoparticles.

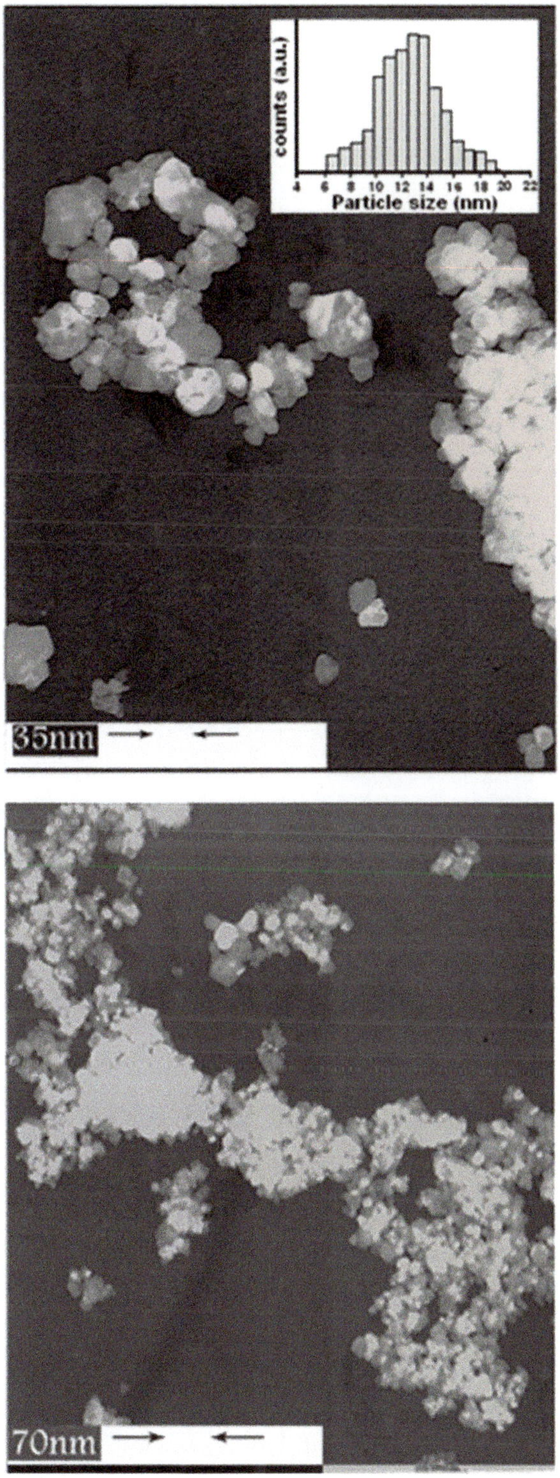

Figure 5 TEM images of the Co_3O_4 nanoparticles in two different magnifications. The inset shows the particle size distribution of Co_3O_4 nanoparticles.

TEM is very close to the average particle size calculated by the Debye-Scherer formula from the XRD pattern.

The EDX spectrum of the product in Figure 6 reveals the presence of only cobalt and oxygen peaks, with no other relevant elements present; the large Si peak can be attributed to the Si wafer upon which the nanoparticles were deposited prior to analysis. The atomic percentages of Co and O were found to be 43.15% and 56.95%, respectively. The atomic ratio of Co and O is 3:3.96, which approaches the theoretical value for Co_3O_4. This observation further confirms that the final product is only highly pure Co_3O_4 nanoparticles.

The surface area of the Co_3O_4 nanoparticles obtained from the decomposition of the complex was measured by the BET method. The specific surface area of the sample was 77.50 m^2/g. Assuming that the nanoparticles are almost spherical, as confirmed by TEM, the surface area can be used to estimate the particle size according to $D_{BET} = 6{,}000/(\rho \times S_{BET})$, where D_{BET} is the diameter of a spherical particle (nm), ρ is the theoretical density of Co_3O_4 (6.05 g/cm^3), and S_{BET} is the specific surface area of Co_3O_4 powder (m^2/g). The particle size calculated from the surface area data is about 12.80 nm, which is in good agreement with the XRD and TEM results.

The optical absorption properties of the as-prepared Co_3O_4 nanoparticles were investigated at room temperature by UV–vis spectroscopy (Figure 7). There are two obvious absorption peaks at 281 and 531 nm. Co_3O_4 is a p-type semiconductor, and its optical bandgap can be obtained from the following equation [16]: $(Ahv)^n = B(hv - E_g)$, where hv is the photon energy (eV), A is the absorption coefficient, B is a constant relative to the material, E_g is the bandgap, and n is either 1/2 for an indirect transition or 2 for a direct transition. The $(Ahv)^2$ versus hv curve for the product is shown in the inset of Figure 7. The value of hv extrapolated to $(Ahv)^2 = 0$ gives the absorption bandgap energy. Two regions with a linear relationship are observed in the ranges of 3.8 to 6.1 and 1.9 to 2.5, respectively, giving two E_g values of 3.2 and 2.3 eV.

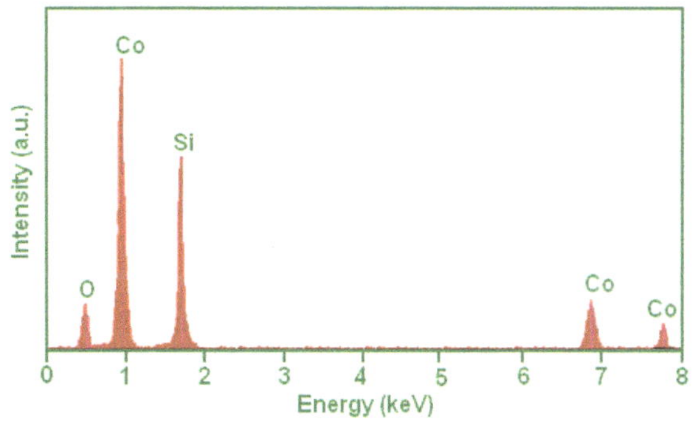

Figure 6 EDX analysis of the Co_3O_4 nanoparticles.

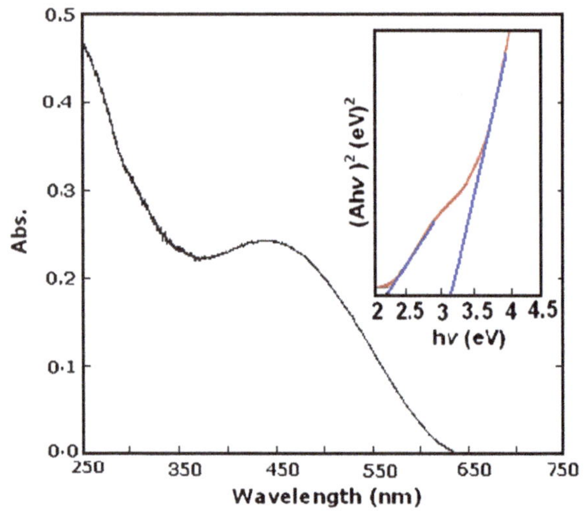

Figure 7 **UV–vis spectrum and $(Ahv)^2$-hv curve (inset) of the Co_3O_4 nanoparticles prepared at 150°C.**

The bandgap of 2.3 eV can be associated with the O^{II-}→Co^{II} charge transfer process (valence to conduction band excitation), while the 3.2 eV bandgap relates to the O^{II-}→Co^{III} charge transfer (with the Co^{III} level located below the conduction band) [51]. As has been investigated in the literatures [27,52], the E_g values of Co_3O_4 nanoparticles are greater than those of bulk Co_3O_4 (E_g=1.77 and 3.17 eV, respectively). The increase in the bandgap of the Co_3O_4 nanoparticles may ascribe to the quantum confinement effects of nanomaterials.

Figure 8 shows the magnetic properties of the Co_3O_4 nanoparticles. The fine shape of the hysteresis loops is a characteristic of a weak ferromagnetic behavior, although bulk Co_3O_4 is antiferromagnetic [53]. From the inset, the coercive field (H_c) and the remanent magnetization (M_r) are estimated to be 0.015 kOe and 0.002 emu/g, respectively. The low coercive fields and remanent magnetizations confirm that the Co_3O_4 nanoparticles have weak ferromagnetic properties. The maximum applied field, 8 kOe, does not saturate the magnetizations which should be attributed to weak ferromagnetic ordering of the spins in the nanoparticles. Co_3O_4 nanoparticles consist of small magnetic domains, each characterized by its own randomly oriented magnetic moment. The total magnetic moment of the nanoparticles is the sum of these magnetic domains coupled by dipolar interactions. The ferromagnetic behavior of the Co_3O_4 nanoparticles can be explained as follows: bulk Co_3O_4 has a normal spinel structure with antiferromagnetic exchange between ions occupying tetrahedral and octahedral sites [53]. It has zero net magnetization due to the complete compensation of sublattice magnetizations. Hence, the change from an antiferromagnetic state for bulk Co_3O_4 to a weakly ferromagnetic state for Co_3O_4 nanoparticles can be ascribed to uncompensated surface spins and/or finite size effects [54,55].

Conclusions

In this work, we presented a simple and low-temperature method of synthesizing spinel-type Co_3O_4 nanoparticles with an average particle size of 13.5 nm through thermolysis of the $[Co(NH_3)_6](NO_3)_2$ at low temperature (150°C). Co_3O_4 nanoparticles are formed from this complex via an explosive redox reaction between NH_3 ligands as the

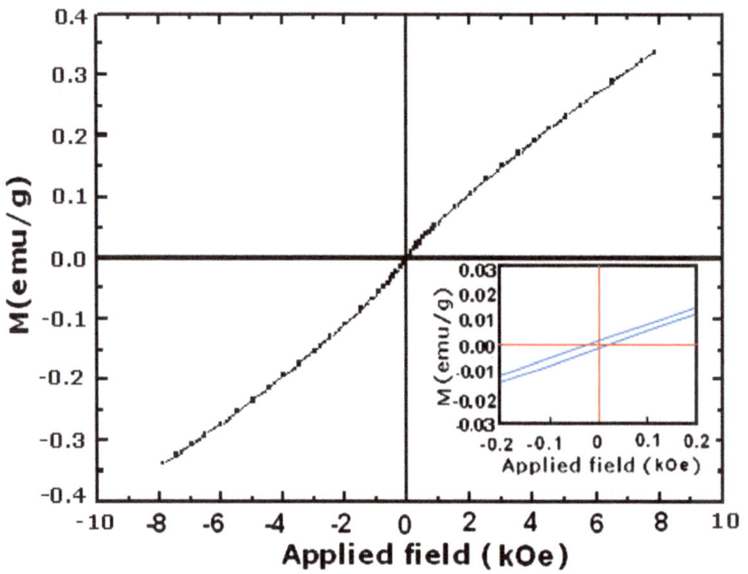

Figure 8 **Magnetization vs. field (*M-H*) of Co_3O_4 nanoparticles prepared from $[Co(NH_3)_6](NO_3)_2$ precursor at 150°C.** The inset shows its expansion of magnetization vs. field near the lower applied field.

reducing agent and the NO_3^- ions acting as the oxidizing agent. This method yields sphere-like Co_3O_4 nanoparticles with weak agglomeration, a narrow size distribution, and weak ferromagnetic behavior. The estimated optical absorption bandgaps of the Co_3O_4 nanoparticles were about 2.3 and 3.2 eV, which are relatively blueshifted, compared with the values for the bulk sample. This approach provides a one-step, general, and inexpensive route for the preparation of Co_3O_4 nanoparticles with high purity for industrial and high-technology applications.

Methods

Preparation of Co_3O_4 nanoparticles

Initially, the $[Co(NH_3)_6](NO_3)_2$ complex was synthesized via a simple reaction of an aqueous solution of $[Co(H_2O)_6](NO_3)_2$ with concentrated ammonia solution according to the reported method [56]. By adding acetone to the reaction mixture, the complex was precipitated as an orange precipitate. The composition of the complex was confirmed by thermal analysis, FT-IR, and elemental analysis: anal. calc. for $[Co(NH_3)_6](NO_3)_2$: Co, 20.68; H, 6.32; N, 39.30; found: Co, 21.10; H, 6.25; N, 39.89. To prepare Co_3O_4 nanoparticles, the $[Co(NH_3)_6](NO_3)_2$ complex was decomposed at selected temperatures for 1 h in an electric furnace under ambient air. The temperatures were selected according to the thermoanalytical data. The decomposition product was collected for characterization.

Characterization techniques

The XRD patterns were recorded by a Rigaku D-max C III X-ray diffractometer (Rigaku Corporation, Shibuya-ku, Japan) using Ni-filtered Cu Kα radiation (λ = 1.5418 Å) in order to determine the phases present in the decomposed samples. Infrared spectra were recorded on a Schimadzu system 160 FT-IR spectrophotometer (Shimadzu Corporation, Kyoto, Japan) using KBr pellets. The Raman spectra were measured on a Spex 1403 Raman spectrometer. Thermal analysis was conducted with a Netzsch STA 409 PC/PG thermal analyzer at a heating rate of 5°C min^{-1} in air. The optical absorption spectrum was recorded on a Shimadzu 1650PC UV–vis spectrophotometer in a wavelength range of 200 to 700 nm at room temperature. The sample for UV–vis studies was well dispersed in distilled water to form a homogeneous suspension by sonication for 25 min. The particle size was determined using a transmission electron microscope (Philips CM10, Philips, Amsterdam, The Netherlands) equipped with an energy dispersive X-ray analyzer. For the TEM measurements, the powders were ultrasonicated in ethanol, and a drop of the suspension was dried on a carbon-coated microgrid. The specific surface area of the product was measured by the BET method using N_2 adsorption-desorption isotherm carried out at −196°C on a surface area analyzer (Micromeritics ASAP 2010, Micromeritics, Norcross, GA, USA). Before each measurement, the sample was degassed at 150°C for 2 h. A vibrating sample magnetometer (Meghnatis Daghigh Kavir Co., Kashan, Iran) was used to measure the magnetic properties of Co_3O_4 nanoparticles.

Competing interests

The authors declare that they have no competing interests.

Authors' contributions

SF participated in the idea of the study, the design of the study, interpretation of the results, and writing the manuscript for publication. KP carried out the synthesis and the physicochemical characterization of the Co_3O_4 nanoparticles and early drafted the manuscript. SS participated in the discussions and interpretation of all characterization results. All authors read and approved the final manuscript.

Authors' information

SF is a professor at the Department of Chemistry, Lorestan University, Iran. He received his BSc degree in Chemistry from Shahid Chamran University in 1991, his MSc degree in Inorganic Chemistry from Tehran University in 1994, and his Ph.D. in Inorganic Chemistry from Isfahan University, Iran, in 2000. His recent research has concentrated on synthesis and characterization of metal and metal oxides nanostructures, and their catalytic applications. KP obtained her BSc and MSc degrees in Chemistry from the Faculty of Science, Lorestan University, Iran, in 2008 and 2012, respectively. During her MSc course, she has been involved in synthesizing cobalt oxide nanostructures through thermal decomposition of some ammine complexes. Her research interest is in the area of nanomaterial preparation. SS obtained her BSc degree in Applied Chemistry in 2009 from the Faculty of Science, Arak University, Iran. She earned her MSc degree in Inorganic Chemistry in 2012 with the thesis entitled 'Preparation and characterization of silver nanoparticles via chemical reduction with hydrogen peroxide (H_2O_2),' from the Faculty of Science, Lorestan University, Iran. Her research area interests include metals and metal oxide nanomaterials.

Acknowledgements

This work was supported by the Lorestan University Research Council and also supported by Iran Nanotechnology Initiative Council (INIC).

References

1. He, JH, Wu, TH, Hsin, CL, Li, KM, Chen, LJ, Chueh, YL: Beak-like SnO_2 nanorods with strong photoluminescent and field-emission properties. Small **2**, 116–120 (2006)
2. Xu, LP, Sithambaram, S, Zhang, YS, Chen, CH, Jin, L, Joesten, R, Suib, SL: Novel urchin-like CuO synthesized by a facile reflux method with efficient olefin epoxidation catalytic performance. Chem. Mater. **21**, 1253–1259 (2009)
3. Hu, CC, Wu, YT, Chang, KH: Low-temperature hydrothermal synthesis of Mn_3O_4 and MnOOH single crystals: determinant influence of oxidants. Chem. Mater. **20**, 2890–2894 (2008)
4. Zhou, J, Ding, Y, Deng, SZ, Gong, L, Xu, NS, Wang, ZL: Three-dimensional tungsten oxide nanowire networks. Adv. Mater. **17**, 2107–2110 (2005)
5. Lou, XW, Deng, D, Lee, JY, Feng, J, Archer, LA: Self-supported formation of needlelike Co_3O_4 nanotubes and their application as lithium-ion battery electrodes. Adv. Mater. **20**, 258–262 (2008)
6. Casas-Cabanas, M, Binotto, G, Larcher, D, Lecup, A, Giordani, V, Tarascon, JM: Defect chemistry and catalytic activity of nanosized Co_3O_4. Chem. Mater. **21**, 1939–1947 (2009)
7. Li, W-Y, Xu, L-N, Chen, J: Co_3O_4 nanomaterials in lithium-ion batteries and gas sensors. Adv. Funct. Mater. **15**, 851–857 (2005)
8. Chou, S-L, Wang, J-Z, Liu, H-K, Dou, S-X: Electrochemical deposition of porous Co_3O_4 nanostructured thin film for lithium-ion battery. J. Power. Sources **182**, 359–364 (2008)
9. Askarinejad, A, Bagherzadeh, M, Morsali, A: Catalytic performance of Mn_3O_4 and Co_3O_4 nanocrystals prepared by sonochemical method in epoxidation of styrene and cyclooctene. Appl. Surface Sci. **256**, 6678–6682 (2010)

10. Li, YG, Tan, B, Wu, YY: Mesoporous Co_3O_4 nanowire arrays for lithium ion batteries with high capacity and rate capacity. Nano Lett. **8**, 265–270 (2008)
11. Mate, VR, Shirai, M, Rode, CV: Heterogeneous Co_3O_4 catalyst for selective oxidation of aqueous veratryl alcohol using molecular oxygen. Catal. Commun. **33**, 66–69 (2013)
12. Makhlouf, SA: Magnetic properties of Co_3O_4 nanoparticles. J. Magn. Magn. Mater. **246**, 184–190 (2002)
13. Baydi, ME, Poillerat, G, Rehspringer, JL, Gautier, JL, Koenig, JF, Chartier, P: A sol–gel route for the preparation of Co_3O_4 catalyst for oxygen electrocatalysis in alkaline medium. J. Solid State Chem. **109**, 281–288 (1994)
14. Chen, Y, Zhang, Y, Fu, S: Synthesis and characterization of Co_3O_4 hollow spheres. Mater. Lett. **61**, 701–705 (2007)
15. Li, L, Chu, Y, Liu, Y, Song, JL, Wang, D, Du, XW: A facile hydrothermal route to synthesize novel Co_3O_4 nanoplates. Mater. Lett. **62**, 1507–1510 (2008)
16. Gu, F, Li, C, Hu, Y, Zhang, L: Synthesis and optical characterization of Co_3O_4 nanocrystals. J. Cryst. Growth **304**, 369–373 (2007)
17. Wang, RM, Liu, CM, Zhang, HZ, Chen, CP, Guo, L, Xu, HB, Yang, SH: Porous nanotubes of Co_3O_4: synthesis, characterization and magnetic properties. Appl. Phys. Lett. **85**, 2080–2082 (2004)
18. Kim, DY, Ju, SH, Koo, HY, Hong, SK, Kangf, YC: Synthesis of nanosized Co_3O_4 particles by spray pyrolysis. J. Alloys Compd. **417**, 254–258 (2006)
19. Mane, AU, Shalini, K, Wohlfart, A, Devi, A, Shivashankar, SA: Strongly oriented thin films of Co_3O_4 deposited on single-crystal MgO(100) by low-pressure, low-temperature MOCVD. J. Cryst. Growth **240**, 157–163 (2002)
20. Xuan, Y, Liu, R, Jia, YQ: Synthesis of a new series of compounds $RE_2Co_{2/3}Nb_{4/3}O_7$ and stability field diagram of $RE_2B_{2/3}'B_{4/3}''O_7$ pyrochlore compounds. Mater. Chem. Phys. **53**, 256–261 (1998)
21. Rumplecker, A, Kleitz, F, Salabas, EL, Schüth, F: Hard templating pathways for the synthesis of nanostructured porous Co_3O_4. Chem. Mater. **19**, 485–496 (2007)
22. Wang, WW, Zhu, YJ: Microwave-assisted synthesis of cobalt oxalate nanorods and their thermal conversion to Co_3O_4 rods. Mater. Res. Bull. **40**, 1929–1935 (2005)
23. Yang, LX, Zhu, YJ, Li, L, Zhang, L, Tong, H, Wang, WW: A facile hydrothermal route to flower-like cobalt hydroxide and oxide. Eur. J. Inorg. Chem. **23**, 4787–4792 (2006)
24. Mohandes, F, Davar, F, Salavati-Niasari, M: Preparation of Co_3O_4 nanoparticles by nonhydrolytic thermolysis of $[Co(Pht)(H_2O)]_n$ polymers. J. Magn. Magn. Mater. **322**, 872–877 (2010)
25. Jiang, J, Li, L: Synthesis of sphere-like Co_3O_4 nanocrystals via a simple polyol route. Mater. Lett. **61**, 4894–4896 (2007)
26. Ren, L, Wang, P, Han, Y, Hu, C, Wei, B: Synthesis of $CoC_2O_4{\cdot}2H_2O$ nanorods and their thermal decomposition to Co_3O_4 nanoparticles. Mater. Phys. Lett. **476**, 78–83 (2009)
27. Kumar, RV, Diamant, Y, Gedanken, A: Synthesis and characterization of nanometer-size transition metal oxides from metal acetates. Chem. Mater. **12**, 2301–2305 (2000)
28. Oh, SW, Bang, HJ, Bae, YC, Sun, Y-K: Effect of calcination temperature on morphology, crystallinity and electrochemical properties of nano-crystalline metal oxides (Co_3O_4, CuO, and NiO) prepared via ultrasonic spray pyrolysis. J. Power. Sources **173**, 502–509 (2007)
29. Lai, T, Lai, Y, Lee, C, Shu, Y, Wang, C: Microwave-assisted rapid fabrication of Co_3O_4 nanorods and application to the degradation of phenol. Catal. Today **131**, 105–110 (2008)
30. Bhatt, AS, Bhat, DK, Tai, C-W, Santosh, MS: Microwave-assisted synthesis and magnetic studies of cobalt oxide nanoparticles. Mater. Chem. Phys. **125**, 347–350 (2011)
31. Wang, X, Chen, XY, Gao, LS, Zheng, HG, Zhang, Z, Qian, YT: One-dimensional arrays of Co_3O_4 nanoparticles: synthesis, characterization, and optical and electrochemical properties. J. Phys. Chem. B **108**, 16401–16404 (2004)
32. Traversa, E, Sakamoto, M, Sadaoka, Y: A chemical route for the preparation of nanosized rare earth Perovskite-type oxides for electroceramic applications. Part. Sci. Technol. **16**, 185–214 (1998)
33. Salavati-Niasari, M, Davar, F, Mazaheri, M: Preparation of ZnO nanoparticles from [bis(acetylacetonato)zinc(II)]–oleylamine complex by thermal decomposition. Mater. Lett. **62**, 1890–1892 (2008)
34. Salavati-Niasari, M, Davar, F: Synthesis of copper and copper(I) oxide nanoparticles by thermal decomposition of a new precursor. Mater. Lett. **63**, 441–443 (2009)
35. Salavati-Niasari, M, Davar, F, Mazaheri, M: Synthesis and characterization of ZnS nanoclusters via hydrothermal processing from [bis(salicylidene)zinc(II)]. J. Alloys Compd. **470**, 502–506 (2009)
36. Davar, F, Fereshteh, Z, Salavati-Niasari, M: Nanoparticles Ni and NiO: synthesis, characterization and magnetic properties. J. Alloys Compd. **476**, 797–801 (2009)
37. Salavati-Niasari, M, Fereshteh, Z, Davar, F: Synthesis of cobalt nanoparticles from [bis(2-hydroxyacetophenato)cobalt(II)] by thermal decomposition. Polyhedron **28**, 1065–1068 (2009)
38. Salavati-Niasari, M, Khansari, A, Davar, F: Synthesis and characterization of cobalt oxide nanoparticles by thermal treatment process. Inorg. Chim. Acta **362**, 4937–4942 (2009)
39. Salavati-Niasari, M, Mir, N, Davar, F: Synthesis and characterization of Co_3O_4 nanorods by thermal decomposition of cobalt oxalate. J. Phys. Chem. Solids **70**, 847–852 (2009)
40. Farhadi, S, Rashidi, N: Preparation and characterization of pure single-phase $BiFeO_3$ nanoparticles through thermal decomposition of the heteronuclear $Bi[Fe(CN)_6]{\cdot}5H_2O$ complex. Polyhedron **29**, 2959–2965 (2010)
41. Farhadi, S, Roostaei-Zaniyani, Z: Simple and low-temperature synthesis of NiO nanoparticles through solid-state thermal decomposition of the hexa(ammine)Ni (II) nitrate, $[Ni(NH_3)_6](NO_3)_2$, complex. Polyhedron **30**, 1244–1249 (2011)
42. Farhadi, S, Safabakhsh, J: Solid-state thermal decomposition of the $[Co(NH_3)_5CO_3]NO_3{\cdot}0.5H_2O$ complex: a simple, rapid and low-temperature synthetic route to Co3O4 nanoparticles. J. Alloys Compd. **515**, 180–185 (2012)
43. Farhadi, S, Pourzare, K: Simple and low-temperature preparation of Co_3O_4 sphere-like nanoparticles via solid-state thermolysis of the $[Co(NH_3)_6](NO_3)_3$ complex. Mater. Res. Bull. **47**, 1550–1556 (2012)
44. Nakamoto, K: Infrared and Raman Spectra of Inorganic and Coordination Compounds, Part B: Applications in Coordination, Organometallic, and Bioinorganic Chemistry, 6th edn. Wiley, New York (2009)
45. Pejova, B, Isahi, A, Najdoski, M: Fabrication and characterization of nanocrystalline cobalt oxide thin films. Mater. Res. Bull. **36**, 161–170 (2001)
46. Klug, HP, Alexander, LE: X-ray Diffraction Procedures, 2nd edn. Wiley, New York (1964)
47. Marinkovic Stanojevic, ZV, Romcevic, N, Stojanovic, B: Spectroscopic study of spinel $ZnCr_2O_4$ obtained from mechanically activated $ZnO–Cr_2O_3$ mixtures. J. Eur. Ceram. Soc. **27**, 903 (2007)
48. Ramana, CV, Massot, M, Julien, CM: XPS and Raman spectroscopic characterization of $LiMn_2O_4$ spinels. Surf. Interface Anal. **37**, 412 (2005)
49. Hadjiev, VG, Iliev, MN, Vergilov, IV: The Raman spectra of Co_3O_4. J. Phys. C. Solid State Phys. **21**, L199 (1988)
50. Cao, B, Cai, W, Duan, G, Li, Y, Zhao, Q, Yu, D: Template-free electrochemical deposition route to ZnO nanoneedle arrays and their optical and field emission properties. Nanotechnology **16**, 2567 (2005)
51. He, T, Chen, DR, Jiao, XL, Wang, YL, Duan, YZ: Solubility-controlled synthesis of high-quality Co_3O_4 nanocrystals. Chem. Mater. **17**, 4023–4030 (2005)
52. Gulino, A, Dapporto, P, Rossi, P, Fragala, I: A novel self-liquid MOCVD precursor for Co_3O_4 thin films. Chem. Mater. **15**, 3748–3752 (2003)
53. Ichiyanagi, Y, Kimishima, Y, Yamada, S: Magnetic study on Co_3O_4 nanoparticles. J. Magn. Magn. Mater. **272–276**, e1245–e1246 (2004)
54. Kodama, RH, Makhlouf, SA, Berkowitz, AE: Growth mechanism and magnon excitation in NiO nanowalls. Phys. Rev. Lett. **79**, 1393–1396 (1997)
55. Ozkaya, T, Baykal, A, Toprak, MS, Koseoglu, Y, Durmus, Z: Reflux synthesis of Co_3O_4 nanoparticles and its magnetic characterization. J. Magn. Magn. Mater. **321**, 2145–2149 (2009)
56. Liszka-skoczylas, M, Mikuli, E, Szklarzewicz, J, Hetmanczyk, J: Thermal behaviour, phase transition and molecular motions in $[Co(NH_3)_6](NO_3)_2$. Thermochim. Acta **496**, 38–44 (2009)

Mechanical properties of multi-walled carbon nanotube/polyester nanocomposites

Mahmoud M Shokrieh[1*], Ali Saeedi[1] and Majid Chitsazzadeh[2]

Abstract

Carbon nanotubes (CNTs), due to their superlative mechanical and physical properties, have shown a high potential to improve properties of polymeric composites. Adding CNTs into polymers at very low weight fractions can improve mechanical properties of the resulting nanocomposites. In the present paper, multi-walled carbon nanotubes (MWCNTs) at different weight ratios (0.05, 0.1, and 0.5 wt.%) were added to polyester. Mechanical stirring and sonication technique were used to achieve good dispersion state of MWCNTs in the polymeric matrix. The results of mechanical tests (tensile and flexural) exhibit improvements of tensile and flexural strengths by 6% and 20%, respectively, at only 0.05 wt.% MWCNT. Improvements in Young's modulus and flexural modulus were also observed. Scanning electron microscopy was employed to determine the dispersion state of nanotubes in the matrix as well as the fracture surface properties.

Keywords: Carbon nanotube, Polyester, Nanocomposites, Mechanical properties

Background

Carbon nanotubes (CNTs) are tiny tubes with diameters of a few nanometers and lengths of several microns made of carbon atoms. Since the discovery of this form of carbon atoms by Iijima [1] in 1991, many attentions have been drawn to use the outstanding physical and chemical properties of CNTs such as high Young's modulus (approximately 1 TPa), tensile strength, and excellent thermal and electrical conductivities [2].

Carbon nanotubes have been used in various fields of applications in recent years due to their high physical, chemical, and mechanical properties [3]. One of these fields is composite materials in which CNTs are added to a matrix not only as reinforcement but also to obtain other physical and chemical properties such as electrical conductivity and corrosion resistance. Carbon nanotubes are specially introduced into polymer matrices like epoxy to fabricate polymer matrix nanocomposites which presents a new generation of composite materials [4-8]. Gojny et al. [4] fabricated double-walled carbon nanotubes/epoxy nanocomposites and reported increases in strength and Young's modulus of the resulting nanocomposites at nanotube content of 0.1 wt.%. Martone et al. [5] investigated the effect of dispersed multi-walled carbon nanotube (MWCNT) on the enhancement of elastic modulus in an epoxy system. Tai et al. [6] reported the enhancement of strength and Young's modulus of phenolic composites reinforced with single-walled carbon nanotubes.

Dispersion of nanofillers plays a very important role in the use of filler properties in polymeric composites. Nanoparticles due to large surface area and mostly high aspect ratio tend to agglomerate greatly which reduces the ability to show their expected properties [9]. A technique to achieve good dispersion of nanoparticles is ultra-sonication which can be used also for CNTs. Applying this method to disperse CNTs in a low-price polymer-like polyester which has good properties such as versatileness, quick curing, and low viscosity leads to fabrication of CNT/polyester composite with enhanced properties. Despite many investigations in CNT/polymer composites as were reviewed by Thostenson et al. [7], only a few works on CNT/polyester composites in the literature are present. Seyhan et al. [10] used amine functionalized and untreated carbon nanotubes to fabricate polyester-based nanocomposites by a three-roll mill dispersion technique. They used a mixture of polyesters without styrene content and characterized the rheological behavior and

* Correspondence: shokrieh@iust.ac.ir
[1]Composites Research Laboratory, Center of Excellence in Experimental Mechanics and Dynamics, School of Mechanical Engineering, Iran University of Science and Technology, Narmak, Tehran 16846-13114, Iran
Full list of author information is available at the end of the article

tensile strength of the resulting composites. The results of their work did not show considerable enhancement in the tensile strength of nanocomposites at very low contents of nanotubes (0.1 wt.%).

In the present study, the effects of adding MWCNT at different weight ratios (0.05, 0.1, and 0.5 wt.%) to the unsaturated polyester resin have been investigated. Mechanical stirring and ultra-sonication technique were used to disperse nanotubes in the matrix. In order to study the effect of addition of nanotubes on the mechanical properties of polyester resin, tensile and flexural tests were conducted. Moreover, scanning electron microscopy was used to determine the properties of fractured surfaces.

Results and discussion

Figure 1a,b,c,d shows the fracture surfaces of samples at 0, 0.05, 0.1, and 0.5 wt.% MWCNT contents. As can be seen, the fracture of neat polyester (Figure 1a) is much smoother than the fracture surfaces of composites, indicating a typical brittle fracture of the polyester, although Figure 1b,c,d shows rougher surfaces due to the addition of more MWCNT particles which force the defects that created the cracks to dodge the CNT particles during the crack propagation.

Figures 2 and 3 present the results of tensile and flexural strengths of MWCNT/polyester composites. According to the reported results, improvements on tensile and flexural strengths of 6% and 20%, respectively, were obtained by the addition of 0.05 wt.% MWCNT. The results show that by increasing the amount of filler content, the tensile strength reaches a peak at 0.1 wt.% and subsequently decreases.

Figures 4 and 5 present the Young's modulus and flexural modulus of the resulting nanocomposites. Unlike the results of tensile and flexural strengths, the optimum amount of filler content for tensile strength is achieved at 0.05 wt.%. The reason for such a phenomenon can be explained by the dispersion state of MWCNT particles.

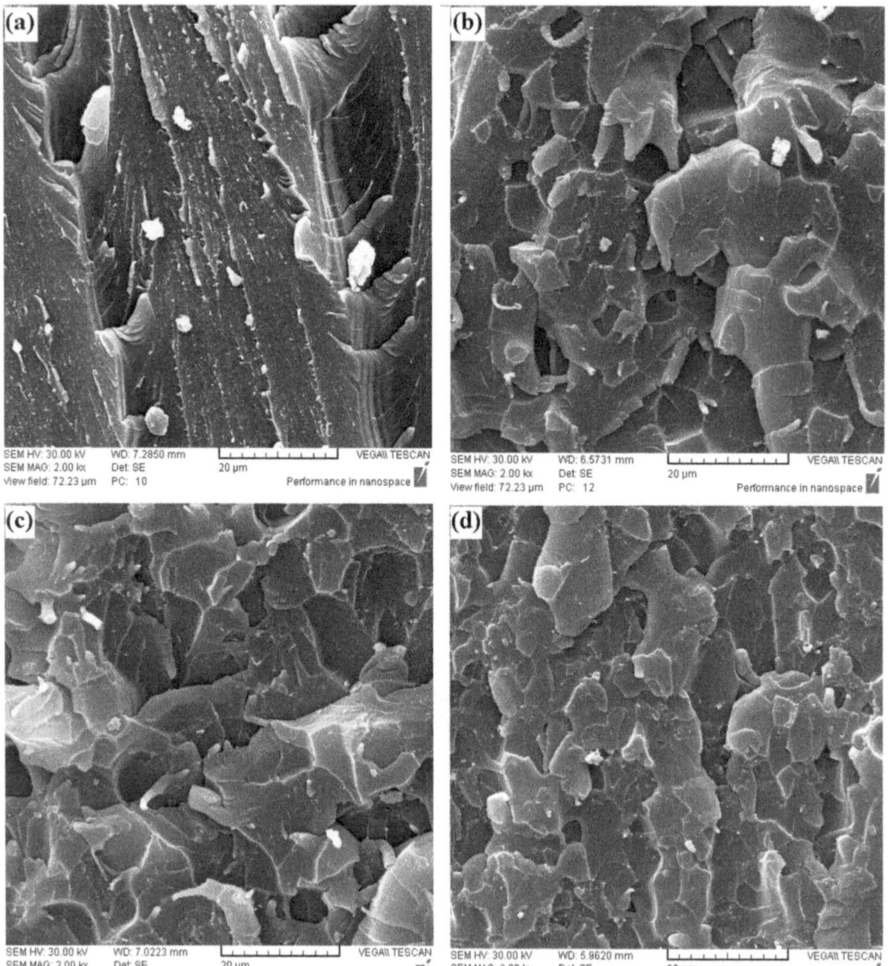

Figure 1 Nanocomposites (a) 0.0, (b) 0.05, (c) 0.1, and (d) 0.5 wt.% MWCNT. Fracture surface of MWCNT/polyester nanocomposites (a) 0.05, (b) 0.1, (c) 0.3, and (d) 0.5 wt.% nanotube contents.

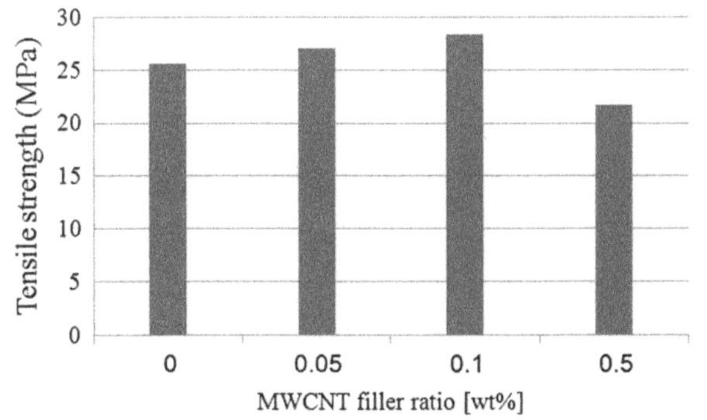

Figure 2 **Tensile strength of MWCNT/polyester nanocomposites in various filler contents.**

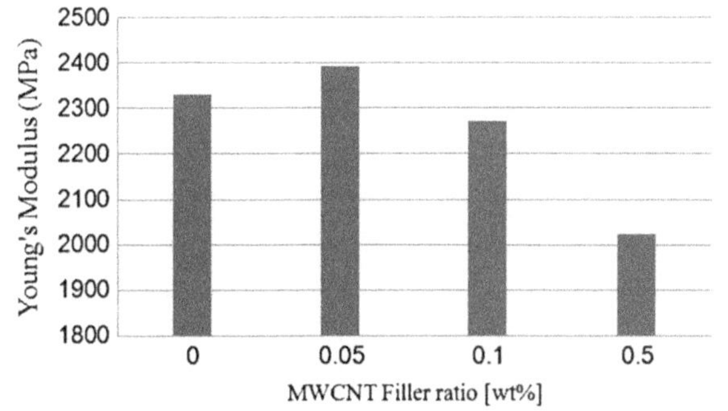

Figure 4 **Young's modulus of MWCNT/polyester nanocomposites in various filler contents.**

Figure 6 shows a scanning electron microscope (SEM) image of dispersion state of MWCNT in a sample containing 0.05 wt.% MWCNT. As depicted in Figure 6, MWCNT particles were fully dispersed in polyester and thus improved its mechanical properties, but at higher weight fractions of nanotubes, the major problem is agglomeration of MWCNT that causes stress concentration and thus decreases the mechanical properties of the nanocomposites. Therefore, it can be assumed that more sonication time or output power may help better dispersion of particles at higher weight fractions, but it should be mentioned that although more sonication leads to a better dispersion of particles, there might be still some agglomerates. Sonication can also decrease the length of nanotubes which reduces the effect of CNT as reinforcement, so increasing the sonication energy necessarily does not enhance the mechanical results. Figure 7 shows the fracture surface of a sample at 0.5 wt.% MWCNT. As can be seen in this figure, a large agglomeration of nanotubes acts as a stress concentration and makes the sample fail at lower stresses.

According to the results shown in Figures 1 and 2, the optimum amount of MWCNT for tensile and flexural moduli was obtained almost at the identical weight fraction for tensile and flexural strengths.

Conclusion

In this paper, mechanical properties of MWCNT/polyester nanocomposites at different weight ratios of nanotubes (0.05, 0.1, and 0.5 wt.%) were investigated. Dispersion of nanotubes in the resin is the key parameter that influences the results of mechanical tests. Scanning electron microscopy demonstrated that MWCNTs were well dispersed in the polyester resin at low contents of nanofillers (0.05 and 0.1 wt.%), and the result did not show significant nanotube aggregates. However, at higher nanotube contents (0.5 wt.%), a large amount of agglomerations are observed that act as stress concentration and reduce the strength of nanocomposites.

The trend of results shows that adding MWCNT at low weight fractions improves tensile and flexural properties of polyester resin. However, more increase in MWCNT weight fraction causes reduction in mechanical properties due to the agglomeration of nanotubes. The optimum amounts of filler for the best properties for tensile strength and flexural strength occur at 0.1 and 0.05 wt.%, respectively.

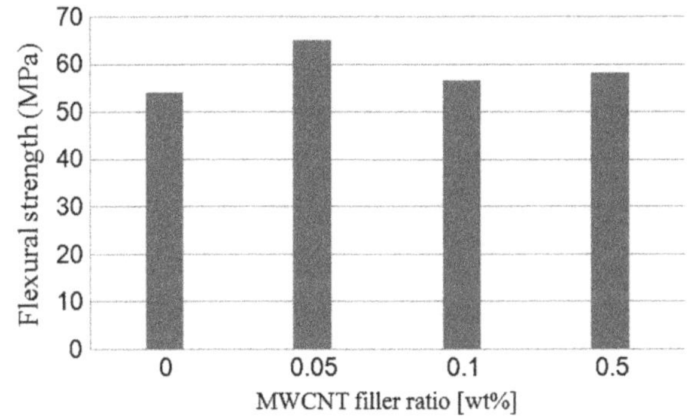

Figure 3 **Flexural strength of MWCNT/polyester nanocomposites in various filler contents.**

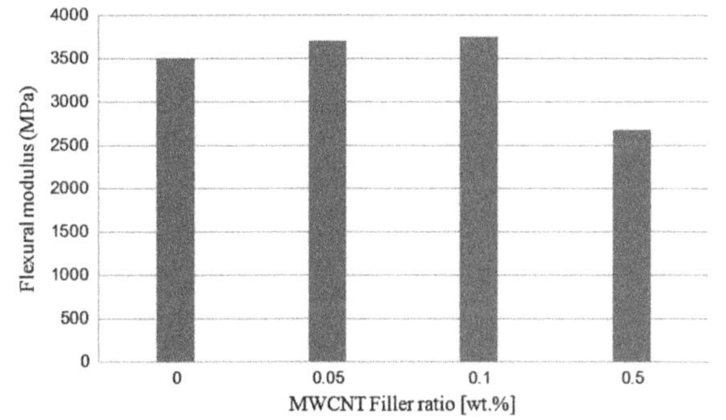

Figure 5 **Flexural modulus of MWCNT/polyester nanocomposites in various filler contents.**

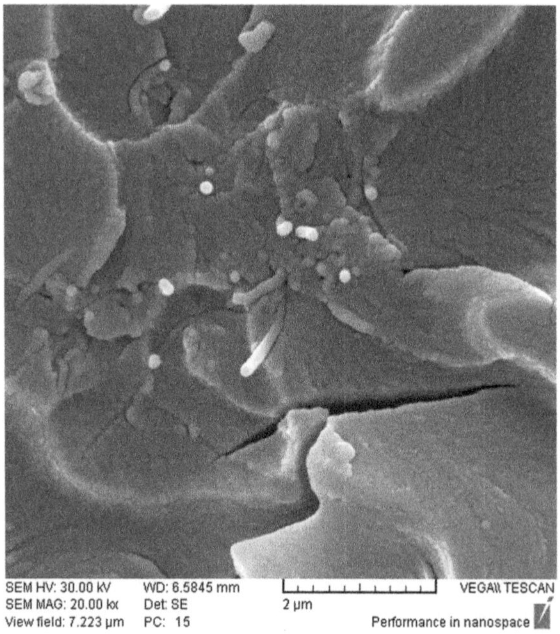

Figure 6 SEM micrograph of fracture surface of a sample containing 0.05 wt.% MWCNT.

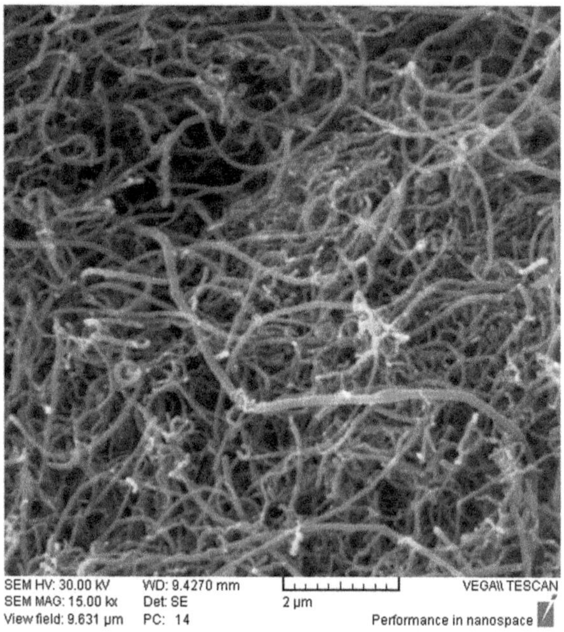

Figure 8 SEM micrograph of supplied MWCNT.

Methods

Materials

The multi-walled carbon nanotubes used in this study were CVD grown with diameters between 10 to 20 nm and lengths between 5 to 15 µm and were supplied by IoLiTec Co. (Germany). Figure 8 depicts a SEM image of carbon nanotubes as supplied. The unsaturated polyester resin was purchased from Boytek Co. (Turkey) and was cured with cobalt naphthenate (8%) and MEKP (Butanox M60, Akzo Nobel Chemicals, Gillingham, Kent, UK) as recommended by the manufacturer.

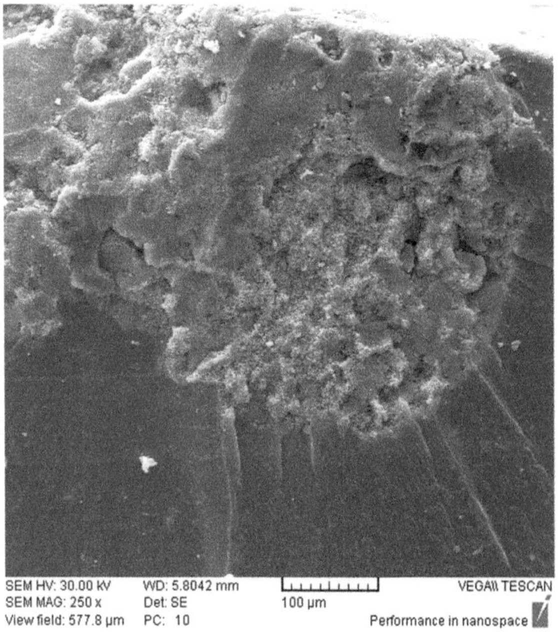

Figure 7 Fracture surface of a sample containing 0.5 wt.% MWCNT agglomeration acts as stress concentration.

Preparation method

Multi-walled carbon nanotube at 0.05, 0.1, and 0.5 wt.% were added to polyester resin and were initially mixed by mechanical stirrer for 20 min at 2,000 rpm. Then, the mixture was sonicated by a 14-mm-diameter probe sonicator, Hielscher UP400S (Tetlow, Germany). The output power of 200 W and sonication time of 45 min were applied to obtain sonication energy of 1 kJ/g of mixture, as the sonication energy should be as much as it can break the agglomerations and disperse the particles in the mixture [11]. However, higher sonication energy leads to damage of the CNTs which might decrease their properties. After the dispersion of MWCNT in polyester resin, the mixture was molded and cured at room temperature followed by a post-cure cycle of 2 h at 80°C.

Characterization

Tensile and flexural tests were performed according to ASTM D3039 [12] and ASTM D790 [13], respectively, to evaluate the effects of adding MWCNT on polyester using the universal testing machine, Santam STM-150 (Tehran, Iran). Tescan Vega II electron microscope (Brno, Czech Republic) was used to investigate the fracture surface and dispersion state of MWCNTs.

Competing interests
The authors declare that they have no competing interests.

Authors' contributions
MMS designed the experiments and revised the manuscript. AS studied the literature and helped MC to fabricate the specimens and perform the tests. Both of them drafted the manuscript. All authors read and approved the final manuscript.

Author details
[1]Composites Research Laboratory, Center of Excellence in Experimental Mechanics and Dynamics, School of Mechanical Engineering, Iran University of Science and Technology, Narmak, Tehran 16846-13114, Iran.
[2]Nanocomposites Research Laboratory, Iran University of Science and Technology, Narmak, Tehran 16846-13114, Iran.

References
1. Iijima, S: Helical microtubules of graphitic carbon. Nature **354**, 56–58 (1991)
2. Coleman, JN, Khan, U, Blau, WJ, Gun'ko, YK: Small but strong: a review of the mechanical properties of carbon nanotube–polymer composites. Carbon **44**, 1624–1652 (2006)
3. Baughman, RH, Zakhidov, AA, de Heer, WA: Carbon nanotubes—the route toward applications. Science **297**, 787–792 (2002)
4. Gojny, FH, Wichmann, MHG, Kopke, U, Fiedler, B, Schulte, K: Carbon nanotube-reinforced epoxy-composites: enhanced stiffness and fracture toughness at low nanotube content. Compos. Sci. Technol **64**, 2363–2371 (2004)
5. Martone, A, Formicola, C, Giordano, M: Reinforcement efficiency of multi-walled carbon nanotube/epoxy nano composites. Compos Sci Tech **70**, 1154–1160 (2010)
6. Tai, NH, Yeh, MK, Peng, TH: Experimental study and theoretical analysis on the mechanical properties of SWNTs/phenolic composites. Composites: Part B **39**, 926–932 (2008)
7. Thostenson, ET, Li, C, Chou, TW: Nanocomposites in context. Compos Sci Tech **65**, 491–516 (2005)
8. Goadagno, L, Vertuccio, L, Sorrentiono, A, Raimondo, M, Naddeo, C, Vittoria, V: Mechanical and barrier properties of epoxy resin filled with multi-walled carbon nanotubes. Carbon **47**, 2419–2430 (2009)
9. Peigney, A, Laurent, CH, Flahaut, E, Bacsa, RR, Rousset, A: Specific surface area of carbon nanotubes and bundles of carbon nanotubes. Carbon **39**, 507–514 (2001)
10. Seyhan, A, Gojny, FH, Tanglu, M, Schulte, K: Critical aspects related to processing of carbon nanotube/unsaturated thermoset polyester nanocomposites. Eur Polym J **43**, 374–379 (2007)
11. Battisti, A, Skordosa, AA, Partridge, IK: Monitoring dispersion of carbon nanotubes in a thermosetting polyester resin. Compos Sci Tech **69**, 1516–1520 (2009)
12. ASTM: ASTM D 3039–00 standard test methods for tensile properties of polymer matrix composite materials, vol 15.03. ASTM, West Conshohocken (2000)
13. ASTM: ASTM D 790–03 standard test methods for flexural properties of unreinforced and reinforced plastics and electrical insulating materials, vol. 8.01. , West Conshohocken (2003)

Optical absorption at one-particle states of charge carriers in semiconductor quantum dots

Pokutnyi Sergey

Abstract

It is shown that the oscillator strengths and dipole moments of the transitions for one-particle states of charge carriers that originated in semiconductor quantum dots take on large values, significantly exceeding the typical values of the corresponding values for semiconductor materials. In the framework of the dipole approach, it is shown that large values of the cross section of light in the studied quasi-zero-dimensional systems enable the use of such nanosystems as new strongly absorbing nanomaterials.

Keywords: Optical absorption; Charge carriers; Semiconductor quantum dots

PACS: 78. 67.Hc; 73. 30. Ly; 78.20. Dj

Correspondence/Findings

Introduction

At present, the optical [1-5] and electro-optical [6-8] properties of quasi-zero-dimensional structures are extensively studied. Such structures commonly consist of spherical semiconductor nanocrystals (the so-called quantum dots (QDs)) with a radius $a \approx 1$ to 10^2 nm grown in semiconductor (or dielectric) matrices. The studies in this field are motivated by the fact that such heterophase systems represent as new promising materials for the development of new components of nonlinear optoelectronics to be used, specifically, for controlling optical signals in optical computers [9-12] or for manufacturing active layers of semiconductor injection lasers [5,13-17].

The optical and electro-optical properties of such quasi-zero-dimensional structures depend on the energy spectrum of a spatially confined electron–hole pair (EHP), i.e., an exciton [1-8]. By the methods of optical spectroscopy, the effects of quantum confinement on the energy spectra of electrons [1,2] and excitons [3,4] were revealed in these heterophase structures.

Previously [18], the conditions for the localization of charge carriers near the spherical interface between two dielectric media were analyzed. In this case, the polarization interaction of a charge carrier with the surface charge induced at the spherical interface, $U(r,a)$, depends on the relative permittivity $\varepsilon = \varepsilon_1/\varepsilon_2$. Here, r is the spacing between the charge carrier and the center of the dielectric particle, a is the radius of the particle, and ε_1 and ε_2 are the permittivities of the surrounding medium and of the dielectric particle embedded in the medium, respectively.

For the charge carriers in motion near the dielectric particle, there are two possibilities:

1. Due to the polarization interaction $U(r,a)$, the carriers can be attracted to the particle surface (to the outer or the inner surface at $\varepsilon < 1$ or $\varepsilon > 1$, respectively), with the formation of outer [19,20] or inner [21] surface states.
2. Due to the polarization interaction $U(r,a)$, the carriers can be, at $\varepsilon < 1$, repelled from the inner surface of the dielectric particle, with the formation of bulk local states inside the particle bulk [22-24]; in this case, the spectrum of the low-energy bulk states is of an oscillatory shape.

It has been shown [18-24] that the formation of the above-mentioned local states is of a threshold-type nature and is possible if the radius of the dielectric particle a is larger than a certain critical radius a_c:

$$a \geq a_c = 6|\beta|^{-1} a_{Bi}, \tag{1}$$

Correspondence: pokutnyi_Sergey@inbox.ru
G. V. Kurdumov Institute for Metal Physics, NAS Ukraine, Vernadsky Blvd., 36, Kiev 03680, Ukraine

where $\beta = (\varepsilon_1 - \varepsilon_2)/(\varepsilon_1 + \varepsilon_2)$ and α_{Bi} is the Bohr radius of a charge carrier in a medium with the permittivity ε_1 $(i = 1, 2)$.

In [14-17], the optical properties of an array of InAs and InSb QDs in GaAs and GaSb matrices and the corresponding operational characteristics of injection lasers, with the active region on the basis of this array, were studied experimentally. In these studies, a large short-wavelength shift of the laser emission line was observed for the array of QDs. In such an array, the energy spectrum of charge carriers is completely discrete [21-24] if the QDs are smaller than $a \approx 1$ to 7 nm in size. In the first-order approximation, the spectrum of such quantum-confined states can be described as a spectrum of a charge carrier in a spherically symmetric well with infinitely high walls.

Until now, there have been no theoretical investigations on optical absorption and scattering at such discrete states in arrays of QDs. To close the gap in this area, here we develop a theory of the interaction of an electromagnetic with one-particle quantum-confined states of charge carriers that originate in the bulk of a semiconductor QD.

In conclusion, we briefly discuss possible physical situations in which the results obtained above can be used for interpreting the experimental data.

Hamiltonian for an electron–hole pair in a quantum dot

We consider a simple model in which a quasi-zero-dimensional system is defined as a neutral spherical semiconductor QD of radius a and permittivity ε_2, embedded in a surrounding with permittivity ε_1. Let an electron (e) and hole (h), whose effective masses are, correspondingly, m_e and m_h, be the motion in this QD. Let the spacing between the electron or hole and the QD center be r_e or r_h, respectively. We assume that the bands for electrons and hole are parabolic. Along with the QD radius a, the characteristic lengths of the problem are a_e, a_h, and a_{ex}, where

$$a_e = \varepsilon_2\hbar^2/m_e e^2, a_h = \varepsilon_2\hbar^2/m_h e^2, a_{ex} = \varepsilon_2\hbar^2/\mu e^2 \quad (2)$$

are the Bohr radii of the electron, hole, and exciton in the semiconductor with the permittivity ε_2, respectively, and $\mu = m_e m_h/(m_e + m_h)$ is the exciton effective mass. All of the characteristic lengths of the problem are much larger than the interatomic spacing α_0:

$$a, a_e, a_h, a_{ex} \gg a_0. \quad (3)$$

This allows us to treat the motion of the electron and hole in the QD in the effective-mass approximation.

In the context of the above-described model and approximations for a quasi-zero-dimensional system, the Hamiltonian of the EHP is

$$H = -\frac{\hbar^2}{2m_e}\Delta_e - \frac{\hbar^2}{2m_h}\Delta_h + V_{eh}(r_e, r_h) + U(r_e, r_h, a) + E_g, \quad (4)$$

where the first two terms in the sum define the kinetic energy of the electron and hole, E_g is the energy bandgap in the bulk (unbounded) semiconductor with the permittivity ε_2, $V_{eh}(\mathbf{r_e},\mathbf{r_h})$ is the energy of the electron Coulomb interaction:

$$V_{eh}(r_e, r_h) = \frac{e^2}{2\varepsilon_2 a}\frac{2a}{\left(r_e^2 - 2r_e, r_h\cos\theta + r_h^2\right)^{1/2}} \quad (5)$$

with the angle θ between the vector $\mathbf{r_e}$ and $\mathbf{r_h}$, and $U(\mathbf{r_e},\mathbf{r_h},a)$ is the energy of the interaction of the electron and hole with the polarization field induced by the electron and hole at the spherical interface between the two media. For arbitrary values of ε_1 and ε_2, the interaction energy $U(\mathbf{r_e},\mathbf{r_h},a)$ can be represented analytically as [18,24]

$$U(r_e, r_h, a) = -\frac{e^2\beta}{\varepsilon_2 a}\frac{1}{\left[(r_e r_h/a^2)^2 - 2(r_e r_h/a^2)\cos\theta + 1\right]^{1/2}} - \frac{e^2\beta}{2(\varepsilon_1+\varepsilon_2)a}\int_0^\infty \frac{dy(a^2/r_h y)^\alpha\Theta(y - a^2/r_h)}{|r_e - y(r_e/r_h)|} - \frac{e^2\beta}{2(\varepsilon_1+\varepsilon_2)a}\int_0^\infty \frac{dy(a^2/r_e y)^\alpha\Theta(y - a^2/r_e)}{|r_h - y(r_h/r_e)|}, \quad (6)$$

where $\Theta(x)$ is the unit step function and

$$\alpha = \frac{\varepsilon_1}{\varepsilon_1 + \varepsilon_2}. \quad (7)$$

The spectrum of charge carriers in a quantum dot

In the bulk of a QD, the electron (hole) energy levels can originate. Their energies are defined as [25]

$$E_{n,l}(a) = \frac{\hbar^2}{2m_{e(h)}a^2}\varphi_{n,l}^2, \quad (8)$$

where the subscripts (n,l) refer to the corresponding quantum size-confined states. Here, n and l are the principal and azimuthal quantum numbers for the electron (hole), respectively, and $\varphi_{n,l}$ are the roots of the Bessel function, i.e., $J_{1+\frac{1}{2}}(\varphi_{n,l}) = 0$. For the quantum-confined levels to originate, it is necessary that in the Hamiltonian (4), the electron (hole) energy $E_{n,l}(a)$ (8) be much larger than the energy of the interaction of the electron (hole) with the polarization field $U(a)$ (6) generated at the spherical QD-dielectric (semiconductor) matrix interface:

$$E_{n,l}(a) = \frac{\hbar^2}{2m_{e(h)}a^2}\varphi_{n,l}^2 \gg U(a) \approx \frac{e^2\beta}{2\varepsilon_2 a}. \quad (9)$$

Condition (9) is satisfied for QDs of radii

$$a \ll a_s^{e(h)} = \frac{\varphi_{n,l}^2}{\beta} a_{e(h)}. \quad (10)$$

At room temperature T_0, the discrete levels of the electron (hole) $E_{n,l}$ (8) in the QD are slightly broadened if the energy separation between the levels is [20-22]

$$\Delta E_{n,l}(a) = E_{n',l} - E_{n,l}(a) \ll kT_0. \quad (11)$$

Taking into account (8), we can rewrite inequality (11) as

$$\frac{\hbar^2}{2m_{e(h)}a^2}\frac{\varphi_{n',l}^2 - \varphi_{n,l}^2}{kT_0} = \eta(a) \ll 1. \quad (12)$$

Formula (8), describing the spectrum of charge carriers in a QD, is applicable to the lowest states (n,l) that satisfy the inequality

$$\Delta E_{n,l}(a) \ll \Delta V_0(a), \quad (13)$$

where $\Delta V_0(a)$ is the depth of the potential well for electrons in the QD. For example, for the CdS QDs whose sizes satisfy inequality (10), the value ΔV_0 is 2.3 to 2.5 eV [26].

If condition (10) is satisfied, we can use, for the electron (hole) wave function in a QD, the wave function of an electron (hole) in a spherical quantum well with infinitely high walls [27]:

$$\Psi_{n,l,m}(r,\theta,\varphi) = Y_{l,m}(\theta,\varphi)\frac{J_{1+1/2}(\varphi_{n,l})\sqrt{2}}{J_{1+3/2}(\varphi_{n,l})a\sqrt{r}}, \quad (14)$$

where $r = r_e$ or $r = r_h$ is the distance of the electron or hole from the QD center, respectively, θ and φ are the azimuthal and polar angles that define the orientation of the radius vector of the electron (hole), respectively, $Y_{l,m}$ are the normalized spherical functions (m is the magnetic quantum number of the electron or hole), and $J_\nu(x)$ are the Bessel functions that can be expressed in terms of the spherical Bessel functions $j_\nu(x)$ [27] as

$$J_{1+1/2}(\varphi_{n,l}) = \sqrt{2/\pi\varphi_{l,n}} j_l(\varphi_{l,n})$$

$$J_{1+3/2}(\varphi_{n,l}) = \sqrt{2/\pi\varphi_{l,n}} j_{l+1}(\varphi_{l,n}). \quad (15)$$

Dipole moments of charge carrier transitions in a quantum dot

In the frequency region corresponding to the above-considered states of charge carriers in the QD bulk, (n,l), the wavelength of light is much larger than the dimensions of these states ($\sim a_e$, a_h), that is the behavior of approximation. In this case, the operator of the dipole moment of the electron (hole) located in the QD bulk is expressed as [28]

$$D(r) = \frac{3\varepsilon_1}{2\varepsilon_1 + \varepsilon_2} D^0(r), D^0(r) = er. \quad (16)$$

To estimate the value of the dipole moment, it is sufficient to consider the transition between the lowest discrete states (8), e.g., between the ground states $|1s = (n = 1, l = 0, m = 0)$ and $|1p = (n = 1, l = 1, m = 0)$. To calculate the matrix element of the dipole moment of the charge carrier transition from the $1s$ state to the $1p$ state, $D_{1,0}(a)$, we assume that the uniform field of the light wave $E(\omega,t)$ is directed only along the axis Z (ω is the wave frequency). In this case, we take the dipole moment $D(r)$ (16) induced by the field $E(\omega,t)$ as the perturbation responsible for such dipole transition. The expression for the dipole moment of the transition $D_{1,0}(a)$ follows from formulas (16) and the expression for the dipole moment of the transition in free space:

$$D_{1,0}^0(a) = 1s|D^0(r)|1p = e1s|r|1p. \quad (17)$$

Taking into account (14) and (15), we can write the wave functions of the $|1s$ and $|1p$ states as

$$|1s = \Psi_{1,0,0}(r,\theta) = Y_{0,0}(\theta)\frac{2}{a^{3/2}}\frac{j_0(\varphi_{0,1}(r/a))}{j_1(\varphi_{0,1})}, \quad (18)$$

$$|1p = \psi_{1,0,0}(r,\theta) = Y_{1,0}(\theta)\frac{2}{\alpha^{3/2}}\frac{j_1(\varphi_{1,1}(r/\alpha))}{j_2(\varphi_{1,1})}. \quad (19)$$

On substitution of (18) and (19) into formula (17) and integration, we obtain the expression for the dipole moment of the transition in free space as follows:

$$D_{1,0}^0(a) = \frac{2\pi\sqrt{6}}{3\varphi_{1,1}j_2(\varphi_{1,1})\left(\varphi_{1,1}^2 - \pi^2\right)} \times \left[\cos\varphi_{1,1} - \frac{\left(3\varphi_{1,1}^2 - \pi^2\right)}{\varphi_{1,1}\left(\varphi_{1,1}^2 - \pi^2\right)}\sin\varphi_{1,1}\right] ea = 0.433ea. \quad (20)$$

Then, according to (20) and (16), the dipole moment of the transition in the QD with the permittivity ε_2 in the surrounding matrix with the permittivity ε_1 is

$$D_{1,0}(a) = \Lambda D^0_{1,0}(a) = \Lambda 0.433ea, \tag{21}$$

where

$$\Lambda = \frac{3\varepsilon_1}{2\varepsilon_1 + \varepsilon_2}. \tag{22}$$

Optical absorption on light at one-particle states of charge carriers in quantum dots

Using the above results for the matrix element of the dipole moment of the transition $D_{1,0}$ (formulas (21) and (22)), we can elucidate the behavior of the semiconductor quasi-zero-dimensional systems on absorbing the energy of the electromagnetic field in the frequency region corresponding to the energies of the quantum-confined states in the QD $E_{n,l}$ (8). The absorption cross section of a spherical QD of radius a can be expressed in terms of the polarizability of QD, $A^{''}(\omega, a)$ A, as [28]

$$\sigma_{\text{abs}}(\omega, a) = 4\pi(\omega/c)A^{''}(\omega, a), \tag{23}$$

where ω is the frequency of the external electromagnetic field.

The polarizability $A^{''}$ can be easily determined if the QD is considered as a single giant ion. Let the radius of QD be a (10). In such QD, the quantum-confined states of charge carriers (n,l) (8) are formed. At room temperature, these states are slightly broadened, satisfying inequality (12). In this case, the polarizability of the charged QD, $A^{''}$, can be expressed in terms of the matrix element of the dipole moment of the transition $D_{1,0}(a)$ (21) between the lowest 1s and 1p states [29,30]:

$$A^{''}(\omega, a) = \frac{e^2}{m_{\text{e(h)}}} \frac{f_{0,1}}{\omega_1^2(a) - \omega^2 - i\omega\Gamma_1(a)}, \tag{24}$$

where

$$f_{0,1} = \frac{2m_{\text{e(h)}}}{\hbar e^2}[\omega_1(a) - \omega_0(a)]\left|D_{1,0}(a)\right|^2 \tag{25}$$

is the oscillator strength of the transition of a charge carrier with the effective mass m_e (or m_h) from the ground 1s state to the 1p state, $\omega_1 = E_{1.1}$ and $\omega_0 = E_{1.0}$ are, correspondingly, the energies of the discrete 1s and 1p levels by formula (8), and $\Gamma_1(a)$ is the width of the 1p level [20,22]. Taking into account formulas (8) and (21), we can express the oscillator strength (25) of the transition as

$$f_{0,1} = \left(\varphi_{1,1}^2 - \pi^2\right)\frac{\Lambda^2 (D^0_{1,0}(a))^2}{e^2 a^2} = 0.967 \times 10^{-1}\Lambda^2 \tag{26}$$

We assume that the frequency ω of the wave of light is far from the resonance frequency ω_1 of the discrete 1p state and, in addition, that the broadening Γ_1 of the 1p level is small, i.e., $\Gamma_1/\omega_1 \ll 1$ [20-22]. Then, for the qualitative estimate of the QD polarizability $A^{''}(\omega, a)$ (24), we obtain, with regard to (8), the following expression:

$$A^{''}(a) = \frac{4f_{0,1}}{\varphi_{1,1}^4}\frac{m_{\text{e(h)}}}{m_0}\left(\frac{a}{a_{\text{B}}}\right)^4 a_{\text{B}}^3, \tag{27}$$

where $a_{\text{B}} = \hbar^2/m_0 e^2$ is the Bohr radius of an electron in free space. Now we write the expression for the cross section of elastic scattering of the electromagnetic wave with frequency ω by the QD of radius a [28] as

$$\sigma_{\text{sc}}(\omega, a) = 2^7 3^{-3}\pi^3(\omega/c)^4\left|A^{''}(a)\right|^2. \tag{28}$$

The theory developed here for QDs is applicable only to the intraband transitions of electrons (holes) whose spectrum $E_{nl}^{e(h)}(a)$ is defined by formula (8). The optical attenuation coefficient, which involves both the absorption and scattering of light by one-particle quantum-confined states of charge carriers in the QD bulk of radius a, can be written, for the concentration of QDs N, as [29,30]

$$\gamma(\omega, a) = N[\sigma_{\text{abs}}(\omega, a) + \sigma_{\text{sc}}(\omega, a)] \tag{29}$$

Formula (29) is applicable to an array of QDs that do not interact with each other. The condition, such that the QDs of radius a and concentration N do not interact with each other, is reduced to the requirement that the spacing between the QDs ($\sim N^{-1/3}$) is much larger than the dimensions of the above-considered one-particle states ($\sim a_{\text{e(h)}}$), i.e.,

$$a_{\text{e(h)}} N^{-1/3} \ll 1. \tag{30}$$

With $a_{\text{e(h)}} \approx 5.0$ nm criterion (30) is satisfied for the concentrations of QDs $N \leq 10^{15}$ cm^{-3} achievable under the experimental conditions of [13-17] for a number of III-V semiconductors.

Comparison of the theory with experiments

Similar to [26], we can assume that under the experimental conditions of [13-17], the annealing of the arrays of InAs and InSb QDs in the GaAs and GaSb matrices at the temperature T = 273 K induces the thermal emission of a light electron, so that a hole alone remains in the QD bulk. In this case, the electron may be localized at a deep trap in the matrix. If the distance d, from this

Table 1 Parameters of quantum-confined states of hole in motion in quantum dots

Quantum dots				Matrices		a_{Bh}	Λ	a	$f_{0,1}$	$D^0_{1,0}$	$D_{1,0} = \Lambda D^0_{1,0}$	η	A	σ_{abs}
Semiconductor	ε_2	$\frac{m_e}{m_0}$	$\frac{m_h}{m_0}$	Semiconductor	ε_1	(nm)		(nm)	(10^{-1})	(D)	(D)	(%)	$(10^{-22}\ cm^3)$	$(10^{-18}\ cm^2)$
InAs	12.5	0.028	0.33	GaAs	12	2.0	0.99	2.0	0.95	6.12	6.04	8.7	9.2	7.5
								3.0		9.2	9.06	19.6	47	38
								4.0		12.2	12.1	35	140	120
InSb	18	0.013	0.18	GaSb	12	5.3	0.94	2.0	0.85	6.12	5.74	4.7	4.6	3.7
								3.0		9.2	8.61	10.6	23	19
								4.0		12.2	11.5	19	74	61
										15.3	14.4	29.4	180	160

The radius a and permittivity ε_2 are embedded in semiconductor matrices with permittivity ε_1. Here, m_e and m_h are the effective masses of an electron and a hole in the QD, respectively; a_{Bh} is the Bohr radius of a hole in the QD; $\Lambda = 3\varepsilon_1/(2\varepsilon_1 + \varepsilon_2)$; $f_{0,1}$ is the oscillator strength of the transition; $D^0_{1,0}$ and $D_{1,0}$ are the dipole moments of the transitions in free space and in the QD, respectively; η is the parameter of the broadening of states at room temperature; A is the polarizability of the QD; and σ_{abs} is the absorption cross section. The unit of measure of the dipole moment is debye (D), (i.e., the dipole moment $D = er$ with $r = 1$ Å (debye)).

trap to the QD center, is large compared to the QD radius a ($d \gg a$), the Coulomb electron–hole interaction $V_{eh}(r_e,r_h)$ (5) in the Hamiltonian (4) can be disregarded. As a result, the one-particle quantum-confined hole states (n,l), with the energy spectrum $E_{n,l}(a)$ described by formula (8), appear in the QD bulk. Now we roughly estimate the cross sections of optical absorption σ_{abs} ((25) and (29)) and σ_{sc} (28) at the quantum-confined hole state in the QDs for the selected $1s \to 1p$ transition under the experimental conditions of [13,17]. For the rough estimation of the cross sections of optical absorption and scattering, we use expressions (23), (27), and (28) on the assumptions that the frequency of the light wave ω is far from the resonance frequency ω_1 of the discrete hole state in the QD and that the broadening Γ_1 of the energy level $E_{1.1} = \omega_1$ (8) is small [20-22] ($\Gamma_1/\omega_1 \ll 1$). In this case, the absorption cross section σ_{abs} and the scattering cross section σ_{sc} take the following forms:

$$\sigma_{abs}(\omega,\alpha) = \frac{16\pi f_{0,1}}{\varphi^4_{1,1}} \frac{\omega}{c} \frac{m_h}{m_0} \left(\frac{\alpha}{\alpha_B}\right)^4 \alpha^3_B, \tag{31}$$

$$\sigma_{sc}(\omega,a) = \frac{1^{11}\pi^3 {f_{0,1}}^2}{3^3 \varphi^8_{1,1}} \left(\frac{\omega}{c}\right)^4 \left(\frac{m_h}{m_0}\right)^2 \left(\frac{a}{a_B}\right)^8 a^6_B. \tag{32}$$

The estimated parameters of the hole states in QDs dispersed in the III-V semiconductors are listed in Table 1. It is worth noting that at room temperature, the quantum-confined hole states are slightly broadened; the parameter $\eta(a)$ (14) does not exceed 35%.

The quasi-zero-dimensional systems considered here are essential nonlinear media with respect to infrared radiation. In fact, the dipole moments of the transitions in QDs of radii $a \approx 20$ to 40 Å are rather large in magnitude: $D_{1,0} \approx 10$ D (see table), being many times larger than the values $D \approx 0.1$ D typical for bulk III-V semiconductors [31,32]. In addition, according to the selection rules for QDs in the electromagnetic field, the dipole transitions between the nearest levels $E_{n,l}$ (8) are allowed [27]. Under these transitions, the azimuthal quantum number changes by unity.

From the estimates presented in the table, it follows that for QDs of radii $a \approx 2.0$ to 4.0 nm, the absorption cross section can be as large as $\sigma_{abs} \approx 10^{16}$ cm^2. This value is 8 orders of magnitude larger than the typical absorption cross sections for atoms [33]. Since the scattering cross sections σ_{sc} (32) under the experimental conditions of [13-17] are negligible compared to the corresponding absorption cross sections σ_{abs} (31) ($\sigma_{sc}/\sigma_{abs} \approx 10^{-12}$), the estimates for σ_{sc} are not included in the table.

Thus, the optical attenuation coefficient γ (31) for QDs is controlled mainly by the absorption process at the quantum-confined states of charge carriers (n,l) (8). If the concentration of QDs of radii a satisfying condition (38) is $N \approx 10^{15}$ cm^{-3}, the optical attenuation coefficient, on absorption at the QDs with the parameters listed in the table, takes the value $\gamma \approx 0.1$ cm^{-1}.

The large optical absorption cross section and attenuation coefficients in the quasi-zero-dimensional systems treated above allow the use of such heterophase structures as new efficient absorbers of electromagnetic waves in a wide wavelength range variable over wide limits in accordance with the nature of materials in contact [34].

Competing interests

The author declares that he has no competing interests.

Acknowledgements

I want to thank M.V. Kozachok for the help in preparing the paper and the editor Sherlyn C. Machica for correcting the paper.

References

1. Ekimov, AI, Onushchenko, AA: Size quantization electrons in microcrystals semiconductor. Pis`ma, Z.: Eksp. Teor. Fiz **40**, 337 (1984) [JETP Lett. 40, 1136 (1984)]
2. Ekimov, AI, Onushchenko, AA, Efros, AL: Size quantization holes in adiabatic potential electron. Pis`ma, Z.: Eksp. Teor. Fiz. **43**, 292 (1986) [JETP Lett. 43, 3766 (1986)]
3. Vandyshev, YV, Dneprovskii, VS, Klimov, VI: Display size quantization levels in spectrum nonlinear absence semiconductor microcrystals. Pis`ma, Z.: Eksp. Teor. Fiz. **53**, 301 (1991) [JETP Lett. 53, 314 (1991)]
4. Agranovich, VM, Benisty, H, Weisbuch, C: Excitons in hybrid organic–inorganic nanostructures. Solid State Commun. **102**, 631 (1997)
5. Alferov, ZI: Progress development in semiconductor nanostructures. Fiz. Tekh. Poluprovodn. (St. Peterburg) **32**, 3 (1998) [Semiconductors **32**, 1 (1998)]
6. Ekimov, AI, Skvortsov, PA, Shubina, TV: Size quantization Stark effect in semiconductor microcrystals. Zh. Tekh. Fiz. **59**, 202 (1989) [Sov. Phys. Tech. Phys. **34**, 371 (1989)]
7. Pokutnyi, SI: Interband absorption of light in quasi-zero-dimensional semiconductor system: theory. J. Nanoscience Lett. **1**, 191 (2011)
8. Pokutnyi, SI: Size quantization Stark effect in semiconductor quantum dots. J. Appl. Phys. **96**, 1115 (2004)
9. Kulchin, Y, Scherbakov, A, Dzyuba, V, Voznesenskiy, S, Mikaelyan, G: Nonlinear-optical properties of heterogeneous liquid nanophase composites based on high-energy-gap Al_2O_3 nanoparticles. Quantum Electronics **38**, 154 (2008)
10. Dzyuba, V, Kulchin, Y, Krasnok, A: Nonlinear refractive index of the dielectric nanocomposites in weak optical fields. Technical Physics Letters (Rus) **36**, 1 (2010)
11. Kulchin, Y, Dzyuba, V, Scherbakov, A: Optical transmittance spectra of nanoparticles. Semiconductors (Rus) **43**, 331 (2009)
12. Dzyuba, V, Kulchin, Y: The variation of the nanoparticles shape and quantum-dimensional states of the quantum dots or nanoparticles. Pacific Science Review **12**, 102 (2010)
13. Gun`ko, NA, Khalfin, VB, Sokolova, ZN, Zegrya, GG: Absorption of light in semiconductor nanosystems. J. Appl. Phys. **84**, 547 (1988)
14. Maksimov, MV, Zhukov, AE, Egorov, A, Kovsh, AR, Alferov, JI: Optical properties of vertically coupled InGaAs quantum dots in a GaAs matrix. Fiz. Tekh. Poluprovodn. (St. Peterburg) **31**, 104 (1997) [Semiconductors **31**, 84 (1997)]
15. Dzyuba, V, Krasnok, A, Kulchin, Y, Dzyuba, I: Nonlinear transmission of light by dielectric nanocomposites. Pacific Science Review **12**, 106 (2010)
16. Zaytsev, SV, Gordeev, YN, Ustinov, VM, Alferov, JI: Study of characteristics of low-threshold heterolasers with clusters of quantum dots. **31**, 539 (1997) [Semiconductors **31**, 452 (1997)]
17. Tsatsul`nicov, AF, Ledentsov, NN, Maksimov, MV, Alferov, JI: Photoluminescence study of InSb quantum dots in GaAs and GaSb matrix. Fiz. Tekh. Poluprovodn. (St. Peterburg) **31**, 68 (1997) [Semiconductors 31, 55 (1997)]
18. Efremov, NA, Pokutniy, SI: Macroscopic local charge states in ultradispersion media. Fiz. Tverd. Tela (Leningrad) **27**, 48 (1985) [Sov. Phys. Solid State **27**, 27 (1985)]
19. Efremov, NA, Pokutnyi, SI: Spectrum of local charge states in ultradispersion media. Fiz. Tverd. Tela (Leningrad) **32**, 2921 (1990) [Sov. Phys. Solid State **32**, 1697 (1990)]
20. Efremov, NA, Pokutnyi, SI: Quasistationary states in quasi-zero-dimensional systems. Fiz. Tverd. Tela (Leningrad) **33**, 2845 (1991) [Sov. Phys. Solid State **33**, 1607 (1991)]
21. Pokutnyi, SI: Theory of macroscopic local single-particle charge states in quasi-zero-dimensional structures: bulk local states. Phys. Status Solidi B **165**, 109 (1991)
22. Pokutnyi, SI: Theory of macroscopic local single-particle charge states in quasi-zero-dimensional structures: surface local states. Phys. Status Solidi B **172**, 573 (1992)
23. Pokutnyi, SI: Volume local states in quasi-zero-dimensional structures. Fiz. Tekh. Poluprovodn. (St. Peterburg) **35**, 257 (1993) [Phys. Solid State **35**, 129 (1993)]
24. Pokutnyi, SI: Macroscopic bulk states of carriers in quasi-zero-dimensional structures. Fiz. Tekh. Poluprovodn. (St. Peterburg) **31**, 1443 (1997) [Phys. Solid State **31**, 1247 (1997)]
25. Efros, AL, Efros, AL: Interzone absorption of light in semiconductor sphere. Fiz. Tekh. Poluprovodn. (Leningrad) **16**, 1209 (1982) [Sov. Phys. Semicond. **16**, 772 (1982)]
26. Grabovskis, V, Dzenis, Y, Ekimov, A: Nonlinear optics of semiconductor-doped glasses. Fiz. Tekh. Poluprovodn. (Leningrad) **31**(7), 272 (1989) [Sov. Phys. Semicond. **31**, 1255 (1989)]
27. Davydov, AS: Quantum Mechanics, 2nd edn. Pergamon, Oxford (1973)
28. Landau, LD, Lifshitz, EM: Electrodynamics of Continuous Media. In: Course of Theoretical Physics, vol. 8, 2nd edn. Pergamon, New York (1973)
29. Pokutnyi, SI: Absorption and scattering of light in quasi-zero-dimensional semiconductor systems. Fiz. Tverd. Tela (St. Peterburg) **39**, 606 (1997). [Phys. Solid State **39**, 528
30. Pokutnyi, SI: Absorption and scattering of light in quasi-zero-dimensional semiconductor systems: dipole moments for the transitions. Fiz. Tverd. Tela (St. Peterburg) **39**, 720 (1997) [Phys. Solid State **39**, 634 (1997)]
31. Petrov, YI: Physics of Small Particles. Nauka, Moscow (1982) [in Russian]
32. Ukhanov, YA: Optical Properties of Semiconductors. Nauka, Moscow (1977) [in Russian]
33. Ashcroft, NW, Mermin, ND: Solid State Physics. Holt, Rinehart and Winston, New York (1979)
34. Agranovich, VM, Ginzburg, VL: Crystal Optics with Spatial Dispersion and Excitons, 2nd edn. Springer, New York (1979)

Phytosynthesis of silver nanoparticles by *Cissus quadrangularis*: influence of physicochemical factors

Mahendran Vanaja, Gnanadhas Gnanajobitha, Kanniah Paulkumar, Shanmugam Rajeshkumar, Chelladurai Malarkodi and Gurusamy Annadurai*

Abstract

Phytosynthesis process of silver nanoparticles using plant extract is simple, cost-effective, and ecofriendly. In this present investigation, we report the green nanoparticles prepared by using stem extract of *Cissus quadrangularis* and assess the physical and chemical factors such as time duration, metal ion concentration, pH, and temperature that play vital role in the nanoparticles synthesis. The maximum synthesis of silver nanoparticles was attained within 1 h, at pH 8 and 1 mM $AgNO_3$ concentration, and 70°C. The nanoparticles obtained are characterized by UV–vis spectroscopy. Silver nanoparticles that were synthesized under these conditions show crystalline nature confirmed by X-ray diffraction and show mostly spherical and some rod and triangle shapes with sizes ranging from 37 to 44 nm, which were characterized by scanning electron microscopy. Fourier transform infrared spectroscopy shows that the functional groups are carboxyl, amine, and phenolic compounds of stem extract which are involved in the reduction of silver ions. Thus, synthesized silver nanoparticles show more antibacterial activity against *Klebsiella planticola* and *Bacillus subtilis*, which was analyzed by disc diffusion method.

Keywords: Stem extract, *Cissus quadrangularis*, XRD, SEM, Silver nanoparticles, Antibacterial activity

Background

Nanotechnology is a multidisciplinary research field that emerges from physical, chemical, and engineering sciences with novel techniques and produces material at nanoscale level [1]. Nanomaterials are the essential building elements of nanotechnology. Nanoparticles are very important and have a distinct property, that is, they exhibit larger surface area to volume ratio [2]. This increase in surface area to volume ratio leads to the change of the properties of nanoparticles better than the bulk particles [3]. Due to their great availability of surface area, they are involved in a many applications such as catalysis [4], drug delivery [5], biosensing [6,7], and optics [8]. The noble metals such as Au, Pt, Pd, and Ag are synthesized by various methods such as physical, chemical, and biological methods. Physically and chemically mediated syntheses necessitate high pressure and temperature, high cost, and toxicity [9]. The involvement of biological entities in the nanoparticle growth is gaining tremendous advantages such as production of anisotropic nanoparticles with size and shape control and intensive energy. Biological methods of nanoparticle synthesis by using microorganisms [10] such as bacteria [11], fungus [12] and algae [13], enzyme [14], and plants both intracellularly and extracellularly show time-consuming and ecofriendly process. Microorganism-assisted nanoparticles synthesis is a time-consuming process which takes 24 to 120 h [15,16]. But green-mediated synthesis of nanoparticles by using plants has more advantages over the micro-mediated synthesis process because it mainly eliminates the maintenance of cell cultures. The utilization of plant materials for biosynthesis of nanoparticles is called green synthesis [17]. Plant-mediated synthesis has been suggested as possibly eco-friendly, cost-effective, and less time-consuming process and do not involve any toxic chemicals [18-20]. The advantage of using plants for the synthesis of nanoparticles is that they are easily available, safe to

* Correspondence: gannadurai@hotmail.com
Environmental Nanotechnology Division, Sri Paramakalyani Centre for Excellence in Environmental Sciences, Manonmaniam Sundaranar University, Alwarkurichi, Tamil Nadu 627412, India

handle, and possess a broad variability of metabolites that may aid in reduction. The time required for 90% of the reduction of silver ions was about 2 or 4 h [21]. The motivation for choosing the plant was its medicinal properties, and capping or reducing agents are available in this plant. The size, shape, stability, and the synthesis of nanoparticles were dependent on the various factors such as capping agents, pH, templates, and temperature. Gericke and Pinches reported that the intracellular particles formation and size of the particles could be controlled by altering the key factors such as pH, temperature, substrate concentration, and exposure time to the substrate [22].

Among the metal nanoparticles, silver has been consumed largely due to their antimicrobial and pharmaceutical applications [23,24]. Silver has long been recognized as having inhibitory effect on microbes present in medical and industrial process [25,26]. The most important application of silver and silver nanoparticles is in medical industry such as topical ointments to prevent infection against burn and open wounds [27]. Medical devices and implants were prepared with silver-impregnated polymers [28] and antibiotics [3,29]. The antimicrobial activity of silver nanoparticles mainly depends on the size property. The small-sized nanoparticles have large surface area to improve the antimicrobial activity in addition to improving chemical stability [30].

Cissus quadrangularis (Linn) has been used by common man in India for the promotion of fracture healing. It is also known as *Vitis quadrangularis* Wall and belongs to family Vitaceae. It is a common perennial climber and is distributed throughout India, particularly in tropical regions [31]. This plant was used in Ayurveda as an alternative anthelmintic, dyspeptic, digestive, tonic, analgesic in eye and ear diseases, and in the treatment of irregular menstruation and asthma. The whole plant is used in oral rehydration; while the leaf, stem, and root extracts of this plant are important in the management of various ailments. In addition, this plant is used in the management of obesity and metabolic disorders [32]. This plant has antioxidant property and free radical-scavenging activity [33].

In this work we report the synthesis of silver nanoparticles by the stem extract of *C. quadrangularis* as a reducing agent and find the effective factors for its synthesis process by varying the pH, temperature, metal ion concentration, and time duration. UV spectrophotometer was used to characterize the synthesized silver nanoparticles.

Results and discussion

Visible observation of nanoparticles synthesis

The formation of silver nanoparticles was preliminarily confirmed by the change of color of the solution (Figure 1). The color change from watery yellow to brown color of the aqueous solution was due to the excitation of the surface plasmon resonance [34] and SPR band which both play an important role in the confirmation of silver nanoparticles formation [35]. It also indicates that reduction of silver ion in the aqueous solution by the stem extract of *C. quadrangularis*.

UV–vis spectrum analysis of influencing the factor in synthesis process

Figure 2 shows UV spectra exposed from the reaction of reduction of silver ions which have polydispersed nanoparticles with broadening peak in the absorbance band at the wavelength of 450 nm. Nanoparticles formation was dependent on the different physical and chemical factors such as metal ion concentration, incubation time, pH, and temperature. These optimized factors affect the silver nanoparticles formation. This study identifies which factors should affect the nanoparticle synthesis.

The first factor is concentration of silver nitrate; the nanoparticles formation was increased while increasing the concentration of silver nitrate (Figure 3). At 1-mM concentration, the band at 410 nm shifts into 450 nm while increasing the silver nitrate concentration. Similar results are obtained using the seed extract of *Jatropha curcas* with different concentrations of silver nitrate [36]. It is because of the bioavailability of functional groups in the 10 ml of stem extract of *C. quadrangularis* is involved in the reduction of silver ion.

Figure 4 shows the factor of time-dependent formation of silver nanoparticles. As the time duration increased, the nanoparticle synthesis also increased. Nanoparticle formation was initiated within 5 min. The completion of nanoparticle synthesis was achieved after 1 h as identified in Figure 4. After 1 h the precipitation of nanoparticles occurred due to the instability of the silver

Figure 1 Color change indicates the formation of silver nanoparticles. (**a**) Silver nitrate solution, (**b**) after the 10 min reduction of silver nitrate, and (**c**) 1 h reaction time.

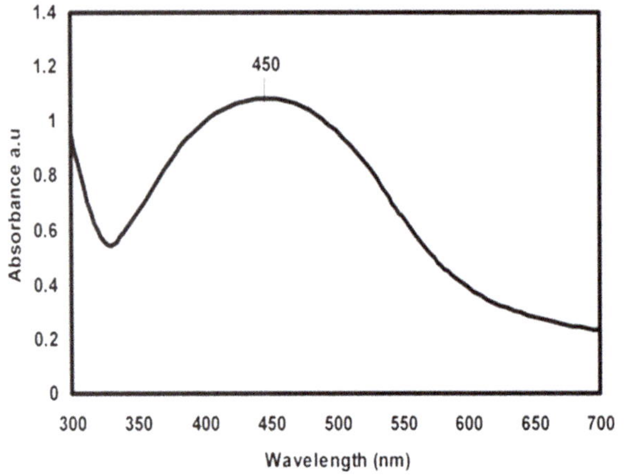

Figure 2 UV spectroscopy of aqueous solution of 1 mM $AgNO_3$ with stem extract. Spectroscopy recorded from the aqueous solution of 1 mM $AgNO_3$ with 10 ml of stem extract at different wavelengths.

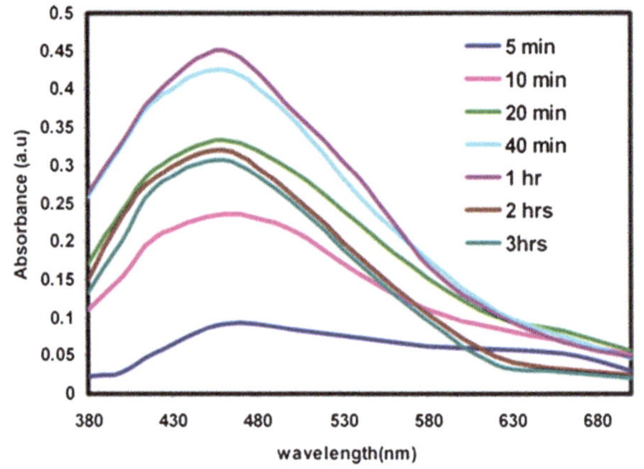

Figure 4 UV spectroscopy recorded from aqueous solution of $AgNO_3$ as a function time at different wavelengths.

nanoparticles. Agglomeration of nanoparticles shows the larger size of nanoparticles. So the optimum time duration for the formation of nanoparticles was 1 h. Earlier studies show it was a long-duration process when using some sources such as fungi which took 72 h in duration for synthesizing the silver nanoparticles [37], but in this study, it is a less time-consuming process. Some of the plants sources also took much longer time duration for silver ion reduction process.

The pH is one of the most important factor for nanoparticle formation. The shape and size of the nanoparticles are dependent on the pH of solution. In this study, the acidic pH 4 shows the peak at 420 nm. The absorbance band was increased while increasing the pH up to 8 due to the excitation of surface plasmon resonance. At pH 8 the absorbance peak was at 450 nm. The acidic pH suppresses the nanoparticle formation. At low pH, agglomeration took place because of the over nucleation and formation of larger nanoparticles. At high pH, a large number of nanoparticles with the small surface area are present due to the bioavailability of functional groups in the stem extract. Nanoparticle formation was increased with increasing pH. In the alkaline conditions, more precipitation or agglomeration occurred due to the instability of silver nanoparticles or due to lack of stabilizing agent. So the alkaline pH 8 was favorable for the nanoparticle formation (Figure 5).

Another one important factor is temperature. In this study nanoparticle synthesis was increased. While increasing the temperature (Figure 6), the absorption peak shifts from 460 nm in the temperature of 30°C; while by increasing the temperature up to 70°C, the band shifts to 440 nm, which is due to the localization of the surface

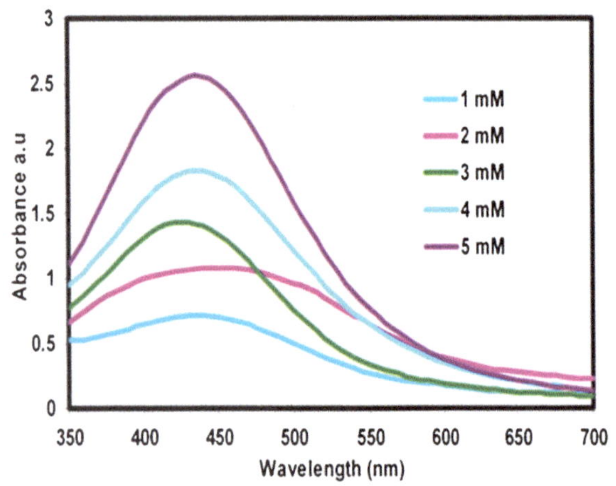

Figure 3 UV spectroscopy recorded from the aqueous solution of $AgNO_3$ at different concentrations at different wavelengths.

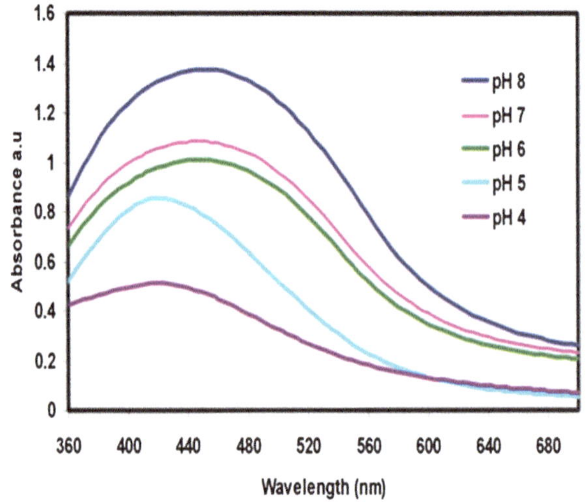

Figure 5 UV spectroscopic analysis shows effect of pH for nanoparticles formation.

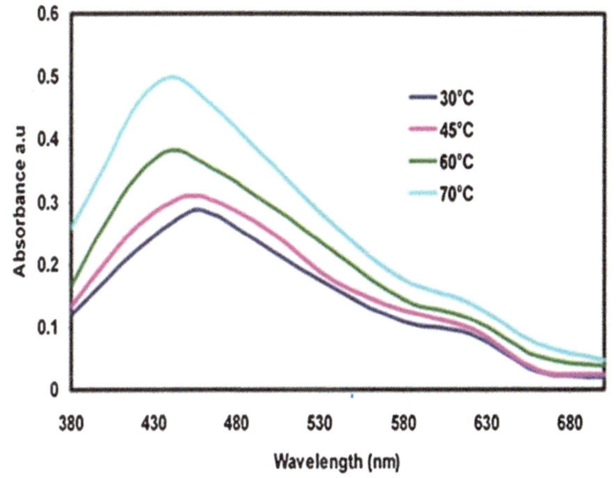

Figure 6 UV spectrum of silver nanoparticles after 1 h of reaction time at different temperatures.

plasmon resonance of the silver nanoparticles [38]. In higher temperature, the reduction occurred very hastily.

Characterization studies

X-ray diffraction

X-ray diffraction (XRD) pattern obtained for the silver nanoparticles was shown in Figure 7. XRD shows that the crystalline structure of silver is face centered cubic. In XRD, silver has similar diffraction profile with intense peaks at 2θ of 38.13°, 44.1°, 64.35°, and 77.88° corresponding to the planes (111), (200), (220), and (311), respectively. This indicates that the synthesized silver nanoparticles by using stem extract of *C. quadrangularis* had crystalline nature. Similar results were obtained by [39] using the *Bacillus cereus* microorganism.

SEM and EDX analysis

Scanning electron microscopy (SEM) image shows (Figure 8) the morphological character of silver nanoparticles synthesized by using extract of *C. quadrangularis.*

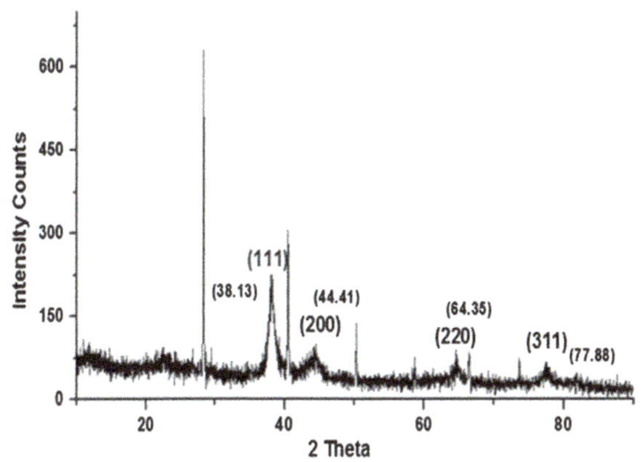

Figure 7 XRD analysis of synthesized silver nanoparticles.

This image shows that the sizes at around 37 to 43 nm with many shapes which are triangle, rod, and spherical are clearly observed. This SEM image also showed the aggregation of the silver nanoparticles.

Energy dispersive X-ray spectroscopy (EDX) shows the chemical analysis of synthesized silver nanoparticles using the *C. quadrangularis* stem extract (Figure 9). The strong signal was obtained at the energy of 3 keV for silver and also some of the weak signals from Cl, K, O, Ca, Mg, and S. The major emission energy at 3 keV indicates that the silver has been correctly identified.

FTIR analysis

FTIR analysis reveals the functional groups of the silver nanoparticles synthesized using stem extract of *C. quadrangularis* as shown in Figure 10. The band at 3,324 cm^{-1} shows O-H stretching vibrations of hydroxyl groups [40], H-bonded alcohols, phenols, or N-H stretching of 1° and 2° amines and amides. The band at 3,195 cm^{-1}appears due to the O-H stretching vibrations of carboxylic acids. A weak band at 2,966 cm^{-1} corresponds to C-H stretch alkanes and O-H stretch carboxylic acids [35]. The peak at 2,856 cm^{-1} shows C-H stretching for alkanes; peaks at 2,297 and 1668 cm^{-1} contribute to C=O stretch α, β-unsaturated aldehydes, and ketones; the peak at 1,596 cm^{-1} shows -C-C- skeletal vibrations [41]. The band at 1,400 cm^{-1} assigned to O-H bend indicates carboxylate [40]; the band 1,193 cm^{-1} shows C-O stretch alcohols and carboxylic acids. The peak at 1,115 cm^{-1} indicates presence of C-N stretch aliphatic amines; the peak at 605 $cm^{-1}$was assigned for C-Cl stretching vibrations of alkyl halides. The functional biomolecules are carboxylic and amine groups involved in the reduction of silver ions, as confirmed by FTIR spectrum.

Assay of antibacterial activity of synthesized silver nanoparticles

Antibacterial activity of synthesized silver nanoparticles against *Bacillus subtilis* and *K. planticola* was assayed by disc diffusion method (Figure 11). The concentration of silver nanoparticles varied from 10 to 50 μl. The zone of inhibition increased while increasing the concentration of silver nanoparticles. The zone of inhibition in diameter was tabulated (Table 1). Silver nanoparticles are very toxic in gram-negative bacteria than the gram-positive bacteria due to the difference in the cell wall of bacteria.

The mechanism of its antibacterial activity was not understood clearly. Some of the researchers explained the possible mechanism of antibacterial activity of silver nanoparticles. Silver nanoparticles attach with the cell wall of bacteria by electrostatic attraction and disrupt the cell permeability and respiration due to the generation of the reactive oxygen species. Silver nanoparticles

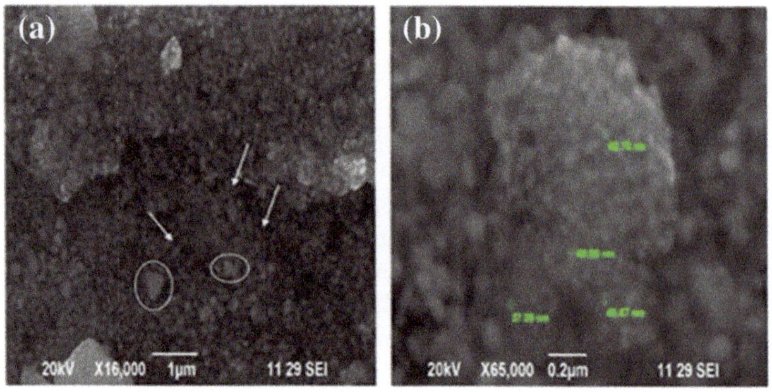

Figure 8 SEM analysis of silver nanoparticles at different magnifications (a) X16000 (b) X65000.

bind with thiol groups of DNA and RNA and affect the protein synthesis of the bacteria [42,43]. Silver nanoparticles form the pits on the cell surface and induce the proton leakage resulting in cell death [44]. The antimicrobial activity of silver nanoparticles depends on the size of the particles. Smaller particles having the larger surface area available for interaction will give more bactericidal effect than the larger particles and since they easily penetrate into the cell [45,46].

Conclusions

This present investigation shows the bioreduction of silver nanoparticles using the medicinal valuable stem extract of *C. quadrangularis* by optimizing the factors for rapid and stabilized nanoparticles synthesis. Nanoparticle synthesis process was controlled by incubation time, metal ion concentration, pH, and temperature. High pH and temperature influence the rapid synthesis of silver nanoparticles. Thus, synthesized silver nanoparticles are mostly spherical as confirmed by SEM. Mainly carboxylic and amine groups may responsible for the reduction process as revealed by FTIR. This green method is simple, rapid, eco-friendly and reliable, and it may have a potential use in the biomedical applications due to its high antibacterial activity.

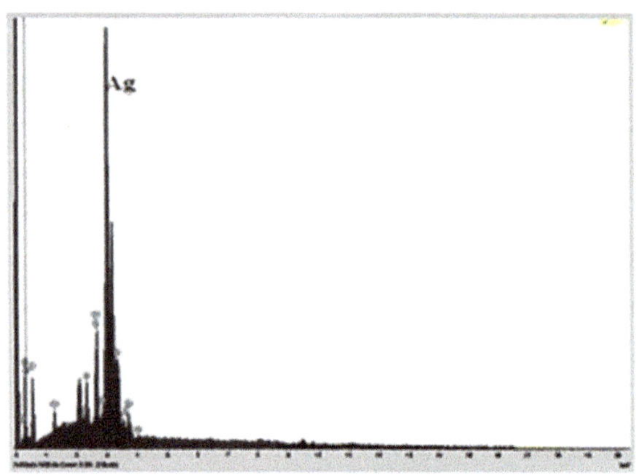

Figure 9 EDAX analysis of silver nanoparticles.

Methods

Preparation of reducing agent

All the chemicals and reagents are purchased from Himedia, Mumbai, India. The stem of *C. quadrangularis* was used as a reducing agent for the preparation of silver nanoparticles, collected from Western Ghats of Tamil Nadu. A 20 g of stem was thoroughly washed with Tween 20 and double-distilled water. The cleaned stem was cut into fine pieces which were boiled in 100 ml of double-distilled water for 5 min and filtered through Whatman no.1 filter paper. Filtered extract was stored at 4°C for 1 week for further experiment use.

Synthesis of silver nanoparticles

A 10 ml of stem extract was added into the 90 ml of aqueous solution of 1 mM of silver nitrate and kept at room temperature. The color change of the solution indicates the formation of silver nanoparticles. The reduced silver nanoparticles were monitored in UV–vis spectrophotometer.

Study of the effect of physicochemical factors in nanoparticles synthesis

To study the effect of silver ion concentration in nanoparticle synthesis, 10 ml of stem extract was added into different concentrations of silver ions (1, 2, 3, 4, and 5 mM). To study the effect of time duration on nanoparticle formation, the reaction solution was incubated at specific time intervals 0, 10, 20, and 40 and 1, 2, 3, 4, and 24 h. To study the effect of pH, experiments were carried out by varying the pH (4, 5, 6, 7, and 8) of the stem extract. The effect of temperature was studied by keeping the reaction solution at different temperatures (30°C, 45°C, 60°C, and 70°C). The influence of these

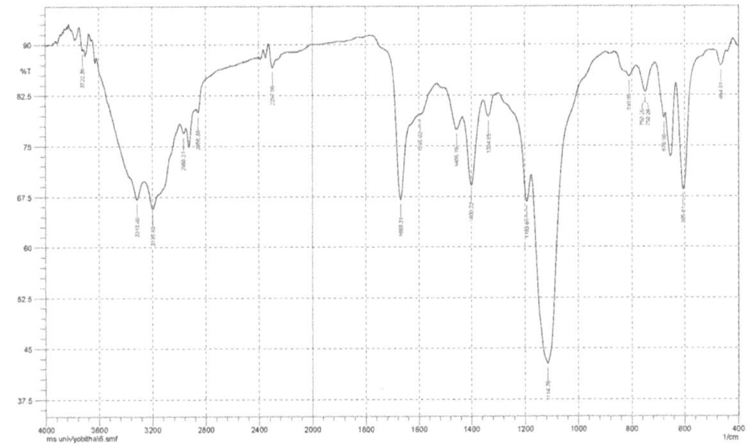

Figure 10 FTIR spectrum of synthesized silver nanoparticles.

physicochemical factors in silver nanoparticles formation was confirmed by UV–vis spectrophotometer at different wavelengths.

Characterization study of synthesized silver nanoparticles

The reduction process of the solution was monitored on a PerkinElmer double-beam UV spectrophotometer (PerkinElmer Inc., Waltham, MA, USA). The reaction of the aliquot solution analyzed at different reaction times in the wavelength ranges between 300 and 700 nm. The crystalline nature of the silver nanoparticles was characterized by X-ray diffraction. The reaction mixture was purified by repeated centrifugation at speed of 10,000 rpm for 10 min, and the pellets were dried at room temperature. The dried powder of silver nanoparticles was characterized by XRD (Philips PW 1830; Royal Philips Electronics, Amsterdam, Netherlands). The morphology and size of the silver nanoparticles were determined by scanning electron microscope (Philip model CM 200). The elemental analysis of silver was carried out by energy disperse analysis X-ray (EDAX). Fourier transform infrared spectroscopy (FTIR) measurements were carried out for dried biomass of leaf extract treated with silver nanoparticles to find out the compound responsible for the synthesis of silver nanoparticles. The FTIR was obtained on a Shimadzu instrument (Shimadzu Co. Ltd., Beijing, China) with the sample as KBR pellet.

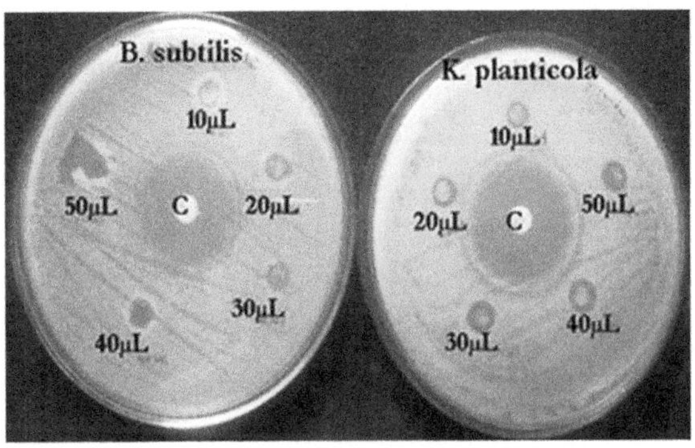

Figure 11 Antibacterial activity of silver nanoparticles synthesized by stem extract of *Cissus quadrangularis*.

Assay of antibacterial activity

Antimicrobial activities of synthesized silver nanoparticles against the two bacterial cultures of *Bacillus subtilis* (3053) and *K. planticola* (2727) (MTCC, Chandigarh, India) were assayed by disk diffusion method. These two bacteria were grown in LB broth for 24 h. Approximately 20 ml of molten and cooled Muller Hinton agar was poured into the Petri dishes. The two tested organisms are swapped over the agar medium, and the silver nanoparticle-containing disks were kept over the medium using sterile forceps. The silver nanoparticle disks were prepared at different concentrations (10, 20, 30, 40, and 50 µL) and incubated at 35°C for 24 h.

Table 1 Antibacterial activity of silver nanoparticles at various concentrations against *B. subtilis* and *K. planticola*

Concentration of silver nanoparticles (µL)	Zone of inhibition (diameter in mm)	
	Bacillus subtilis	*Klebsiella planticola*
10	8	9
20	9	10
30	9	9
40	11	12
50	11	13

Competing interests
The authors declare that they have no competing interests.

Authors' contributions
MV and GG carried out the nanoparticles synthesis. KP carried out the manuscript preparation. SR and CM carried out the antimicrobial activity. All authors read and approved the final manuscript.

Author's information
MV completed B.Sc. in Microbiology and M.Sc. Environmental Biotechnology in Sri Paramakalyani Centre for Environmental Sciences, Manonmaniam Sundaranar University, Alwarkurichi, Tirunelveli. She obtained her Ph.D. degree under the guidance of GA in the field of nanotechnology. Her research interests include ecofriendly synthesis of nanoparticles and its application in the environment. GG completed M.Sc. in Biotechnology in Periyar University and obtained her Ph.D. degree in Nanotechnology under the guidance of GA. Her research interests include green-meditated synthesis of nanoparticles and its agricultural applications for controlling plant diseases. KP received his B.Sc. (2005) and M.Sc. in Biotechnology (2007) from Madurai Kamaraj University and Periyar University, India, respectively. He is currently working to obtain his Ph.D. degree at Manonmaniam Sundaranar University in India. He is interested in the green and immobilized microbe-mediated synthesis process of nanoparticles and nanocomposites and their biomedical and textile industry applications. SR completed M.Sc. Biotechnology in Periyar University and obtained his Ph.D. degree in Nanotechnology under the guidance of GA. He published seven research articles in nanoparticles synthesis using algae. He is interested in the metallic nanoparticles synthesis using algae and algae-derived compounds and their potential applications in the biomedical field. CM obtained her M.Sc. in Environmental Biotechnology in Sri Paramakalyani Centre for Environmental Sciences, Manonmaniam Sundaranar University, Alwarkurichi, Tirunelveli. She achieved her Ph.D. degree under the guidance of GA in the field of nanotechnology. Her research interests include synthesis of semiconductor nanomaterials and its biomedical applications. GA received his M.Sc. in Applied Chemistry (1992) and Ph.D. in Environmental Biotechnology (1997) from Anna University, India. He had ten years (1999 to 2008) post-doctoral experiences from the National Taiwan University in Taiwan, National Institute of Advanced Industrial Science and Technology in Japan, and National Central University in Taiwan. He has received many research awards from Indian and other country governments. He is an associate editor in five international journals. At present, he is an associate professor of the Environmental Biotechnology and the leader of Environmental Nanobiotechnology Division at Sri Paramakalyani Centre for Environmental Sciences, Manonmaniam Sundaranar University, India. His main research interest is on biosynthesis of nanoparticles and nanomaterials, nanobiocatalyst, and environmental chemistry.

Acknowledgment
Authors gratefully acknowledge the DST-FIST-sponsored program of the Department of Science Technology, New Delhi, India for funding the research development (ref. no. S/FST/ESI-101/2010) and in carrying out of this work.

References
1. Schmid, G: Nanoparticles from theory to applications. Wiley-VCH, Weinheim (2004)
2. Annamalai, A, Babu, ST, Jose, NJ, Sudha, D, Lyza, CV: Biosynthesis and characterization of silver and gold nanoparticles using aqueous leaf extraction of *Phyllanthusamarus* Schum & Thonn. World Appl Sci J **13**, 1833–1840 (2011)
3. Raimondi, F, Scherer, GG, Kotz, R, Wokaun, A: Nanoparticles in energy technology: examples from electochemistry and catalysis. Angew Chem Int Ed **44**, 2190–2209 (2005)
4. Tsang, C, Caps, V, Paraskevas, I, Chadwick, D, Thompsett, D: Magnetically separable, carbon-supported nanocrystals for the manufacture of fine chemical. Angew Chem **116**, 5763–5767 (2004)
5. Baptista, PV: Cancer nanotechnology - prospects for cancer diagnostics and therapy. Curr Canc Ther Rev **5**(2), 80–88 (2009)
6. Huang, CC, Yang, Z, Lee, KH, Chang, HT: Synthesis of highly fluorescent gold nanoparticles for sensing mercury (II). Angew Chem Int Ed **46**, 6824–6828 (2007)
7. Jain, PK, Lee, KS, El-Sayed, IH, El-Sayed, MA: Calculated absorption and scattering properties of gold nanoparticles of different size, shape, and composition: applications in biological imaging and biomedicine. J Phys Chem B **110**, 7238–7248 (2006)
8. Murphy, CJ, Sau, TK, Gole, AM, Orendorff, CJ, Gao, J, Gou, L, Hunyadi, SE, Li, T: Anisotropic metal nanoparticles: synthesis, assembly, and optical applications. J Phys Chem B **109**, 13857–13870 (2005)
9. Sulochana, S, Krishnamoorthy, P, Sivaranjani, K: Synthesis of silver nanoparticles using leaf extract of *Andrographis paniculata*. J Pharmacol Toxicol **7**(5), 251–258 (2012)
10. Schultz, S, Smith, DR, Mock, JJ, Schultz, DA: Single-target molecule detection with nonbleaching multicolor optical immunolabels. Proc Natl Acad Sci **97**, 996–1001 (2000)
11. Shahverdi, AR, Fakhimi, A, Shahverdi, HR, Minaian, SA: Synthesis and effect of silver nanoparticles on the antibacterial activity of different antibiotics against *Staphylococcus aureus* and *Escherichia coli*. Nanomed Nanotechnol Biol Med **3**, 168–171 (2007)
12. Shaligram, NS, Bule, M, Bhambure, R, Singhal, RS, Singh, SK, Szakacs, G, Pandey, A: Biosynthesis of silver nanoparticles using aqueous extract from the compactin producing fungal strain. Proc Biochem **44**, 939–943 (2009)
13. Singaravelu, G, Arockiamary, JS, Ganesh Kumar, V, Govindaraju, K: A novel extracellular synthesis of monodisperse gold nanoparticles using marine alga, *Sargassum wightii* Greville. Colloid Surf B: Biointer **57**, 97–101 (2007)
14. Nair, B, Pradeep, T: Coalescence of nanoclusters and formation of submicron crystallites assisted by *Lactobacillus* strains. Cryst Growth Des **2**, 293–298 (2002)
15. Ahmad, A, Mukherjee, P, Senapati, S, Mandal, D, Khan, DI, Kumar, R, Sastry, M: Extracellular biosynthesis of silver nanoparticles using the fungus *Fusarium oxysporum*. Colloids Surf B Biointerfaces **28**, 313–318 (2003)
16. Ahmad, A, Senapati, S, Khan, MI, Kumar, R, Ramani, R, Srinivas, V, Sastry, M: Intracelluar synthesis of gold nanoparticles by a novel alkalotolerant actinomycete *Rhodococcus* species. Nanotechnology **14**, 824–825 (2003)
17. Sathishkumar, FM, Sneha, K, Won, SW, Cho, CW, Kim, S, Yun, YS: *Cinnamon zeylanicum* bark extract and powder mediated green synthesis of nano-crystalline silver particles and its bactericidal activity. Colloid Surf B: Biointerf **73**, 332–338 (2009)
18. Vigneshwaranm, N, Ashtaputre, NM, Varadarajan, PV, Nachane, RP, Paralikar, KM, Balasubramanya, RH: Biological synthesis of silver nanoparticles using the fungus *Aspergillus flavus*. Mater Lett **61**, 1413–1418 (2007)
19. Mohanpuria, P, Rana, NK, Yadav, SK: Biosynthesis of nanoparticles: technological concepts and future applications. J Nanopart Res **10**, 507–517 (2008)
20. Shankar, SS, Rai, A, Ahmad, A, Sastry, MJ: Rapid synthesis of Au, Ag, and bimetallic Au core-Ag shell nanoparticles using Neem (*Azadirachta indica*) leaf broth. J Colloid Interface Sci **275**, 496–502 (2004)
21. Chandran, SP, Chaudhary, M, Pasricha, R, Ahmad, A, Sastry, M: Synthesis of gold nanotriangles and silver nanoparticles using *Aloe vera* plant extract. Biotechnol Prog **22**, 577–583 (2006)
22. Gericke, M, Pinches, A: Microbial production of gold nanoparticles. Gold Bull **39**, 22–28 (2006)
23. Kim, YK, Lee, YS, Jeong, DH, Cho, MH, et al.: Antimicrobial effect of silver nanoparticles. Nanomed **3**, 95–101 (2007)
24. Choi, O, Deng, KK, Kim Jr, N, Ross, L, Rao, YS, Hu, Z: The inhibitory effects of silver nanoparticles, silver ions, and silver chloride colloids on microbial growth. Water Res **42**, 3066–3074 (2008)
25. Jose, RM, Jose, LEC: The bactericidal effect of silver nanoparticles. Alejandra Nanotech **16**, 2346–2353 (2005)
26. Lok, C, Ho, R, Chen, Q, He, W, Yu, H, Sun, PK, Tam, J, Chiu, C, Che, CM: Silver-nanoparticles: partial oxidation and antibacterial activities. J Biol Inorg Chem **12**, 527–534 (2007)
27. Ip, M, Lui, SL, Poon, VKM, Lung, I, Burd, A: Antimicrobial activities of silver dressings: an *in vitro* comparison. J Med Microbiol **55**, 59–63 (2006)
28. Silver, S: Bacterial silver resistance: molecular biology and use and misuses of silver compounds. FEMS Microbiol Rev **27**, 341–353 (2003)
29. Gajbhiye, M, Kesharwani, J, Ingle, A, Gade, A, Rai, M: Fungus-mediated synthesis of silver nanoparticles and their activity against pathogenic fungi in combination with fluconazole. Nanomed: Nanotechnol Biol Med **5**, 382–386 (2009)

30. Toshikazu, T: Antimicrobial agent composed of silica-gel with silver complex. Inorg Mat **6**, 505–511 (1999)
31. Mishra, G, Srivastava, S, Nagori, BP: Pharmacological and therapeutic activity of *Cissus quadrangularis*: an overview. Inter J Pharm Tech Res **2**, 1298–1310 (2010)
32. Mallika, J, Shyamala, CSD: *In vitro* and *in vivo* evaluation of free radical scavenging potential of *Cissus quadrangularis*. Afri J of Biomed Res **8**, 95–99 (2005)
33. Oben, J, Kuate, D, Agbor, G, Momo, C, Talla, X: The use of a *Cissus quadrangularis* formulation in the management of weight loss and metabolic syndrome. Lipids Health Dis **5**, 24 (2006)
34. Krishnaraj, C, Jagan, EG, Rajasekar, S, Selvakumar, P, Kalaichelvan, PT, Mohan, N: Synthesis of silver nanoparticles using *Acalypha indica* leaf extracts and its antibacterial activity against water borne pathogens. Colloids Surf B Biointerfaces **76**, 50–56 (2010)
35. Rajeshkumar, S, Kannan, C, Annadurai, G: Synthesis and characterization of antimicrobial silver nanoparticles using marine brown seaweed *Padina tetrastromatica*. Drug Invention Today **4**, 511–513 (2012)
36. Bar, H, Bhui, DK, Sahoo, GP, Sarkar, P, De, SP, Misra, A: Green synthesis of silver nanoparticles using latex of *Jatropha curcas*. Colloid Surf A **339**, 134–139 (2009)
37. Banu, A, Rathod, V, Ranganath, E: Silver nanoparticle production by *Rhizopus stolonifer* and its antibacterial activity against extended spectrum β-lactamase producing (ESBL) strains of *Enterobacteriaceae*. Mater Res Bull **46**, 1417–1423 (2011)
38. Fayaz, AM, Balaji, K, Kalaichelvan, PT, Venkatesan, R: Fungal based synthesis of silver nanoparticles - an effect of temperature on the size of particles. Colloid Surf B **74**, 123–126 (2009)
39. Ganesh Babu, MM, Gunasekaran, P: Production and structural characterization of crystalline silver nanoparticles from *Bacillus cereus* isolate. Colloid Surf B: Biointerf **74**, 191–195 (2009)
40. Rastogi, L, Arunachalam, J: Sunlight based irradiation strategy for rapid green synthesis of highly stable silver nanoparticles using aqueous garlic (*Allium sativum*) extract and their antibacterial potential. Mater Chem Phy **129**, 558–563 (2011)
41. Bansal, V, Rautray, D, Ahmad, A, Sastry, M: Biosynthesis of zirconia nanoparticles using the fungus *Fusarium oxysporum*. J Mater Chem **14**, 3303–3305 (2004)
42. Kulthong, K, Srisung, S, Boonpavanitchaku, K, Kangwansupamonkon, W, Maniratanachote, R: Determination of silver nanoparticle release from antibacterial fabrics into artificial sweat. Particle Fibre Toxicol **7**, 1–8 (2010)
43. Feng, QL, Wu, J, Chen, GQ, Cui, FZ, Kim, TN, Kim, JO: A mechanistic study of the antibacterial effect of silver ions on *Escherichia coli* and *Staphylococcus aureus*. J Biomed Mater Res **52**, 662–668 (2000)
44. Dibrov, P, Dzioba, J, Gosink, KK, Hase, CC: Chemiosmotic mechanism of antimicrobial activity of Ag(+) in *Vibrio cholerae*. Antimicrob Agents Chemother **46**, 2668–2670 (2002)
45. Panacek, A, Kvytek, L, Prucek, R, Kolar, M, Vecerova, R, Pizurova, N, Virender, KS, Nevecna, T, Zboril, R: Silver colloid nanoparticles: synthesis, characterization, and their antibacterial activity. J Phys Chem B **110**, 16248–16253 (2006)
46. Shrivastava, S, Bera, T, Roy, A, Singh, G, Ramachandran, P, Das, D: Characterization of enhanced antibacterial effect of novel silver nanoparticle. Nanotechnology **18**, 1–9 (2007)

Removal of ethidium bromide by carbon nanotube in aqueous solution: isotherms, equilibrium mechanism studies, and its comparison with nanoscale of zero valent iron as adsorbent

Fahimeh Najafi[1*], Mehdi Norouzi[2], Karim Zare[3,4] and Ali Fakhri[5]

Abstract

The adsorption of ethidium bromide (EtBr) by single-walled carbon nanotubes (SWCNTs) and nanoscale of zero valent iron (NZVI) were investigated to assess its possible use as adsorbents. The effect of various factors, namely initial adsorbate concentration, adsorbent dosage, and contact time, were studied to identify adsorption capacity of SWCNTs and NZVI surfaces. The experiment demonstrated the maximum EtBr which was obtained at 5 min to attain equilibrium for SWCNTs and NZVI surfaces. Adsorption data were modeled with the Langmuir, Freundlichand, Temkin isotherms. Langmuir adsorption model was used for the mathematical description of the adsorption equilibrium, and the equilibrium data fitted very well with this model for both surfaces as adsorbents. The study showed that SWCNTs and NZVI surfaces could be used as new and efficient adsorbent materials for the removal of EtBr from aqueous solution. Also, the result showed that the SWCNTs were more effective than NZVI in the removal of EtBr from aqueous solution.

Keywords: Ethidium bromide; Single-walled carbon nanotube; NZVI; Removal; Isotherms; Kinetics

Background

Ethidium bromide (EtBr; 3,8-diamino-6-phenyl-5-ethyl-phenanthridinium bromide), commonly used in research laboratories as a stain for the visualization of nucleic acids in electrophoresis gels, is a toxic chemical and a potent mutagen [1]. When used in nucleic acid staining, ethidium bromide fluoresces a red-orange to pink color under ultraviolet light and with increased fluorescence when bound to double-stranded DNA. While it is not specifically regulated as a hazardous waste, the mutagenic properties may present health hazards and disposal concerns if it is not managed properly in the laboratory [2]. Adsorption has been found to be an effective and economical method with high potential for the removal, recovery, and recycle of pollution from the wastewater [3]. In this decade, the use of single-walled carbon nanotubes (SWCNTs) and nanoscale of zero valent iron (NZVI) as a reactive medium for the treatment of toxic chemicals is one of the most significant techniques and has attracted a lot of attention because the iron metal is of low cost, low toxicity, is easy to get, and has good effectiveness and ability to degrading contaminants. Laboratory studies have demonstrated that SWCNTs and NZVI surfaces as adsorbents can effectively transform chlorinated solvents, organochlorine pesticides, organic dyes and heavy metals [4-11] into nontoxic forms. Nanoscale of zero valent iron has drawn great attention as an inexpensive and environmentally friendly strong reducing agent [12,13]. Addition of NZVI to oxygen-containing water results in oxidation of organic compounds [14]. Therefore, extensive efforts have been directed to assess the potential application of NZVI for the decomposition of organic pollutants in the wastewater. SWCNTs, ever since their discovery, have attracted extensive attention due to their unique properties. They have shown potential applications in many areas, such as conductive and high-strength composites. With the significant increase in production and use of SWCNTs, one major concern is the

* Correspondence: fahimnajafi@gmail.com
[1]Department of Chemistry, Roudehen Branch, Islamic Azad University, Roudehen, Iran
Full list of author information is available at the end of the article

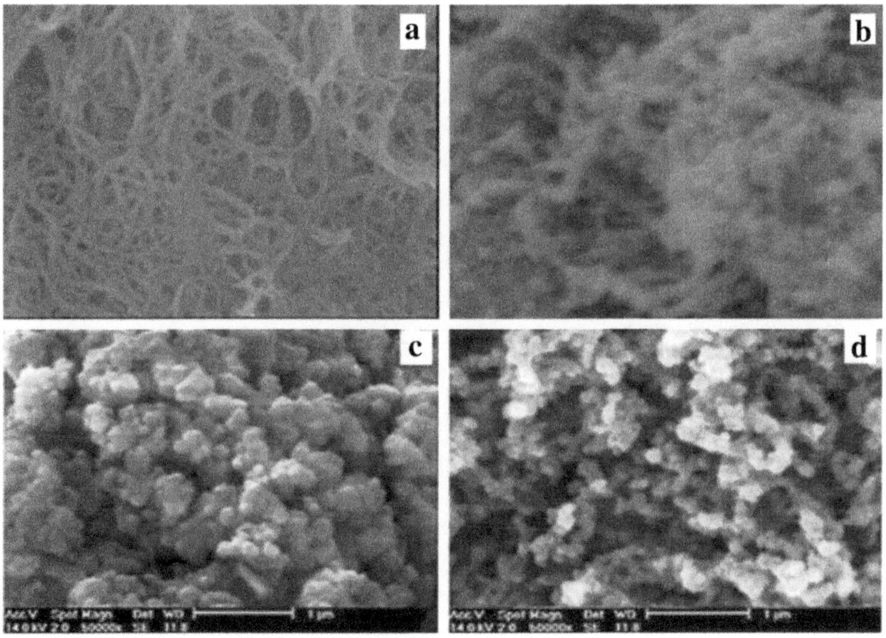

Figure 1 SEM images of SWCNT and NZVI surfaces before and after reaction. Before reaction (**a**,**c**, respectively) and after reaction (**b**,**d**, respectively) with EtBr molecules in aqueous solution.

health and environmental risks posed by these nanomaterials once they are released to the environment. Evidences for potential risks of SWCNTs to humans and ecosystems are accumulating rapidly [15]. Due to their highly hydrophobic surfaces, SWCNTs exhibit strong adsorption affinities to synthetic organic chemicals, such as polycyclic aromatic hydrocarbons. As a consequence, toxicity of SWCNTs may be further enhanced by adsorption of toxic chemicals. Therefore, understanding of SWCNT interactions is critical for the environmental of both SWCNTs and toxic pollutants as well as for the applications of SWCNTs as potential adsorbents [16].

The objective of this research is to investigate the adsorption kinetics and isotherm models of EtBr removal by SWCNT and NZVI surfaces. Equilibrium data can be used to predict the rate at which the target contaminant is removed from aqueous solutions, and equilibrium adsorption isotherms are used to quantify the adsorptive capacity of an adsorbent. Results from this study can be used to assess the utility of SWCNT and NZVI surfaces for EtBr removal, in particular EtBr adsorption, at the field scale.

Results and discussion

Characterization of the adsorbent

Figure 1 shows the SEM image of freshly synthesized SWCNTs (a,b) and iron nanoparticles (c,d). It can be observed that the iron particles are in the form of nanospheres, which exist in contact with each other and form chains having diameters of 50 to 100 nm [17].

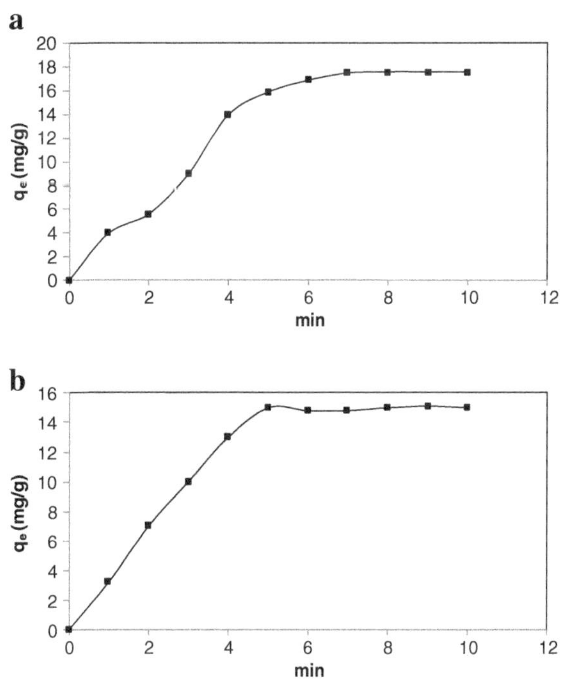

Figure 2 The effect of contact time on removal of EtBr by (a) SWCNT and (b) NZVI surfaces. (a) Initial concentration, 20 mg/L; adsorbent dosage, 30 mg; and temperature = 20°C ± 0.1°C. **(b)** Initial concentration, 20 mg/L; adsorbent dosage, 30mg; and temperature = 20°C ± 0.1°C.

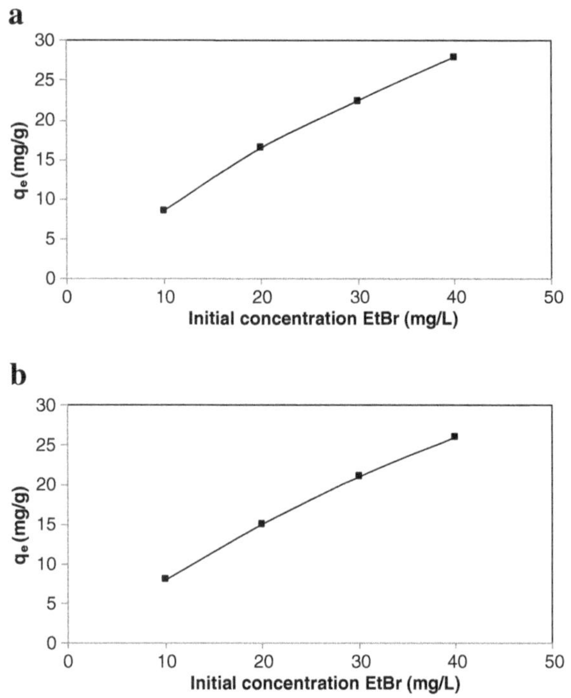

Figure 3 Effect of initial concentration on removal of EtBr by (a) SWCNT and (b) NZVI surfaces. (a) Contact time, 5 min; adsorbent dosage, 30 mg; and temperature = 20°C ± 0.1°C. **(b)** Contact time, 5 min; adsorbent dosage, 30 mg; and temperature = 20°C ± 0.1°C.

Figure 1b,d showed that the EtBr molecules removed by SWCNT and NZVI surfaces were covered in aqueous solution.

Effect of contact time

In these experiments, initial concentration of EtBr molecules on SWCNT and NZVI surfaces was 20 mg/L (Figure 2). The effects of contact time (0 to 10 min) were investigated at 20 mg/50 mL sample dosage. The extent of removal of EtBr by SWCNT and NZVI surfaces was found to increase and has reached its maximum with the increase in contact time. The time at which the maximum percentage removal of EtBr occurs is fixed as the contact time (Figure 2a for the removal of SWCNTs and Figure 2b for the removal of NZVI).

Effect of initial EtBr concentration

The effect of initial concentration (10 to 40 mg/L) of EtBr adsorption on to SWCNTs and NZVI surfaces are presented in Figure 3 (Figure 3a for the removal of SWCNTs and Figure 3b for the removal of NZVI). A rapid removal is observed at the initial stages, and it then proceeds slowly until equilibrium is reached. This may be due to the availability of the number of vacant adsorption sites at the initial stage. The equilibrium adsorption capacity of SWCNT and NZVI surfaces were decreased as there is increasing concentration from 10 to 40 mg/L. At lower EtBr concentration, the available adsorption sites are relatively high, and consequently, the EtBr species can easily find the accessible adsorption sites for both adsorbents (SWCNT and NZVI surfaces).

Effect of adsorbent dosage

From Figure 4 (Figure 4a for the removal of SWCNTs and Figure 4b for the removal of NZVI), we see that the optimum dose of SWCNT and NZVI surfaces for the EtBr were 30 mg/50 mL. Though at 40 mg/50 mL, there is a slight increase in adsorption value, but if we get nearly the same result as we get at an adsorbent dosage of 50 mg/50 mL, then going for 70 mg/50 mL will be expensive and there will be loss of adsorbent. It is obvious, with increasing the amount, what the active sites are for adsorption of three dyes.

Effect of temperature

To study the effect of temperature on the removal of EtBr, the experiments were carried out at temperatures varying from 283 to 313 K. It is revealed from Figure 5 (Figure 5a for the removal of SWCNTs and Figure 5b

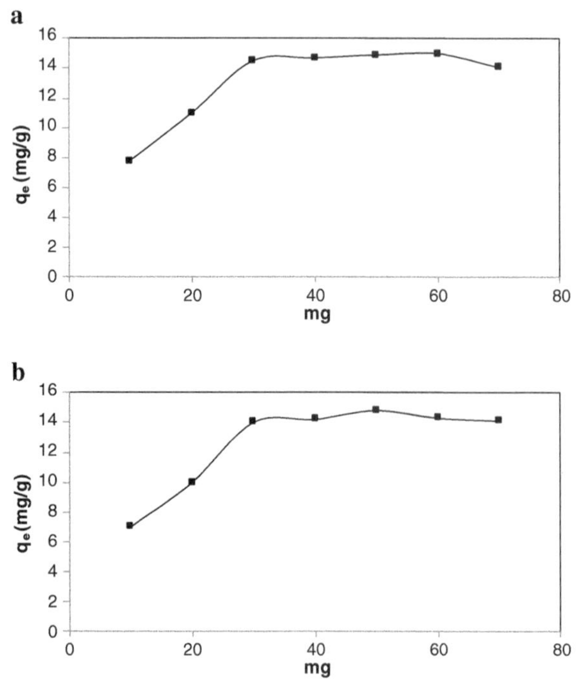

Figure 4 Effect of adsorbent dosage on removal of EtBr by (a) SWCNT and (b) NZVI surfaces. (a) Initial concentration, 20 mg/L; contact time, 5 min; and temperature = 20°C ± 0.1°C. **(b)** Initial concentration, 20 mg/L; contact time, 5 min; and temperature = 20°C ± 0.1°C.

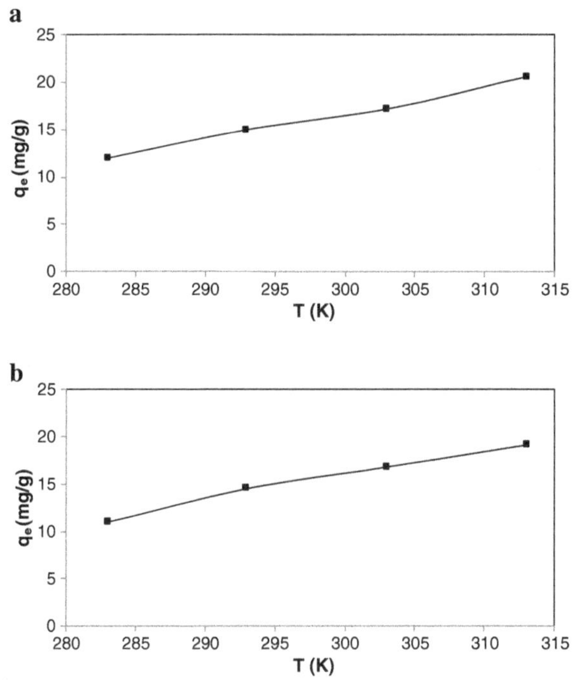

Figure 5 Effect of temperature on the removal of EtBr by (a) SWCNT and (b) NZVI surfaces. (a) Initial concentration, 20 mg/L; contact time, 5 min; and adsorbent dosage, 30 mg. **(b)** Initial concentration, 20 mg/L; contact time, 5 min; and adsorbent dosage, 30 mg.

for the removal of NZVI), that the removal of EtBr with temperatures would increase the mobility of the EtBr and produce a swelling effect within the internal structure of adsorbent thus enabling the large molecules of EtBr to penetrate further. The greater removal of EtBr due to increasing temperature may have more interaction between the adsorbate and adsorbent.

Adsorption isotherms

Equilibrium isotherm equations are used to describe the experimental adsorption data. In this study, three adsorption models, that is the Langmuir, Freundlich, and Temkin isotherms, were used to describe the reactive dye equilibrium. The Langmuir isotherm [18] which is valid for monolayer sorption onto a completely homogeneous surface with a finite number of identical sites and with negligible interaction between adsorbed molecules is given by Equation 1:

$$q_e = \frac{q_m K_L C_e}{(1 + K_L C_e)}, \tag{1}$$

where q_e (mg/g) and C_e (mg/L) are the amount of adsorbed EtBr per unit mass of adsorbent and an adsorbed EtBr concentration in solution at equilibrium, respectively. q_m is the maximum amount of the EtBr per unit mass of adsorbent to form a complete monolayer on the surface

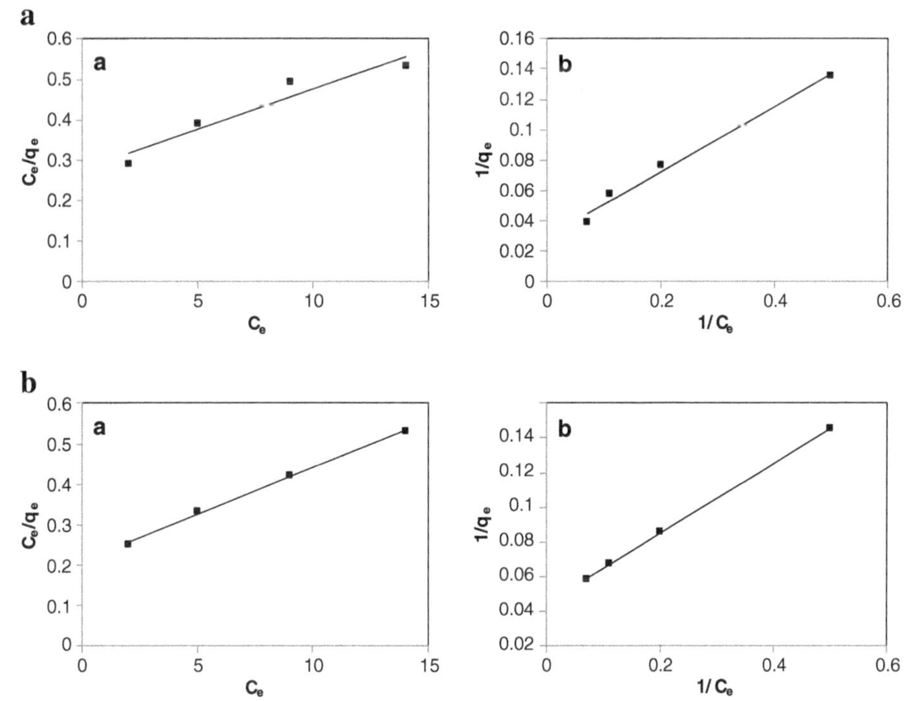

Figure 6 Langmuir plots. (a) Langmuir plots (Type 1A; Type 2B) for EtBr removal by SWCNTs. **(b)** Langmuir plots (Type 1A; Type 2B) for EtBr removal by NZVI.

bound at high C_e (mg/g), and K_L is a constant related to the affinity of the binding sites (L/mg). The two linear transformations of the Langmuir isotherm is often expressed as Equation 2 (Type1) and Equation 3 (Type 2):

$$\frac{C_e}{q_e} = \frac{1}{K_L q_m} + \frac{C_e}{q_m}, \tag{2}$$

$$\frac{1}{q_e} = \frac{1}{q_m} + \frac{1}{K_L q_m C_e}, \tag{3}$$

The essential characteristics of the Langmuir equation can be expressed in terms of a dimensionless separation factor, R_L, defined by [19]:

$$R_L = \frac{1}{1 + KC_0}, \tag{4}$$

where C_0 is the initial EtBr concentration (mg/L). R_L values indicate the type of isotherm to be irreversible ($R_L = 0$), favorable ($0 < R_L < 1$), linear ($R_L = 1$), or unfavorable ($R_L > 1$). The empirical Freundlich isotherm based on sorption on a heterogeneous surface is given by Equation 5 [20]:

$$q_e = K_F C_e^{1/n}, \tag{5}$$

or can be written in linearized form (Equation 6) as:

$$lnq_e = lnK_F + \frac{1}{n} lnC_e, \tag{6}$$

where K_F and n are the Freundlich constants characteristic of the system. K_F and n are indicators of adsorption capacity and adsorption intensity, respectively. q_e is the equilibrium EtBr concentration in solid phase (mg/g), and C_e is the equilibrium EtBr concentration in liquid phase (mg/L). Temkin isotherm contains a factor that explicitly takes into the account the adsorbing species-adsorbent interactions. The Temkin equation is given as [21]:

$$q_e = \frac{RT}{b} ln\ K_T C_e, \tag{7}$$

which can be linearized as:

$$q_e = B_T lnK_T + B_T lnC_e, \tag{8}$$

where $B_T = RT/b$, T (K) is the absolute temperature, R (8.314 J/mol) is the universal gas constant, K_T (per milligram) is the equilibrium binding constant, and b (J/mol) is related to the heat of adsorption. The isotherm constants K_T and B_T are calculated from the slope and intercept of the q_e versus lnC_e plot. The plots of linearized Langmuir, Freundlich, and Temkin isotherms are illustrated in Figures 6, 7, and 8, respectively. Parameters of the three isotherms were calculated and listed in Table 1. Based on the values of R^2, it can be concluded that the Langmuir isotherm best fits the equilibrium data. The

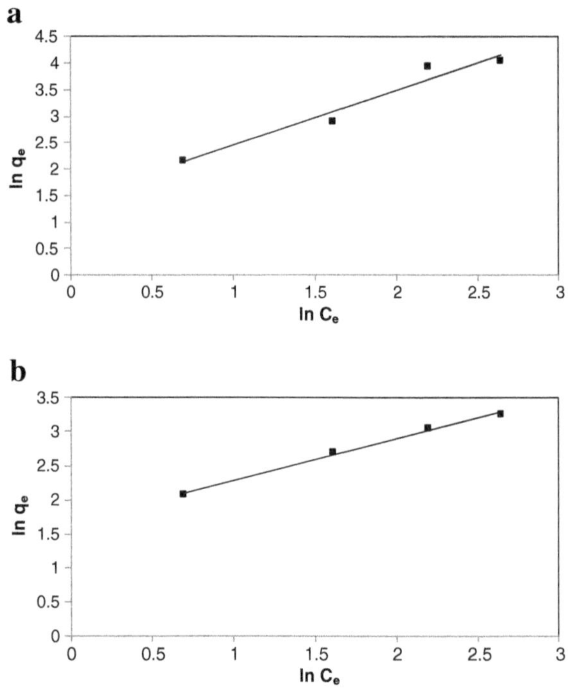

Figure 7 Freundlich plots. **(a)** Freundlich plots for EtBr adsorption by SWCNTs. **(b)** Freundlich plots for EtBr adsorption by NZVI.

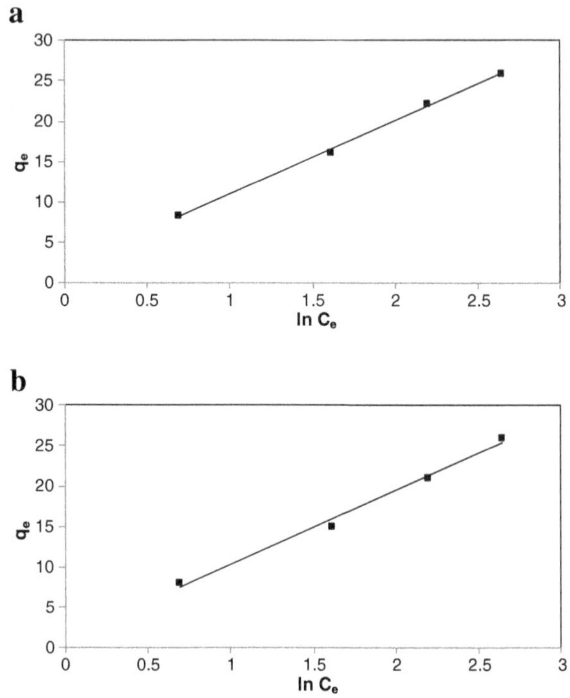

Figure 8 Temkin plots. **(a)** Temkin plots for EtBr adsorption by SWCNTs. **(b)** Temkin plots for EtBr adsorption by NZVI.

Table 1 Isotherm parameters for the adsorption of EtBr by SWCNTs

Langmuir (Type 1)				Langmuir (Type 2)				Freundlich			Temkin		
K_L (L/g)	q_m (mg/g)	R^2	R_L	K_L (L/g)	q_m (mg/g)	R^2	R_L	K_F (mg/g)/(mg/L)$^{1/n}$	n	R^2	K_T (m/g)	b (J/mol)	R^2
0.2001	45.111	0.9994	0.5	0.1186	42.160	0.9994	0.6	5.4286	1.733	0.9955	1.1370	270.01	0.9990

maximum adsorption capacity (q_m) of adsorbent calculated from Langmuir isotherm defines the total capacity of the adsorbent for EtBr. The values of R_L calculated from Equation 4 are listed in Table 2. The R_L values are between 0 and 1, and the adsorption process is favorable. The fact that the Langmuir isotherm fits the experimental data very well may be due to the homogeneous distribution of active sites on the SWCNTs and NZVI surface because application of the Langmuir equation involves the assumption that the surface were homogeneous.

Conclusions

The removal of ethidium bromide by SWCNTs and nanoparticle zero valent iron were studied in a batch system with respect to the temperature, adsorbent dose, and contact time. Removal of EtBr from aqueous solution was possible using SWCNTs and NZVI surfaces. It was seen that maximum adsorption takes place after 5 to 7 min. The optimum doses of adsorbents were 30 mg for NZVI. The Langmuir, Freundlich, and Temkin isotherm models provided the best fits to predict the adsorption for the EtBr by SWCNTs and NZVI surfaces. From the above results, we conclude that the adsorbent NZVI was effective in removing EtBr from aqueous solution. Also, the removal of EtBr by SWCNT surfaces was more effective than that of NZVI surfaces.

Methods

Materials

Ethidium bromide was supplied by Merck, Darmstadt, Germany (maximum purity available). All solutions were prepared with deviations of less than ±0.1% from the desired concentrations. SWCNTs (Armchair (6,6), purity >95; diameter, 1 to 2 nm; length, 5 to 30 nm; surface area, 400 m^2/g; and manufacturing method, catalytic chemical vapor deposition).

Preparation of NZVI particles

NZVI particles were prepared by liquid-phase reduction method. All solvents were degassed and saturated for 30 min with N_2 before use. NZVI was synthesized by adding 1 M $NaBH_4$ solution into 0.5 M $FeCl_3$ solution during vigorous stirring under N_2 atmosphere. The mixture's color turned from red brown to light yellow and then eventually to black. At the same time, the mixture gradually produced more black grain particles in the three-neck flask. Ferric iron (Fe^{+3}) was reduced to elemental iron according to the following reaction:

$$4Fe^{+3} + 3BH_4 + 9H_2O \rightarrow 4Fe^0(s) + 3H_2BO_3^-(aq) + 12H^+ + 6H_2$$

Then, black NZVI particles were vacuum-filtered and washed with deionized water and 1:1 (*V/V*) ethanol/acetone; doing so prevented the NZVI from oxidizing, and then, resulting gray-black solid was dried under nitrogen atmosphere before use.

Batch equilibrium studies

Adsorption experiments were carried out by agitating 50 mg of SWCNTs and NZVI for each individually in aqueous solution with 30 ml of EtBr solution of desired concentration for each surfaces (SWCNTs and NZVI) and pH at 180 rpm, 25°C in a thermostated mechanical shaker (ORBITEK, Chennai). Concentration of EtBr was estimated spectrophotometrically by monitoring the absorbance at 274 nm using a UV–vis spectrophotometer (Hitachi model 8543, Chiyoda-ku, Japan).

The effect of adsorbent dosage was studied by agitating 30 mL of EtBr solutions with different adsorbent doses (10 to 70 mg) at equilibrium time. The amounts of EtBr adsorbed by the adsorbent were calculated using the following equation:

$$\frac{(C_0 - C_t)V}{W}, \tag{9}$$

where C_0 is the initial EtBr concentration, C_t is the EtBr concentration (mg/L) at any time, V is the volume of solution (L), and W is the mass of the adsorbents (g).

Table 2 Isotherm parameters for the adsorption of EtBr by NZVI

Langmuir (Type 1)				Langmuir (Type 2)				Freundlich			Temkin		
K_L (L/g)	q_m (mg/g)	R^2	R_L	K_L (L/g)	q_m (mg/g)	R^2	R_L	K_F (mg/g)/(mg/L)$^{1/n}$	n	R^2	K_T (m/g)	b (J/mol)	R^2
0.1106	43.290	0.9981	0.4	0.1186	42.160	0.9991	0.5	5.3275	1.626	0.9940	1.1261	264.94	0.9911

Competing interests
The authors declare that they have no competing interests.

Authors' contributions
FN carried out the result analysis, participated to draft the manuscript and in the manuscript elaboration, carried out the experiments, and obtained most of the experimental images. MN, KZ, and AF coordinated the project, discussed the results, and helped to draft the manuscript. All authors read and approved the final manuscript.

Acknowledgements
The authors would like to thank the Islamic Azad University Roudehen Branch for their financial support.

Author details
[1]Department of Chemistry, Roudehen Branch, Islamic Azad University, Roudehen, Iran. [2]Department of Virology, School of Public Health and Institute of Health Research, Tehran University of Medical Sciences, Tehran, Iran. [3]Department of Chemistry, Science and Research Branch, Islamic Azad University, Tehran, Iran. [4]Department of Chemistry, Shahid Beheshti University, Evin, Tehran, Iran. [5]Department of Chemistry, Shahre-Qods Branch, Islamic Azad University, Tehran, Iran.

References

1. Lunn, G, Sansone, EB: Ethidium bromide: destruction and decontamination of solutions. Anal Biochem **162**, 453–458 (1987)
2. Bountiff, L, Levantis, P, Oxford, J: Electrophoretic Analysis of Nucleic Acids Isolated from Scrapie-infected Hamster Brain. J Gen Virol **77**, 2371–2378 (1996)
3. Bailey, SE, Olin, TJ, Bricka, RM, Adrian, DD: A review of potentially low-cost sorbents for heavy metals. Water Res **33**, 2469–2479 (1999)
4. Uzum, C, Shahwan, T, Eroglu, AE, Lieberwirth, I, Scott, TB, Hallam, KR: Application of zero-valent iron nanoparticles for the removal of aqueous Co^{2+} ions under various experimental conditions. Chem Eng J **144**, 213–220 (2008)
5. Cantrell, KJ, Kaplan, DI, Wietsma, TW: Zero-valent iron for the in situ remediation of selected metals in groundwater. J Hazard Mater **42**, 201–212 (1995)
6. Cho, HH, Park, JW: Sorption and reduction of tetrachloroethylene with zero valent iron and amphiphilic molecules. Chemosphere **64**, 1047–1052 (2006)
7. Dombek, T, Dolan, E, Schultz, J, Klarup, D: Rapid reductive dechlorination of atrazine by zero-valent iron under acidic conditions. Environ Pollut **111**, 21–27 (2001)
8. Ghauch, A, Rima, J, Amine, C, Martin-Bouyer, M: Rapid treatment of water contamined with atrazine and parathion with zero-valent iron. Chemosphere **39**, 1309–1315 (1999)
9. Noubactep, C: Characterizing the discoloration of methylene blue in Fe^0/H_2O systems. J Hazard Mater **166**, 79–87 (2009)
10. Noubactep, C, Kurth, AMF, Sauter, M: Evaluation of the effects of shaking intensity on the process of methylene blue discoloration by metallic iron. J Hazard Mater **169**, 1005–1011 (2009)
11. Noubactep, C: Characterizing the reactivity of metallic iron upon methylene blue discoloration in Fe^0/MnO_2/H_2O systems. J Hazard Mater **168**, 1613–1616 (2009)
12. Li, XQ, Elliott, DW, Zhang, WX: Zero-Valent Iron Nanoparticles for Abatement of Environmental Pollutants: Materials and Engineering Aspects. Crit. Rev. Solid State Mater. Sci. **31**, 111–122 (2006)
13. Noubactep, C: A Critical review on the process of contaminant removal in fe0–h2o systems. Environ. Technol. **29**, 909–920 (2008)
14. Noradoun, CE, Cheng, IF: EDTA Degradation Induced by Oxygen Activation in a Zerovalent Iron/Air/Water System. Environ. Part Sci Technol **39**, 7158–7163 (2005)
15. Wang, H, Zhou, A, Peng, F, Yu, H, Yang, J: Mechanism study on adsorption of acidified multi-walled carbon nanotubes to Pb(II). J Colloid Interface Sci **316**, 277–283 (2007)
16. Lu, C, Chiu, H, Liu, C: Removal of zinc(II) from aqueous solution by purified carbon nanotubes: Kinetics and equilibrium studies. Ind Eng Chem Res **45**, 2850–2855 (2006)
17. Feng, J, Lim, TT: Iron-mediated reduction rates and pathways of halogenated methanes with nanoscalePd/Fe: Analysis of linear free energy relationship. Chemosphere **66**, 1765–1774 (2007)
18. Langmuir, I: The adsorption of gases on plane surfaces of glass, mica and platinum. J Am Chem Soc **40**, 1361–1368 (1918)
19. Mahmoodi, NM, Arami, M: Modeling and sensitivity analysis of dyes adsorption onto natural adsorbent from colored textile wastewater. J Appl Polym Sci **109**, 4043–4048 (2008)
20. Freundlich, HMF: Uber die adsorption in I osungen (adsorption in solution). J Phys Chem **57**, 385–470 (1906)
21. Moradi, O, Zare, K, Monajjemi, M, Yari, M, Aghaie, H: The studies of equilibrium and thermodynamic adsorption of Pb(II), Cd(II) and Cu(II) ions from aqueous solution on SWCNTs and SWCNT-COOH surfaces. Fullerenes, Nanotubes, and Carbon Nanostructures **18**, 285–302 (2010)

38

Alteration in protein profile of *Pseudomonas aeruginosa* (PTSOX4) coated with magnetic Fe_3O_4 nanoparticles

Fazeleh Bahrampour, Jamshid Raheb* and Zohreh Rabiei

Abstract

One of the remarkable features of bacterial species is their capacity for rapid growth when the appropriate environmental condition for growth is provided. Some bacteria, during their growth period, encounter stress factors in their natural environments, such as limitation in growth bioavailability, heat shock, heavy metal, etc. One stress factor not studied is the effect of magnetic Fe_3O_4 nanoparticles on bacterial growth rate. The effect of magnetic Fe_3O_4 nanoparticles on the protein profiles of genetically engineered bacterial strain *Pseudomonas aeruginosa* (PTSOX4), a strain with biological desulfurization characteristic, was investigated. The magnetic Fe_3O_4 nanoparticles were synthesized using co-sedimentation method, and their morphology was observed by scanning electron microscopy (SEM). The topography of magnetic Fe_3O_4 nanoparticles was detected by X-ray diffraction, and the average nanoparticle size measured was 40 to 50 nm. The bacterial cells were coated with magnetic nanoparticles, and the SEM electrographs of the bacterial cells indicated that the nanoparticles were uniformly coated on the cell surface. Proteins from both uncoated and coated bacterial cells were extracted by sonication and subjected to two-dimensional sodium dodecyl sulfate polyacrylamide gel electrophoresis. Some novel protein bands appeared in the protein profiles of coated bacterial cells; however, some protein bands disappeared. The two-dimensional gel electrophoresis results highlighted the presence of two different polypeptide groups, with molecular weights of 30 to 56 kDa and 56 to 65 kDa.

Keywords: Desulfurization; Nanoparticle; Protein profile; Polypeptide; *Pseudomonas aeruginosa*

Background

The synthesis of magnetic monodisperse nanoparticles has attracted great interest in the field of nanotechnology, and the investigation is developed with several types of iron oxides that have been carried out in the field of magnetic nanoparticles (NPs). The application of small iron oxide (NPs) *in vitro* diagnostics has been investigated for nearly half a century [1,2]. Nanotechnology has developed to such an extent that it has become possible to fabricate, characterize, and specially tailor the functional properties of NPs for biomedical and diagnostic applications [3,4]. At present, a number of different chemical and biological methods have been reported for magnetite Fe_3O_4 NPs [5]. The study on chemical co-precipitation has been widely used to produce magnetite NPs due to its facility and large-volume capability [6,7]. This method involves co-precipitation to synthesize iron oxides from aqueous Fe^{2+} and Fe^{3+} salt solutions by adding a base [8,9] as follows: $Fe^{2+} + 2Fe_3^+ + 8OH^- \rightarrow Fe_3O_4 + 4H_2O$.

Nano-sized particles have physical and chemical properties that are characteristic of neither the atom nor the bulk counterparts [10]. Among various types of NPs, magnetic iron oxide NPs have been explored increasingly by researchers for their applications in drug delivery, magnetic resonance imaging, tissue repair, hyperthermia, and cell separation [11,12]. In the field of nano-sized magnetic NPs (the single domain with diameter of about 5 to 20 nm), magnetite is a very promising candidate since its biocompatibility has already been proven [13,14]. In fact, two main types of iron oxides, magnetite (Fe_3O_4) and maghemite (γ-Fe_2O_3), are widespread in the environment and have been used in various biomedical applications [3,15]. Maghemite could

* Correspondence: jam@nigeb.ac.ir
National Institute of Genetic Engineering and Biotechnology, Tehran free way 15 Km, Pajouhesh Blv., P.O. Box 14925/16, Tehran 1417863171, Iran

be considered as a Fe(II)-deficient magnetite, and it is formed by the topotactic oxidation of magnetite [7]. The study on magnetic Fe_3O_4 NPs is of increasing interest because of its special magnetic properties and low toxicity [8,16]. Moreover, it has been found to have various biomedical applications. On the other hand, bacterial genomes are ideal model systems for the development and application of proteomic technology because they are relatively small organisms [4]. A combination of proteomic technologies was used in this study for the global analysis of protein expression and regulation in the Gram-negative bacteria *Pseudomonas aeruginosa* (PTSOX4).

The term proteomics was coined in 1995 by Marc Wilkins as a complement of proteins expressed by a genome. In the simplest term, it is the study on proteome, the collection of proteins in an organism [17]. Proteins are organic compounds that consist of 20 common amino acids joined by peptide bonds [18].

All amino acids have a central carbon atom to which a hydrogen atom, an amino group (NH_2), and a carboxyl group (COOH) are attached in common [19]. In fact, sequences of amino acids fold to generate compact domains (three-dimensional structures: secondary, tertiary, and quaternary) from linear chains (primary structure) [19]. In protein molecules, electrostatic interactions, Van der Waals forces, hydrogen bonds, and hydrophobic interactions play an important role in defining and stabilizing the three-dimensional structure and adsorption of protein molecules [17,18]. In addition, many proteins have carbohydrate-binding sites on their surface [20]. They play an important role in bacterial adhesion and recognition of pathogens by specific surface carbohydrates of the immune system [20]. In fact, proteins are one group of recognizable organic matter from the environment [21]. Proteins play an important role in bacterial transport in natural environments. Therefore, it is important to understand the role of proteins in bacteria [21]. On the other hand, proteomics offers the most effective approach to identify protein markers specific to epidemic strains that could serve as targets for novel therapeutics [22,23]. Proteomics can have practical applications through the identification of proteins that may be potential targets for the biotechnology industry and through expansion of our understanding on the physiological action of these proteins [24]. Sodium dodecyl sulfate polyacrylamide gel electrophoresis (SDS-PAGE) is an electrophoretic method for the separation of polypeptides according to their molecular weights. The technique is performed in polyacrylamide gels containing sodium dodecyl sulfate. SDS-PAGE is the most widely used in analytical methods to resolve separate components of a protein mixture. The technique is also a powerful tool for estimating the molecular weights of proteins [25]. In the present study, using proteomic strategies to achieve these aims includes (a) analysis by SDS-PAGE and (b) two-dimensional (2-D) gel electrophoresis.

Results and discussion

Determination of nanoparticle size for bacterial cell coating

Scanning electron microscopy (SEM) was used to characterize the nanoparticle size suitable for bacterial cell coating. The nanoparticle diameter in the range of 45 to 55 nm was determined directly from the SEM images, as shown in (Figure 1). An average diameter of 40 nm was selected for nanoparticles since it is small enough to have super paramagnetic characteristics. The SEM electrophotograph of the cells coated with Fe_3O_4 nanoparticles is shown in (Figure 2), in which it is clearly demonstrated that the Fe_3O_4 nanoparticles were adsorbed on the surface of bacterial cells.

Structural characterization of the magnetic Fe_3O_4 nanoparticles

The X-ray diffraction (XRD) pattern (Figure 3) of the Fe_3O_4 nanoparticles prepared under standard conditions revealed diffraction peaks at (220), (311), (400), (422), (511), (440), etc., which were characteristic peaks of Fe_3O_4 in a crystal form with a cubic spinel structure. It is clear that only the phase of Fe_3O_4 is detectable and there was no other phase such as $Fe(OH)_3$ or Fe_2O_3, as usual products in a chemical co-precipitation procedure (Figure 4). It was also found that particle size was quite small from the relatively wide half-peak breadth. According to the XRD pattern, the average diameter could be evaluated by the Scherrer equation [9].

SDS-PAGE analysis of total bacterial cell protein

SDS-PAGE was applied for total cell protein analysis. The total proteins include the control and those treated with Fe_3O_4 nanoparticle, and the marker is demonstrated in

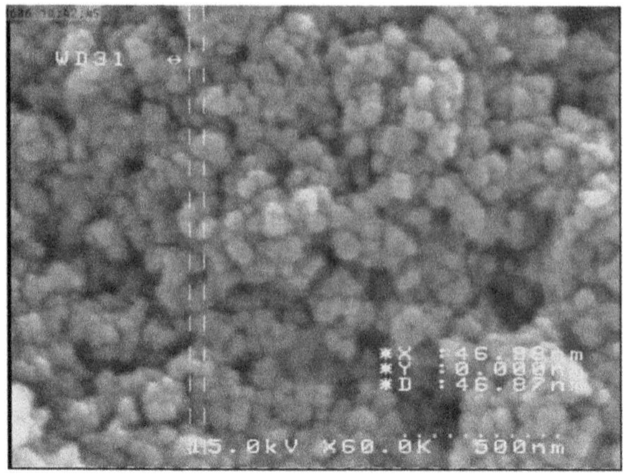

Figure 1 SEM images (a total of 100 particles were measured). Mean size is 47.22 ± 0.96 nm (SD).

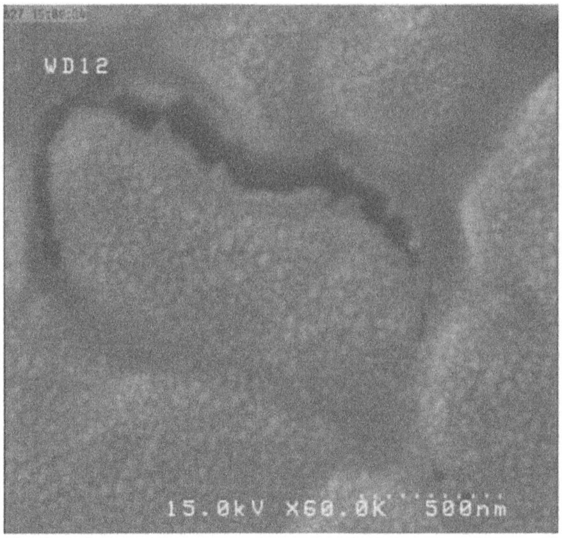

Figure 2 SEM of the surfaces of cells coated with Fe_3O_4 nanoparticles.

Figure 5. The results indicated the presence of two different polypeptide groups, with molecular weights of 30 to 45 kDa and 17 to 30 kDa. The polypeptide with the molecular weight of 78 kDa was more upregulated than the others. Moreover, the effects of Fe_3O_4 nanoparticles on the cell surface of the bacteria could be considered as the main reason for the elimination of polypeptide available per bacterial cells to yield a suitable level of cell coating.

2-D gel electrophoretic analysis of total bacterial cell protein

In order to define protein expression and regulation in *P. aeruginosa* grown under Fe_3O_4 nanoparticle stress, proteomic technology was utilized. The total cell protein of *P. aeruginosa* was extracted by a 7-cm Immobiline DryStrip (GE Life Sciences, Chalfont St. Giles, England), pH 3 to 10. The Ettan IPGphor isoelectric focusing system (Figure 6) and treatment sample (Figure 7) showed complete differentiation of protein quantity in the control and treatment samples, which was distinguished by the white area. The standard ladder (Bio-Rad, Hercules, CA, USA) used in the 2-D technique was determined between 97 and 21.5 kDa.

Ishikawa and coworkers have purified an iron-binding protein from *Campylobacter jejuni*, cloned and sequenced the dps gene encoding this protein, and investigated the role of the dps gene in hydrogen peroxide stress resistance [26,27]. In this work, the protein was purified from *P. aeruginosa* to investigate the role of iron oxide nanoparticles in stress resistance.

Proteins can be quantitatively precipitated from a variety of sources without interference from detergents, chaotropes, and other common reagents used to solubilize protein (Figure 6). Recovery is generally greater than 50%. The procedure does result in spot gain or loss or changes in spot position relative to untreated samples (Figure 7). The precipitated proteins are easily re-suspended in a 2-D sample solution. On the other hand, it showed that in the SDS-PAGE, some bands disappeared, but majority of the protein banding patterns in both were stable, which introduce the effect of Fe_3O_4 nanoparticles on the bacterial protein profile. Many proteins were differentially expressed between each mutant strain situation.

Conclusions

High resistance to environmental stress factors is one of the excellent characteristics of bacterial responses. Stimulation of these factors in different bacteria is triggered for some production, leading to the formation of a series of proteins

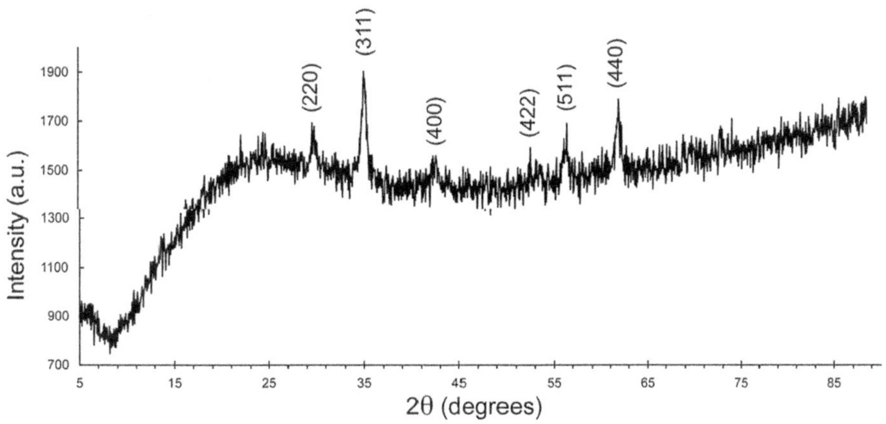

Figure 3 X-ray diffraction pattern of the nanoparticles. The numbers in parenthesis give the Miller indices of pure Fe_3O_4 (NBS, 1976) assigned to the observed peaks.

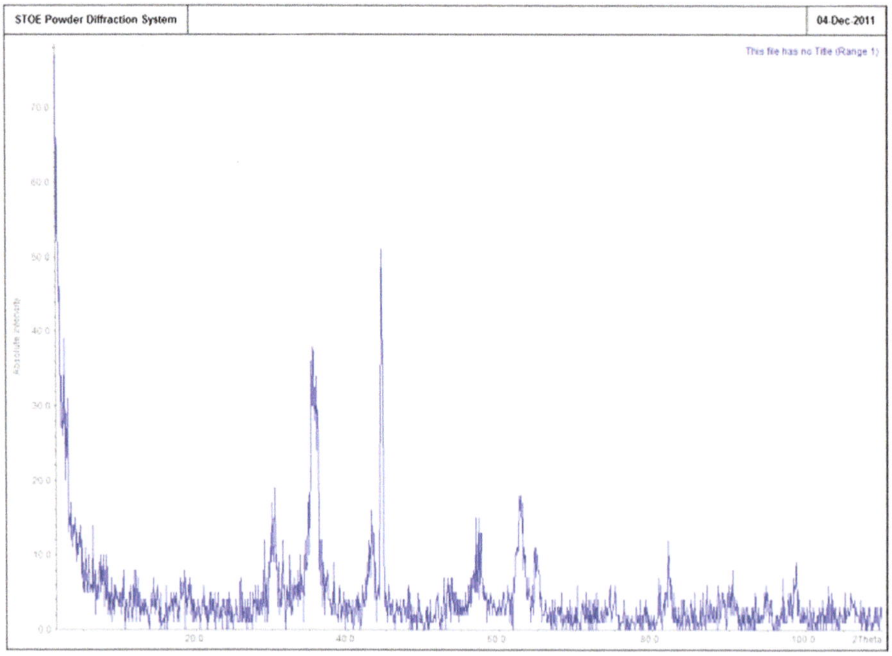

Figure 4 XRD pattern indicating the sole existence of inverse cubic spinel phase of Fe_3O_4. The average size was about 40 nm.

in response to the environmental stress factors such as nanoparticles. Magnetic nanoparticles with a super paramagnetic behavior have very low toxicity on living cells [28]. However, these nanoparticles as an environmental stress factor cause the production of a number of proteins in response to resist the toxicity of these nanoparticles. This mechanism occurs due to the development of a more resistant state in most bacterial cells. Exposure to nonlethal levels of magnetic nanoparticles also increases the resistance of actively growing cells to subsequent lethal level of the same and, occasionally, other stressors. This enhanced resistance which is developed during exposure to the magnetic nanoparticle requires newly synthesized proteins.

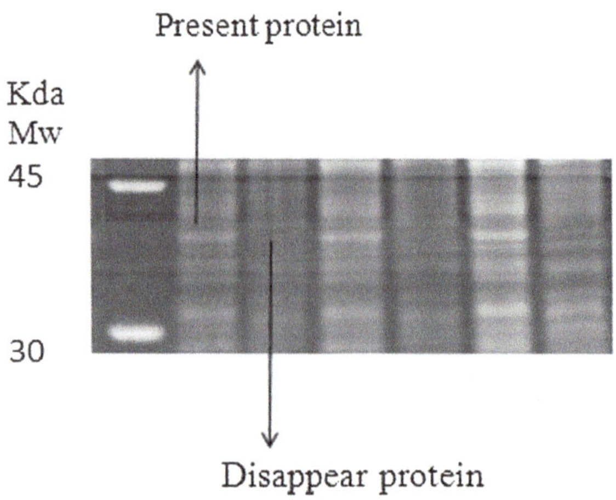

Figure 5 Data obtained from SDS-PAGE showing the separation of bacterial protein. Column 1, ladder; columns 2, 4 and 6 represent the uncoated bacteria as control, and columns 3, 5 and 7 represent coated bacteria.

In *Vibrio furnissii*, in response to some environmental factors, cell mass and rRNA decreased rapidly while the total cell counts remained stable. This reduction in cell mass is seen as miniaturization of cell, similar to the phenomena observed in other *Vibrio* species [29]. The overall rate of proteins synthesized in *Vibrio* upon exposure to one of the stressors after 24 h was about 10% of that of cells growing exponentially in a complex medium, and the global rate of protein synthesis was reduced by 95% in *Escherichia coli* [30]. In both cases, the proteins being synthesized were those associated with one environmental factor and the development of resistance to other factors in these bacteria. The rate of total protein synthesis decreased immediately at the onset of stress, and after 48 h, it was less than 1% of the synthesis rate during growth [31,32].

In the experimental reported here, the total number of proteins seen in the profiles of *P. aeruginosa* (PTSOX4) cells run on SDS-PAGE showed several changes over an exposure period to magnetic nanoparticles. The changes observed depend on the time period of exposure, and after a longer time, the changes were more considerable. No measurements of the total amount of proteins in the cells or the overall rate of the protein synthesis were made in this experiment. However, it is evident from an

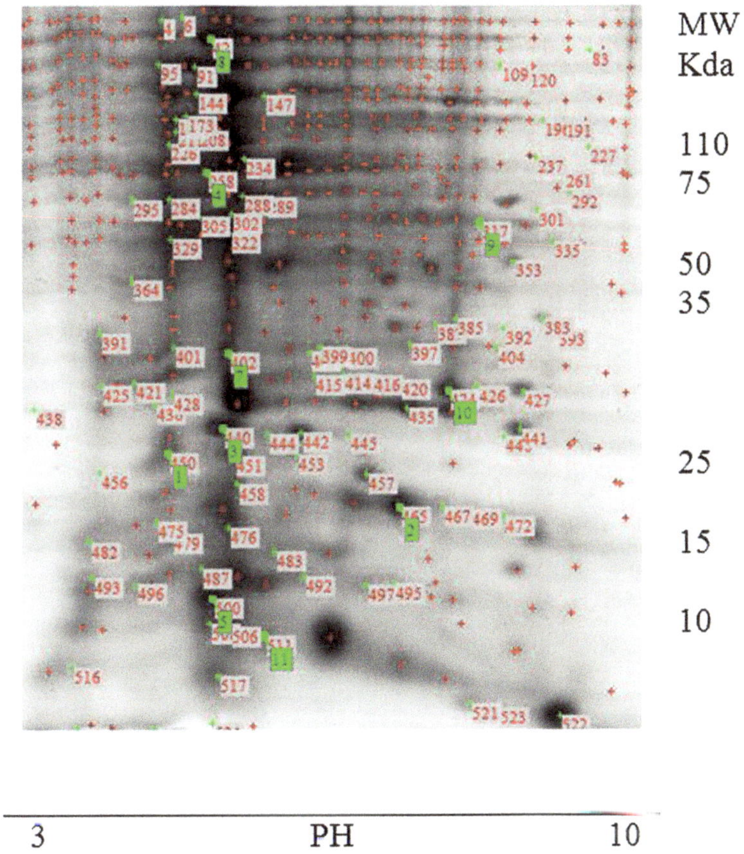

Figure 6 2-D protein of *P. aeruginosa*. Extracted by a 7-cm Immobiline DryStrip, pH 3 to 10, Ettan IPGphor isoelectric focusing system. Protein profiles of uncoated bacterial cells analyzed by Melani software.

examination of the gels that the total amount of cellular protein decreased and that there were reproducible changes observed in the protein profile of cells when exposed to the nanoparticles. As SDS-PAGE only analyze the proteins based on molecular weight, it is impossible to identify many proteins which may have a different isoelectric point but the same molecular weight. Many of the bands seen on this gel could be due to the multiple proteins. Consequently, two-dimensional gel electrophoresis was used to demonstrate in more detail the changes in the total protein profiles of cells under different stresses. The study on cells coated with nanoparticles showed that a number of proteins were eliminated. These proteins are responsible for overall integrity, protection against oxidative stress, and change in osmotic pressure. Previously, we showed that these nanoparticles are less toxic to PTSOX4 bacteria. We suggest that the coating of Fe_3O_4 NPs has no effects on the osmotic equilibrium of the membrane. Some report that the synthesis of some proteins is most critical when the stress effect response in *E. coli* occurred in the first period of incubation and that these proteins play a major role in the response to long-term effects of environmental factors [33-35].

Methods

Materials

Bacterial strain and medium

P. aeruginosa (PTSOX4) and *E. coli* DH5α were provided by the National Institute of Genetic Engineering and Biotechnology and have the ability to convert dibenzothiophene (DBT) to 2-hydroxy-biphenyl and sulfate. The bacteria were grown on a sulfur-free culture medium (bifidus selective medium (BSM) composed of 2.44 g KH_2PO_4, 5.47 g Na_2HPO_4, 0.2 g $MgCl_2.6H_2O$, 0.001 g $CaCl_2.2H_2O$, 0.001 g $FeCl_3.6H_2O$, 0.004 g $MnCl_2.4H_2O$, and 2 mL glycerol in 1 L deionized water. In addition, DBT solution was added to obtain the final solution of 100 ppm L^{-1}. *Pseudomonas* strain was grown in BSM at 30°C. *E. coli* strains were grown in Luria-Bertani (LB) medium (Difco, Becton, Dickinson and Company, Franklin Lakes, NJ, USA) at 37°C.

Methods

Nanoparticle synthesis

The synthesis of magnetic nanoparticles was done by the following method. We dissolved 0.045 g $FeCl_2.4H_2O$ and $FeCl_3.6H_2O$ in 150 mL deionized water, previously

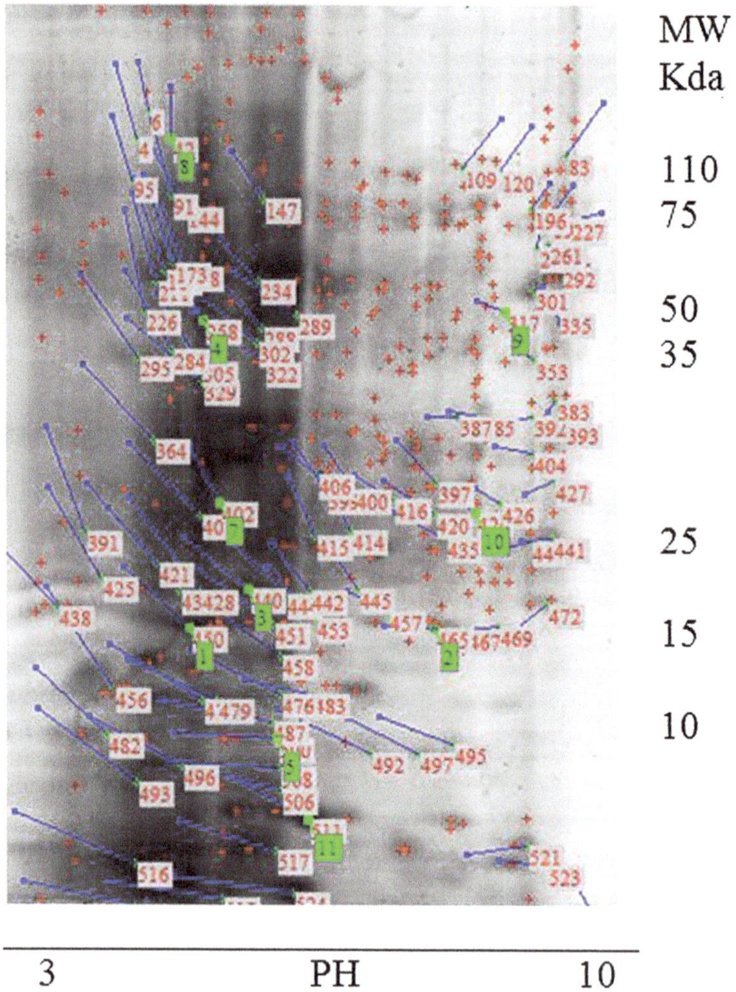

Figure 7 2-D protein of *P. aeruginosa* coated with iron oxide nanoparticles. Extracted by a 7-cm Immobiline DryStrip, pH 3 to 10, Ettan IPGphor isoelectric focusing system. Protein profiles of coated bacterial cells analyzed by Melani software.

acidified by 1 mL HCl (37%), with mechanical stirring at 1,100 rpm and 65°C, then quickly added NH_4OH (1 M) until the pH reached 11. To the resulting suspension, 0.09 g glycine was added over a period of 10 min. After 20 min, the magnetic precipitate was separated by a centrifuge (4,000 rpm). The sample was washed two times and dried at 80°C by vacuum drying.

Methods of NP characterization

In contrast, the surface coated with nanoparticles is more freely aligned with the external field than the uncoated surface . The repulsive force between hydrophobic surfactant molecules coated on individual particles can prevent them from agglomeration.

Structure of the crystal

A crystallographic study on magnetic Fe_3O_4 NPs was performed with an X-ray powder diffractometer using a CuKα (λ = 0.15406 nm) target, graphite monochromator. The working voltage was 40 kV, and the working electrical current was 30 mA with a scan speed of 8°C min^{-1}.

Cell growth and coating

The bacterial cells were grown in 100 mL LB until the mid-exponential growth phase, harvested by centrifugation at 1,400×*g* for 10 min, and concentrated to obtain 5.6 mg of dry cells per liter. The cells were coated with magnetic nanoparticles through the following method: 10 mL of a suspension containing 100 μg mL^{-1} Fe_3O_4 NPs in water solution was mixed with 100 mL of the cell suspension in LB at a final concentration of 0.5 mM. The ratio of nanoparticle mass to biomass was measured, which was 1.78 *w*/*w* and considered appropriate to make sufficient particles.

Protein extraction and purification

Bacterial cells were grown in a suitable liquid medium for 24 h at 30°C. The cells were harvested by centrifugation and sonicated in 1 mM Tris buffer (pH 8.0). The sonicated cells were centrifuged at 1,500×*g* for 20 min at 4°C. The pellet was discarded, and the supernatant was centrifuged at 48,400×*g* in a JA20 rotor (Beckman Coulter, Brea, CA, USA) for 60 min at 4°C to harvest the membrane proteins and leave the soluble proteins in the supernatant. The membrane proteins were re-suspended in 3 mL Tris buffer (pH 8.0) and stored at −20°C.

The supernatant was concentrated to about 5 mL using an Amicon (Millipore Co., Billerica, MA, USA) ultrafiltration apparatus. The filters were able to retain proteins with molecular weight greater than 10,000 Da. An equal amount of saturated ammonium sulfate was added to the concentrated supernatant fraction to give a final ammonium sulfate concentration of 50%, which was left overnight to allow the proteins to precipitate. The precipitate was pelleted by centrifugation at 35,000×*g* for 30 min, re-suspended in 5 mL of sterile distilled water, and dialyzed overnight against at least 4 L of distilled water to remove any ammonium sulfate from the samples.

Two-dimensional gel electrophoresis

This technique separated proteins according to their isoelectric point in the first dimension and according to their molecular weight in the second dimension by SDS-PAGE. These two unrelated parameters give the maximum resolution of the proteins present in a complex mixture.

Approximately 75 μg of bacterial protein was placed in a rehydration buffer that was immobilized by a dry strip (pH 3 to 10 for 16 h) and then placed in an Ettan IPGphor isoelectric focusing system (92.5 kV h for 21 h; GE Life Sciences). After that, the equilibrated IPG strips for the second-dimension gel between the two glass plates were applied. The IPG plastic backing was placed against one of the glass plates in room temperature on one side of the strip with the filter paper and protein marker.

Rehydration buffer

The buffer was prepared for one IPG strip of 7 cm, 50 μL of buffer including 12 mg (8 M) urea, 0.075 mg thiourea, 2 mg 4% CHAPS, 2.5 μL 100% (1 M) DTT, 0.2 μL 1% IPG buffer (ampholines), 20 μL Bromophenol blue, and dionized water up to 125 μL.

Equilibration buffer

The following buffer was applied as an equilibration buffer in this work: Tris (1 M), pH 8.8 5 mL, 36.03 g urea, 30 mL 100% glycerol, 20 mL 10% SDS, 200 μL 1% Bromophenol blue, and deionized water up to 100 mL.

SDS polyacrylamide gel

Gel electrophoresis was carried out in a vertical SDS polyacrylamide gel containing 10% SDS, with stacking gel of 6% acrylamide and a resolving gel of 12% acrylamide. Electrophoresis was performed for 2 to 3 h at 100 mA. The standard protein markers (Bio-Rad) were included. Gels were fixed and stained with Coomassie blue as previously described. Photographs of each gel were taken by a gel documentation system (LabScan TM 6.0, Swiss Institute of Bioinformatics, Geneva, Switzerland).

Competing interests

The authors declare that they have no competing interests.

Authors' contributions

JR and FB carried out the nanobiotechnology and proteomics studies.
JR, FB and ZR participated in writing the manuscript.

Authors' information

FB is a PhD student at the National Institute of Genetic Engineering and Biotechnology. JR and ZR are academic staff and associate professor at the National Institute of Genetic Engineering and Biotechnology.

Acknowledgements

The authors gratefully acknowledge the financial support of this work by the National Institute of Genetic Engineering and Biotechnology (NIGEB).

References

1. Guobin, S, Xing, JM, Zhang, HY, Liu, H: Biodesulfurization of dibenzothiophene by microbial cells coated with magnetite nanoparticles. Appl. Environ. Microbiol. **71**, 4497 (2005)
2. Hu, C, Gao, Z, Yong, X: Fabrication and magnetic properties of Fe_3O_4 octahedra. Chem. Phys. Lett. **429**, 513 (2006)
3. Moghimi, SM, Hunter, ACH, Murray, JC: Long-circulating and target-specific nanoparticles: theory to practice. Pharm. Rev. **53**, 283 (2001)
4. Gygi, SP, Corthals, GL, Zhang, Y, Rochon, Y, Aebersold, R: Evaluation of two-dimensional gel electrophoresis-based proteome analysis technology. Proc. Natl. Acad. Sci. USA. **97**, 9390 (2009)
5. Montazeri, H, Amani, A, Shahverdi, HR, Haratifar, E, Shahverdi, AR: Separation of the defect-free Fe_3O_4-Au core/shell fraction from magnetite-gold composite nanoparticles by an acid wash treatment. J. Nanostruc. Chem. **3**, 25 (2013)
6. Theil, EC: Ferritin structure, gene regulation, and cellular function in animals, plants, and microorganisms. Annu. Rev. Biochem. **56**, 289 (1987)
7. Tartaj, P, González, T, Serna, CJ: Single-step nanoengineering of silica coated maghemite hollow spheres with tunable magnetic properties. Adv. Mater. **13**, 1620 (2001)
8. Carpenter, EE: Iron nanoparticles as potential magnetic carriers. J. Magn. Magn. Mater. **225**, 17 (2001)
9. Cornell, RM, Schertmann, U: Iron Oxides in the Laboratory: Preparation and Characterization, vol. 185, p. 8723. VCH, Weinheim (1991)
10. Babes, L, Denizot, B, Tanguy, G, LeJeune, JJ, Jallet, P: Synthesis of iron oxide nanoparticles used as MRI contrast agents: a parametric study. Coll. Int. Sci. **212**, 474 (1999)
11. Gupta, AK, Gupta, M: Synthesis and surface engineering of iron oxide nanoparticles for biomedical applications. Biomath. **26**, 3995 (2005)
12. Gupta, AK, Gupta, M: Synthesis and surface engineering of superparamagnetic iron oxide nanoparticles for drug delivery and cellular targeting. In: Ravi Kumar, MNV (ed.) Handbook of Particulate Drug Delivery. American Scientific, Valencia (2007)
13. Panyam, J, Labhasetwar, V: Biodegradable nanoparticles for drug and gene delivery to cells and tissue. Adv. Drug. Del. Rev. **55**, 329 (2003)
14. Mahmoudi, M, Simchi, A, Milani, AS, Stroeve, P: Cell toxicity of superparamagnetic iron oxide nanoparticles. Collo. Inter. Sci. **336**, 510 (2009)
15. Hafmann-Amtenbrik, M, Rechenberg, B, Hafmann, H: Superparamagnetic nanoparticles for biomedical applications. Trans. World Res. Netw. **37**, 661 (2009)

16. Gupta, AK, Well, S: Surface-modified superparamagnetic nanoparticles for drug delivery: preparation, characterization and cytotoxicity studies. IEEE Trans Nanobiosci. **3**, 66 (2004)
17. Mansur, HS, Lobato, ZP, Orefice, RL, Vasconcelos, WL, Oliveira, C, Machado, LJ: Surface functionalization of porous glass networks: effects on bovine serum albumin and porcine insulin immobilization. Biomacromolecules **1**, 789 (2000)
18. Darby, NJ, Creighton, TE: Protein Structure. Oxford University Press, Oxford (1993)
19. Branden, C, Tooze, J: Introduction to Protein Structure. Garland, New York (1999)
20. Castro, LB, Kappl, M, Petri, DF: Adhesion forces between hybrid colloidal particles and concanavalin A. Langmuir **22**, 3757 (2006)
21. Fabiano, M, Marrale, D, Misic, C: Bacteria and organic matter dynamics during a bioremediation treatment of organic-rich harbour sediments. Mar. Pollut. Bull. **46**, 1164 (2003)
22. Nouwens, AS, Willcox, MD, Walsh, BJ, Cordwell, SJ: Proteomic comparison of membrane and extracellular proteins from invasive (PAO1) and cytotoxic (6206) strains of *Pseudomonas aeruginosa*. Proteomics **21**, 325 (2002)
23. Nouwens, AS, Walsh, BJ, Cordwell, SJ: Application of proteomics to *Pseudomonas aeruginosa*. Adv. Biochem. Eng. Biotechnol. **83**, 117 (2003)
24. Han, MJ, Lee, SY, Koh, ST, Noh, SG, Han, WH: Biotechnological applications of microbial proteomes. J. Biotechnol. **15**, 341 (2010)
25. Wai, S, Takata, NT, Takade, A, Hamasaki, N, Amako, K: Purification and characterization of ferritin from *Campylobacter jejuni*. Arch. Microbiol. **164**, 1 (1995)
26. Ishikawa, T, Mizunoe, Y, Kawabata, SI, Takade, A, Harada, M, Wai, SN, Yoshida, SI: The iron-binding protein Dps confers hydrogen peroxide stress resistance to *Campylobacter jejuni*. J. Bacteriol. **185**, 1010 (2003)
27. Hong, RY, Pan, TT, Li, HZ: Microwave synthesis of magnetic Fe_3O_4 nanoparticles used as a precursor of nanocomposites and ferrofluids. J. Magn. Magn. Mater. **303**, 60 (2006)
28. Samanta, B, Yan, H, Fischer, NO, Shi, J, Jerry, DJ, Rotello, VM: Protein-passivated Fe_3O_4 nanoparticles: low toxicity and rapid heating for thermal therapy. J. Mater. Chem. **18**, 1204 (2008)
29. Kramer, JK, Singleton, FL: Variations in rRNA content of marine *Vibrio spp.* during starvation-survival and recovery. App. Environ. Microbiol **58**, 201 (1992)
30. Albertson, NH, Nystrom, T, Kjelleberg, S: Macromolecular synthesis during recovery of the marine *Vibrio sp. S14* from starvation. J. G. Microbiol **136**, 2201 (1990)
31. Flardh, K, Cohen, PS, Kjelleberg, S: Ribosomes exist in large excess over the apparent demand for protein synthesis during carbon starvation in marine *Vibriosp-strain CCUG 15956*. J. Bacteriol. **174**, 6780 (1992)
32. Nowack, B, Bucheli, TD: Occurrence, behavior and effects of nanoparticles in the environment. Environ. Pollut. **150**, 5 (2007)
33. Reeve, CA, Bockman, AT, Matin, A: Role of protein degradation in the survival of carbon-starved *Escherichia coli* and *Salmonella typhimurium*. J. Bacteriol. **157**, 758 (1984)
34. Lim, CH: The effect of environmental factors on the physiology of *Aeromonashydrophila* in lake water. PhD thesis. University of Warwick (1995)
35. Marrow, JB, Arange, CP, Helbrook, RD: Association of quantum dot nanoparticles with *Pseudomonas aeruginosa* biofilm. Environ. Qual. **39**, 1934 (2010)

Seaweed-mediated synthesis of gold nanoparticles using *Turbinaria conoides* and its characterization

Shanmugam Rajeshkumar[1,2], Chelladurai Malarkodi[1], Gnanadhas Gnanajobitha[1,2], Kanniah Paulkumar[1], Mahendran Vanaja[1], Chellapandian Kannan[2*] and Gurusamy Annadurai[1*]

Abstract

The synthesis of metal nanoparticles using algae has been unexplored, but it is a more biocompatible method than the other biological methods. Metal nanoparticle synthesis using algae extract shows rapid and non-toxic process which resulted to nano sizes having the greatest potential for biomedical applications. In this investigation, we studied the green synthesis of gold nanoparticles using the algae extract of *Turbinaria conoides*. Green synthesis of gold nanoparticles was preliminarily confirmed by color changing from yellow to dark pink in the reaction mixture, and the broad surface plasmon resonance band was centered at 520 to 525 nm which indicates polydispersed nanoparticles. Transmission electron microscopy and selected-area electron diffraction analysis show the morphology and crystalline structure of synthesized gold nanoparticles with the size range of 6 to 10 nm. The four strong diffraction peaks were observed by X-ray diffraction; it confirmed the crystalline nature of synthesized gold nanoparticles. The carboxylic, amine, and polyphenolic groups were associated with the algae-assisted synthesized gold nanoparticles which was confirmed using Fourier transform-infrared spectroscopy. This study eliminates the use of chemical substances as reducing and stabilizing agent. Because it has natural several constituents which are fucoidan and polyphenolic substances, it does a dual function as both reducing and stabilizing agent for nanoparticles. Thus, algae-mediated synthesis process of biomedically valuable gold nanoparticles is a one-spot, facile, convenient, large-scaled, and eco-friendly method.

Keywords: Green synthesis, Gold nanoparticles, *Turbinaria conoides*, TEM, Seaweed

Background

Nanotechnology is anticipated to be the beginning of many of the main technological innovations of this century. An essential area of research in nanoscience and nanotechnology deals with the synthesis of nanoparticles of different chemical compositions, dimensions, and controlled monodispersity [1,2]. Currently, there is a growing necessity to improve the environmentally benign nanoparticle synthetic processes that are gratis from toxic chemicals in the synthesis procedure [3]. As the nanorevolution persisted to unfold and it is very important to manufacturing, the nanoparticles entrenched end products including environmentally beneficial and non-polluting technologies. Chemical and physical synthesis methods lead to the presence of some deadly chemical species adsorbed on the surface that may have unfavorable effects in medical applications [4]. In chemical-mediated synthesis of nanoparticles, some toxic substances could be used as reducing and stabilizing agent for stopping the agglomeration. As a result, researchers in the field of the synthesis of nanoparticles and assembly have turned to green system [5]. The use of eco-friendly materials like bacteria, fungi, plants, and parts of plant materials and algae for the synthesis of gold nanoparticles (AuNPs) offers several benefits of good environment and compatibility for biomedical applications as they do not use noxious chemicals in the synthesis procedure [6]. Green-mediated synthesis of metal nanoparticles provides

* Correspondence: chellapandiankannan@gmail.com; annananoteam@gmail.com
[2]Department of Chemistry, Manonmaniam Sundaranar University, Tirunelveli, Tamilnadu 627012, India
[1]Environmental Nanotechnology Division, Sri Paramakalyani Centre for Environmental Sciences, Manonmaniam Sundaranar University, Alwarkurichi, Tamilnadu 627412, India

progression over chemical and physical methods as it is economical and environment friendly, and in this method, there is no need to use high pressure, energy, temperature, and toxic chemicals [7]. Algae are otherwise called bionanofactories because they synthesized nanoparticles with high stability, are easy to handle, and eliminate cell maintenance [8]. Algae are the naturally available plant, which are important source of phytochemicals involved in the production of metallic nanoparticles. Recently, gold nanoparticle synthesized using the extract of algae such as *Sargassum wightii* [9], *Turbinaria conoides* [10], *Laminaria japonica* [11], and *Stoechospermum marginatum* [12] was reported.

Gold is one of the precious, inert, and less toxic metals, and it is utilized for curing various diseases. Gold nanoparticles play a vital role in nanobiotechnology as biomedicine because of convenient surface bioconjugation with biomolecular probes and remarkable plasmon-resonant optical properties [13-15]. Gold nanoparticles have an important function in the delivery of nucleic acids, proteins, gene therapy, *in vivo* delivery, targeting, etc. [16].

Among the marine sources, the macroalgae (seaweeds) occupy a significant place as a source of biomedical compounds. The compounds derived from macroalgae are reported to have a broad range of biological activities such as antibacterial [17], anticoagulant [18], and antifouling activity [19]. Seaweeds have been used since ancient times and have an exclusive place in traditional medicine of maritime nation as vermifuges, aesthetics, and antibiotics in the treatment of cough, wounds, gout, goiter, hypertension, venereal diseases, cancer, and a variety of other sickness [20]. *T. conoides* is the brown algae coming under the order of Fucales and in the family of Sargassaceae. The components found in such are a major amount sterols, some are fucosterol, and different molecules containing vinyl and ethyl cholesterol types, cyclohexane [21], and some sulfated polysaccharides such as fucoidan, neutral glucan, and guluronic and mannuronic acid residues containing alginic acid [22], rhamnose, fucose, arabinose, xylose, mannose, galactose, glucose, and uronic acids [23], providing a medicinal value for the brown algae.

This present investigation deals with a simple, eco-friendly synthesis of gold nanoparticles by the reduction of aqueous $AuCl_4$ into nanoparticles using the extract of marine algae *T. conoides*. The optical property of synthesized gold nanoparticles was characterized using a ultraviolet–visible (UV–vis) spectrophotometer; morphological characters was studied by performing TEM; crystalline nature was analyzed using selected-area electron diffraction (SAED) and X-ray diffraction (XRD) patterns; and functional molecules involved in the reduction process were comprehensively studied using Fourier transform-infrared (FTIR) spectroscopy.

Results and discussion

Visual inspection

The reduction of gold ions into gold nanoparticles was visually identified by color change from yellow to dark pink during the exposure of algae extract into the gold ion solution. Generally, gold ion exhibits yellow color in distilled water. Since the algae extract was added into the gold ion complex solution, it started to change the color from yellow to pink which indicates the reduction of gold ions into gold nanoparticles. Variation in the color change was observed; it mainly depends on the reaction time and phytochemical components of the algae extract. In this observation, dark pink color was formed at 1-h time of incubation, and the completion of reduction process that occurred at 24 h was identified by the precipitation of nanoparticles at the bottom of the conical flask (Figure 1a,b). These color changes happen because of the excitation of surface plasmon vibrations with the gold nanoparticles [24]. Quite interestingly, Dhamotharan et al. reported gold nanoparticles synthesis started at 2 h using algae biomass of *Padina tetrastromatica*. Some other biological resources such as fungi [9,25,26], bacteria [27], and plants [28] were used for synthesis of the gold nanoparticles by reduction process; they took time to complete the reduction which was about 24 to120 h.

UV–vis spectrophotometer analysis

UV–vis spectroscopy is an important technique to determine the fabrication and stability of metal nanoparticles in an aqueous solution. Figure 1c shows the UV–vis spectrograph of the colloidal gold solution that has been recorded as the function of time. Initially, the broad surface plasmon resonance (SPR) band for gold nanoparticles occurred at 520 nm. The broad peak indicates the presence of nanoparticles with a large size distribution. After 1 h, increases in intensity occurred as a role of the time of reaction, the SPR band was observed at 525 nm (Figure 1c). This band blue shift occurred due to longitudinal excitation of surface plasmon resonance of the nanoparticles [29]. The plasmon resonance bands of gold nanoparticles are broad with an absorption tail in the longer wavelength that extends well into the near-infrared region attributing the excitation of the in-plane SPR and indicates considerable anisotropy in the shape of Au nanoparticles. According to the Mie theory, the small size and spherical shape of nanoparticles which were formed in the reaction mixture were identified by forming a single SPR band for gold nanoparticles [30] and were confirmed by TEM. The optical absorbance was increased while increasing the incubation time indicates increasing nanoparticle synthesis. Absorbance intensity and increased nanoparticles are directly proportional to the incubation time. The conversion of gold ions into gold nanoparticles was found 90% within

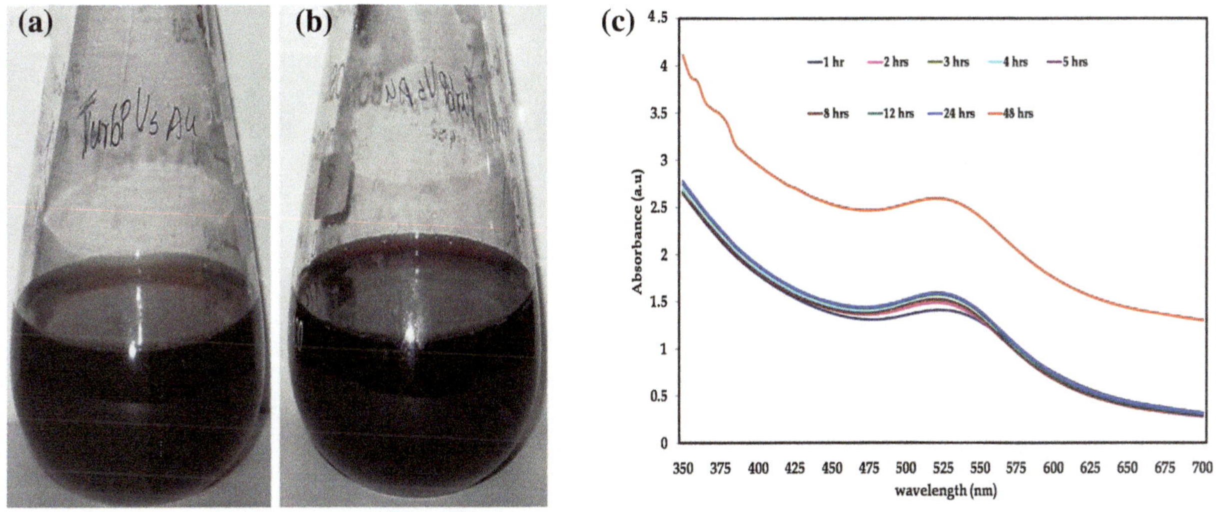

Figure 1 Green synthesis of gold nanoparticles. (**a**) Color change at 1 h of incubation, (**b**) dark purple pink at 24-h time of incubation, and (**c**) UV–vis absorption spectrum for gold nanoparticles synthesized using marine alga T. conoides.

1 h, identified by the absorbance intensity of reaction mixture. Similarly, Ghodake and Lee [11] reported that the SPR band was located at 530 nm for gold nanoparticles synthesized using brown algae *L. japonica*.

Transmission electron microscope

The structure and size of the nanoparticles were shown in Figure 2a,b,c. The TEM images of *T. conoides*-synthesized gold nanoparticles show mostly large and small spherical-, triangle-, and pseudo-spherical-shaped particles. The spherical and undefined shapes are also found in the TEM images which indicated that these particles are synthesized at the beginning of the reaction, i.e., at 6-to18-h incubation. After 24-h incubation, the particles are aggregated and form a bulk structure. The low absorbance was observed after 24 h in the UV–vis spectrum which also proved the aggregation of the particles. Due to the absence of stabilizing agent in the brown algal extracts, the particles are aggregated with each other and form a bulk structure. The average size of resulting particles is about 6 to 10 nm. When compared to the previous reports of Noruzi et al. [31] and Raghunandan et al. [32], herein, the algal-mediated gold nanoparticle shape are varied as spherical, triangle, and pseudo-spherical. Similarly, Ghodake and Lee [11] synthesized gold nanoparticles in the size range between 15 and 20 nm. The SAED pattern also suggested that the synthesized gold nanoparticles are crystalline in nature. The result of SAED pattern coincided with the result of XRD. The appearance of rings is attributed to set of diffraction planes (1 1 1), (2 0 0), (2 2 0), and (3 1 1) of fcc gold (Figure 2d). Similar SAED pattern was obtained using the algae [32].

X-ray diffraction assay

The green-synthesized nanoparticle in crystalline nature was clearly analyzed using XRD patterns [6]. Figure 3 shows the XRD patterns of the AuNPs prepared using the algal extract and its simulated solution. The XRD patterns revealed that AuNPs corresponded to the crystalline gold fcc phase. The diffraction peaks obtained at $2\theta = 38.36°$ (1 1 1), 44.13° (2 0 0), 64.78° (2 2 0), and 77.98° (3 1 1) are identical with those reported for the standard gold metal (Au^0) (Joint Committee on Powder Diffraction Standards-JCPDS, USA). The presence of these four intense peaks corresponding to the nanoparticles was in agreement with the Bragg's reflections of gold identified with the diffraction pattern [33]. Thus, the XRD pattern suggests that the gold nanoparticles were essentially crystalline nature. Hence, the simulated solution exhibited tremendous performance on the synthesis of AuNPs as that of the algae *T. conoides* extract. The XRD pattern of *T. conoides*-derived AuNPs shows some unidentified peaks that if raised with standard Au peaks reveals the association of algal biomolecules with the synthesized AuNPs. Similar result of XRD for gold nanoparticles synthesized using brown algae *S. marginatum*was found by Rajathi et al. [12].

FTIR analysis

FTIR measurement was carried out for functional molecules or constituents in the *T. conoides* responsible for the reduction of gold ions to nanoparticles and stabilizing the nanoparticles. The gold nanoparticles synthesized from *T. conoides* exhibit a lot of biomolecules which were involved in the synthesis of AuNP process (Figure 4).

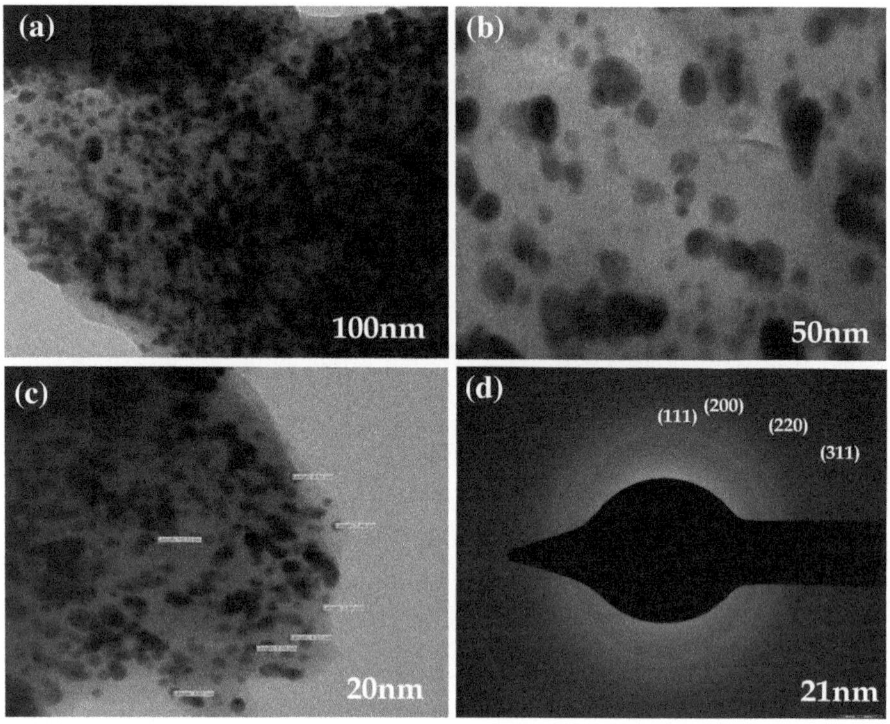

Figure 2 TEM images of AuNPs synthesized using *T. conoides* at different scale bars show spherical-shaped nanoparticles. The (**a**) 100 nm, (**b**) 50 nm, (**c**) 20 nm, and (**d**) SAED pattern of AuNPs.

FTIR shows the presence of different functional groups, which give rise to the well-known signatures in the IR region of electromagnetic spectrum. The strong and broad band observed at 3,414 cm^{-1} indicates the presence of polyphenolic O-H group and primary amine O-H band [34], C-H stretching vibrations of alkanes group was formed at the absorption band at 2,924 cm^{-1},a narrow band at 1,638 cm^{-1} indicates the presence of amide I, 1,418 cm^{-1} corresponding to C-C stretching aromatic ring, C-O stretching carboxylic acid group assigned at 1,243 cm^{-1},the narrow peak at 1,030 $cm^{-1}$is because of the presence of C-N stretching vibrations of aliphatic amines of proteins [35],and 813 $cm^{-1}$is assigned to S-O stretching of sulfonates. The weak bands at 639 and 586 cm^{-1} correspond to alkyl halides. This study also confirms that the carbonyl group from amino acids or proteins has stronger ability to bind metal so that the proteins or enzymes could most possibly cap the metal nanoparticles

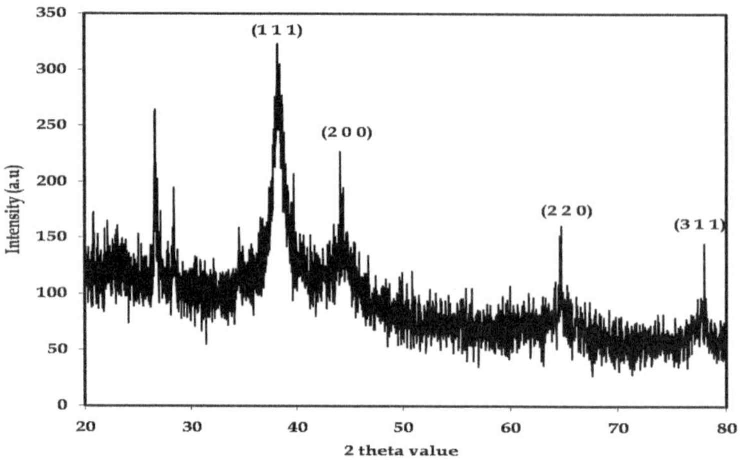

Figure 3 XRD spectra of gold nanoparticles synthesized using *T. conoides* algal extract.

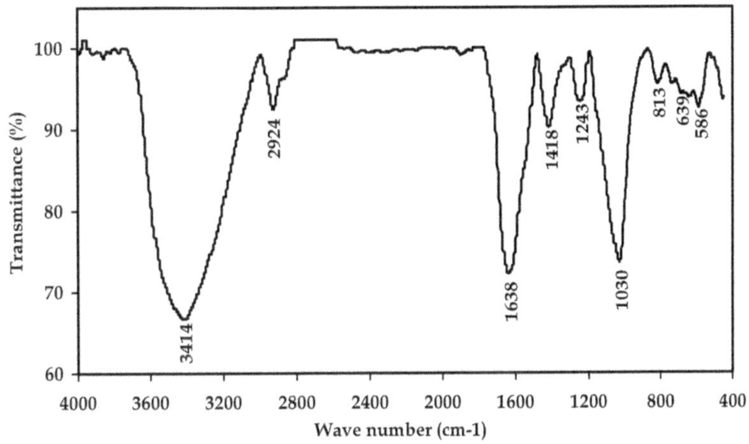

Figure 4 **FTIR spectra of *T. conoides*-synthesized gold nanoparticles.**

to prevent the agglomeration of the particles. This evidence reveals that sulfonated polysaccharide compounds and amide linkage proteins molecules could possibly be involved in the reduction of gold ions to nanoparticles and stabilization of gold nanoparticles in the aqueous medium. Previously, similar report has been presented [11,36]. Thangaraju et al. [37] reported that the presence of carboxylic, amine, phosphate, and hydroxyl functional groups is involved in the reduction of silver ions in the algae extract of *Sargassum polycystum*.

Polysaccharide and protein biomolecules in the algae extract do a dual function as reducing the gold ions and stabilizing the gold nanoparticles. Algal-mediated gold nanoparticle synthesis mechanism was proposed to involve electrostatic interactions between gold anions and algal functional groups. The $AuCl_4$ anion binds to positively charged functional groups, such as amino groups (NH_2), on the algal surface. After 40 and 50 min, algal extracts reduced Au (III) to gold nanoparticles. Hydroxyl groups present in the brown algal polysaccharides were involved in the bioreduction of Au (III) ions into Au (0) [38]. The hydroxyl groups are present on the secondary metabolites, polysaccharides, and monosaccharides [22,23] of the algal material which might be involved in the synthesis of gold nanoparticles [9].

Conclusions

In conclusion, the gold nanoparticle was synthesized using the algae extract of *T. conoides* as this seaweed is an easily available marine source. The synthesized gold nanoparticles were preliminarily examined by the color change of reaction mixture to dark pink. Furthermore, UV–vis spectrophotometer shows that the strong band at 520 to 525 nm indicates small spherical-shaped polydispersed nanoparticles, which were also confirmed by TEM, and the size ranges from 6 to 10 nm. Four intense peaks shown in the diffraction analysis indicate the crystalline structure of synthesized gold nanoparticles. The polyphenolic and amide linkages in the proteins or enzymes present in the algae extract may be responsible for the reduction of gold ions to gold nanoparticles that was identified using FTIR spectrum. This green synthesis is an efficient, environmentally reliable, less time-consuming, and single-step process, and it reveals that algae extract as a competent resources for manufacturing nontoxic biomimetic and biocompatible gold nanoparticles.

Methods

Chemicals

Gold chloride was obtained from Hi Media, Mumbai, India. All glasswares have been washed with distilled water and dried in oven before use. Dried algae of *T. conoides* have been collected from the Gulf of Mannar, South India.

Preparation of algae extract

The dried algae of *T. conoides* were washed several times with distilled water to remove the waste materials. Algae extract used for the synthesis was prepared from 5 g of thoroughly washed seaweed in a 500-mL Erlenmeyer flask and boiled in 100 mL double-distilled water for 15 min. Filtered algae extract was stored at −15°C for further use, being usable for several weeks.

Green synthesis of gold nanoparticles

To prepare gold nanoparticles, 10 mL of algae extract was added into 90 mL of 1 mM aqueous solution of gold chloride. The reduction was started at 10 min in room temperature, resulting in a color changing of solution indicating the formation of AuNPs. The bioreduction of gold ions in the solution was monitored periodically by measuring the UV–vis spectra (350 to 700 nm) of the solutions.

Characterization studies

UV–vis spectroscopy measurements (Lambda double-beam UV-Spectrophotometer, Perkin Elmer, Waltham, MA, USA) was carried out as a function of time of the reaction at room temperature operated at a resolution of 1nm. The reduction of gold ions was confirmed by qualitative testing of supernatant, obtained after centrifugation. Transmission electron microscopic analysis was done using a TEM (CM200, PHILIPS, Amsterdam, The Netherlands). Three microliters of the sample was placed on the carbon-coated copper grid, making a thin film of sample on the grid, and an extra sample was removed using the cone of a blotting paper and kept in a grid box sequentially. XRD measurements of film of the green-synthesized and washed gold nanoparticle solution cast onto glass slides were done on a Bruker AXS D8 Advance X-Ray Powder Diffractometer (Bruker AXS, Inc., Madison, WI, USA) operating at a voltage of 40 kV and current of 30 mA. The FTIR spectra of algae extract-synthesized AuNPs were analyzed which discussed for the possible function groups for the formation of AuNPs. The gold nanoparticle pellet obtained after centrifugation was redispersed in water and washed (centrifugation and redispersion) with distilled water for three times. Finally, the pellet was redispersed in water prior to FTIR (RX1 spectrophotometer, Perkin Elmer, Waltham, MA, USA) analysis.

Competing interests

The authors declare that they have no competing interests.

Authors' contributions

SR carried out the nanoparticles synthesis and characterization. CM, MV, and GG carried out the sample collection. KP carried out the manuscript preparation. GA and CK supervised the research work and corrected the manuscript. All authors read and approved the final manuscript.

Authors' information

SR completed M.Sc. Biotechnology in Periyar University and obtained his Ph.D. degree in Nanotechnology under the guidance of GA. He published 11 research articles in nanoparticles synthesis. He is interested in the metallic nanoparticles synthesis using algae and algae-derived compounds and their potential applications in the biomedical field. CM completed B.Sc. in Zoology and M.Sc. Environmental Biotechnology in SPKCES, Manonmaniam Sundaranar University, Alwarkurichi, Tirunelveli. She achieved her Ph.D. degree under the guidance of GA in the field of nanotechnology. Her research interests include synthesis of semiconductor nanomaterials and metal nanoparticles and its biomedical applications. MV obtained her M.Sc. in Environmental Biotechnology in Sri Paramakalyani Centre for Environmental Sciences, Manonmaniam Sundaranar University, Alwarkurichi, Tirunelveli. She obtained her Ph.D. degree under the guidance of GA in the field of Nanotechnology. Her research interests include eco-friendly synthesis of nanoparticles and its application in the environment. KP received his B.Sc. (2005) and M.Sc. in Biotechnology (2007) from Madurai Kamaraj University and Periyar University, India, respectively. He is currently working to obtain his Ph.D. degree at Manonmaniam Sundaranar University in India. He is interested in the green and immobilized microbe-mediated synthesis process of nanoparticles and nanocomposites and their biomedical and textile industry applications. GG completed M.Sc. in Biotechnology in Periyar University and obtained her Ph.D. degree in Nanotechnology under the guidance of GA. Her research interests include green-meditated synthesis of nanoparticles and its agricultural applications for controlling plant diseases. CK had done his master's degree and doctoral degree in the discipline of Chemistry. His research interests are in the field of Green Chemistry, Environmental Chemistry, and synthesis of nanoparticles. Currently, he works as an Associate Professor in MSU, Tirunelveli. GA received his M.Sc. in Applied Chemistry (1992) and Ph.D. in Environmental Biotechnology (1997) from Anna University, India. He had 10 years (1999 to 2008) post-doctoral experiences from the National Taiwan University in Taiwan, National Institute of Advanced Industrial Science and Technology in Japan, and National Central University in Taiwan. He has received many research awards from Indian and other country governments. He is an associate editor in five international journals. At present, he is an Associate Professor of the Environmental Biotechnology and the leader of Environmental Nanobiotechnology Division at SPKCES, MS University, India. His main research interest is on biosynthesis of nanoparticles and nanomaterials, nanobiocatalyst, and environmental chemistry.

Acknowledgments

The authors gratefully acknowledge the DST-FIST sponsored programme, Department of Science Technology, New Delhi, India, for funding the research development (Ref no. S/FST/ESI-101/2010) to carry out this work.

References

1. Armendariz, V: Size controlled gold nanoparticle formation by *Avena sativa* biomass. J. Nanoparticle Res. **6**, 377–382 (2004)
2. Taleb, C, Pettai, M, Pileni, P: Nanoparticles and nanostructured films preparation characterization. Appl. Chem. **22**, 1203–1203 (1998)
3. Shankar, SS, Rai, A, Ahmad, A, Sastry, M: Rapid synthesis of Au, Ag, and bimetallic Au core Ag shell nanoparticles using Neem (*Azadirachta indica*) leaf broth. J. Colloid Interface Sci. **275**, 496–502 (2004)
4. Nadagouda, MN: Varma, RS: Green synthesis of silver and palladium nanoparticles at room temperature using coffee and tea extract. Green Chem. **10**(8), 859–862 (2008)
5. Ahmad, A, Senapati, S, Islam Khan, M, Kumar, R, Sastry, M: Extracellular biosynthesis of monodisperse gold nanoparticles by a novel extremophilic actinomycete, *Thermomonospora* sp. Langmuir **19**, 3550–3553 (2003)
6. Singh, M, Kalaivani, R, Manikandan, S, Sangeetha, N, Kumaraguru, AK: Facile green synthesis of variable metallic gold nanoparticle using *Padina gymnospora*, a brown marine macroalga. Appl. Nanosci. **3**(2), 145–151 (2013)
7. Rajeshkumar, S, Kannan, C, Annadurai, G: Green synthesis of silvernanoparticles using marine brown algae *Turbinaria conoides* and its antibacterial activity. Int. J. Pharm. Bio. Sci. **3**(4), 502–510 (2012)
8. Song, JY, Kim, BS: Rapid biological synthesis of silver nanoparticles using plant leaf extracts. Bioprocess Biosyst. Eng. **32**, 79–84 (2009)
9. Singaravelu, G, Arockiyamari, J, Ganesh Kumar, V, Govindaraju, K: A novel extracellular biosynthesis of monodisperse gold nanoparticles using marine algae, *Sargassum wightii* Greville. Colloid Surf B: Biointerf. **57**, 97–101 (2007)
10. Vijayaraghavan, K, Mahadevan, A, Sathishkumar, M, Pavagadhi, S, Balasubramanian, R: Biosorption and subsequent bioreduction of trivalent aurum by a brown marine alga *Turbinaria conoides*. Chem. Eng. J. **167**, 223–227 (2011)
11. Ghodake, G, Lee, DS: Biological synthesis of gold nanoparticles using the aqueous extract of the brown algae *Laminaria japonica*. J. Nanoelectron Optoe. **6**, 1–4 (2011)
12. Rajathi, FAA, Parthiban, C, Ganesh Kumar, V, Anantharaman, P: Biosynthesis of antibacterial gold nanoparticles using brown alga, *Stoechospermum marginatum* (kützing). Spectrochim. Acta Part A **99**, 166–173 (2012)
13. Daniel, MC, Astruc, D: Gold nanoparticles: assembly, supramolecular chemistry, quantum-size-related properties, and applications toward biology, catalysis, and nanotechnology. Chem. Rev. **104**, 293–346 (2004)
14. Wu, C-C, Chen, D-H: Facile green synthesis of gold nanoparticles with gum arabic as a stabilizing agent and reducing agent. Gold Bull **43**, 234–239 (2010)
15. Kreibig, U, Vollmer, M: Optical properties of metal clusters. Springer-Verlag, Berlin (1995)
16. Tiwari, PM, Vig, K, Dennis, VA, Singh, SR: Functionalized gold nanoparticles and their biomedical applications. Nanomaterials **1**, 31–63 (2011)
17. Selvin, J, Huxley, AJ, Lipton, AP: Immunomodulatory potential of marine secondary metabolites against bacterial diseases of shrimp. Aquaculture **230**, 241–248 (2004)
18. Lipton, AP: Jose, J: Anticoagulant and immune enhancing activities of marine macroalgae explored. Spectrum - ICAR News **12**(4), 8–6 (2006)

19. Selvin, J, Lipton, AP: Development of a rapid mollusc foot adherence bioassay for detecting potent antifouling bioactive compounds. Curr. Sci. **83**, 735–737 (2002)
20. South, GR, Whittick, A: Introduction to phycology, pp. 21–25. Oxford press, London (1987)
21. Kumar, V, Fausto, N, Abbas, A: Robbins & Cotran pathologic basis of disease, 7th edn, pp. 914–7. Saunders (2003)
22. Nabanita, C, Ghosh, T, Sinha, S, Kausik, C, Karmakar, P, Bimalendu, R: Polysaccharides from *Turbinaria conoides*: structural features and antioxidant capacity. Food Chem. **118**(3), 823–829 (2010)
23. Shyamali, S, Silva, D, Savitri Kumar, N: Carbohydrate constituents of the marine algae of Sri Lanka part III: composition of the carbohydrates extracted from the brown seaweed *Turbinaria conoides*. J. National Sci. Council Sri Lanka **16**(2), 201–208 (1988)
24. Shankar, SS, Ahmad, A, Sastry, M: Geranium leaf assisted biosynthesis of silver nanoparticles. Biotechnol. Prog. **19**, 1627–31 (2003)
25. Mukherjee, P, Senapati, S, Mandal, D, Ahmad, A, Khan, MI, Kumar, R, Sastry, M: Extracellular synthesis of gold nanoparticles by the fungus *Fusarium oxysporum*. Chem. Bio. Chem. **3**, 461–463 (2002)
26. Ahmad, A, Senapati, S, Khan, MI, Kumar, R, Ramani, R, Srinivas, V, Sastry, M: Intracellular synthesis of gold nanoparticles by a novel alkalotolerant actinomycete *Rhodococcus* species. Nanotechnol. **14**, 824–825 (2003)
27. He, S, Guo, Z, Zhang, Y, Zhang, S, Wang, J, Gu, N: Biosynthesis of gold nanoparticles using the bacteria *Rhodopseudomonas capsulate*. Mater. let. **61**, 3984–3987 (2007)
28. Annamalai, A, Christina, VLP, Sudha, D, Kalpana, M, Lakshmi, PTV: Green synthesis, characterization and antimicrobial activity of Au NPs using *Euphorbia hirta* L. leaf extract. Colloids Surf. B. Biointerfaces **108**, 60–65 (2013)
29. Kamat, PV, Flumiani, M, Hartland, GV: Picosecond dynamics of silver nanoclusters. Photoejection of electrons and fragmentation. J. Phys. Chem. B. **102**, 3123–3128 (1998)
30. Smitha, SL, Philip, D, Gopchandran, KG: Green synthesis of gold nanoparticles using *Cinnamomum zeylanicum* leaf broth. Spectrochim. Acta Part A **74**, 735–739 (2009)
31. Noruzi, M, Zare, D, Khoshnevisan, K, Davoodi, D: Rapid green synthesis of gold nanoparticles using Rosa hybrid petal extract at room temperature. Spectrochimica Acta Part A **79**(5), 1461–1465 (2011)
32. Raghunandan, D, Basavaraja, S, Mahesh, B, Balaji, S, Manjunath, SY, Venkataraman, A: Biosynthesis of stable polyshaped gold nanoparticles from microwave-exposed aqueous extracellular anti-malignant guava (*Psidium guajava*) leaf extract. Nano Biotechnol. **5**(1–4), 34–41 (2009)
33. Saxena, A, Tripathi, RM, Singh, P: Biological Synthesis of silver nanoparticles by using onion (*Allium cepa*) extract and their antibacterial activity. Digest. J. Nanomater Biostruct. **5**, 427–432 (2010)
34. Devi, JS, BhimbA, BV, Ratnam, K: *Invitro* anticancer activity of silver nanoparticles synthesized using the extract of *Gelidiella sp*. Int. J. Pharm. Pharm. Sci. **4**, 710–715 (2012)
35. Gole, A, Dash, C, Ramakrishnan, V, Sainkar, SR, Mandale, AB, Rao, M: Pepsin-gold colloid conjugates: preparation, characterization and enzymatic activity. Langmuir. **17**, 1674–9 (2001)
36. Xie, J, Lee, JY, Wang, DIC, Ting, YP: Identification of active biomolecules in the high-yield synthesis of single-crystalline gold nanoplates in algal solutions. Small **3**, 672–682 (2007)
37. Thangaraju, N, Venkatalakshmi, RP, Arulvasu, C, Pandian, K: Synthesis of silver nanoparticles and the antibacterial and anticancer activities of the crude extract of *Sargassum polycystum* C. Agardh, Nano Biomed. Eng. **4**(2), 89–94 (2012)
38. Vijayaraghavan, GR, David, S, Bermudez-Allende, M, Sarwat, H: Imaging-guided parenchymal liver biopsy: how we do it. J. Clin. Imaging Sci. **1**, 30 (2011)

Room temperature photo-induced, Eu^{3+}-doped IGZO transparent thin films fabricated using sol–gel method

Rajagopalan Krishnan[1], Jagannathan Thirumalai[1*] and Rathinam Chandramohan[2]

Abstract

Red-emitting Eu^{3+}-doped indium gallium zinc oxide (IGZO) transparent thin films were fabricated using sol–gel method under UV (254-nm irradiation) in nitrogen atmosphere (inside the glove box) and thermally annealed at 500°C for 1 h. Structure, morphology, composition, and optical properties of the materials were examined using X-ray diffraction analysis, scanning electron microscope, X-ray photoelectron spectroscopy, and photoluminescence, respectively.

Keywords: Oxides, Annealing, Photoelectron spectroscopy, Luminescence

Background

Transparent semiconducting oxides have received greater attention because of their wide range of applications in active matrix electronics, displays, sensor arrays, and X-ray detectors [1-5]. It is well established that the indium gallium zinc oxide (IGZO) is transparent to visible light and has higher electron mobility than the conventional amorphous semiconductors, (e.g., amorphous silicon) and has higher switching speed, better electronic stability, more transparent displays, and more flexible electronics [6,7]. During recent years, solution-based synthesis materials have got more interest due to their excellent optical transparency and as new prospects for low-cost, high-throughput, easy-compositional modification and as printable and transparent electronics devices [1,8-10].

For a long time, the conventional sol–gel method is used to grow thin films by the coating of wet precursors followed by thermal annealing at higher temperatures for their decomposition and oxidation [8]. Recently, Kim et al. [1] have developed a new photochemical activation induced by deep-ultraviolet light from a low-pressure mercury lamp in an inert atmosphere (to prevent reactive ozone formation) to achieve high degrees of sol–gel condensation and film densification in amorphous metal oxide semiconductor systems. Unfortunately, there are no reports available on the rare-earth-doped IGZO solution-based synthesis. Normally, rare-earth materials are the noble candidates for optically active ions in semiconductor materials because they show many absorption and fluorescence transitions in the almost visible region. Rare-earth materials have additional significant features in contrast to other optically active ions as well: the wavelengths of their emission and absorption transitions are moderately unresponsive to host materials; they have longer lifetimes of energy states, higher quantum efficiency and good color rendering index [11,12], which lead to their excellent performance in many optoelectronic applications. Recently, Muth et al. [13,14] have developed the transparent europium-doped IGZO thin film for electroluminescent (TFEL) device using the integrated circuits that were fabricated using pulsed laser deposition (PLD). To the best of our knowledge, no systematic study has been carried out on the rare-earth-doped IGZO using sol–gel thin films.

With this aim, we prepared rare-earth-doped IGZO using sol–gel thin films. Our synthesis work on IGZO:Eu^{3+} has a similar route to that of Kim et al. [1], but there is a slight modification in the UV annealing treatment which is explained below. In this work, we have prepared the transparent IGZO:Eu^{3+} sol–gel thin films under UV (254-nm irradiation) [under nitrogen atmosphere (all the UV curing treatments were carried out inside the glove box)] and thermally annealed (500°C, 1 h)

* Correspondence: thirumalaijg@gmail.com
[1]Department of Physics, B.S. Abdur Rahman University, Vandalur, Chennai 600048, Tamil Nadu, India
Full list of author information is available at the end of the article

for comparative investigation. The selection of high-energy UV lamp (254 nm) is the alternative source for chemical cleavage and rearrangement of disordered M–O–M networks under room temperature UV treatment instead of high-temperature annealing, which results in the preparation of high-quality IGZO:Eu^{3+} thin films. The structure and phase purity of the samples were analyzed using X-ray diffraction pattern. The scanning electron microscope (SEM) images of IGZO:Eu^{3+} film confirm that the smaller nanocrystalline grains are homogenously dispersed in the elaborated thin films. The elemental and compositional analysis using X-ray photoelectron spectroscopy (XPS) confirms the presence of In, Ga, Zn, O, and Eu. The room temperature photoluminescence emission properties of IGZO:Eu^{3+} films were studied in detail.

Results and discussion

Structure and morphology of IGZO:Eu^{3+} thin films

In Figure 1, the IGZO:Eu^{3+} films that are post-annealed at 150°C show an amorphous phase, while the film with the annealing temperature at 450°C exhibits a less polycrystalline phase with (009) and (104) peaks as investigated by X-ray diffraction (XRD) measurement. Subsequently, the as-spun films were ultraviolet-treated (254 nm, 90 min) under N_2 atmosphere which also exhibits the similar behavior. No other additional peaks of impurity were observed. Furthermore, the SEM images of IGZO:Eu^{3+} film confirm that the nanocrystalline grains are homogenously dispersed in the elaborated thin films (thickness of 50 to 70 nm) with an average size of 10 nm as seen in Figure 2a,b.

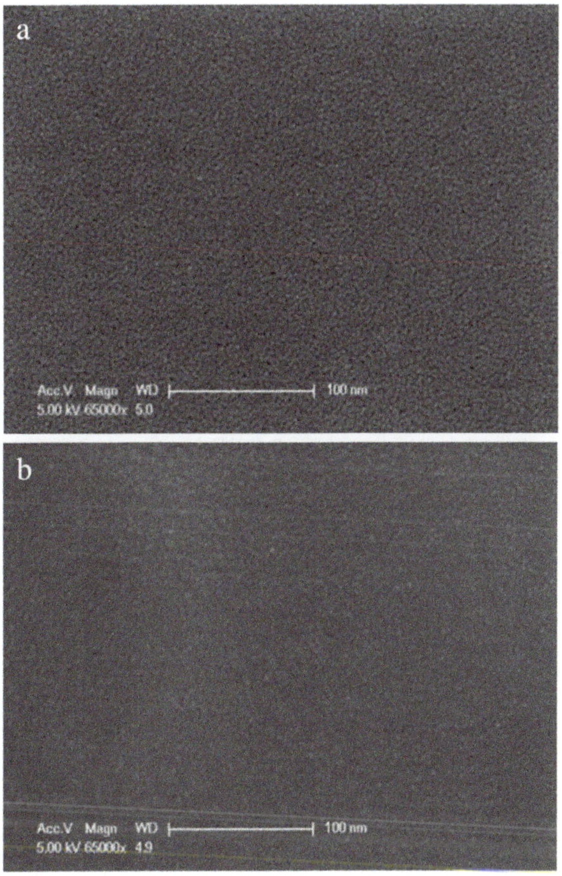

Figure 2 SEM image of (a) thermal and (b) UV-annealed IGZO:Eu^{3+} thin films.

X-ray photoelectron spectroscopy

Figure 3 shows the XPS spectra of as-synthesized IGZO:Eu^{3+} thin films [UV-irradiated and thermally annealed

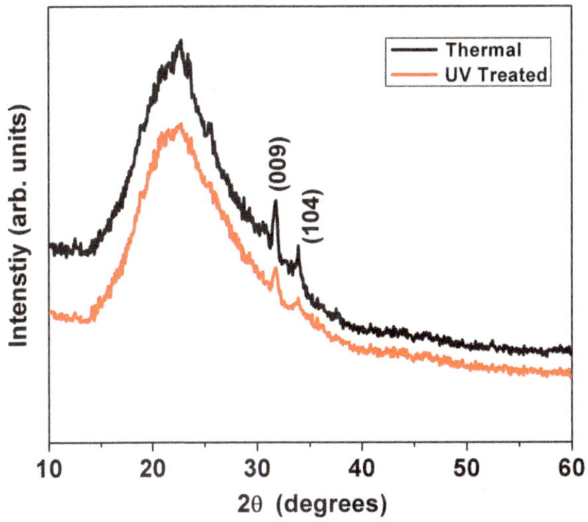

Figure 1 X-ray diffraction of thermal and UV-annealed IGZO:Eu^{3+} thin films.

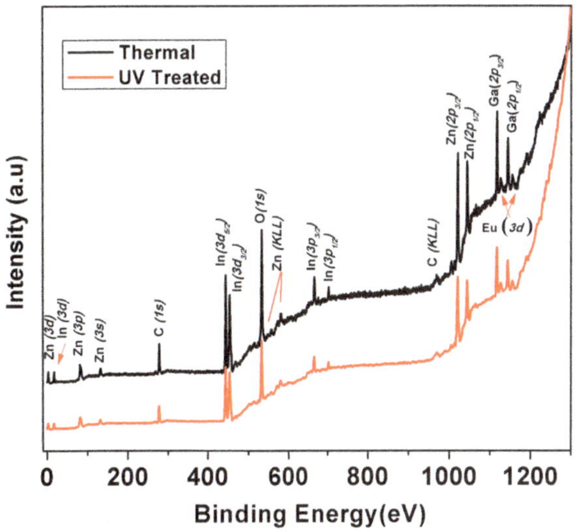

Figure 3 X-ray photoelectron spectroscopy of thermal and UV-annealed IGZO:Eu^{3+} thin films.

(at 450°C)]. From the XPS analysis, it is observed that the presence of the core levels of In, Ga, Zn, O, and Eu in IGZO:Eu^{3+}thin films obtained are in the ranges of 100 to 300 eV (Zn 3*d* at 3 eV; In 3*d* at 17 eV; Zn 3*p* at 81.5 eV; Zn 3 s at 133 eV), 300 to 700 eV (In $3d_{5/2}$ at 444 eV; In $3d_{3/2}$ 452 eV; O 1 s 530.7 eV; In $3p_{3/2}$ at 665 eV; In $3p_{1/2}$ at 703 eV), and 800 to 1,200 eV (Zn $2p_{3/2}$ at 1,022 eV; Zn $2p_{1/2}$ 1,044 eV; Ga $2p_{3/2}$ at 1,117 eV; Ga $2p_{1/2}$ 1,144 eV; Eu $3d_{5/2}$ at 1,125.6 eV; Eu $3d_{3/2}$ at 1,152 eV). The binding energy data of IGZO:Eu^{3+} thin films are calibrated using carbon (1 s, 283.5 eV) as the reference. The line shape and peak positions are in good agreement with the literature [15].

Photoluminescence studies of IGZO:Eu^{3+} thin films

As can be seen in Figure 4, the photoluminescence (PL) emission spectra show the Stark splitting pattern of $^5D_0 \rightarrow {}^7F_J$ (J = 1–4), intra-configurational *f-f* electronic transitions of Eu^{3+}-doped IGZO UV, and thermally annealed thin films. Upon excitation with 254-nm UV irradiation, the emission spectrum is dominated by the hypersensitive red emission, showing a transition $^5D_0 \rightarrow {}^7F_2$ (due to electric dipole transition) which is stronger than $^5D_0 \rightarrow {}^7F_1$ (magnetic dipole transition). These transitions are known to be hypersensitive to the crystal structure and chemical surroundings. The presence of electric dipole transition confirmed that Eu^{3+} ions were located at sites without inversion symmetry (C_{3v} symmetry). The other transitions of $^5D_0 \rightarrow {}^7F_1$, $^5D_0 \rightarrow {}^7F_3$, and $^5D_0 \rightarrow {}^7F_4$ were relatively very weak for all the samples. From the emission spectrum, it is seen that the transition $^5D_0 \rightarrow {}^7F_2$ is predominating for all the samples and it indicates that higher energy levels were coupled to ligand to metal stretching vibrations. The presence of strong luminescent intensity from trivalent europium ion (Eu^{3+}) in the IGZO matrix has intricate energy levels with energy gaps of various magnitudes. The optical property of the IGZO:Eu^{3+} thin film is in good agreement with the literature [12,13], which was obtained by PLD. The result envisages an effective route to synthesize IGZO:Eu^{3+} thin film and has a potential application in TFEL display devices, flexible transistor, and solar cell systems.

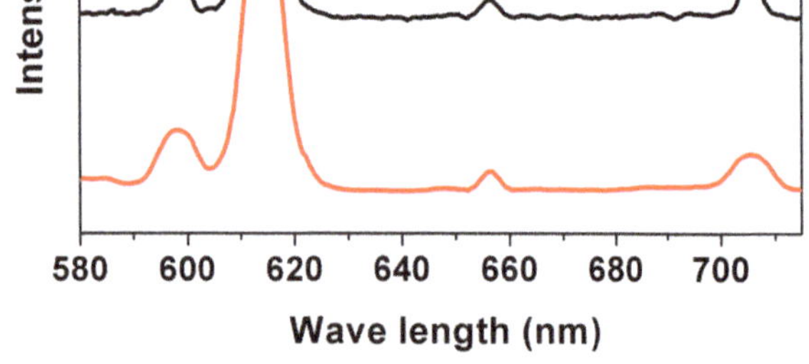

Figure 4 Photoluminescence emission spectra of thermal and UV-annealed IGZO:Eu^{3+} thin films.

Conclusions

In summary, we have fabricated amorphous IGZO:Eu^{3+} thin films on glass substrates by using a spin-coating method and studied the structure, composition, and optical property dependences of the UV and thermally annealed IGZO:Eu^{3+} thin films. The elemental and compositional analysis (XPS) confirms the presence of In, Ga, Zn, O, and Eu. The photoluminescence properties of IGZO:Eu^{3+} were thoroughly investigated and these materials have potential application in the display devices.

Methods

Synthesis of rare-earth (Eu^{3+})-doped IGZO thin film

All chemicals were purchased from Sigma-Aldrich (St. Louis, MO, USA). The precursor solution of IGZO:Eu^{3+} was prepared by dissolving zinc acetate dihydrate ($Zn(CH_3CO_2)_2{\cdot}2H_2O$), gallium nitrate hydrate ($Ga(NO_3)_3{\cdot}xH_2O$), and indium nitrate hydrate ($In(NO_3)_3{\cdot}xH_2O$) in 2-methoxyethanol (2-ME) solvent. The precursors were mixed at an millimolar concentration of In/Ga/Zn/Eu = 0.045:0.0125:0.0275:0.02 in the final oxide films. Then, 2.0 M of monoethanolamine (MEA) was added as a sol stabilizer, and acetic acid CH_3COOH was dropped to make a homogeneous solution while stirring. After vigorously stirring at 80°C for 5 h, the IGZO:Eu^{3+} sol was subsequently aged for 1 day before spin coating. The IGZO:Eu^{3+} thin films were spin-coated on the substrates (0.7-mm-thick glass) with spin rotation speeds of 3,000 rpm for 30 s. The light-induced photochemical treatment was performed using a high-density ultraviolet treatment (254 nm) in N_2 atmosphere inside the glove box. Upon UV annealing process, the radiant thermal energy of UV lamp increases the surface temperature of the substrate to 110°C to 120°C. The output energy intensity of the lamp was 20 to 23 mW cm^{-2}. The total energy delivered to the sample surface is calculated to be approximately 125 J cm^{-2}. For comparative investigation, the as-spun films were thermally annealed at 500°C for 1 h.

Characterization

The XRD patterns were recorded using Philips X'Pert PRO PanAlytical diffractometer (PanAlytical Instruments, Almelo, the Netherlands) with CuKα radiation (λ = 0.15406 nm) at a scanning rate of 5° min^{-1}. The SEM measurements were carried out using FESEM JSM-7600 F. The XPS measurements were performed using a LAS-3000 surface analysis system (RIBER, Paris, France) employing AlKα X-rays (1,486.6 eV, width 0.5 eV). Room temperature PL spectra were recorded using Horiba-Jobin Yvon Fluromax-4P bench-top spectrofluorometer (HORIBA Instruments, Kyoto, Japan).

Competing interests
The authors declare that they have no competing interests.

Authors' contributions
RK synthesized the materials and carried out all the characterization studies. JT participated in the sequence alignment, drafted the manuscript, and conceived of the study and participated in its design and coordination. RC participated in the sequence alignment. All authors read and approved the final manuscript.

Acknowledgments
The authors acknowledge gratefully the CECRI (CSIR) in Karaikudi, IGCAR in Kalpakkam, and CUSAT in Cochin for extending their instrumentation facilities for characterization.

Author details
[1]Department of Physics, B.S. Abdur Rahman University, Vandalur, Chennai 600048, Tamil Nadu, India. [2]Department of Physics, Sree Sevugan Annamalai College, Devakottai 630303, Tamil Nadu, India.

References

1. Kim, YH, Heo, JS, Kim, TH, Park, SJ, Yoon, MH, Kim, J, Oh, MS, Yi, GR, Noh, YY, Park, SK: Flexible metal-oxide devices made by room temperature photochemical activation of sol–gel films. Nature **489**, 128–133 (2012)
2. Fortunato, EMC, Barquinha, PMC, Pimentel, ACMBG, Gonçalves, AMF, Marques, AJS, Martins, RFP, Pereira, LMN: Wide-bandgap high-mobility ZnO thin-film transistors produced at room temperature. Appl. Phys. Lett. **85**, 2541–2543 (2004)
3. Kim, MG, Kanatzidis, MG, Facchetti, A, Marks, TJ: Low-temperature fabrication of high performance metal oxide thin film electronics via combustion processing. Nature. Mater. **10**, 382–388 (2011)
4. Dehuff, NL, Kettenring, ES, Hong, D, Chiang, HQ, Wager, JF, Hoffman, RL, Park, CH, Keszler, DA: Transparent thin-film transistors with zinc indium oxide channel layer. J. Appl. Phys **97**, 064505–064505-5 (2005)
5. Kamiya, T, Nomura, K, Hosono, H: Present status of amorphous In-Ga-Zn-O thin-film transistors. Sci. Technol. Adv. Mater. **11**, 044305–044323 (2010)
6. Yang, SH, Bak, JY, Yoon, SM, Ryu, MK, Oh, H, Hwang, CS, Kim, GH, Park, SHK, Jang, J: Low-temperature processed flexible In-Ga-Zn-O thin-film transistors exhibiting high electrical performance. Electron. Dev. Lett **32**, 1692–1694 (2011)
7. Kim, YH, Han, MK, Han, JI, Park, SK: Effect of metallic composition on electrical properties of solution-processed indium-gallium-zinc-oxide thin film transistors. IEEE Trans. Electron. Dev. **57**, 1009–1014 (2010)
8. Hwang, SY, Lee, JH, Woo, CH, Lee, JY, Cho, HK: Effect of annealing temperature on the electrical performances of solution-processed InGaZnO thin film transistors. Thin. Sol. Film. **519**, 5146–5149 (2011)
9. Imai, H: Ultraviolet (UV) irradiation. In: Sakka, S (ed.) Handbook of Sol–gel Science and Technology: Processing, Characterization and Applications vol. 1, pp. 639–650. Kluwer, Boston (2005)
10. Kim, YH, Han, JI, Park, SK: Effect of Zn/tin composition ratio on the operational stability of solution-processed zinc tin oxide thin film transistors. IEEE Electron. Device. Lett. **33**, 50–52 (2012)
11. Thirumalai, J, Chandramohan, R, Divakar, R, Mohandas, E, Sekar, M, Parameswaran, P: Eu^{3+} doped gadolinium oxysulfide (Gd_2O_2S) nanostructures - synthesis and optical and electronic properties. Nanotechnology **19**, 395703 (2008)
12. Krishnan, R, Thirumalai, J, Shameem Banu, IB, John, P: A: Rugby-ball-shaped $Na_{0.5}La_{0.5}MoO_4:Eu^{3+}$ 3D architectures: synthesis, characterization, and their luminescence behavior. J. Nanostruct. Chem **3**, 14 (2013)
13. Wellenius, P, Suresh, A, Muth, JF: Bright, low voltage europium doped gallium oxide thin film electroluminescent devices. Appl. Phys. Lett. **92**, 021111–021113 (2008)
14. Wellenius, P, Suresh, A, Luo, H, Lunardi, LM, Muth, JF: An amorphous indium-gallium-zinc-oxide active matrix electroluminescent pixel. J. Display Tech. **5**, 438–445 (2009)
15. Moulder, JF, Stickle, WF, Sobol, PE, Bomben, KD: Handbook of X-ray Photoelectron Spectroscopy. Perkin-Elmer Corporation, Eden Prairie (1992)

Permissions

All chapters in this book were first published in JNC, by Springer; hereby published with permission under the Creative Commons Attribution License or equivalent. Every chapter published in this book has been scrutinized by our experts. Their significance has been extensively debated. The topics covered herein carry significant findings which will fuel the growth of the discipline. They may even be implemented as practical applications or may be referred to as a beginning point for another development.

The contributors of this book come from diverse backgrounds, making this book a truly international effort. This book will bring forth new frontiers with its revolutionizing research information and detailed analysis of the nascent developments around the world.

We would like to thank all the contributing authors for lending their expertise to make the book truly unique. They have played a crucial role in the development of this book. Without their invaluable contributions this book wouldn't have been possible. They have made vital efforts to compile up to date information on the varied aspects of this subject to make this book a valuable addition to the collection of many professionals and students.

This book was conceptualized with the vision of imparting up-to-date information and advanced data in this field. To ensure the same, a matchless editorial board was set up. Every individual on the board went through rigorous rounds of assessment to prove their worth. After which they invested a large part of their time researching and compiling the most relevant data for our readers.

The editorial board has been involved in producing this book since its inception. They have spent rigorous hours researching and exploring the diverse topics which have resulted in the successful publishing of this book. They have passed on their knowledge of decades through this book. To expedite this challenging task, the publisher supported the team at every step. A small team of assistant editors was also appointed to further simplify the editing procedure and attain best results for the readers.

Apart from the editorial board, the designing team has also invested a significant amount of their time in understanding the subject and creating the most relevant covers. They scrutinized every image to scout for the most suitable representation of the subject and create an appropriate cover for the book.

The publishing team has been an ardent support to the editorial, designing and production team. Their endless efforts to recruit the best for this project, has resulted in the accomplishment of this book. They are a veteran in the field of academics and their pool of knowledge is as vast as their experience in printing. Their expertise and guidance has proved useful at every step. Their uncompromising quality standards have made this book an exceptional effort. Their encouragement from time to time has been an inspiration for everyone.

The publisher and the editorial board hope that this book will prove to be a valuable piece of knowledge for researchers, students, practitioners and scholars across the globe.

List of Contributors

Farhad Shirini
Department of Chemistry, College of Science, the University of Guilan, PO Box 1914, Rasht 41335, Iran

Manouchehr Mamaghani
Department of Chemistry, College of Science, the University of Guilan, PO Box 1914, Rasht 41335, Iran

Seyyed Vahid Atghia
Department of Chemistry, College of Science, the University of Guilan, PO Box 1914, Rasht 41335, Iran

Ibrahim M El-Sherbiny
Zewail University, Zewail City of Science and Technology, 6th October City, Giza 12588, Egypt
Chemistry Department, Faculty of Science, Mansoura University, Mansoura ET-35516, Egypt

Ehab Salih
Biological Advanced Materials, Physics Department, Faculty of Science, Mansoura University, Mansoura ET-35516, Egypt

Fikry M Reicha
Biological Advanced Materials, Physics Department, Faculty of Science, Mansoura University, Mansoura ET-35516, Egypt

Moayad Hossaini Sadr
Chemistry Department, Faculty of Science, Azarbaijan Shahid Madani University, Tabriz 53714-161, Iran

Hafezeh Nabipour
Chemistry Department, Faculty of Science, Azarbaijan Shahid Madani University, Tabriz 53714-161, Iran

Farshad Taleshi
Department of Applied Science, Qaemshahr Branch, Islamic Azad University, Qaemshahr 163, Iran

Ali A Hosseini
Department of Physics, University of Mazandaran, Babolsar 47416-95447, Iran

Zahra Saadi
Department of Chemical engineering, Faculty of engineering, South Tehran Branch, Islamic Azad University, P.O. Box 11365-4435, Tehran, Iran

Reyhane Saadi
Department of Chemical engineering, Faculty of engineering, South Tehran Branch, Islamic Azad University, P.O. Box 11365-4435, Tehran, Iran

Reza Fazaeli
Department of Chemical engineering, Faculty of engineering, South Tehran Branch, Islamic Azad University, P.O. Box 11365-4435, Tehran, Iran

Hassan Keypour
Faculty of Chemistry, Bu-Ali Sina University, Hamedan 65174, Iran

Mohammad Noroozi
Faculty of Chemistry, Bu-Ali Sina University, Hamedan 65174, Iran
Center for Research and Development of Petroleum Technologies at Kermanshah, Research Institute of Petroleum Industry (RIPI), Kermanshah, Iran

Alimorad Rashidi
Nanotechnology Research Center, Research Institute of Petroleum Industry (RIPI), Tehran, Iran

Chelladurai Malarkodi
Environmental Nanotechnology Division, Sri Paramakalyani Centre for Environmental Sciences, Manonmaniam Sundaranar University, Alwarkurichi, Tamil Nadu 627 412, India

Shanmugam Rajeshkumar
Environmental Nanotechnology Division, Sri Paramakalyani Centre for Environmental Sciences, Manonmaniam Sundaranar University, Alwarkurichi, Tamil Nadu 627 412, India

Mahendran Vanaja
Environmental Nanotechnology Division, Sri Paramakalyani Centre for Environmental Sciences, Manonmaniam Sundaranar University, Alwarkurichi, Tamil Nadu 627 412, India

Kanniah Paulkumar
Environmental Nanotechnology Division, Sri Paramakalyani Centre for Environmental Sciences, Manonmaniam Sundaranar University, Alwarkurichi, Tamil Nadu 627 412, India

Gnanadhas Gnanajobitha
Environmental Nanotechnology Division, Sri Paramakalyani Centre for Environmental Sciences, Manonmaniam Sundaranar University, Alwarkurichi, Tamil Nadu 627 412, India

Gurusamy Annadurai
Environmental Nanotechnology Division, Sri Paramakalyani Centre for Environmental Sciences, Manonmaniam Sundaranar University, Alwarkurichi, Tamil Nadu 627 412, India

Heydar Maddah
Shahrood Branch, Islamic Azad University, Shahrood 3619995161, Iran

Mahdokht Rezazadeh
Young Research Club, Saveh Branch, Islamic Azad University, Saveh 39187/ 366, Iran

Mojtaba Maghsoudi
Sama-Saveh Branch, Islamic Azad University, Saveh 39187/366, Iran

Syamak NasiriKokhdan
Kashan University, Kashan 87314, Iran

Michail O Danilov
V.I. Vernadskii Institute of General and Inorganic Chemistry of the Ukrainian NAS, prospekt Palladina 32-34, 03680, Kyiv 142, Ukraine

Ivan A Slobodyanyuk
V.I. Vernadskii Institute of General and Inorganic Chemistry of the Ukrainian NAS, prospekt Palladina 32-34, 03680, Kyiv 142, Ukraine

Igor A Rusetskii
V.I. Vernadskii Institute of General and Inorganic Chemistry of the Ukrainian NAS, prospekt Palladina 32-34, 03680, Kyiv 142, Ukraine

Gennadii Ya Kolbasov
V.I. Vernadskii Institute of General and Inorganic Chemistry of the Ukrainian NAS, prospekt Palladina 32-34, 03680, Kyiv 142, Ukraine

Rajendra Bhosale
D.B.F. Dayanad College of Arts and Science, Solapur, Maharashtra 413002, India

Sidram Pujari
D.B.F. Dayanad College of Arts and Science, Solapur, Maharashtra 413002, India

Gajanan Muley
Department of Physics, Sant Gadge Baba Amravati University, Amravati, Maharashtra 444602, India

Bathuwell Pagare
Department of Chemistry, Shri Anand College of Science, Pathardi, Ahmednagar, Maharashtra 414102, India

Anil Gambhire
Department of Chemistry, Shri Anand College of Science, Pathardi, Ahmednagar, Maharashtra 414102, India

Hojatollah Montazeri
Department of Medical Biotechnology, School of Advanced Medical Technologies, Tehran University of Medical Sciences, Tehran 1417614411, Iran

Amir Amani
Department of Medical Biotechnology, School of Advanced Medical Technologies, Tehran University of Medical Sciences, Tehran 1417614411, Iran

Hamid Reza Shahverdi
Department of Materials Engineering, University of Tarbiat Modares, 14115-143, Tehran, Iran

Emad al din Haratifar
Department of Materials Engineering, University of Tarbiat Modares, 14115-143, Tehran, Iran

Ahmad Reza Shahverdi
Biotechnology Research Center, Faculty of Pharmacy, Tehran University of Medical Sciences, Tehran 1417614411, Iran
Pharmaceutical Sciences Research Center, School of Pharmacy, Tehran University of Medical Sciences, Tehran 417614411, Iran

Sajjad Sedaghat
Department of Chemistry, Shahre-Qods Branch, Islamic Azad University, Tehran, Iran

Fariba Golbaz
Department of Chemistry, Shahre-Qods Branch, Islamic Azad University, Tehran, Iran

M Raghasudha
Department of Chemistry, Jayaprakash Narayan College of Engineering, Mahabubnagar 509001 Andhra Pradesh, India

D Ravinder
Department of Physics, Nizam College, Basheerbagh Osmania University, Hyderabad 500001 Andhra Pradesh, India

P Veerasomaiah
Department of Chemistry, Osmania University, Hyderabad 500007 Andhra Pradesh, India

Abbas Rahdar
Department of Physics, Faculty of Science, University of Zabol, Zabol, Iran

Mohammad Barshan-Tashnizi
Department of Life Science Engineering (LSE), Faculty of New Sciences and Technologies (FNST), University of Tehran, Tehran 14174, Iran

Bahman Ebrahimi-Hoseinzadeh
Department of Life Science Engineering (LSE), Faculty of New Sciences and Technologies (FNST), University of Tehran, Tehran 14174, Iran

Parandis Daneshgar
Institute of Biochemistry and Biophysics, University of Tehran, Tehran 14174, Iran

Parviz Norouzi
Center of Excellence in Electrochemistry, University of Tehran, Tehran 14174, Iran

Mohamad M Ayad
Department of Chemical and Petrochemical Engineering, Egypt-Japan University of Science and Technology, Borg El Arab, Alexandria 21934, Egypt
Department of Chemistry, Faculty of Science, University of Tanta, Tanta 31111, Egypt

Ahmed Abu El-Nasr
Department of Chemistry, Faculty of Science, University of Tanta, Tanta 31111, Egypt

Bahniman Ghosh
Microelectronics Research Center, University of Texas at Austin, 10100, Burnet Road, Bldg. 160, Austin, TX 78758, USA

Shoubhik Gupta
Department of Electrical Engineering, Indian Institute of Technology, Kanpur, 208016, India

Smriti Kumari
Department of Electrical Engineering, Indian Institute of Technology, Kanpur, 208016, India

Akshaykumar Salimath
Department of Electrical Engineering, Indian Institute of Technology, Kanpur, 208016, India

Alireza Keshavarz
Department of Physics, Shiraz University of Technology, Shiraz, Iran

Zohreh Parang
Department of Physics, Shiraz branch, Islamic Azad University, Shiraz, Iran

Ahmad Nasseri
Department of Physics, Shiraz University of Technology, Shiraz, Iran

Rajagopalan Krishnan
Department of Physics, B.S. Abdur Rahman University, Vandalur, Chennai, Tamilnadu 600048, India

Jagannathan Thirumalai
Department of Physics, B.S. Abdur Rahman University, Vandalur, Chennai, Tamilnadu 600048, India

Ithrees Basha Shameem Banu
Department of Physics, B.S. Abdur Rahman University, Vandalur, Chennai, Tamilnadu 600048, India

Anthuvan John Peter
Department of Physics, B.S. Abdur Rahman University, Vandalur, Chennai, Tamilnadu 600048, India

Abdul Kareem Thottoli
PG & Research Department of Physics, Kongunadu Arts and Science College, GN Mills PO, Coimbatore, Tamilnadu 641 029, India

Anu Kaliani Achuthan Unni
PG & Research Department of Physics, Kongunadu Arts and Science College, GN Mills PO, Coimbatore, Tamilnadu 641 029, India

Ramin Karimian
Applied Biotechnology Research Center, Baqiyatallah University of Medical Sciences, Tehran, Iran

Farideh Piri
Department of Chemistry, Faculty of Science, Zanjan University, P.O. Box 45195-313, Zanjan, Iran

Ali Asghar Safari
Department of Chemistry, Faculty of Sciences, Zanjan Branch, Islamic Azad University, Zanjan, Iran

Seyed Javad Davarpanah
Applied Biotechnology Research Center, Baqiyatallah University of Medical Sciences, Tehran, Iran

Mohammad Mirjalili
Department of Textile Engineering, Yazd Branch, Islamic Azad University, Yazd, Iran

Mina Abbasipour
Department of Textile Engineering, Yazd Branch, Islamic Azad University, Yazd, Iran

Abdollah MortezaAli
Laboratory of Surface Physics, Department of Physics, Alzahra University, Tehran 19938, Iran

Saeideh Ramezani Sani
Department of Mechanic, Islamic Azad University, Roudehen Branch, Roudehen, Iran

Behzad Zeynizadeh
Department of Chemistry, Faculty of Science, Urmia University, Urmia 57159-165, Iran

Mostafa Karimkoshteh
Department of Chemistry, Faculty of Science, Urmia University, Urmia 57159-165, Iran

Dariush Robati
Department of Chemistry, Eslamshahr Branch, Islamic Azad University, Eslamshahr, Iran

Yoshio Kobayashi
Department of Biomolecular Functional Engineering, College of Engineering, Ibaraki University, 4-12-1 Nakanarusawa-cho, Hitachi, Ibaraki 316-8511, Japan

Hikaru Morimoto
Department of Biomolecular Functional Engineering, College of Engineering, Ibaraki University, 4-12-1 Nakanarusawa-cho, Hitachi, Ibaraki 316-8511, Japan

Tomohiko Nakagawa
Department of Nano-Medical Science, Graduate School of Medicine, Tohoku University, Seiryo-machi, Aoba-ku, Sendai 980-8575, Japan

Yohsuke Kubota
Department of Nano-Medical Science, Graduate School of Medicine, Tohoku University, Seiryo-machi, Aoba-ku, Sendai 980-8575, Japan

Kohsuke Gonda
Department of Nano-Medical Science, Graduate School of Medicine, Tohoku University, Seiryo-machi, Aoba-ku, Sendai 980-8575, Japan

Noriaki Ohuchi
Department of Nano-Medical Science, Graduate School of Medicine, Tohoku University, Seiryo-machi, Aoba-ku, Sendai 980-8575, Japan
Department of Surgical Oncology, Graduate School of Medicine, Tohoku University, Seiryo-machi, Aoba-ku, Sendai 980-8574, Japan

Mohammad Ghorbanpour
Chemical Engineering Department, University of Mohaghegh Ardabili, Ardabil, Iran

Cavus Falamaki
Chemical Engineering Department, Amirkabir University of Technology, P.O. Box 15875-4413 Tehran, Iran

Saeed Farhadi
Department of Chemistry, Lorestan University, Khoramabad 68135-465, Iran

Jalil Safabakhsh
Department of Chemistry, Lorestan University, Khoramabad 68135-465, Iran

Parisa Zaringhadam
Department of Chemistry, Lorestan University, Khoramabad 68135-465, Iran

Biswadip Banerji
Department of Chemistry, CSIR-Indian Institute of Chemical Biology, 4, Raja S.C. Mullick Road, Kolkata 700032, India

Sumit Kumar Pramanik
Department of Chemistry, CSIR-Indian Institute of Chemical Biology, 4, Raja S.C. Mullick Road, Kolkata 700032, India

Uttam Pal
Department of Structural Biology and Bioinformatics, CSIR-Indian Institute of Chemical Biology, 4, Raja S.C. Mullick Road, Kolkata 700032, India

Nakul Chandra Maiti
Department of Structural Biology and Bioinformatics, CSIR-Indian Institute of Chemical Biology, 4, Raja S.C. Mullick Road, Kolkata 700032, India

Kolsoum Pourzare
Department of Chemistry, Lorestan University, Khoramabad 68135-465, Iran

Shokooh Sadeghinejad
Department of Chemistry, Lorestan University, Khoramabad 68135-465, Iran

Mahmoud M Shokrieh
Composites Research Laboratory, Center of Excellence in Experimental Mechanics and Dynamics, School of Mechanical Engineering, Iran University of Science and Technology, Narmak, Tehran 16846-13114, Iran

Ali Saeedi
Composites Research Laboratory, Center of Excellence in Experimental Mechanics and Dynamics, School of Mechanical Engineering, Iran University of Science and Technology, Narmak, Tehran 16846-13114, Iran

Majid Chitsazzadeh
Nanocomposites Research Laboratory, Iran University of Science and Technology, Narmak, Tehran 16846-13114, Iran

Pokutnyi Sergey
G. V. Kurdumov Institute for Metal Physics, NAS Ukraine, Vernadsky Blvd., 36, Kiev 03680, Ukraine

Fahimeh Najafi
Department of Chemistry, Roudehen Branch, Islamic Azad University, Roudehen, Iran

Mehdi Norouzi
Department of Virology, School of Public Health and Institute of Health Research, Tehran University of Medical Sciences, Tehran, Iran

Karim Zare
Department of Chemistry, Science and Research Branch, Islamic Azad University, Tehran, Iran
Department of Chemistry, Shahid Beheshti University, Evin, Tehran, Iran

Ali Fakhri
Department of Chemistry, Shahre-Qods Branch, Islamic Azad University, Tehran, Iran

Fazeleh Bahrampour
National Institute of Genetic Engineering and Biotechnology, Tehran freeway 15 Km, Pajouhesh Blv., P.O. Box 14925/16, Tehran 1417863171, Iran

Jamshid Raheb
National Institute of Genetic Engineering and Biotechnology, Tehran freeway 15 Km, Pajouhesh Blv., P.O. Box 14925/16, Tehran 1417863171, Iran

Zohreh Rabiei
National Institute of Genetic Engineering and Biotechnology, Tehran freeway 15 Km, Pajouhesh Blv., P.O. Box 14925/16, Tehran 1417863171, Iran

Chellapandian Kannan
Department of Chemistry, Manonmaniam Sundaranar University, Tirunelveli, Tamilnadu 627012, India

Rajagopalan Krishnan
Department of Physics, B.S. Abdur Rahman University, Vandalur, Chennai 600048, Tamil Nadu, India

Jagannathan Thirumalai
Department of Physics, B.S. Abdur Rahman University, Vandalur, Chennai 600048, Tamil Nadu, India

Rathinam Chandramohan
Department of Physics, Sree Sevugan Annamalai College, Devakottai 630303, Tamil Nadu, India